T0238628

Lecture Notes in Computer Sc

Commenced Publication in 1973
Founding and Former Series Editors:
Gerhard Goos, Juris Hartmanis, and Jan van Leeuwen

Sebastian Link   Henri Prade (Eds.)

Foundations of Information and Knowledge Systems

6th International Symposium, FoIKS 2010
Sofia, Bulgaria, February 15-19, 2010
Proceedings

Volume Editors

Sebastian Link
School of Information Management
The Victoria University of Wellington
Wellington, New Zealand
E-mail: sebastian.link@vuw.ac.nz

Henri Prade
Institut de Recherche en Informatique de Toulouse
CNRS
University of Toulouse III
Toulouse, France
E-mail: prade@irit.fr

Library of Congress Control Number: 2010920282

CR Subject Classification (1998): G.2, F.4.1, I.2.3, I.2.4, D.3

LNCS Sublibrary: SL 3 – Information Systems and Application, incl. Internet/Web and HCI

ISSN 0302-9743
ISBN-10 3-642-11828-3 Springer Berlin Heidelberg New York
ISBN-13 978-3-642-11828-9 Springer Berlin Heidelberg New York

springer.com

Printed in Germany

Typesetting: Camera-ready by author, data conversion by Scientific Publishing Services, Chennai, India
Printed on acid-free paper SPIN: 12990597 06/3180 5 4 3 2 1 0

Preface

This volume contains the articles presented at the 6th International Symposium on Foundations of Information and Knowledge Systems (FoIKS 2010) which was held in Sofia, Bulgaria during February 15–19, 2010.

The FoIKS symposia provide a biennial forum for presenting and discussing theoretical and applied research on information and knowledge systems. The goal is to bring together researchers with an interest in this subject, share research experiences, promote collaboration and identify new issues and directions for future research.

FoIKS 2010 solicited original contributions dealing with any foundational aspects of information and knowledge systems. This included submissions that apply ideas, theories or methods from specific disciplines to information and knowledge systems. Examples of such disciplines are discrete mathematics, logic and algebra, model theory, information theory, complexity theory, algorithmics and computation, statistics, and optimization.

Previous FoIKS symposia were held in Pisa (Italy) in 2008, Budapest (Hungary) in 2006, Vienna (Austria) in 2004, Schloss Salzau near Kiel (Germany) in 2002, and Burg/Spreewald near Berlin (Germany) in 2000. FoIKS took up the tradition of the conference series Mathematical Fundamentals of Database Systems (MFDBS), which initiated East–West collaboration in the field of database theory. Former MFDBS conferences were held in Rostock (Germany) in 1991, Visegrad (Hungary) in 1989, and Dresden (Germany) in 1987.

The FoIKS symposia are a forum for intense discussions. Speakers are given sufficient time to present their ideas and results within the larger context of their research. Furthermore, participants are asked in advance to prepare a first response to a contribution of another author.

Suggested topics for FoIKS 2010 included, but were not limited to:

- Database design: formal models, dependencies and independencies
- Dynamics of information: models of transactions, concurrency control, updates, consistency preservation, belief revision
- Information fusion: heterogeneity, views, schema dominance, multiple source information merging, reasoning under inconsistency
- Integrity and constraint management: verification, validation, consistent query answering, information cleaning
- Intelligent agents: multi-agent systems, autonomous agents, foundations of software agents, cooperative agents, formal models of interactions, logical models of emotions
- Knowledge discovery and information retrieval: machine learning, data mining, formal concept analysis and association rules, text mining, information extraction

- Knowledge representation, reasoning and planning: non-monotonic formalisms, probabilistic and non-probabilistic models of uncertainty, graphical models and independence, similarity-based reasoning, preference modeling and handling, argumentation systems
- Logics in databases and AI: classical and non-classical logics, logic programming, description logic, spatial and temporal logics, probability logic, fuzzy logic
- Mathematical foundations: discrete structures and algorithms, graphs, grammars, automata, abstract machines, finite model theory, information theory, coding theory, complexity theory, randomness
- Security in information and knowledge systems: identity theft, privacy, trust, intrusion detection, access control, inference control, secure Web services, secure Semantic Web, risk management
- Semi-structured data and XML: data modeling, data processing, data compression, data exchange
- Social computing: collective intelligence and self-organizing knowledge, collaborative filtering, computational social choice, Boolean games, coalition formation, reputation systems
- The Semantic Web and knowledge management: languages, ontologies, agents, adaption, intelligent algorithms
- The WWW: models of Web databases, Web dynamics, Web services, Web transactions and negotiations

The call for papers resulted in the submission of 50 full articles. A rigorous refereeing process saw each submitted article referred by at least three international experts. The 13 articles judged best by the Program Committee were accepted for long presentation. In addition, six articles were accepted for short presentation. This volume contains versions of these articles that have been revised by their authors according to the comments provided in the reviews. After the conference, authors of a few selected articles were asked to prepare extended versions of their articles for publication in a special issue of the journal *Annals of Mathematics and Artificial Intelligence.*

We would like to thank all authors who submitted articles and all conference participants for the fruitful discussions. We are grateful to Philippe Balbiani, Jan Paredaens, and Dimiter Vakarelov, who presented invited talks at the conference. We would also like to thank the members of the Program Committee and external referees for their timely expertise in carefully reviewing the submissions. A special thank-you goes to Markus Kirchberg for his outstanding work as FoIKS Publicity Chair. Finally, we wish to express our appreciation to Stefan Dodunekov and his team for being our hosts and for the wonderful days in Sofia.

February 2010

Sebastian Link
Henri Prade

Conference Organization

Program Committee Chairs

Sebastian Link	Victoria University of Wellington, New Zealand
Henri Prade	Université de Toulouse, France

Program Committee

Leila Amgoud	University of Toulouse, France
Lyublena Antova	Cornell University, USA
Marcelo Arenas	Pontificia Universidad Católica de Chile, Chile
Salem Benferhat	University of Lens, France
Jonathan Ben-Naim	University of Toulouse, France
Leopoldo Bertossi	Carleton University, Canada
Philippe Besnard	University of Toulouse, France
Joachim Biskup	University of Dortmund, Germany
Piero A. Bonatti	University of Naples "Federico II", Italy
Gerhard Brewka	University of Leipzig, Germany
Balder ten Cate	INRIA, France
Jan Chomicki	University at Buffalo, USA
Samir Chopra	City University of New York, USA
Marina De Vos	University of Bath, UK
Michael I. Dekhtyar	Tver State University, Russia
James P. Delgrande	Simon Fraser University, Canada
Jürgen Dix	Clausthal University of Technology, Germany
Stefan Dodunekov	Bulgarian Academy of Sciences, Bulgaria
Thomas Eiter	Vienna University of Technology, Austria
Ronald Fagin	IBM Almaden Research Center, San Jose, USA
Victor Felea	"Al.I. Cuza" University of Iasi, Romania
Flavio Ferrarotti	University of Santiago de Chile and Yahoo! Research Latin America, Chile
Floris Geerts	The University of Edinburgh, UK
Lluis Godo	Artificial Intelligence Research Institute (IIIA - CSIC), Spain
Edward Hermann Haeusler	Pontifícia Universidade Católica, Brazil
Joe Halpern	Cornell University, USA
Sven Hartmann	Clausthal University of Technology, Germany
Stephen J. Hegner	Umeå University, Sweden

Andreas Herzig	University of Toulouse, France
Eyke Hüllermeier	University of Marburg, Germany
Anthony Hunter	University College London, UK
Yasunori Ishihara	Osaka University, Japan
Ulrich Junker	ILOG, France
Gyula O.H. Katona	Alfréd Rényi Institute, Hungarian Academy of Sciences, Hungary
Gabriele Kern-Isberner	University of Dortmund, Germany
Hans-Joachim Klein	University of Kiel, Germany
Henning Koehler	The University of Queensland, Australia
Phokion G. Kolaitis	University of California, Santa Cruz and IBM Almaden Research Center, USA
Sébastien Konieczny	University of Lens, France
Gerhard Lakemeyer	RWTH Aachen University, Germany
Jérôme Lang	University of Paris 9, France
Mark Levene	Birkbeck University of London, UK
Thomas Lukasiewicz	University of Oxford, UK
Sebastian Maneth	NICTA and University of New South Wales, Australia
Pierre Marquis	University of Artois, France
Carlo Meghini	Institute of Information Science and Technologies, Italy
Leora Morgenstern	New York University, USA
Wilfred S.H. Ng	Hong Kong University of Science and Technology, Hong Kong
Juliana Peneva	New Bulgarian University, Bulgaria
Patrice Perny	University of Paris 6, France
Attila Sali	Alfréd Rényi Institute, Hungarian Academy of Sciences, Hungary
Vladimir Sazonov	University of Liverpool, UK
Francesco Scarcello	Universita' degli Studi della Calabria, Italy
Torsten Schaub	University of Potsdam, Germany
Klaus-Dieter Schewe	Information Science Research Centre, New Zealand
Karl Schlechta	Université de Provence, France
Dietmar Seipel	University of Würzburg, Germany
Guillermo R. Simari	Universidad Nacional del Sur, Argentina
Margarita Spiridonova	Bulgarian Academy of Sciences, Bulgaria
Nicolas Spyratos	University of Paris-South, France
Letizia Tanca	Politecnico di Milano, Italy
Bernhard Thalheim	University of Kiel, Germany
Miroslaw Truszczynski	University of Kentucky, USA
José María Turull-Torres	Massey University Wellington, New Zealand

Jan Van den Bussche	Universiteit Hasselt, Belgium
Wiebe van der Hoek	University of Liverpool, UK
Dirk Van Gucht	Indiana University, USA
Victor Vianu	University of California San Diego, USA
Millist Vincent	University of South Australia, Australia
Irina B. Virbitskaite	Russian Academy of Sciences, Russia
Evgenii E. Vityaev	Russian Academy of Sciences, Russia
Peter Vojtas	Charles University, Czech Republic
Jef Wijsen	University of Mons-Hainaut, Belgium
Mary-Anne Williams	University of Technology, Sydney, Australia
Masatoshi Yoshikawa	Kyoto University, Japan

External Referees

Loredana Afanasiev	Universiteit van Amsterdam, The Netherlands
Foto Afrati	National Technical University of Athens, Greece
Diego Arroyuelo	Yahoo! Research Latin America, Chile
Pablo Barcelo	University of Chile, Chile
Nils Bulling	Clausthal University of Technology, Germany
Michael Fink	Vienna University of Technology, Austria
Olivier Gauwin	INRIA, France
Rita Hartel	University of Paderborn, Germany
Thomas Krennwallner	Vienna University of Technology, Austria
Aurelien Lemay	University of Lille, France
Jörg Pührer	Vienna University of Technology, Austria
Luigi Sauro	University of Naples "Federico II", Italy
Qing Wang	University of Otago, New Zealand
Marco Zaffalon	Dalle Molle Institute for Artificial Intelligence, Switzerland

Local Arrangements Chair

Stefan Dodunekov	Bulgarian Academy of Sciences, Bulgaria

Publicity Chair

Markus Kirchberg	Institute for Infocomm Research, A*STAR, Singapore

Table of Contents

Invited Talks

Regular Articles

Tools and Techniques in Qualitative Reasoning about Space

Philippe Balbiani

CNRS — Université de Toulouse
Institut de recherche en informatique de Toulouse
118 ROUTE DE NARBONNE, 31062 TOULOUSE CEDEX 9, France
Philippe.Balbiani@irit.fr

As a subfield of artificial intelligence, qualitative reasoning is about that kind of knowledge representation languages and automated deduction methods that is used by scientists and engineers when a precise quantitative description of the physical bodies is not available or when a complete quantitative calculation of their relationships is not feasible. A special area of qualitative reasoning is concerned with the qualitative aspects of representing and reasoning about spatial entities. Applications of qualitative spatial reasoning (QSR) can be found in natural language processing [1], spatial information systems [8], etc. They have given rise to numerous knowledge representation languages and automated deduction methods for space.

One particular formalism, namely *RCC*, has been more deeply considered: the region connection calculus [10]. The fundamental approach of *RCC* takes extended entities such as spatial regions as the primary notions. The basis of RCC is one binary relation C read as "connects with". If one thinks of regions as being regular subsets of a topological space S, then $C(x, y)$ will hold when at least one point in S is incident both in x and y. Other binary predicates relating pairs of regions can be defined in terms of C: "partially overlaps", "is part of", etc. A theory comprising a set of 8 jointly exhaustive and pairwise distinct relations, namely $RCC8$, has been introduced and developed within the context of constraint satisfaction. Some of the computational properties of $RCC8$ have been established by Renz and Nebel [11] who proved that reasoning with $RCC8$ is NP-complete and who considered a maximal tractable fragment of the region connection calculus.

Other works in spatial reasoning have also addressed other aspects of space such as size [7], direction [9], etc. Intended as a short introductory course for students who wish to study QSR deeply, this talk will give the necessary background on some of the numerous knowledge representation languages and automated deduction methods for space that have been proposed so far: rectangle algebra [2], reasoning with lines [3], egg-yolk approaches [4], contact algebras [5,6], topological and size information [7], cardinal directions [9]. It does not presuppose training in mathematics or logic.

S. Link and H. Prade (Eds.): FoIKS 2010, LNCS 5956, pp. 1–2, 2010.

References

1. Asher, N., Vieu, L.: Toward a geometry of common sense: a semantics and a complete axiomatization of mereotopology. In: Proceedings of the 14th International Joint Conference on Artificial Intelligence, pp. 846–852. Morgan Kaufmann, San Francisco (1995)
2. Balbiani, P., Condotta, J.-F., Fariñas del Cerro, L.: A model for reasoning about bidimensional temporal relations. In: Proceedings of the 6th International Conference on Knowledge Representation and Reasoning, pp. 124–130. Morgan Kaufmann, San Francisco (1998)
3. Challita, K.: Reasoning with lines in the Euclidean space. In: Proceedings of the 21st International Joint Conference on Artificial Intelligence. Association for the Advancement of Artificial Intelligence, pp. 462–467 (2009)
4. Cohn, A., Gotts, N.: The 'egg-yolk' representation of regions with indeterminate boundaries. In: Geographic Objects with Indeterminate Boundaries, pp. 171–188. Taylor & Francis, Abington (1996)
5. Dimov, G., Vakarelov, D.: Contact algebras and region-based theory of space: a proximity approach — I. Fundamenta Informaticæ, 209–249 (2006)
6. Düntsch, I., Winter, M.: A representation theorem for Boolean contact algebras. Theoretical Computer Science 347, 498–512 (2005)
7. Gerevini, A., Renz, J.: Combining topological and size information for spatial reasoning. Artificial Intelligence 137, 1–42 (2002)
8. Laurini, R., Thompson, D.: Fundamentals of Spatial Information Systems. Academic Press, London (1992)
9. Ligozat, G.: Reasoning about cardinal directions. Journal of Visual Languages and Computing 9, 23–44 (1998)
10. Randell, D., Cui, Z., Cohn, A.: A spatial logic based on regions and connection. In: Proceedings of the 3rd International Conference on Knowledge Representation and Reasoning, pp. 165–176. Morgan Kaufmann, San Francisco (1992)
11. Renz, J., Nebel, B.: On the complexity of qualitative spatial reasoning: a maximal tractable fragment of the region connection calculus. Artificial Intelligence 108, 69–123 (1999)
12. Stell, J.: Boolean connection algebras: a new approach to the region-connection calculus. Artificial Intelligence 122, 111–136 (2000)

A Simple but Formal Semantics for XML Manipulation Languages

Jan Paredaens[1] and Jan Hidders[2]

[1] University of Antwerp, Antwerp, Belgium
jan.paredaens@ua.ac.be
[2] University of Technology, Delft, The Netherlands
a.j.h.hidders@tudelft.nl

Abstract. XML is a modern format that is nowadays used to store many documents, expecially on the Web. Since a set of documents can be considered as a semi-structured dataset or database, we want to query these documents. Moreover in many applications we need to transform documents into other documents, containing the same information, but with a different structure. Finally, most documents are represented in HTML on the Web, which can be seen as XML-documents.

We discuss sublanguages of XPath, XQuery and XSLT. The latter are well-used manipulation languages for XML that were developed during the last decade in which we can express these queries and transformations. XPath is a simple language that enables us to navigate through a document. XQuery is a powerful query language for XML and XSLT is a transformation language. These three languages are conceptually totally different. For each of them we have defined an upward compatible sublanguage respectively MiXPath, MiXQuery and MiXSLT. These are three manipulation languages for XML, whose semantics is defined in a formal, uniform, compact and elegant way.

These three languages will enable us later on to investigate more easily certain aspects such as the expressive power of certain types of expressions found in XPath, XQuery and XSLT, the expressive power of recursion and possible syntactical restrictions that let us control this power, the complexity of deciding equivalence of expressions for purposes such as query optimization, the functional character in comparison with functional languages such as LISP and ML, the role of XPath 1.0 and 2.0 in XQuery in terms of expressive power and query optimization, and the relationship between queries on XML and the classical well-understood concept of generic database queries. The contribution of MiXPath, MiXQuery and MiXSLT is their relatively simple syntax and semantics that is appropriate both for educational and research purposes. Indeed, we are convinced that these languages have a number of interesting properties, that can be proved formally, and that can be transposed to XPath, XQuery and XSLT.

The semantics of MiXPath, MiXQuery and MiXSLT use the same data model, called *a store*, which is the formal representation of a sequence of XML-documents

S. Link and H. Prade (Eds.): FoIKS 2010, LNCS 5956, pp. 3–5, 2010.

and fragments of XML-documents. The actual value for each variable and parameter and some local values are defined in the *environment*. After defining the syntax of each language we give the formal semantics for each syntactic rule.

MiXPath. MiXPath is a simple language that enables us to navigate through a document. It has a Unix-like syntax and is not closed. It is upward compatible with XPath 1.0, that is defined by W3C, and whose complete definition can be found on "`http://www.w3.org/TR/xpath/`".

MiXPath is a language whose expressions have as input an XML document and a node, and as output a sequence of nodes of that document or atomic values. Each expression consists of a sequence of *steps* that contain each an *axis* that generates a sequence of nodes and of a *node-test* and a *predicate* that are both filters on the generated node sequence. Typical syntax rules are:

$$\begin{array}{ll} \langle \mathit{Step} \rangle & \rightarrow (\text{ ``.''} \mid \text{``..''} \mid (\langle \mathit{Axis} \rangle \text{ ``::''})^? \langle \mathit{Node\text{-}test} \rangle) \\ & \quad (\text{ ``[''} (\langle \mathit{P\text{-}Pred} \rangle \mid \langle \mathit{Position} \rangle) \text{ ``]''})^* \\ \langle \mathit{Node\text{-}test} \rangle & \rightarrow \mathit{Name} \mid \text{``*''} \mid \text{``}\mathtt{node()}\text{''} \end{array}$$

They describe the general form of a step and of a node-test, their semantics are expressed by the following rules, where $St, En \vdash_P exp \Rightarrow (St', v)$ indicates that given the store St and the environment En the evaluation of the expression exp results in the value v and the new store St':

$$\frac{St, En \vdash_P \mathtt{self::node()} \Rightarrow (St', v)}{St, En \vdash_P \mathtt{.} \Rightarrow (St', v)} \qquad \frac{St, En \vdash_P \mathtt{parent::node()} \Rightarrow (St', v)}{St, En \vdash_P \mathtt{..} \Rightarrow (St', v)}$$

$$\frac{St, En \vdash_P \mathtt{child::}\ nt \Rightarrow (St', v)}{St, En \vdash_P nt \Rightarrow (St', v)}$$

$$\frac{St, En \vdash_P ax \Rightarrow (St', [n_1, \ldots, n_m]) \qquad v = [n_i \mid \lambda(n_i) = name]}{St, En \vdash_P ax\mathtt{::}name \Rightarrow (St', v)}$$

$$\frac{St, En \vdash_P ax \Rightarrow (St', [n_1, \ldots, n_m]) \qquad v = [n_i \mid \lambda(n_i) \neq \mathtt{text()}]}{St, En \vdash_P ax\mathtt{::}\ \mathtt{*} \Rightarrow (St', v)}$$

$$\frac{St, En \vdash_P ax \Rightarrow (St', v)}{St, En \vdash_P ax\mathtt{::}\ \mathtt{node()} \Rightarrow (St', v)}$$

$$\frac{\begin{array}{c} St, En \vdash_P st \Rightarrow (St_0, [n_1, \ldots, n_m]) \qquad En' = En[\mathbf{m} \mapsto m] \\ \{\ St_{i-1}, En'[\mathbf{x} \mapsto n_i][\mathbf{k} \mapsto i] \vdash_P pprp \Rightarrow (St_i, b_i)\ \}_{1 \leq i \leq m} \qquad v = [n_i | b_i = \mathbf{true}] \end{array}}{St, En \vdash_P st\ \mathtt{[}\ pprp\ \mathtt{]} \Rightarrow (St_m, v)}$$

MiXQuery. MiXQuery is a query language for XML documents. It is upward compatible with XQuery, that is defined by W3C, and whose complete definition can be found on "`http://www.w3.org/TR/xquery`". MiXQuery is a versatile functional query language, whose queries have as input zero or more XML documents and have as output a sequence of one or more XML documents, nodes or atomic values. MiXQuery is Turing complete and recursive.

The typical syntactic entity is here the *flower expression* whose syntax and formal semantics is partially given by

$$\langle \mathit{ForExpr}\rangle \rightarrow \text{“}\mathtt{for}\text{”}\ \langle \mathit{Var}\rangle\ (\ \text{“}\mathtt{at}\text{”}\ \langle \mathit{Var}\rangle\ )?\ \text{“}\mathtt{in}\text{”}\ \langle \mathit{Q\text{-}Expr}\rangle\ \text{“}\mathtt{return}\text{”}\ \langle \mathit{Q\text{-}Expr}\rangle$$

$$\frac{\begin{array}{c} St, En \vdash_Q qe \Rightarrow (St_0, [x_1, \ldots, x_m]) \\ \{\ St_{i-1}, En[\mathbf{v}(name) \mapsto x_i][\mathbf{v}(name') \mapsto i] \vdash_Q qe' \Rightarrow (St_i, v_i)\ \}_{1 \le i \le m} \end{array}}{St, En \vdash_Q \mathtt{for}\ \$name\ \mathtt{at}\ \$name'\ \mathtt{in}\ qe\ \mathtt{return}\ qe' \Rightarrow (St_m, v_1 \circ \ldots \circ v_m)}$$

MiXSLT. MiXSLT is a language for transformations that originated from the world of stylesheets and is mainly used to transform XML-documents into HTML-documents. It is upward compatible with XSLT, that is defined by W3C, and whose complete definition can be found on “`http://www.w3.org/TR/xslt/`”. A stylesheet or program is a set of functions, called templates. There are two kinds of templates, those who can be *called* in a traditional way, and those who can be *applied* using a match method that is rather complicated. A MiXSLT stylesheet is syntactically an XML-document. The syntax of the Apply-templates-expression is

⟨*Apply-templates*⟩ → “`<xsl:apply-templates`”
  “`select = "`”⟨*S*⟩-Expr“`"`” (“`mode = `”*String*)? “`>`”
  ⟨*With-param*⟩*
  “`</xsl:apply-templates>`”

Let I be an input-document and S be a script or a part of a stylesheet. The semantics are based on the environment-generating function $\nu_{S,I}$, that transforms an environment and an XML-fragment into an environment, and the fragment-generating function $\Phi_{S,I}$ that transforms an environment and an XML-fragment into an XML-fragment. We finally give the semantics of the rule above:

Let $p_1 \ldots p_k$ be with-parameter elements. Let $\Phi_{S,I}((\mathbf{v}, \mathbf{x}), p_i) =< s_i, v_i >, 1 \le i \le k$. Let $(St, (\mathbf{v}, \mathbf{x})) \vdash_S S\text{-}Exp \Rightarrow [n^1, \ldots, n^l]$, with all n^j nodes. For each $j, 1 \le j \le l$, let the set of names of the parameters of $\mathbf{TemplateToApply}_S(St, n^j, md)$ be a subset of $\{s_1 \ldots s_k\}$ and let $et_1^j \ldots et_{m^j}^j$ be its body. Let $En^j = (\bot[s_1 \mapsto v_1] \ldots [s_k \mapsto v_k], n^j), 1 \le j \le l$, then:

$\nu_{S,I}((\mathbf{v}, \mathbf{x}),$`<xsl:apply-templates select = "`$S\text{-}Expr$`" mode = "`$md$`">`$p_1 \ldots p_k$
`</xsl:apply-templates>`$) = (\mathbf{v}, \mathbf{x})$
$\Phi_{S,I}((\mathbf{v}, \mathbf{x}),$`<xsl:apply-templates select = "`$S\text{-}Expr$`" mode = "`$md$`">`$p_1 \ldots p_k$
`</xsl:apply-templates>`$) = \Phi_{S,I}(En^1, et_1^1 \ldots et_{m^1}^1) \ldots \Phi_{S,I}(En^l, et_1^l \ldots et_{m^l}^l)$

Algorithmic Definability and Completeness in Modal Logic
(Extended Abstract)*

Dimiter Vakarelov

Department of Mathematical Logic,
Faculty of Mathematics and Computer Science,
Sofia University blvd James Bourchier 5,
1164 Sofia, Bulgaria
dvak@fmi.uni-sofia.bg

One of the nice features of modal languages is that sometimes they can talk about abstract properties of the corresponding semantic structures. For instance the truth of the modal formula $\Box p \Rightarrow p$ in the Kripke frame (W, R) is equivalent to the reflexivity of the relation R. Using a terminology from modal logic [13], we say that the condition of reflexivity – $(\forall x)(xRx)$, is a first-order equivalent of the modal formula $\Box p \Rightarrow p$, or, that the formula $\Box p \Rightarrow p$ is first-order definable by the condition $(\forall x)(xRx)$. More over, adding the formula $\Box p \Rightarrow p$ to the axioms of the minimal modal logic K we obtain a complete logic with respect to the class of reflexive frames and the completeness proof can be done by the well known in modal logic canonical method (such formulas are called canonical). Let us note that definability and completeness are some of the good features in the applications of modal logic, and hence it is important to have algorithmic methods for establishing such properties. In our talk we will describe several algorithmic approaches to this problem.

The first algorithmic result of such a kind is the famous Sahlqvist theorem [9] for first-order definability and completeness in modal logic. It describes by a direct syntactic definition a large effective class of modal formulas (subsequently called Sahlqvist formulas) which are first-order definable and canonical, and more over it presents an algorithm for computing their first-order equivalents. For a long time the class of Sahlqvist formulas was considered as the optimal syntactically defined class with these two properties. Sahlqvist theorem was recently generalized in several ways extending considerably the original Sahlqvist class for the most general modal languages [8,10]. This method will be called syntactical, because it gives explicit syntactical definition of the class of formulas on which definability algorithm works. Extending in such a way the Sahlqvist class has some unpleasant features, the definitions of the extension are quite complicate and the formulas in the extension have numerous syntactic limitations.

* This work is partially supported by Contract 63/2009, Sofia University.

S. Link and H. Prade (Eds.): FoIKS 2010, LNCS 5956, pp. 6–8, 2010.

Another algorithmic method, related only to modal definability, uses the fact that finding first-order equivalents of modal formulas is a special case of second-order quantifier elimination (see the recent book [6]). In the literature there are two basic algorithms for this purpose – SCAN, introduced by Gabbay and Ohlbach and DLS, introduced by Doherty, Łukaszewicz, and A. Szałas (see [6] for their description). Both algorithms are designed to eliminate monadic second-order quantifiers from formulas α of the form $(\exists P_1 \ldots \exists P_n)A$, where A is a Skolemized first-order formula. To apply these algorithms to modal formulas, the later are first translated into the second-order monadic logic. SCAN is a semidecidable procedure which eliminates the second-order quantifiers using resolution techniques, while DLS is a transformation algorithm based on a special lemma of Ackermann. Both algorithms use also skolemization and reverse skolemization (called also unskolemization). These two algorithms are incomparable, but both succeed on Sahlqvist formulas. DLS was also extended to works on some formulas definable in the extension of first-order logic with least fixpoints (such formulas are called μ-definable).

Recently another algorithm for computing first-order equivalents of modal formulas has been proposed [1,2,3], called SQEMA (**S**econd-Order **Q**uantifier **E**limination for **M**odal formulas using **A**ckermann's lemma). SQEMA is based on a modal version of the Ackermann Lemma, works directly on modal formulas and hence does not use skolemization and unskolemization. SQEMA succeeds with all Sahlqvist formulas and all polyadic Sahlqvist formulas (called inductive formulas in [2,3]). An implementation of SQEMA working online is given in [7]. In [4] an extension of SQEMA is given, based on special substitutions which succeeds also on all complex Sahlqvist formulas introduced in [10,4]. All versions of SQEMA have another nice property, which is not known to be true for SCAN and DLS, namely, all modal formulas on which SQEMA succeeds are both first-order definable and canonical. In this way SQEMA can be considered also as a result of Sahlqvist kind, because the algorithms also defines effectively in an indirect way a class of modal formulas which are both canonical and first-order definable. The result of [4] can be considered also as one of the largest known extensions of the Sahlqvist theorem. In [5] another extension of SQEMA is given with the property that all formulas for which the algorithm succeeds are μ-definable. This extension succeeds on all complex recursive modal formulas introduced in [4] and can be considered as a generalization of the definability part of the Sahlqvist theorem. It is also an algorithm which succeeds on the largest known class of μ-definable modal formulas.

A third direction in algorithmic definability has an algebraic flavor. It is based on an observation that modal definability in some cases can be reduced to the problem of solving equations in modal algebras over Kripke frames by means of some algebraic generalizations of Ackermann Lemma (see [11,12]). The Ackermann Lemma in this approach presents a necessary and sufficient first-order condition on the modal structure which guaranties existence of a solution of the considered system of equations. In [11,12] several generalizations of the Ackermann Lemma are given making possible to solve definability problem for large classes of first-order or μ-definable formulas. In some sense this approach is equivalent to the approach given by some versions of SQEMA, because it gives a new mathematical interpretation of SQEMA and a new intuition of the procedures on which SQEMA is based.

References

1. Conradie, W., Goranko, V., Vakarelov, D.: Elementary canonical formulae: a survey on syntactic, algorithmic, and model-theoretic aspects. In: Schmidt, R., Pratt-Hartmann, I., Reynolds, M., Wansing, H. (eds.) Advances in Modal Logic, vol. 5, pp. 17–51. Kings College, London (2005)
2. Conradie, W., Goranko, V., Vakarelov, D.: Algorithmic correspondence and completeness in modal logic I: The core algorithm SQEMA. Logical Methods in Computer Science 2(1:5) (2006)
3. Conradie, W., Goranko, V., Vakarelov, D.: Algorithmic correspondence and completeness in modal logic II. Polyadic and hybrid extensions of the algorithm SQEMA. Journal of Logic and Computation 16, 579–612 (2006)
4. Conradie, W., Goranko, V., Vakarelov, D.: Algorithmic Correspondence and Completeness in Modal Logic. III. Extensions of the Algorithm SQEMA with Substitutions. Fundam. Inform. 92(4), 307–343 (2009)
5. Conradie, W., Goranko, V., Vakarelov, D.: Algorithmic correspondence in modal logic. In: V: Recursive extensions of the algorithm SQEMA (2009) (submitted)
6. Gabbay, D.M., Schmidt, R., Szałas, A.: Second-Order Quantifier Elimination: Foundations, Computational Aspects and Applications. Studies in Logic, vol. 12. College Publications (2008)
7. Georgiev, D., Tinchev, T., Vakarelov, D.: SQEMA - an algoritm for computing first-order equivalents in modal logic: a computer realization. In: Pioneers of Bulgarian Mathematics, International Conference dedicated to Nicola Obrechkoff and Lubomir Tschakaloff, Sofia, Abstracts, July 8-10 (2006)
8. Goranko, V., Vakarelov, D.: Elementary Canonical Formulae: Extending Sahlqvist Theorem. Annals of Pure and Applied Logic 141(1-2), 180–217 (2006)
9. Sahlqvist, H.: Correspondence and completeness in the first and second-order semantics for modal logic. In: Kanger, S. (ed.) Proc. of the 3rd Scandinavial Logic Symposium, Uppsala 1973, pp. 110–143. North-Holland, Amsterdam (1975)
10. Vakarelov, D.: Modal definability in languages with a finite number of propositional variables, and a new extention of the Sahlqvist class. In: Balbiani, P., Suzuki, N.-Y., Wolter, F., Zakharyaschev, M. (eds.) Advances in Modal Logic, vol. 4, pp. 495–518. King's College Publications, London (2003)
11. Vakarelov, D.: Modal definability, solving equations in modal algebras and generalization of the Ackermann lemma. In: Proceedings of 5th Panhellenic Logic Symposium, Athens, July 25-28, pp. 182–189 (2005)
12. Vakarelov, D.: A recursive generalizations of Ackermann lemma with applications to μ-definability. In: Kaouri, G., Zahos, S. (eds.) 6th Panhellenic Logic Symposium PLS 2007, Volos, Greece, July 5-8, pp. 133–137 (2007) (Extended abstracts)
13. van Benthem, J.F.A.K.: Modal Logic and Classical Logic. Bibliopolis, Napoli (1983)

A Probabilistic Temporal Logic That Can Model Reasoning about Evidence

Dragan Doder[1], Zoran Marković[2], Zoran Ognjanović[2], Aleksandar Perović[3], and Miodrag Rašković[2]

[1] Faculty of Mechanical Engineering, Kraljice Marije 16, 11120 Belgrade, Serbia
ddoder@mas.bg.ac.rs

[2] Mathematical Institute of Serbian Academy of Sciences and Arts, Kneza Mihaila 36, 11000 Belgrade, Serbia
{zoranm,zorano,miodragr}@mi.sanu.ac.rs

[3] Faculty of Transportation and Traffic Engineering, Vojvode Stepe 305, 11000 Belgrade, Serbia
pera@sf.bg.ac.rs

Abstract. The aim of the paper is to present a sound, strongly complete and decidable probabilistic temporal logic that can model reasoning about evidence.

1 Introduction

The present paper offers a solution of the problem proposed by Halpern and Pucella in [5], in which they presented a first order logic for reasoning about evidence and left as an open problem the existence of complete propositional logic that can model evidence.

In short, we have constructed propositional counterparts of both logics developed in [5] and proved strong completeness theorems for both of them. From the technical point of view, we have modified some of our earlier developed completion techniques presented in [8,9]. The addition of weight for evidence, next and until operator to the standard probabilistic formalism introduced by Fagin, Halpern and Megiddo in the early nineties (see [1]) induced some technical difficulties, so we have to somewhat extensively modify techniques presented in [8,9].

Concerning decidability of the introduced systems, it can be straightforwardly derived form the corresponding results presented in [5,8], so we have omitted repetition of the argumentation presented there.

There are many relevant papers concerning evidence, temporal and probabilistic logics, rendering almost impossible to credit all of them in one place. Therefore, we have decided to include only those references that are absolutely essential for this paper. For temporal logic we refer the reader to [2]; for probabilistic logic, we refer the reader to [1,3,6,7]; for evidence, we refer the reader to [5,10,11].

The rest of the paper is organized as follows: in Section 2 we present some basic concepts concerning evidence; in Section 3 we present a propositional counterpart

S. Link and H. Prade (Eds.): FoIKS 2010, LNCS 5956, pp. 9–24, 2010.

of the first formal system introduced in [5]; in Section 4 we present a propositional counterpart of the second formal system introduced in [5], with additional until operator; concluding remarks are in the final section.

2 Preliminaries

Halpern and Fagin [4] have suggested that evidence can be seen as a function from prior beliefs to beliefs after making an observation (posterior beliefs). Let us consider the following example.

One coin is randomly chosen from a box that contains also double-headed coins. The coin is tossed, and it lands heads. After that observation we examine two possible hypotheses: the coin is fair, or it is double-headed.

Intuitively, the probability of hypotheses depends on:

- the prior probabilities of the hypotheses (i.e. the proportion of double-headed coins in the box);
- in what extend the observations support the hypotheses.

The second item is formalized by the weight of evidence - the function which assigns a number from the unit interval to every observation and hypothesis. In this section, we give a short overview of how evidence is modeled and formalized, in the case of one or more observations.

Let $\mathcal{P}$ be the set of propositional letters and let $For_{\mathcal{P}}$ be corresponding set of formulas. A function $\mu : For_{\mathcal{P}} \longrightarrow [0,1]$ which satisfies

1. $\mu(\phi) = 1$, whenever ϕ is a tautology;
2. $\mu(\phi) = 0$, whenever ϕ is a contradiction;
3. $\mu(\phi) = \mu(\psi)$, whenever $\phi \leftrightarrow \psi$ is a tautology;
4. $\mu(\phi \vee \psi) = \mu(\phi) + \mu(\psi) - \mu(\phi \wedge \psi)$

will be called a (finitely additive) probability measure.

Let $H = \{h_1, \ldots, h_m\}$ be the set that represents mutually exclusive and exhaustive hypotheses and $O = \{o_1, \ldots, o_n\}$ be the set of possible observations.

For the hypothesis h_i, let μ_i be a likelihood function on O, i.e., the function which satisfies

- $\mu_i : O \longrightarrow [0,1]$;
- $\mu_i(o_1) + \ldots + \mu_i(o_n) = 1$.

We assume that for every $o \in O$ there is $i \in \{1, \ldots, m\}$ such that $\mu_i(o) > 0$. An evidence space is a tuple $\mathtt{E} = \langle H, O, \mu_1, \ldots, \mu_m \rangle$. For an evidence space $\mathtt{E}$, we define a weight function $w_{\mathtt{E}}$ by the following conditions:

- $w_{\mathtt{E}} : O \times H \longrightarrow [0,1]$;
- $w_{\mathtt{E}}(o_i, h_j) = \frac{\mu_j(o_i)}{\mu_1(o_i)+\ldots+\mu_m(o_i)}$.

Halpern and Pucella provided a characterization of weight functions, see [5].

Theorem 1. *Let $H = \{h_1, \ldots, h_m\}$ and $O = \{o_1, \ldots, o_n\}$, and let f be a real-valued function, $f : O \times H \longrightarrow [0, 1]$. Then there exists an evidence space*

$$\mathtt{E} = \langle H, O, \mu_1, \ldots, \mu_m \rangle$$

such that $f = w_{\mathtt{E}}$ iff f satisfies the following properties:

1. *$f(o_i, h_1) + \ldots + f(o_i, h_m) = 1$, for every $i \in \{1, \ldots, n\}$.*
2. *There exists $x_1, \ldots, x_n > 0$ such that, for all $j \in \{1, \ldots, m\}$, $x_1 f(o_1, h_j) + \ldots + x_n f(o_n, h_j) = 1$.*

Moreover, if 1. and 2. are satisfied, then the likelihood functions μ_i, $i \in \{1, \ldots, m\}$ are defined by

$$\mu_j(o_i) = \frac{f(o_i, h_j)}{x_j}.$$

Dempster's rule of combination combines probability distributions ν_1 and ν_2 on $\mathcal{H}$ in the following way: for every measurable $H \subseteq \mathcal{H}$

$$(\nu_1 \oplus \nu_2)(H) = \frac{\sum_{h \in H} \nu_1(h)\nu_2(h)}{\sum_{h \in \mathcal{H}} \nu_1(h)\nu_2(h)}.$$

Let μ be a probability measure on the set $For(H)$ of propositional formulas over H, which satisfies

$$\mu(h_1) + \ldots + \mu(h_m) = 1.$$

Since hypotheses are mutually exclusive, μ should satisfy $\mu(h_i \wedge h_j) = 0$, for $i \neq j$. Then $\mu(h_1 \vee \ldots \vee h_m) = \mu(h_1) + \ldots + \mu(h_m)$, so the the fact that hypotheses are exhaustive may be expressed by the equality $\mu(h_1) + \ldots + \mu(h_m) = 1$.

Note that for any $\phi \in For(H)$, there exists $\phi' \in For(H)$ of the form $\bigvee_{i \in I} h_i$, for some $I \subseteq \{1, \ldots, m\}$, such that $\mu(\phi) = \mu(\phi')$.

Indeed, it is obvious if $\phi \in H$; suppose that $\mu(\phi_1) = \mu(\phi_1')$ and $\mu(\phi_2) = \mu(\phi_2')$, where ϕ_1' is of the form $\bigvee_{i \in I_1} h_i$ and ϕ_2' is of the form $\bigvee_{i \in I_2} h_i$.

Then, $\mu(\phi_1 \wedge \phi_2) = \mu((\phi_1 \wedge \phi_2)')$ and $\mu(\neg\phi_1) = \mu((\neg\phi_1)')$ for $(\phi_1 \wedge \phi_2)' = \bigvee_{i \in I_1 \cap I_2} h_i$ and $(\neg\phi_1)' = \bigvee_{i \in \{1,\ldots,m\} \setminus I_1} h_i$.

In [5] Halpern and Pucella noticed that, for each observation o for which $\mu_1(o) + \ldots + \mu_m(o) > 0$, $w_{\mathtt{E}}(o, h_1) + \ldots + w_{\mathtt{E}}(o, h_m) = 1$ holds, so there is a unique probability measure on $For(H)$ which is an extension of $w_{\mathtt{E}}(o, \cdot)$, such that hypotheses are mutually exclusive. Hence, we will also denote that measure with $w_{\mathtt{E}}(o, \cdot)$. Informally, we may assume that elements of $For(H)$ are subsets of H.

If $\mathtt{E} = \langle H, O, \mu_1, \ldots, \mu_m \rangle$ is an evidence space, we define

$$\mathtt{E}^* = \langle H, O^*, \mu_1^*, \ldots, \mu_m^* \rangle$$

as follows:

- $O^* = \{\langle o^1 \ldots, o^k \rangle | k \in \omega, o^i \in O\}$.

- $\mu_i^* : O \longrightarrow [0,1]$ is defined by

$$\mu_i^*(\langle o^1 \dots, o^k \rangle) = \mu_i(o^1) \cdots \mu_i(o^k).$$

It is shown in [5] that $w_{E^*}(\langle o^1 \dots, o^k \rangle, \cdot) = w_E(o^1, \cdot) \oplus \cdots \oplus w_E(o^k, \cdot)$.

Informally, $w_{E^*}(\langle o^1 \dots, o^k \rangle, h)$ is the weight that hypothesis h is true, after observing $o^1 \dots, o^k$. We will also use the following equality in an axiomatization of our temporal logic:

$$w_{E^*}(\langle o^1 \dots, o^k \rangle, h_i) = \frac{w_{E^*}(o^1, h_i) \cdots w_{E^*}(o^k, h_i)}{w_{E^*}(o^1, h_1) \cdots w_{E^*}(o^k, h_1) + \cdots + w_{E^*}(o^1, h_m) \cdots w_{E^*}(o^k, h_m)}.$$

3 Logic for Reasoning about Evidence, Prior and Posterior Probabilities

3.1 Syntax and Semantics

Let V be a set of propositional letters. The set $For(V)$ of propositional formulas over V is defined inductively as the smallest set containing propositional letters from V and closed under formation rules: if α and β are propositional formulas, then $\neg\alpha$ and $\alpha \wedge \beta$ are propositional formulas. The connectives $\vee, \rightarrow$ and $\leftrightarrow$ are introduced in the usual way. $\top$ denotes an arbitrary tautology and $\bot$ denotes an arbitrary contradiction.

Let $H = \{h_1, \dots, h_m\}$, $O = \{o_1, \dots, o_n\}$ and $C = \{c_1, \dots, c_n\}$.

Definition 1. *We define the set Term of all probabilistic terms recursively as follows:*

- $Term(0) = \{P_0(\alpha), P_1(\alpha) | \alpha \in For(H)\} \cup \{w(o,h) | o \in O, h \in H\} \cup C \cup \{0,1\}$.
- $Term(n+1) = Term(n) \cup \{(\mathtt{f} + \mathtt{g}), (\mathtt{f} \cdot \mathtt{g}), (-\mathtt{f}) | \mathtt{f}, \mathtt{g} \in Term(n)\}$.
- $Term = \bigcup_{n=0}^{\infty} Term(n)$. □

Probabilistic terms will be denoted by $\mathtt{f}, \mathtt{g}$ and $\mathtt{h}$, possibly with indices. Furthermore, we introduce the usual abbreviations : $\mathtt{f} + \mathtt{g}$ is $(\mathtt{f} + \mathtt{g})$, $\mathtt{f} + \mathtt{g} + \mathtt{h}$ is $((\mathtt{f} + \mathtt{g}) + \mathtt{h})$, $\mathtt{f} \cdot \mathtt{g}$ is $(\mathtt{f} \cdot \mathtt{g})$ and $(\mathtt{f} \cdot \mathtt{g}) \cdot \mathtt{h} = \mathtt{f} \cdot (\mathtt{g} \cdot \mathtt{h})$. Similarly, $-\mathtt{f}$ is $(-\mathtt{f})$, $\mathtt{f} - \mathtt{g}$ is $(\mathtt{f} + (-\mathtt{g}))$, $2 = 1 + 1$, $3 = 2 + 1$, $2\mathtt{f} = \mathtt{f} + \mathtt{f}$ and so on.

Definition 2. *A basic probabilistic formula is any formula of the form*

$$\mathtt{f} \geqslant 0.$$

The set For of formulas is the smallest set containing basic probabilistic formulas, observations and hypotheses that is closed under Boolean connectives $\neg$ *and* $\wedge$. □

Formulas will be denoted by ϕ, ψ and θ, possibly with indices. The other Boolean connectives are introduced as in the propositional case. To simplify notation, we define the following abbreviations:

- $\mathtt{f} \leqslant 0$ is $-\mathtt{f} \geqslant 0$.
- $\mathtt{f} > 0$ is $\neg(\mathtt{f} \leqslant 0)$.
- $\mathtt{f} < 0$ is $\neg(\mathtt{f} \geqslant 0)$.
- $\mathtt{f} = 0$ is $\mathtt{f} \leqslant 0 \wedge \mathtt{f} \geqslant 0$.
- $\mathtt{f} \neq 0$ is $\neg(\mathtt{f} = 0)$.
- $\mathtt{f} \geqslant \mathtt{g}$ is $\mathtt{f} - \mathtt{g} \geqslant 0$. Similarly are defined $\mathtt{f} \leqslant \mathtt{g}$, $\mathtt{f} > \mathtt{g}$, $\mathtt{f} < \mathtt{g}$, $\mathtt{f} = \mathtt{g}$ and $\mathtt{f} \neq \mathtt{g}$.

We may assume that rational numbers are also terms. For example, the formula $\frac{1}{3}\mathtt{f} \geqslant \frac{1}{2}\mathtt{g}$ is abbreviation for $2\mathtt{f} - 3\mathtt{g} \geqslant 0$.

A *model* $\mathcal{M}$ is any tuple $\langle \mathtt{E}, \mu, o, h, d_1, \ldots, d_n \rangle$ such that:

- $\mathtt{E} = \langle H, O, \mu_1, \ldots, \mu_m \rangle$ is an evidence space.
- μ is a finitely additive probability measure on $For(H)$, such that

$$\mu(h_1) + \ldots + \mu(h_m) = 1.$$

- $o \in O$ is an observation.
- $h \in H$ is an hypothesis.
- $d_1, \ldots, d_n$ are positive real numbers such that

$$d_1 w_{\mathtt{E}}(o_1, \mathrm{h}_j) + \ldots + d_n w_{\mathtt{E}}(o_n, h_j) = 1$$

  holds (the existence of such numbers is provided by Theorem 1).

It was mentioned before that we may assume that elements of $For(H)$ are subsets of H. This fact will allow us to apply Dempster's Rule of Combination in the following definition.

Definition 3. *Let $\mathcal{M} = \langle \mathtt{E}, \mu, o, h, d_1, \ldots, d_n \rangle$ be any model. We define the satisfiability relation $\models$ recursively as follows:*

- *For $h' \in H$, $\mathcal{M} \models h'$ if $h' = h$.*
- *For $o' \in O$, $\mathcal{M} \models o'$ if $o' = o$.*
- *$\mathcal{M} \models \mathtt{f} \geqslant 0$ if $\mathtt{f}^{\mathcal{M}} \geqslant 0$, where $\mathtt{f}^{\mathcal{M}}$ is recursively defined in the following way:*
  - $0^{\mathcal{M}} = 0$, $1^{\mathcal{M}} = 1$.
  - $c_i^{\mathcal{M}} = d_i$.
  - $P_0(\phi)^{\mathcal{M}} = \mu(\phi)$, $\phi \in For(H)$.
  - $w(o, h)^{\mathcal{M}} = w_{\mathtt{E}}(o, h)$.
  - $P_1(\phi)^{\mathcal{M}} = (\mu \oplus w_{\mathtt{E}}(o, \cdot))(\phi)$, $\phi \in For(H)$.
  - $(\mathtt{f} + \mathtt{g})^{\mathcal{M}} = \mathtt{f}^{\mathcal{M}} + \mathtt{g}^{\mathcal{M}}$.
  - $(\mathtt{f} \cdot \mathtt{g})^{\mathcal{M}} = \mathtt{f}^{\mathcal{M}} \cdot \mathtt{g}^{\mathcal{M}}$.
  - $(-\mathtt{f})^{\mathcal{M}} = -(\mathtt{f}^{\mathcal{M}})$.
- *$\mathcal{M} \models \neg\phi$ if $\mathcal{M} \not\models \phi$.*
- *$\mathcal{M} \models \phi \wedge \psi$ if $\mathcal{M} \models \phi$ and $\mathcal{M} \models \psi$.* □

A formula ϕ is *satisfiable* if there is a $\mathcal{M}$ such that $\mathcal{M} \models \phi$. A formula ϕ is *valid* if it is satisfied in every model. The set T of formulas is *satisfiable* if there is a model $\mathcal{M}$ such that $\mathcal{M} \models \phi$ for all $\phi \in T$.

3.2 Axiomatization

The formal system presented below contains six groups of axioms and two inference rules. Propositional axioms provide syntactical verification of tautology instances and substitution of provably equal terms in formulas. Probabilistic axioms guarantee that probability weights $P(\alpha)$ actually behave as finitely additive probabilities on formulas. Axioms about hypothesis and evidence formally express a setting in which agent perceive exactly one observation and decides a single hypothesis that yields it. Axioms about commutative ordered rings formally provide the usual manipulations with terms (commutativity, associativity etc). Finally, axioms about evidence provide that w actually behaves like a weight function.

Propositional axioms

A1. $\tau(\phi_1, \ldots, \phi_n)$, where $\tau(p_1, \ldots, p_n) \in For_C$ is any propositional tautology.
A2. $\mathtt{f} = \mathtt{g} \rightarrow (\phi(\ldots, \mathtt{f}, \ldots) \rightarrow \phi(\ldots, \mathtt{g}, \ldots))$

Probabilistic axioms ($i \in \{0, 1\}$)

A3. $P_i(\alpha) \geqslant 0$.
A4. $P_i(\top) = 1$.
A5. $P_i(\alpha) = P_i(\beta)$, whenever $\alpha \leftrightarrow \beta$ is a propositional tautology.
A6. $P_i(\alpha \vee \beta) = P_i(\alpha) + P_i(\beta) - P_i(\alpha \wedge \beta)$.

Axioms about hypotheses

A7. $h_1 \vee \ldots \vee h_m$.
A8. $h_i \rightarrow \neg h_j$, for all $i, j \in \{1, \ldots, m\}$, $i \neq j$.

Axioms about observations

A7. $o_1 \vee \ldots \vee o_n$.
A8. $o_i \rightarrow \neg o_j$, for all $i, j \in \{1, \ldots, n\}$, $i \neq j$.

Axioms about commutative ordered rings

A10. $0 < 1$.
A11. $\mathtt{f} + \mathtt{g} = \mathtt{g} + \mathtt{f}$.
A12. $(\mathtt{f} + \mathtt{g}) + \mathtt{h} = \mathtt{f} + (\mathtt{g} + \mathtt{h})$.
A13. $\mathtt{f} + 0 = \mathtt{f}$.
A14. $\mathtt{f} - \mathtt{f} = 0$.
A15. $\mathtt{f} \cdot \mathtt{g} = \mathtt{g} \cdot \mathtt{f}$.
A16. $\mathtt{f} \cdot (\mathtt{g} \cdot \mathtt{h}) = (\mathtt{f} \cdot \mathtt{g}) \cdot \mathtt{h}$.
A17. $\mathtt{f} \cdot 1 = \mathtt{f}$.
A18. $\mathtt{f} \cdot (\mathtt{g} + \mathtt{h}) = (\mathtt{f} \cdot \mathtt{g}) + (\mathtt{f} \cdot \mathtt{h})$.

A19. $\mathtt{f} \geqslant \mathtt{f}$.
A20. $\mathtt{f} \geqslant \mathtt{g} \lor \mathtt{g} \geqslant \mathtt{f}$.
A21. $(\mathtt{f} \geqslant \mathtt{g} \land \mathtt{g} \geqslant \mathtt{h}) \rightarrow \mathtt{f} \geqslant \mathtt{h}$.
A22. $\mathtt{f} \geqslant \mathtt{g} \rightarrow \mathtt{f} + \mathtt{h} \geqslant \mathtt{g} + \mathtt{h}$.
A23. $(\mathtt{f} \geqslant \mathtt{g} \land \mathtt{h} > 0) \rightarrow \mathtt{f} \cdot \mathtt{h} \geqslant \mathtt{g} \cdot \mathtt{h}$.
A24. $(\mathtt{f} \geqslant \mathtt{g} \land \mathtt{h} < 0) \rightarrow \mathtt{f} \cdot \mathtt{h} \leqslant \mathtt{g} \cdot \mathtt{h}$.

Axioms about evidence

A26. $w(o,h) \geqslant 0$, $o \in O$, $h \in H$.
A27. $w(o,h_1) + \ldots + w(o,h_m) = 1$, $o \in O$.
A28. $o \rightarrow P_0(h)w(o,h) = P_1(h)(P_0(h_1)w(o,h_1) + \ldots + P_0(h_m)w(o,h_m))$, $o \in O$, $h \in H$.
A29. $c_1 > 0 \land \ldots \land c_n > 0 \land c_1 w(o_1,h_1) + \ldots + c_n w(o_n,h_1) = 1 \land \ldots \land c_1 w(o_1,h_m) + \ldots + c_n w(o_n,h_m) = 1$.

Inference rules

R1. From ϕ and $\phi \rightarrow \psi$ infer ψ.
R2. From the set of premises

$$\{\phi \rightarrow \mathtt{f} \geqslant -n^{-1} \mid n = 1,2,3,\ldots\}$$

infer $\phi \rightarrow \mathtt{f} \geqslant 0$.

Definition 4. *A formula ϕ is deducible from a set T of sentences ($T \vdash \phi$) if there is an at most countable sequence of formulas $\phi_0, \phi_1, \ldots, \phi$, such that every ϕ_i is an axiom or a formula from the set T, or it is derived from the preceding formulas by an inference rule. A formula ϕ is a theorem ($\vdash \phi$) if it is deducible from the empty set. A set T of formulas is consistent if there is at least one formula from For that is not deducible from T, otherwise T is inconsistent.*

A consistent set T of sentences is said to be maximally consistent if for every $\phi \in For$, either $\phi \in T$ or $\neg\phi \in T$. A set T is deductively closed if for every $\phi \in For$, if $T \vdash \phi$, then $\phi \in T$. □

In our logic, the length of inference may be any successor ordinal lesser than the first uncountable ordinal ω_1.

3.3 Completeness

Theorem 2. *Every consistent set T of formulas can be extended to a maximal consistent set, i.e., the consistent set T^* which satisfies the following condition:*

for each $\phi \in For$, either $\phi \in T^*$, or $\neg\phi \in T^*$.

Proof. The proof is pretty much the same as for Theorem 5 given below, so we have omitted it. □

For the completion T^*, we define a *canonical model* $\mathcal{M}^* = \langle \mathtt{E}, \mu, o, h, d_1, \ldots, d_n \rangle$ as follows:

- $d_i = \sup\{r \in [0,1] \cap \mathbb{Q} \mid T^* \vdash c_i \geqslant r\}$.
- Evidence space $\mathtt{E} = \langle H, O, \mu_1, \ldots, \mu_m \rangle$ is defined by sets H and O and likelihood functions $\mu_j : O \longrightarrow [0,1]$ defined by

$$\mu_j(o_i) = \frac{w_{\mathtt{E}}(o_i, h_j)}{d_j},$$

  where $w_{\mathtt{E}}(o_i, h_j) = \sup\{r \in [0,1] \cap \mathbb{Q} \mid T^* \vdash w(o_i, h_j) \geqslant r\}$.
- $\mu(\phi) = \sup\{r \in [0,1] \cap \mathbb{Q} \mid T^* \vdash P_0(\phi) \geqslant r\}$.
- h is the unique hypothesis such that $T^* \vdash h$.
- o is the unique observation such that $T^* \vdash o$.

Lemma 1. *$\mathcal{M}^*$ is a model.*

Proof. We need to prove that μ is a probability measure, i.e., that:

1. $\mu(\top) = 1$.
2. $\mu(\phi) = \mu(\psi)$, whenever $\phi \leftrightarrow \psi$ is a tautology.
3. $\mu(\phi \vee \psi) = \mu(\phi) + \mu(\psi) - \mu(\phi \wedge \psi)$.

The first and the second item are immediate, so we will prove 3. Using the fact that $\mathbb{Q}$ is dense in $\mathbb{R}$, we may chose increasing sequence $\underline{a}_0 < \underline{a}_1 < \underline{a}_2 < \cdots$ and decreasing sequence $\overline{a}_0 < \overline{a}_1 < \overline{a}_2 < \cdots$ in $\mathbb{Q}$ such that $\lim \underline{a}_n = \lim \overline{a}_n = \mu(\phi)$. By the definition of μ and completeness of T^*, we obtain

$$T^* \vdash P_0(\phi) \geqslant \underline{a}_n \wedge P_0(\phi) < \overline{a}_n$$

for all n. Similarly, we may chose increasing sequences $(\underline{b}_n)_{n\in\omega}$ and $(\underline{c}_n)_{n\in\omega}$, and decreasing sequences $(\overline{b}_n)_{n\in\omega}$ and $(\overline{c}_n)_{n\in\omega}$ in $\mathbb{Q}$, such that $\lim \underline{b}_n = \lim \overline{b}_n = \mu(\psi)$ and $\lim \underline{c}_n = \lim \overline{c}_n = \mu(\phi \wedge \psi)$.

Using axioms about commutative ordered rings, we have

$$T^* \vdash \underline{a}_n + \underline{b}_n - \overline{c}_n \leqslant P_0(\phi) + P_0(\psi) - P_0(\phi \wedge \psi) < \overline{a}_n + \overline{b}_n - \underline{c}_n$$

for all n. Since $\vdash P_0(\phi \vee \psi) = P_0(\phi) + P_0(\psi) - P_0(\phi \wedge \psi)$, we have that

$$T^* \vdash \underline{a}_n + \underline{b}_n - \overline{c}_n \leqslant P_0(\phi \vee \psi) < \overline{a}_n + \overline{b}_n - \underline{c}_n$$

for all n. Finally, from

$$\mu(\phi \vee \psi) = \sup\{r \mid T^* \vdash P_0(\phi \vee \psi) \geqslant r\}$$

and

$$\lim \underline{a}_n + \underline{b}_n - \overline{c}_n = \lim \overline{a}_n + \overline{b}_n - \underline{c}_n = \mu(\phi) + \mu(\psi) - \mu(\phi \wedge \psi),$$

we obtain $\mu(\phi \vee \psi) = \mu(\phi) + \mu(\psi) - \mu(\phi \wedge \psi)$.

Next, we will prove equality $d_1 w_{\mathtt{E}}(o_1, h_1) + \ldots + d_n w_{\mathtt{E}}(o_n, h_1) = 1$. As above, we chose increasing sequences of rational numbers $(\underline{a}^i_k)_{k\in\omega}$ and $(\underline{b}^i_k)_{k\in\omega}$, and decreasing sequences of rational numbers $(\overline{a}^i_k)_{k\in\omega}$ and $(\overline{b}^i_k)_{k\in\omega}$, $i \in \{1, \ldots, n\}$, such that $\lim \underline{a}^i_k = \lim \overline{a}^i_k = d_i$ and $\lim \underline{b}^i_k = \lim \overline{b}^i_k = w_{\mathtt{E}}(o_i, h_1)$ $(i \in \{1, \ldots, n\})$.

Now, the equality is immediate consequence of the following facts:

- $T^* \vdash \underline{a}_k^1 \underline{b}_k^1 + \ldots + \underline{a}_k^n \underline{b}_k^n \leqslant 1$, for all k.
- $T^* \vdash \overline{a}_k^1 \overline{b}_k^1 + \ldots + \overline{a}_k^n \overline{b}_k^n > 1$, for all k.
- $\vdash c_1 w(o_1, h_1) + \ldots + c_n w(o_n, h_1) = 1$.
- $\lim \underline{a}_k^1 \underline{b}_k^1 + \ldots + \underline{a}_k^n \underline{b}_k^n = \lim \overline{a}_k^1 \overline{b}_k^1 + \ldots + \overline{a}_k^n \overline{b}_k^n = d_1 w_{\mathtt{E}}(o_1, h_1) + \ldots + d_n w_{\mathtt{E}}(o_n, h_1)$.

Similarly, we may show that the other conditions of Theorem 1 are satisfied, so $w_{\mathtt{E}}$ is a weight function and it defines likelihood functions μ_i.

Finally, it is obvious from axioms about hypotheses and observations (and completeness of T^*) that there is the unique h such that $T^* \vdash h$ and the unique o such that $T^* \vdash o$. □

Theorem 3 (Strong completeness theorem). *Every consistent set T of formulas is satisfiable.*

Proof. By theorem 2 ,we can extend a consistent set of formulas T to a maximal consistent set T^* and define a model $\mathcal{M}^*$, as above. We have to prove that for every formula ϕ, $\mathcal{M}^* \models \phi$ iff $\phi \in T^*$. The proof is by the induction on the complexity of formulas. We omit the obvious base cases when the formula is an observation or a hypothesis, as well as the cases when it is Boolean combination in the induction step.

Let $\mathtt{f} \geqslant 0 \in T^*$. Using the axioms for ordered commutative rings, we can show that

$$\vdash \mathtt{f} = r_1 \mathtt{g}_1 + \ldots + r_{n_\mathtt{f}} \mathtt{g}_{n_\mathtt{f}},$$

for some $n_\mathtt{f} \in \omega$, where each $\mathtt{g}_i$ is of the form $\mathtt{g}_i = \mathtt{h}_1 \cdots \mathtt{h}_{n_i}$, for some $\mathtt{h}_j \in Term(0) \setminus \{0,1\}$.

Note that $\vdash \mathtt{g}_i \geqslant 0$ and $\mathtt{h}_i^{\mathcal{M}^*} = \sup\{r \in [0,1] \cap \mathbb{Q} \mid T^* \vdash \mathtt{h}_i \geqslant r\}$.

Using increasing sequences of rational numbers $(\underline{a}_k^i)_{k \in \omega}$, and decreasing sequences of rational numbers $(\overline{a}_k^i)_{k \in \omega}$, such that $\lim \underline{a}_k^i = \lim \overline{a}_k^i = \mathtt{h}_i^{\mathcal{M}^*}$, one may show, as in the proof of the previous theorem, that

$$\vdash \underline{a}_k^1 \cdots \underline{a}_k^{n_i} \leqslant \mathtt{g}_i < \overline{a}_k^1 \cdots \overline{a}_k^{n_i},$$

for all $k \in \omega$ and $i \in \{1, \ldots, n_\mathtt{f}\}$, and consequently

$$\mathtt{g}_i^{\mathcal{M}^*} = \sup\{r \in [0,1] \cap \mathbb{Q} \mid T^* \vdash \mathtt{g}_i \geqslant r\}.$$

Without the loss of generality, suppose that $T^* \vdash r_i \geqslant 0$, for $1 \leqslant i \leqslant m_\mathtt{f}$, and $T^* \vdash r_i < 0$, for $m_\mathtt{f} < i \leqslant n_\mathtt{f}$. Once again, using increasing sequences of rational numbers $(\underline{b}_k^i)_{k \in \omega}$, and decreasing sequences of rational numbers $(\overline{b}_k^i)_{k \in \omega}$, such that $\lim \underline{b}_k^i = \lim \overline{a}_k^i = \mathtt{g}_i^{\mathcal{M}^*}$, we have

$$\vdash r_1 \underline{b}_k^1 + \ldots + r_{m_\mathtt{f}} \underline{b}_k^{m_\mathtt{f}} + r_{m_\mathtt{f}+1} \overline{b}_k^{m_\mathtt{f}+1} + \ldots + r_{n_\mathtt{f}} \overline{b}_k^{n_\mathtt{f}} \leqslant \mathtt{f},$$

and

$$\vdash \mathtt{f} < r_1 \overline{b}_k^1 + \ldots + r_{m_\mathtt{f}} \overline{b}_k^{m_\mathtt{f}} + r_{m_\mathtt{f}+1} \underline{b}_k^{m_\mathtt{f}+1} + \ldots + r_{n_\mathtt{f}} \underline{b}_k^{n_\mathtt{f}},$$

for all $k \in \omega$.

Finally,

$$\mathtt{f}^{\mathcal{M}^*} = \sup\{r \in [0,1] \cap \mathbb{Q} \mid T^* \vdash \mathtt{f} \geqslant r\},$$

so $\mathtt{f}^{\mathcal{M}^*} \geqslant 0$ or, equivalently, $\mathcal{M}^* \models \mathtt{f} \geqslant 0$.

For the other direction, let $\mathcal{M}^* \models \mathtt{f} \geqslant 0$. If $\mathtt{f} \geqslant 0 \notin T^*$, by the construction of T^*, there is a positive integer n such that $\mathtt{f} < -n^{-1} \in T^*$. Reasoning as above, we can prove that $\mathtt{f}^{\mathcal{M}^*} \geqslant 0$, a contradiction. So, $\mathtt{f} \geqslant 0 \in T^*$. □

4 Temporal Logic That Can Reason about Evidence

In this section we introduce a temporal logic that can deal with sequences of observations made over the time. We assume that the flow of the time is isomorphic to natural numbers. The language of our logic contains the "next" operator $\bigcirc$ and the "until" operator U.

4.1 Syntax, Semantics and Axiomatization

We define the set $Term$ of all probabilistic terms recursively, similarly as in previous section; the difference is that now we don't need two probability operators, since posterior probability P_1 is now represented as the probability in the next time step.

- $Term(0) = \{P(\alpha) | \alpha \in For(H)\} \cup \{w(o,h) | o \in O, h \in H\} \cup C \cup \{0,1\}$.
- $Term(n+1) = Term(n) \cup \{(\mathtt{f}+\mathtt{g}), (\mathtt{f}\cdot\mathtt{g}), (-\mathtt{f}) | \mathtt{f},\mathtt{g} \in Term(n)\}$.
- $Term = \bigcup_{n=0}^{\infty} Term(n)$.

The set For of formulas is defined recursively as the smallest set that satisfies the following conditions:

- Expressions of the form $f \geqslant 0$, $f \in Term$ (basic probabilistic terms), observations and hypotheses are formulas.
- If ϕ and ψ are formulas, then $\neg\phi$, $\phi \wedge \psi$, $\bigcirc\phi$ and $\phi U \psi$ are formulas.

In order to simplify notation, we define abbreviations (for both terms and formulas) as in previous section. Also, $\bigcirc^0\phi$ is ϕ and $\bigcirc^{n+1}\phi$ is $\bigcirc(\bigcirc^n\phi)$. If T is a set of formulas, then $\bigcirc T$ denotes $\{\bigcirc\phi | \phi \in T\}$ and $\bigcirc^{-1}T$ denotes $\{\phi | \bigcirc\phi \in T\}$.

Furthermore, we introduce temporal operators F (sometimes) and G (always):

- $F\phi$ is $\top U \phi$.
- $G\phi$ is $\neg F \neg \phi$.

An example of the formula is

$$o \wedge w(o,h) \geqslant r \wedge G(P(h) > 0) \rightarrow F(o \rightarrow P(h) \geqslant s)$$

which can be read as "if o is observed, the weight of evidence of o for h is at least r and if the probability of h is always positive, then, sometimes in the future, probability of h will be at least s, if o is observed".

Adopting semantics from the previous section, we define a model $\overline{\mathcal{M}}$ as an infinite sequence $\langle \mathcal{M}_0, \mathcal{M}_1, \mathcal{M}_2, \ldots \rangle$, such that

$$\mathcal{M}_k = \langle \mathtt{E}^*, \mu, h, d_1, \ldots, d_n, o^1, o^2, \ldots, o^k \rangle$$

(specially, $\mathcal{M}_0 = \langle \mathtt{E}^*, \mu, h, d_1, \ldots, d_n \rangle$).

For a model $\overline{\mathcal{M}} = \langle \mathcal{M}_0, \mathcal{M}_1, \mathcal{M}_2, \ldots \rangle$ we define the satisfiability relation $\models$ recursively:

- $\mathcal{M}_k \models h'$ if $h' = h$.
- $\mathcal{M}_k \models o'$ if $o' = o^k$.
- $\mathcal{M} \models \mathtt{f} \geqslant 0$ if $\mathtt{f}^{\mathcal{M}} \geqslant 0$, where $\mathtt{f}^{\mathcal{M}}$ is recursively defined in the following way:
  - $0^{\mathcal{M}_k} = 0$, $1^{\mathcal{M}_k} = 1$, $c_i^{\mathcal{M}_k} = d_i$.
  - $P(\phi)^{\mathcal{M}_k} = \mu \oplus w_{\mathtt{E}^*}(\langle o^1, \ldots, o^k \rangle, \cdot)(\phi)$, $\phi \in For(H)$.
  - $w(\langle o^{i_1}, \ldots, o^{i_k} \rangle, h')^{\mathcal{M}_k} = w_{\mathtt{E}^*}(\langle o^{i_1}, \ldots, o^{i_k} \rangle, h')$.
  - $(\mathtt{f} + \mathtt{g})^{\mathcal{M}_k} = \mathtt{f}^{\mathcal{M}_k} + \mathtt{g}^{\mathcal{M}_k}$.
  - $(\mathtt{f} \cdot \mathtt{g})^{\mathcal{M}_k} = \mathtt{f}^{\mathcal{M}_k} \cdot \mathtt{g}^{\mathcal{M}_k}$.
  - $(-\mathtt{f})^{\mathcal{M}_k} = -(\mathtt{f}^{\mathcal{M}_k})$.
- $\mathcal{M}_k \models \neg\phi$ if $\mathcal{M}_k \not\models \phi$.
- $\mathcal{M}_k \models \phi \wedge \psi$ if $\mathcal{M}_k \models \phi$ and $\mathcal{M}_k \models \psi$.
- $\mathcal{M}_k \models \bigcirc\phi$ if $\mathcal{M}_{k+1} \models \phi$.
- $\mathcal{M}_k \models \phi U \psi$ if there is $l \in \omega$ such that $\mathcal{M}_{k+l} \models \psi$, and for every $l' \in \omega$ such that $l' < l$, $\mathcal{M}_{k+l'} \models \phi$.

Note that U is a strong version of until operator, i.e., if $\phi U \psi$ holds, then ψ must hold in some future time instant.

A set of formulas T is satisfiable if there is a model $\overline{\mathcal{M}}$ and $k \in \omega$ such that $\mathcal{M}_k \models \phi$ holds for every formula $\phi \in T$. A formula ϕ is satisfiable if the set ϕ is satisfiable. A formula ϕ is valid, if for every model $\overline{\mathcal{M}}$ and every $k \in \omega$ $\mathcal{M}_k \models \phi$ holds.

Axiomatization

We will modify axioms from the previous section; the axiomatization includes propositional axioms and axioms about observations, hypotheses and commutative ordered rings, as well as probabilistic axioms, with the difference that we drop indices, since there is only one probabilistic operator P.

Axioms about evidence

Axioms A26, A27, and A29 hold. Axiom A28 needs to be replaced with the following axiom:

A30. $\bigcirc(P(h) \geqslant r) \rightarrow P(h)w(o,h) \geqslant r(P(h_1)w(o,h_1) + \ldots + P(h_m)w(o,h_m))$, $o \in O$, $h \in H$, $r \in [0,1] \cap \mathbb{Q}$.

A31. $w(o^1,h) \cdots w(ob^k,h) =$
$w(\langle o^1, \ldots, o^k \rangle, h)(w(o^1,h_1) \cdots w(o^k,h_1) + \cdots + w(o^1,h_m) \cdots w(o^k,h_m))$.

Temporal axioms

A32. $\bigcirc(\phi \to \psi) \to (\bigcirc\phi \to \bigcirc\psi)$.
A33. $\neg \bigcirc \phi \leftrightarrow \bigcirc\neg\phi$.
A34. $\phi U \psi \leftrightarrow \psi \vee (\phi \wedge \bigcirc(\phi U \psi))$.
A35. $\phi U \psi \to F\psi$.
A36. $\phi \leftrightarrow \bigcirc\phi$, $\phi \in For(H)$.
A37. $\mathtt{f} \geqslant 0 \leftrightarrow \bigcirc(\mathtt{f} \geqslant 0)$, if f does not contain an occurrence of P.

Inference rules

R1. From ϕ and $\phi \to \psi$ infer ψ.
R2. From ϕ infer $\bigcirc\phi$, if ϕ is a theorem.
R3. From the set of premises $\{\phi \to \bigcirc^m\mathtt{f} \geqslant -n^{-1} \mid n = 1,2,3,\ldots\}$ infer $\phi \to \bigcirc^m\mathtt{f} \geqslant 0$ (for any $m \in \omega$).
R4. From the set of premises $\{\phi \to \bigcirc^n\psi \mid n = 1,2,3,\ldots\}$ infer $\phi \to G\psi$.

We list some properties of the temporal part of the above axiomatization:

Lemma 2. *1.* $\vdash \bigcirc(\phi \wedge \psi) \leftrightarrow (\bigcirc\phi \wedge \bigcirc\psi)$.
2. $\vdash \bigcirc(\phi \vee \psi) \leftrightarrow (\bigcirc\phi \vee \bigcirc\psi)$.
3. If $T \vdash \phi$, *then* $\bigcirc T \vdash \bigcirc\phi$.
4. $\{\phi, \bigcirc\phi, \ldots, \bigcirc^{n-1}\phi, \bigcirc^n\psi\} \vdash \phi U \psi$.

4.2 Completeness

Using a straightforward induction on the length of the inference, one can easily prove the following lemma.

Lemma 3. *The above axiomatization is sound with respect to the class of models.*

Theorem 4 (Deduction theorem). *Suppose that T is an arbitrary set of formulas and that $\phi, \psi \in For$. Then, $T \cup \{\phi\} \vdash \psi$ implies $T \vdash \phi \to \psi$.*

Proof. We will use the transfinite induction on the length of the inference. The case when ψ is a theorem is standard, as well as the case when we apply inference rule R1. The case when ψ is obtained by application of R2. is immediate - rule R2. can be applied to the theorems only, so ϕ, ψ and $\phi \to \psi$ are theorems.

Assume that $T \cup \{\phi\} \vdash \psi$, where ψ is obtained by the inference rule R3. Then ψ is of the form $\theta \to \bigcirc^m\mathtt{f} \geqslant 0$, and $T \cup \{\phi\} \vdash \theta \to \bigcirc^m\mathtt{f} \geqslant -n^{-1}$, for all $n \in \omega$. By the induction hypothesis, we obtain $T \vdash \phi \to (\theta \to \bigcirc^m\mathtt{f} \geqslant -n^{-1})$, or, equivalently, $T \vdash (\phi \wedge \theta) \to \bigcirc^m\mathtt{f} \geqslant -n^{-1}$, for all $n \in \omega$. Finally, using R3. we have $T \vdash (\phi \wedge \theta) \to \bigcirc^m\mathtt{f} \geqslant 0$, so $T \vdash \phi \to \psi$.

Suppose that $T \cup \{\phi\} \vdash \psi \to G\theta$ is obtained by the inference rule R4. Then $T \cup \{\phi\} \vdash \psi \to \bigcirc^n\theta$, for all $n \in \omega$. Similarly as above, $T \vdash (\phi \wedge \psi) \to \bigcirc^n\theta$, for all $n \in \omega$. Hence, using R4. we obtain $T \vdash (\phi \wedge \psi) \to G\theta$, or, equivalently, $T \vdash \phi \to (\psi \to G\theta)$. □

Theorem 5. *Every consistent set T of formulas can be extended to a maximal consistent set.*

Proof. Suppose that $For = \{\phi_i \mid i = 0, 1, 2, 3, \ldots\}$. We define a completion T^* of T recursively:

1. $T_0 = T$.
2. If ϕ_i is consistent with T_i, then $T_{i+1} = T_i \cup \{\phi_i\}$.
3. If ϕ_i is not consistent with T_i, then:
   (a) If ϕ_i has the form $\psi \to \bigcirc^m \mathtt{f} \geqslant 0$, then
   $$T_{i+1} = T_i \cup \{\psi \to \bigcirc^m \mathtt{f} < -n^{-1}\},$$
   where n is a positive integer such that T_{i+1} is consistent (the existence of such n is provided by Deduction theorem; if we suppose that $T_i \cup \{\phi \to \bigcirc^m \mathtt{f} < -n^{-1}\}$ is inconsistent for all n, we can conclude that
   $$T_i \vdash \phi \to \bigcirc^m \mathtt{f} \geqslant -n^{-1}$$
   for all n. By R3., $T_i \vdash \phi \to \bigcirc^m f \geqslant 0$, so T would be inconsistent).
   (b) Otherwise, if ϕ_i has the form $\psi \to G\theta$, then
   $$T_{i+1} = T_i \cup \{\psi \to \neg \bigcirc^n \theta\},$$
   where n is a positive integer such that T_{i+1} is consistent (if we suppose that $T_i \cup \{\psi \to \neg \bigcirc^n \theta\}$ is inconsistent for all n, it follows by Deduction theorem that $T_i \vdash \neg(\psi \to \neg \bigcirc^n \theta)$, for all n. Using propositional axioms, we obtain
   $$T_i \vdash \psi \to \bigcirc^n \theta,$$
   for all n, so, by R4., $T_i \vdash \psi \to G\theta$, which contradicts the assumption).
   (c) Otherwise, $T_{i+1} = T_i$.
4. $T^* = \bigcup_{n \in \omega} T_n$.

Obviously, each T_i is consistent. Let us prove that T^* is maximal, i.e., for each $\phi \in For$, either $\phi \in T^*$ or $\neg\phi \in T^*$. Let $\phi = \phi_i$ and $\neg\phi = \phi_j$. If both $\phi \notin T^*$ and $\neg\phi \notin T^*$, then, by construction of T^* and Deduction theorem we obtain $T_i \vdash \neg\phi$ and $T_j \vdash \phi$. If n is positive integer such that $n > i, j$, then $T_n \vdash \phi \wedge \neg\phi$, so T_n would be inconsistent; a contradiction.

Next, we will show that T^* is deductively closed, i.e., that $T^* \vdash \phi$ implies $\phi \in T^*$. Since any axiom is consistent with any consistent set, each instance of any axiom is in $T*$, so it is enough to prove that T^* is closed under inference rules R1.-R4.

R1: Let $\{\phi, \phi \to \psi\} \subseteq T^*$, and suppose that $\phi = \phi_i$, $\psi = \phi_j$ and $\neg\psi = \phi_k$. If $\neg\psi \in T^*$, then for any positive integer n such that $n > i, j, k$, $T_n \vdash \psi \wedge \neg\psi$. Since T_n is consistent, $\neg\psi \notin T^*$, so by maximality of T^*, $\psi \in T^*$.

R2: Let $\vdash \phi$, and $\neg\bigcirc\phi = \phi_i$. If $\neg\bigcirc\phi \in T^*$, then $T_{i+1} \vdash \neg\bigcirc\phi \wedge \bigcirc\phi$ ($\vdash \phi$ implies $\vdash \bigcirc\phi$), so T_{i+1} would be inconsistent. By maximality of T^*, $\bigcirc\phi \in T^*$.

R3: Let $\phi_{l_n} = (\phi \rightarrow \bigcirc^m \mathtt{f} \geqslant -n^{-1}) \in T^*$, for all $n \in \omega$. If $\phi_j = \phi \rightarrow \bigcirc^m \mathtt{f} \geqslant 0 \notin T^*$, then, by maximality of T^*, $\neg(\phi \rightarrow \bigcirc^m \mathtt{f} \geqslant 0) \in T_i$, for some $i \in \omega$. Consequently, $T_i \vdash \phi$. By the construction of T^*, $\phi \rightarrow \bigcirc^m \mathtt{f} < -m^{-1}) \in T_{j+1}$, for some $m \in \omega$. If k is positive integer such that $k > i, j, l_m$, then $T_k \vdash \phi$, $T_k \vdash \bigcirc^m \mathtt{f} < -m^{-1}$ and $T_k \vdash \bigcirc^m \mathtt{f} \geqslant -m^{-1}$. Hence, T_k would be inconsistent.

R4: Let $\phi_{l_n} = (\phi \rightarrow \bigcirc^n \psi) \in T^*$, for all $n \in \omega$. If $\phi_j = \phi \rightarrow G\psi \notin T^*$, then $\neg(\phi \rightarrow G\psi) \in T_i$, for some $i \in \omega$ (so, $T_i \vdash \phi$). By the construction of T^*, $\phi \rightarrow \neg \bigcirc^m \psi \in T_{j+1}$, for some $m \in \omega$. If k is positive integer such that $k > i, j, l_m$, then $T_k \vdash \phi$, so $T_k \vdash \bigcirc^m \psi \wedge \neg \bigcirc^m \psi$; a contradiction.

Finally, T^* is consistent: if $T^* \vdash \bot$, then, by deductive closeness of T^*, $\bot \in T^*$, so $\bot \in T_i$, for some $i \in \omega$; a contradiction. □

Lemma 4. *If T^* is maximal consistent set of formulas, then $\bigcirc^{-1}T^*$ is maximal consistent.*

Proof. If $\bigcirc^{-1}T^* = \{\phi | \bigcirc\phi \in T^*\}$ is not maximal, there exists a formula ϕ such that both $\phi \notin \bigcirc^{-1}T^*$ and $\neg\phi \notin \bigcirc^{-1}T^*$. Then $\bigcirc\phi \notin T^*$ and $\bigcirc\neg\phi \notin T^*$,so, by axiom A33., $\neg \bigcirc \phi \notin T^*$, which is in contradiction with the maximality of T^*.

If $\bigcirc^{-1}T^*$ is inconsistent, there exists a formula ϕ such that $\bigcirc^{-1}T^* \vdash \phi \wedge \neg\phi$. Since $T^* = \bigcirc(\bigcirc^{-1}T^*)$, by lemma 2 we obtain $T^* \vdash \bigcirc(\phi \wedge \neg\phi)$ and, by the same lemma and axiom A33., $T^* \vdash \bigcirc\phi \wedge \neg \bigcirc \phi$; a contradiction. □

Using theorem 5 and lemma 4, for a given consistent theory T we can build a model $\overline{\mathcal{M}}$ as follows:

- First we extend T to the maximal consistent set T_0^*.
- For any positive integer n, let $T_n^* = \bigcirc^{-1}T_{n-1}^*$.
- For any $n \in \omega$, let $\mathcal{M}'_n = \langle \mathtt{E}^*, \mu, h, d_1, \ldots, d_n, o^n \rangle$ be the canonical model of the maximal consistent theory T_n^*, defined similarly as in the subsection 3.3 (with the difference that $\mu(\phi) = \sup\{r \in [0,1] \cap \mathbb{Q} \mid T^* \vdash P(\phi) \geqslant r\}$, since there is only one probability operator in this logic).
- $\mathcal{M}_n = \langle \mathtt{E}^*, \mu, h, d_1, \ldots, d_n, o^1, \ldots, o^n \rangle$, where

$$\mathcal{M}'_k = \langle \mathtt{E}^*, \mu, h, d_1, \ldots, d_n, o^1, \ldots, o^k \rangle,$$

  for all $k \leqslant n$.
- $\overline{\mathcal{M}} = \langle \mathcal{M}_0, \mathcal{M}_1, \mathcal{M}_2, \ldots \rangle$.

The model $\overline{\mathcal{M}}$ is well defined, since axioms A36. and A37. ensure that hypothesis and weight function don't depend on a time moment. Moreover, axiom A30. ensures that the measure μ in the next time moment changes in accordance with the observation in that moment.

The proof of The completeness theorem is based on the proof of theorem 3. We prove that for every formula ϕ, $\mathcal{M}_i \models \phi$ iff $\phi \in T_i^*$, using the induction on the complexity of the formulas. We only have to examine two additional cases in the induction - when the formula is of the form $\bigcirc\phi$ and when it is of the form $\phi U \psi$:

- $\bigcirc\phi \in T_i^*$ iff $\phi \in T_{i+1}^*$ iff (by the induction hypothesis) $\mathcal{M}_{i+1} \models \phi$ iff $\mathcal{M}_i \models \bigcirc\phi$.
- If $\phi U \psi \in T_i^*$, then $F\psi \in T_i^*$, by axiom A35. Let us chose minimal n such that $\bigcirc^n \psi \in T_i^*$. We analyze two cases:
  1. If $n = 0$ then $\psi \in T_i^*$, so $\mathcal{M}_i \models \psi$ (by the induction hypothesis), and obviously $\mathcal{M}_i \models \phi U \psi$.
  2. Let $n > 0$. From $\phi U \psi \in T_i^*$ and axiom A34. we obtain:
     $\psi \vee (\phi \wedge \bigcirc(\phi U \psi)) \in T_i^*$,
     $\psi \vee (\phi \wedge \bigcirc(\psi \vee (\phi \wedge \bigcirc(\phi U \psi)))) \in T_i^*$, and, by lemma 2,
     $\psi \vee (\phi \wedge (\bigcirc\psi \vee (\bigcirc\phi \wedge \bigcirc^2(\phi U \psi)))) \in T_i^*$. Continuing, we have
     $\psi \vee (\phi \wedge (\bigcirc\psi \vee (\bigcirc\phi \wedge \ldots \wedge (\bigcirc^{n-1}\psi \vee (\bigcirc^{n-1}\phi \wedge \bigcirc^n(\phi U \psi)))))) \in T_i^*$.
     Since $\bigcirc^k \psi \notin T_i^*$, for all $k < n$, by maximality of T_i^* it follows that $\bigcirc^k \phi \in T_i^*$, for all $k < n$. Besides, $\bigcirc^n \psi \in T_i^*$. By the induction hypothesis, $\mathcal{M}_{i+k} \models \phi$, for all $k < n$ and $\mathcal{M}_{i+n} \models \psi$, so $\mathcal{M}_i \models \phi U \psi$.

  Conversely, suppose that $\mathcal{M}_i \models \phi U \psi$. Then $\mathcal{M}_{i+n} \models \psi$, for some $n \in \omega$ and $\mathcal{M}_{i+k} \models \phi$, for every $k < n$. It follows from the induction hypothesis that $\psi \in T_{i+n}^*$ and $\phi \in T_{i+k}^*$, for every $k < n$. Consequently, $\bigcirc^n \psi \in T_i^*$ and $\bigcirc^k \phi \in T_i^*$, for every $k < n$. By lemma 2, $\phi U \psi \in T_i^*$.

Thus, we proved the following theorem.

Theorem 6 (Strong completeness theorem). *Every consistent set T of formulas has a model.*

5 Conclusion

As we have mentioned in the introduction, the main contribution of this paper is a solution of the problem formulated by Halpern and Pucella in [5]. Our propositional systems are infinitary, since it is the only nontrivial way to obtain real valued strongly complete propositional logic that can model probability, evidence and temporal operators.

Namely, if we remove the Archimedean rule R2, the obtained system would be incomplete, since

$$T = \{P(p) > 0\} \cup \{P(p) \leqslant 2^{-n} \mid n \in \mathbb{N}\}$$

will be a consistent unsatisfiable theory in it. The key property of the Archimedean rule is the fact that it is a syntactical way to ensure that no nonstandard measure (i.e. measure that can have a proper infinitesimals in its range) can be a model for our systems.

Acknowledgement

This work is supported by grant 144013 of Serbian ministry of science through Mathematical Institute of Serbian Academy of Sciences and Arts.

References

1. Fagin, R., Halpern, J., Megiddo, N.: A logic for reasoning about probabilities. Information and Computation 87(1-2), 78–128 (1990)
2. Gabbay, D., Hodkinson, I., Reynolds, M.: Temporal logic. In: Mathematical Foundations and Computational Aspects, vol. 1. Clarendon Press, Oxford (1994)
3. Godo, L., Marchioni, E.: Coherent conditional probability in a fuzzy logic setting. Logic Journal of the IGPL 14(3), 457–481 (2006)
4. Halpern, J., Fagin, R.: Two views of belief: belief as generalized probability and belief as evidence. Artificial Intelligence 54, 275–317 (1992)
5. Halpern, J., Pucella, R.: A logic for reasoning about evidence. Journal of Artificial Intelligence Research 26, 1–34 (2006)
6. Nilsson, N.: Probabilistic logic. Artificial Intelligence 28, 71–87 (1986)
7. Ognjanović, Z., Rašković, M., Marković, Z.: Probability Logics. In: Ognjanović, Z. (ed.) Logic in Computer Science, Mathematical Institute of Serbian Academy of Sciences and Arts, pp. 35–111 (2009) ISBN 978-86-80593-40-1
8. Ognjanović, Z.: Discrete linear-time probabilistic logics: completeness, decidability and complexity. J. Log. Comput. 16(2), 257–285 (2006)
9. Perović, A., Ognjanović, Z., Rašković, M., Marković, Z.: A probabilistic logic with polynomial weight formulas. In: Hartmann, S., Kern-Isberner, G. (eds.) FoIKS 2008. LNCS, vol. 4932, pp. 239–252. Springer, Heidelberg (2008)
10. Shafer, G.: A Mathematical Theory of Evidence. Princeton University Press, Princeton (1976)
11. Shafer, G.: Belief functions and parametric models (with commentary). Journal of the Royal Statistical Society, Series B 44, 322–352 (1982)

An Algorithm for Generating Nash Stable Coalition Structures in Hedonic Games

Helena Keinänen

Helsinki University of Technology, Faculty of Information and Natural Sciences

Abstract. In this paper, we consider a problem of generating Nash stable solutions in coalitional games. In particular, we present an algorithm for constructing the set of all Nash stable coalition structures from players' preferences in a given additively separable hedonic game. We show the correctness and completeness of the algorithm. Our experiments with several classes of hedonic games demonstrate the usefulness and practical efficiency of the algorithm.

1 Introduction

An important issue in multi-agent knowledge systems is to generate partitioning of self-interested agents into cooperating coalitions which are stable in the sense that no agent has an incentive to deviate from its coalition. Application areas of multi-agent coalition formation include, e.g., automated negotiation and electronic commerce [1], distributed vehicle routing [2], and multi-sensor networks [3]. Efficient computation of all stable solutions in these kinds of settings is crucial in order to find and compare viable allocations of agents or resources.

Hedonic games [4] are coalitional games where all players only give their preferences over coalitions they may belong to. Hedonic games provide a suitable framework to model agents which are self-interested but willing to cooperate. An important solution concept in hedonic games is Nash stability [5,6]. Essentially, solving a hedonic game with respect to this solution concept amounts to partitioning all players into coalition structures, i.e. disjoint coalitions, where no-one has an incentive to deviate from the coalition she belongs to. Despite hedonic games have received much research attention recently (see, e.g., [7] for the most recent results), no algorithms have been presented in the literature for the problem of generating all Nash stable coalition structures of a hedonic game.

We contribute to the literature by introducing a novel algorithm which generates all Nash stable coalition structures of a hedonic game with additively separable preferences. In particular, the presented algorithm is useful in order to compare properties of different Nash stable coalition structures. The algorithm is based on a new Theorem 2 concerning hedonic games, which effectively allows to detect unstable coalitions, and thus the algorithm avoids unnecessary calculation on unstable coalition structures. We show through numerical experiments that the new algorithm can often generate all Nash stable coalition structures by checking only a very small portion (up to 0.0012%) of the all possible solutions of a given game.

S. Link and H. Prade (Eds.): FoIKS 2010, LNCS 5956, pp. 25–39, 2010.

In the field of economics, an extensive research has been directed at hedonic games with the focus on models and stability concepts (see, e.g.,[8,5,6,9,10]). In a hedonic game, the existence of a Nash stable coalition structure is guaranteed with restrictions on players preferences [5,6]. However, with generic preferences the set of Nash stable coalition structures may be empty. Deciding whether there exists a Nash stable partitioning of players in general settings is NP-complete [11]. Also, the problem of deciding existence of a Nash stable partitioning of players with additively separable preferences is NP-complete [12]. Recently, hedonic coalition games have been applied in multi-agent settings and an algorithm for the core membership checking has been proposed in [7].

This paper is structured as follows. First, in Section 2 we present the notations and definitions of the hedonic games together with the problem of Nash stable coalition structure generation. Then, in Section 3 we prove important properties of hedonic games which are useful in the design of our algorithm. In Section 4, we give the formal description of the algorithm for Nash stable coalition structure generation, and show its correctness, completeness and complexity. Finally, In Section 5, we experimentally evaluate the algorithm, and demonstrate the effectiveness of our approach. Finally, in Section 6 we conclude.

2 Hedonic Games

We consider coalitional games with a finite set of players $N = \{1, 2, \ldots, n\}$. A coalition S is a subset of the players $S \subseteq N$. For all players $i \in N$, the set of coalitions which contain i is denoted by $\mathcal{A}^i = \{S \subseteq N \,|\, i \in S\}$. A coalition structure CS over N is a partition of N into mutually disjoint coalitions; that is, for all coalitions $S, S' \in CS$ with $S \neq S'$ we have $S \cap S' = \emptyset$ and $\bigcup_{S \in CS} = N$. Given a player $i \in N$ and a coalition structure CS over N, $CS(i)$ denotes the coalition in CS with the coalition member i. By $\mathcal{C}(N)$ we denote the collection of all coalition structures over N.

For all players $i \in N$, we define a preference relation $\succeq_i$ over $\mathcal{A}^i$. The players preferences are *purely hedonic* in the sense that players order only those coalitions where they are members themselves. The preferences of a player $i \in N$ over $\mathcal{C}(N)$ are completely determined by the preference relation of i such that, for all coalition structures CS, $CS' \in \mathcal{C}(N)$, i weakly prefers CS to CS' if and only if $CS(i) \succeq_i CS'(i)$.

A *hedonic game* G is a tuple

$$G = \langle N, \succeq \rangle$$

where $N = \{1, 2, \ldots, n\}$ is a finite set of players and $\succeq = (\succeq_1, \succeq_2, \ldots, \succeq_n)$ is a profile of preference orders. Given a game $G = \langle N, \succeq \rangle$ we say that G is *additively separable*, if for all players $i \in N$ there exists a valuation function $v_i : N \rightarrow \mathbb{R}$ characterizing $\succeq_i$ such that for all $S, S' \in \mathcal{A}^i$:

$$S \succeq_i S' \text{ iff } \sum_{j \in S} v_i(j) \geq \sum_{j \in S'} v_i(j).$$

We restrict to additively separable hedonic games because their preference profiles have a concise representation of size $|N| \times |N|$. For every $i \in N$ and for every $S \in \mathcal{A}^i$, we shall use $v_i(S)$ as an abbreviation for $\sum_{j \in S} v_i(j)$. Without loss of generality we assume that $v_i(i) = 0$, since the value $v_i(i)$ has no effect on preference relation $\succeq_i$.

There are several sub-classes of additively separable hedonic games such as *aversion to enemies* and *symmetric* games. These are defined as follows. In hedonic games with additively separable preferences, players' coalition partners can be either friends, enemies or neutral partners. A player $j \in N$ is called a friend (enemy) of a player $i \in N$ if and only if $v_i(j) > 0$ ($v_i(j) < 0$). If $v_i(j) = 0$, then the player i considers the player j as a neutral partner. An aversion to enemies game is based on preferences $\succeq$ where, for all players $i \in N$, $v_i(\cdot) \in [-n, 1]$ and $v_i(i) = 0$. A symmetric game is based on preferences where, for all players $i, j \in N$, $v_i(j) = v_j(i)$ holds.

Hedonic games with aversion to enemies and symmetric preferences are interesting sub-cases because they both serve as good examples of hedonic games with structured preference profiles. Aversion to enemies preferences can be used to model coalition formation situations where players divide the others into desirable and undesirable coalition partners. Symmetric preferences can be used to model coalition formation situations where the existence of a Nash stable outcome is required.

In this paper, we focus on the notion of Nash stability in hedonic games. A coalition structure CS is *Nash stable*, if for all $i \in N$

$$CS(i) \succeq_i S_k \cup \{i\}$$

for all $S_k \in CS \cup \{\emptyset\}$. In a Nash stable coalition structure it is guaranteed that no player has an incentive to join another coalition, given that the other players do not deviate from their coalitions. We are concerned with the following problem.

Problem 1 (Nash Stable Coalition Structure Generation). Given a hedonic game $G = \langle N, \succeq \rangle$ the Nash stable coalition structure generation problem is to generate all Nash stable coalition structures of G.

The problem of deciding whether there exists a Nash stable coalition structure in a hedonic game is already known to be NP-complete [11,12]. We now turn to consider properties of hedonic games which are important in solving the Problem 1.

3 Properties of Hedonic Games

In this section we consider properties of hedonic games that are important in generating the set of all Nash stable coalition structures. Previously, it is shown in [5] that there always exists a Nash stable coalition structure in a hedonic game with additively separable preferences which satisfy the symmetry condition $v_i(j) = v_j(i)$ for all $i, j \in N$. Here, we prove bounds on the size of the set of all Nash stable coalition structures, which gives the asymptotic lower bound for Problem 1.

Theorem 1. *The following statement is true for any set of players $N = \{1, 2, \ldots, n\}$: there are additively separable preferences $\succeq = (\succeq_1, \succeq_2, \ldots, \succeq_n)$ such that every coalition structure over N, i.e. every member of $\mathcal{C}(N)$, is Nash stable in $G = \langle N, \succeq \rangle$.*

Proof. We prove the statement with induction on the number n of players in the game G.

When $n = 1$ the statement is true, since $\mathcal{C}(\{1\})$ consists of the only coalition structure $\{\{1\}\}$ which is obviously Nash stable.

So, consider some $k \geq 1$ for which the statement holds. We show that it also holds when $n = k+1$. By the induction hypothesis, with players' preferences $(\succeq_1, \succeq_2, \ldots, \succeq_k)$, every coalition structure in $\mathcal{C}(\{1, 2, \ldots, k\})$ is Nash stable. Now, consider an arbitrary coalition structure $CS \in \mathcal{C}(\{1, 2, \ldots, k\})$. Then, CS can be extended by adding the player $k + 1$, and we have the resulting coalition structure CS' with $k + 1$-players. For all players $1 \leq j \leq k$, let $v_j(k + 1) = 0$. Furthermore, let $v_{k+1}(j) = 0$ for all players $1 \leq j \leq k + 1$. Given these hedonic preferences $(\succeq_1, \succeq_2, \ldots, \succeq_k, \succeq_{k+1})$, no player $i \in \{1, 2, \ldots, k\}$ has an incentive to deviate from $CS'(i)$ since $v_i(CS(i)) = v_i(CS'(i))$. Furthermore, we have $v_{k+1}(CS'(k + 1)) = 0$. This means that the statement holds for every coalition structure in $\mathcal{C}(\{1, 2, \ldots, k + 1\})$, which finalizes the induction step of the proof. □

Intuitively, Theorem 1 follows also from the observation that if all players are totally indifferent between all coalitions, then all coalition structures are Nash stable.

Sandholm et al. [13] show that the number of coalition structures is $O(n^n)$ (with n players). More precisely, it is known that the exact number of coalition structures for an n-player hedonic game is given by the so-called Bell number $B(n)$ defined as $B(n) = \sum_{k=1}^{n} Z(n, k)$ where $Z(n, k) = kZ(n-1, k) + Z(n-1, k-1)$ is a Stirling number of the second kind. The Stirling number gives the number of coalition structures where the players are distributed into k coalitions such that $k \leq n$. For instance, notice that e.g. $B(15) = 1\ 382\ 952\ 545$, and already for a 20-player game the number of coalition structures is of astronomical size.

On the one hand, by Theorem 1 all algorithms for Problem 1 with n-players must have the time complexity $O(B(n)) = O(n^n)$. Whenever the set of Nash stable coalition structures equals $\mathcal{C}(\{1, 2, \ldots, n\})$, every algorithm to solve Problem 1 must use at least $O(B(n)) = O(n^n)$ operations. On the other hand, usually a given hedonic game has only a few Nash stable coalition structures (as experimentally shown in our problem analysis in Sect. 5.2). These facts make Problem 1 computationally difficult and in hedonic games with a large number of players an exhaustive search strategy is neither feasible nor effective in order to generate all Nash stable coalition structures. Therefore, we devise a complete algorithm which usually searches only a small subset of $\mathcal{C}(N)$, but still satisfies the requirement that all Nash stable coalition structures shall be generated.

Our algorithm is based on an observation that, in a Nash stable coalition structure, players with certain kinds of preferences can never be members of the same coalition. In particular, we have the following theorem which is a useful technical result for the design of our algorithm.

Theorem 2. *Let* $G = \langle N, \succeq \rangle$ *be an additively separable hedonic game. Let* $enemies(i, G)$ *denote the set of enemies for player* i *in* G *defined as follows:*

$$enemies(i, G) = \{j \in N \mid v_i(j) < 0\}.$$

For all players $i \in N$ *and for all* $S \subseteq enemies(i, G)$*:*

$$\sum_{l \in S} v_i(l) + \sum_{l \in N, v_i(l) > 0} v_i(l) < 0$$

implies that there does not exist any Nash stable coalition structure $CS \in \mathcal{C}(N)$ *where player* i *and players in* S *are all members of the same coalition.*

Proof. For the sake of a contradiction, suppose that there is a Nash stable coalition structure $CS \in \mathcal{C}(N)$ where i and all players in S are members of the same coalition $CS(i)$. Then, supposing that the inequality $\sum_{l \in S} v_i(l) + \sum_{l \in N, v_i(l) > 0} v_i(l) < 0$ holds, we show that the player i has an incentive to deviate from $CS(i)$. Let us partition the coalition $CS(i)$ into three disjoint subsets: $\{i\}$, S and $CS(i) \setminus (\{i\} \cup S)$. Consider the preferences of the player i on these subsets. It is clear that $v_i(\{i\}) = 0$ and $v_i(S) < 0$. Also, we know by the assumption of Nash stable coalition structures that $v_i(CS(i) \setminus (\{i\} \cup S)) \leq \sum_{l \in N, v_i(l) > 0} v_i(l)$ holds for the third of the subsets. It follows that $v_i(\{i\}) + v_i(S) + v_i(CS(i) \setminus (\{i\} \cup S)) < 0$. Thus, $v_i(CS(i)) < 0$ holds. Recall that $v_i(\{i\}) = 0$ holds. Therefore, $v_i(CS(i)) < v_i(\{i\})$ holds, too. Clearly, i has an incentive to deviate from the coalition $CS(i)$ to the coalition $\{i\}$, and the coalition structure CS cannot be Nash stable. □

Intuitively, the meaning of the inequality in the above theorem is that in a Nash stable coalition structure of a hedonic game with additively separable preferences a player can never be a member of a coalition where resides too many enemies for the player. In the following section we introduce an algorithm based on Theorem 2.

4 Algorithm for Generating Nash Stable Coalition Structures

The formal description of our algorithm is given in Alg. 1. Our algorithm is a variation of an anytime, breadth-first search (BFS) algorithm for optimal coalition structure generation for characteristic function games [13,14] (see the merge algorithm in Sect. 3 of [14]). The BFS merge algorithm in [13,14] essentially constructs a so-called *coalition structure graph* whose nodes are coalition structures in $\mathcal{C}(N)$, and it is based on an approach in which all coalition structures are processed one at a time in the order of the BFS.

Our variation of the algorithm corresponds to the special case of the BFS on a directed acyclic graph where coalition structures in $\mathcal{C}(N)$ are nodes and, for all

Algorithm 1. GenerateNashStableCoalitionStructures(G)

1: enqueue $\{\{1\}, ..., \{n\}\}$;
2: $workset := \emptyset$;
3: **while** the queue is non-empty **do**
4: dequeue a coalition structure CS and examine it:
5: **if** for some player $i \in N$: $\sum_{j \in CS(i), v_i(j)<0} v_i(j) + \sum_{j \in N, v_i(j) \geq 0} v_i(j) < 0$ **then**
6: do nothing with CS;
7: **else**
8: **if** CS is Nash stable **then**
9: print CS;
10: **end if**
11: let $merges(CS) = \begin{cases} \emptyset & \text{if } |CS| = 1, \\ \{(((CS \setminus S) \setminus S') \cup (S \cup S')) | S, S' \in CS \wedge S \neq S'\} & \text{otherwise.} \end{cases}$
12: **for** all $CS' \in merges(CS)$ **do**
13: **if** $CS' \notin workset$ **then**
14: enqueue CS';
15: $workset := workset \cup \{CS'\}$;
16: **end if**
17: **end for**
18: **end if**
19: **if** $|CS| > 1$ **then**
20: let CS'' be front of the queue;
21: **if** $|CS| > |CS''|$ **then**
22: $workset := \emptyset$;
23: **end if**
24: **end if**
25: **end while**

$CS, CS' \in \mathcal{C}(N)$, there is a directed edge (CS, CS') iff CS' is obtained from CS by merging exactly two coalitions of CS. The algorithm starts the BFS from the root node $\{\{1\}, \{2\}, \ldots, \{n\}\}$, iterates the while-loop over a subset of coalition structures in $\mathcal{C}(N)$, and enqueues each encountered coalition structure at most once in the course of the BFS. In addition, our variation of the BFS algorithm is distinguished from the existing algorithm [14,13] as follows:

(i) it processes and stores in memory the coalition structures on one BFS level at a time, instead of building and storing the whole coalition structure graph
(ii) once the algorithm detects a coalition structure violating Theorem 2, it is not enqueued at all, thus, avoiding also the processing of its all descendants.

Let us explain the execution of Alg. 1 in more detail. The algorithm takes as its input a hedonic game $G = \langle N, \succeq \rangle$ with players $N = \{1, 2, \ldots, n\}$. It starts by putting the coalition structure $\{\{1\}, \{2\}, \ldots, \{n\}\}$, which consists of singleton coalitions, onto the queue (line 1), and then sets a working storage *workset* as empty (line 2). The *workset* is used for the purpose of ensuring that no coalition structure is enqueued more than once in the course of the execution of the algorithm (lines 13, 15 and 19–24). The essential functionality of the algorithm is to iteratively process coalition structures from the queue (line 4), and to examine whether they contain coalitions violating Theorem 2 (line 5)[1]. If a dequeued coalition structure violates Theorem 2, then we forget it and do not process it at all (line 6). The coalition structures not violating Theorem 2 are further processed as follows (line 7). It is first checked whether or not the coalition structure under consideration is Nash stable (according to the definition given in Sect. 2), and Nash stable coalition structures are printed out (lines 8–10). Then, all of the BFS successors of the coalition structure under consideration are calculated and enqueued, if not already placed before in the queue (lines 11–17). Essentially, the BFS successors are all coalition structures that can be obtained by merging exactly two coalitions to form a single coalition (line 11). Finally, the algorithm terminates when the last coalition structure form the queue has been examined (lines 3 and 23).

The following theorem establishes the correctness of Alg. 1.

Theorem 3. *Upon termination the algorithm has printed out every Nash stable coalition structure of the input hedonic game.*

Proof. Suppose that there is a Nash stable CS which has not been printed out until the algorithm converges. Then, by the assumption and by the construction of the algorithm there must be a coalition structure CS' such that:

(i) the condition of the first if-clause in line 5 holds for the CS', and
(ii) either the $CS = CS'$ or CS can be obtained from CS' by merging coalitions.

Otherwise, CS must have been placed in the queue while executing the algorithm, and finally CS would have been printed out in the second if-clause in lines 8–9 before termination.

By (i), there is some player i and some $C \subset CS'(i)$ s.t.

$$\sum_{l \in C} v_i(l) + \sum_{l \in N, v_i(l) > 0} v_i(l) < 0$$

holds. Clearly, if $CS = CS'$, then by Theorem 2 CS cannot be Nash stable. Furthermore, if $CS \neq CS'$ and CS can be obtained from CS' via a number of coalition merges, then $C \subset CS(i)$ must hold too. It follows by Theorem 2 that CS cannot be Nash stable. Thus, all Nash stable coalition structures has been printed out whenever the execution of the algorithm terminates. □

The worst-case time complexity of Alg. 1 is as follows.

Theorem 4. *Given an n-player hedonic game, the algorithm converges after $O(n^n)$ and $\omega(n^{n/2})$ iterations of the while-loop, and each iteration requires time $O(2^n)$.*

Proof. The size of $\mathcal{C}(N)$ is $O(n^n)$ and $\omega(n^{n/2})$. The while-loop (lines 3–23) is run for each element in $\mathcal{C}(N)$ at most once. For a single element in $\mathcal{C}(N)$, one

[1] Notice that the algorithm can be easily speed up considerably at the level of implementation, e.g., by computing and storing the second sum in line 5 once for each i.

iteration of the while-loop obviously takes time $O(2^n)$ because by the end of i'th iteration the *workset* consists of all coalition structures of size $n - i + 1$, and clearly, for $i \approx n/2$, this set will be exponential in size. □

By Theorem 1 no other algorithm can solve the problem faster than $O(n^n)$, thus our algorithm is line with the asymptotic lower bound for the problem. In the next section, we show experimentally that the implementation of the algorithm performs quite well in practice, and we demonstrate that the above theoretical asymptotic bound of Theorem 4 is rarely met in practice.

Notice that the space requirement of Alg. 1 is only $O(Z(n, \lceil n/2 \rceil))$ where n is the number of agents and Z is the Stirling number of the second kind (see Sect. 3 for the definition). This space requirement is clearly significantly less than $O(|\mathcal{C}(\{1, 2, \ldots, n\})|)$ which would be required to store all coalition structures. In the next section, our experimental results indicate that in practice Alg. 1 uses typically orders of magnitudes less memory than $O(Z(n, \lceil n/2 \rceil))$.

5 Experimental Results

Alg. 1 proposed in this paper has been implemented in the C programming language [15]. To the best of our knowledge, the proposed algorithm is the first algorithm in the literature to generate all Nash stable coalition structures of a hedonic game. There is no other algorithms that can be used for a numerical comparison in this setting. Thus, for the purposes of this paper several problem instance generators for hedonic games have been implemented to evaluate our algorithm.

5.1 Benchmarks and Setup

As benchmarks we use three different hedonic game classes from [8,5,10], namely generic additively separable, symmetric and aversion to enemies games (see Sect. 2 for the definitions of these classes). Problem instances are generated uniformly at random as follows:

1. A random real valued matrix v of size $|N| \times |N|$ is produced to represent the players' additively separable preferences $v_i : N \to \mathbb{R}$. The i-th line of matrix v represents player's $i \in N$ values for the other players inclusion in the same coalition with i. For every player $i \in N$, and for every coalition $S \in \mathcal{A}^i$ containing i, i's value $v_i(S)$ for coalition S is $\Sigma_{j \in S} v[i, j]$.
2. We iterate over each player's preference list as follows. For player $i \in N$ and another player $j \in N \setminus \{i\}$ in i's preference list, we generate a random real number between -1 and 1 in the generic case, and a random real number either $-|N|$ or 1 in the aversion to enemies case. In the symmetric case, it is also ensured that $v[i, j] = v[j, i]$ always holds.
3. For all $i \in N$, the diagonal values $v[i, i]$ is set constantly 0.

We have instrumented the source code of the implementation of Alg. 1 in order to measure the run-times needed to solve the problems. In particular, the run-times were measured as the number of checked coalition structures before termination. All of the experiments were run with a 2.13GHz Intel Celeron CPU running on Linux with sufficient RAM memory (i.e., 1000 MB) and the experiments reported in the following sub-sections took in excess of 3 weeks CPU time.

5.2 Benchmark Analysis

To investigate the density of Nash stable coalition structures over $\mathcal{C}(N)$, we generated 300 instances of 10-player games, 100 instances per each generic, symmetric and aversion to enemies game classes. Notice that density of Nash stable outcomes is a feature of the problem, and is algorithm independent. We run Alg. 1 on the 300 instances to count the Nash stable coalition structures.

Figure 1 shows the density of the Nash stable coalition structures for the different preference profiles. In case of generic additively separable preferences, we observe that there are not more than 3 Nash stable coalition structures in an instance, and 99 instances of 100 has less than 3 Nash stable coalition structures. Moreover, only 15% of the instances have a Nash stable coalition structure, and mostly there is only one Nash stable coalition structure. In case of symmetric preferences, the number of Nash stable coalition structures is larger.

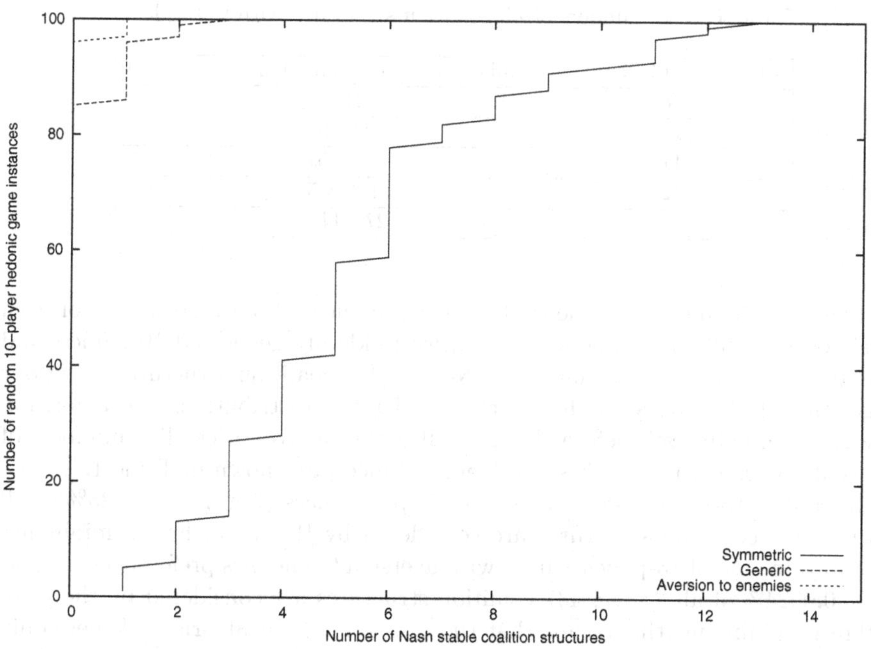

Fig. 1. Portion of Nash stable coalition structures in randomly generated 10-player hedonic games

Recall that symmetric preferences ensure that there always exists at least one Nash stable coalition structure in each game. However, the number of Nash stable coalition structures is still relatively small. There are not more than 13 Nash stable coalition structures in an instance. In case of aversion to enemies preferences, we observe only four instances of 100 having a Nash stable coalition structure.

We see that as the preference distribution changes from the aversion to enemies to the symmetric case, the proportion of Nash stable coalition structures per instance slightly increases. We observe, however, that for all hedonic game classes the numbers of Nash stable solutions are always very small compared to the number of all coalition structures which is 115 975.

5.3 Results

In the first series of experiments we study, for all considered game classes, the run-times of our algorithm on benchmarks with game sizes varying from 9 to 13 players. The aim is to investigate how the run-times of the algorithm increase compared to the number of all coalition structures, as the game sizes grow. That is, in order to evaluate the improvement brought by Theorem 2 (see line 5 of Alg. 1 respectively), we conduct experiments to compare the two variants of Alg. 1 with and without line 5.

Table 1. The numbers of all coalition structures with 9–13 players

Number of players $\|N\|$	Number of all coalition structures $\|\mathcal{C}(N)\|$
9	21 147
10	115 975
11	678 570
12	4 213 597
13	27 644 437

Figures 2, 3 and 4 show the runtime performances of our algorithm. For each preference profile and game size listed, we randomly generated 10 games, and measured run-times to calculate the Nash stable coalition structures using our algorithm. It can be seen that, for all preference distributions the algorithm always terminates well before checking all coalition structures. The numbers of all coalition structures in these problem instances are shown in Table 1.

In case of 13-player games with generic preferences (Fig. 2), only 23% of all 27 644 437 coalition structures are considered by the algorithm in minimum. Notably, in case of 13-player games with aversion to enemies preferences (Fig. 4) only 0.0024% of all 27 644 437 coalition structures are considered by the algorithm in minimum (this means that the run-time of our algorithm is here only 670 coalition structures). This seems to be due to the fact that, in hedonic games with aversion to enemies preferences, players assign large negative values to their enemies, and consequently one enemy for some coalition member is enough to

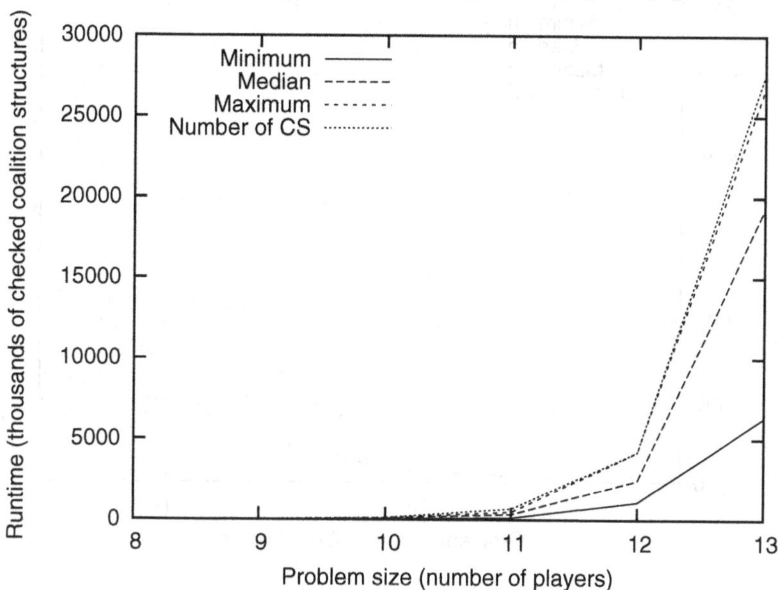

Fig. 2. Run-times for hedonic games with generic preferences

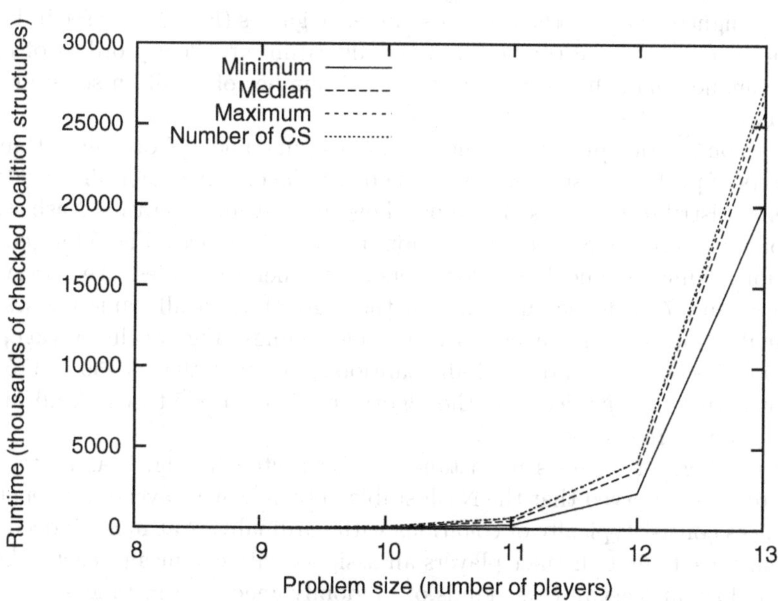

Fig. 3. Run-times for hedonic games with symmetric preferences

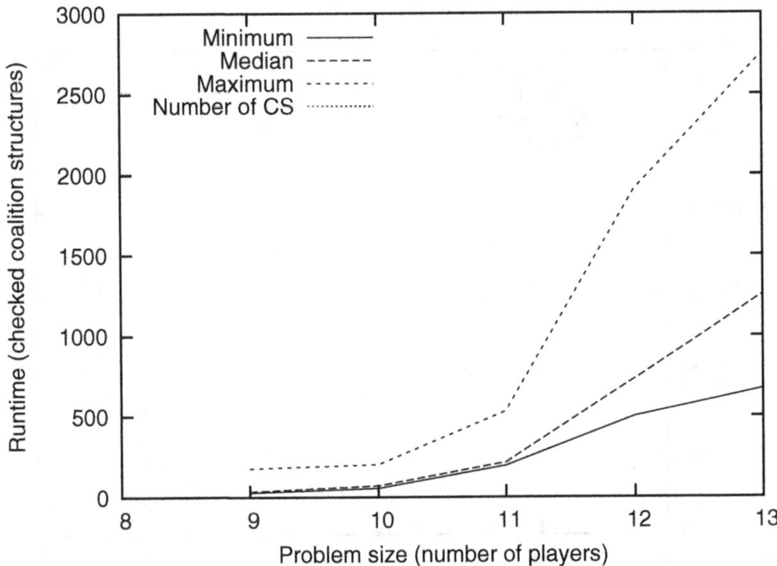

Fig. 4. Run-times for hedonic games with aversion to enemies preferences. Note the scale is here different than in Figs. 2 and 3.

turn the coalition structure violating Theorem 2. We observe that our algorithm performs slightly worse in the class of symmetric games (Fig. 3). We see in Fig. 4 that the increases in the run-times of the algorithm are clearly orders of magnitudes smaller than the increase in the total number of coalition structures in the games.

Our second series of experiments measures run-times of our algorithm on larger sets of problem instances consisting of 10-player games. For all considered preference distributions, Figs. 5, 6 and 7 show the costs of generating Nash stable coalition structures on a set of 100 random problem instances: Fig. 5 for generic, Fig. 6 for symmetric and Fig. 7 for aversion to enemies games. We observe in Figs. 5, 6 and 7 that the run-times of the algorithm on all game classes are very small. Notably, with aversion to enemies games (Fig. 7), in average, the algorithm checks only 0.0012% of all coalition structures, that is 136 of 115 975 coalition structures. Notice that the algorithm never needs to check all of the coalition structures of the games.

The numbers of enemies in instances are reflected in Fig. 7 as a step-like structure. It is observed that the Nash stable outcomes with aversion to enemies preferences consist typically of coalitions with cardinality 2 at most; indeed, the probability that three distinct players all assign positive values to each other is relatively low, at least in these kinds of randomly generated instances.

We observe that the approach presented here relies on a heuristic, which allows one to avoid considering considerable amount of coalition structures. Clearly, the heuristic is in its best whenever most (or at least some) of the values that agents

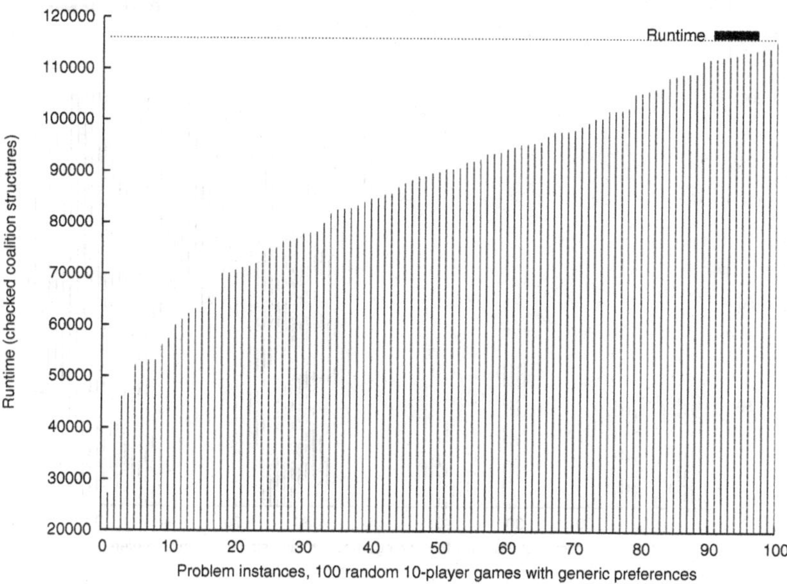

Fig. 5. Run-times for 100 10-player hedonic games with generic preferences. The horizontal line in shows the total number of coalition structures in a 10-player game.

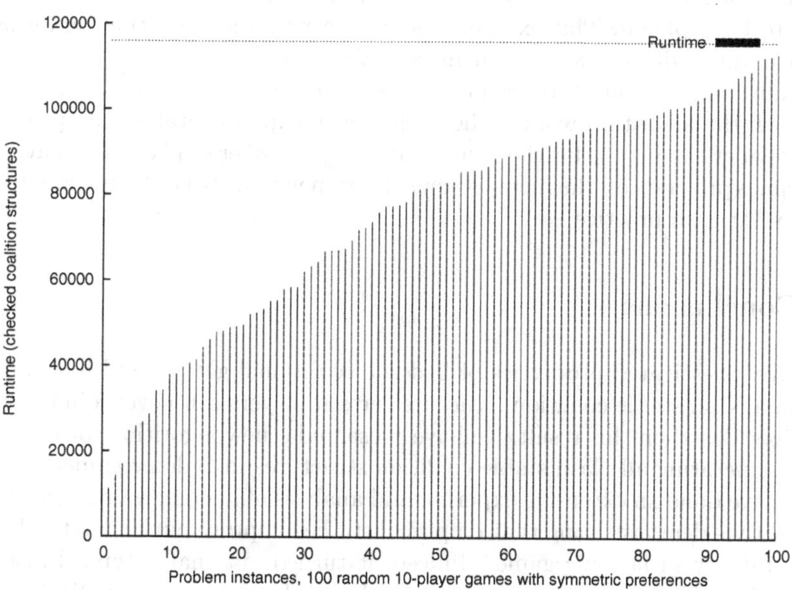

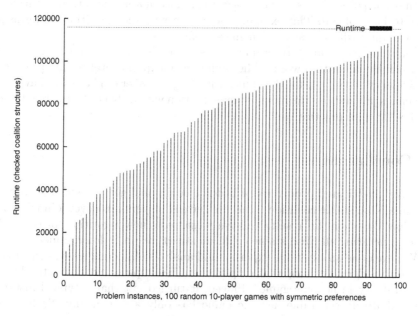

Fig. 6. Run-times for 100 10-player hedonic games with symmetric preferences. The horizontal line in shows the total number of coalition structures in a 10-player game.

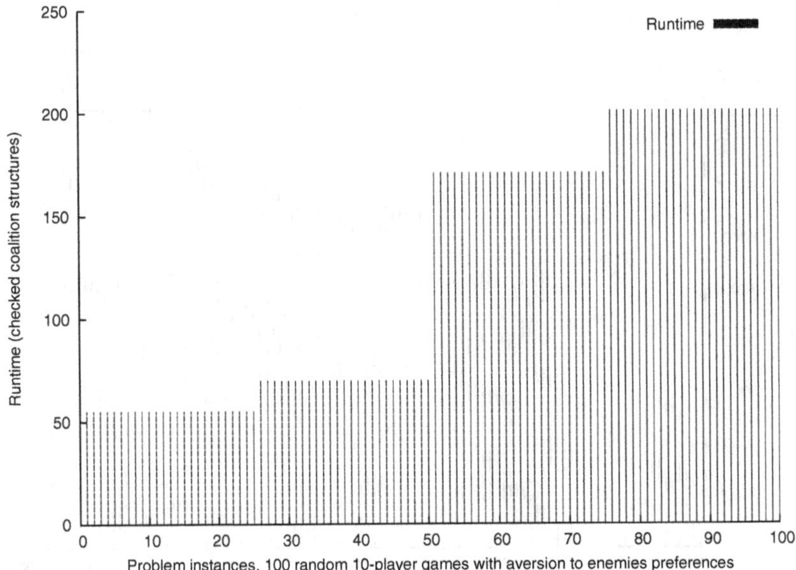

Fig. 7. Run-times for 100 10-player hedonic games with aversion to enemies preferences. Note the scale is here different than in Figs. 5 and 6.

assign to each other are negative; otherwise, unfortunately the entire search space needs to be explored. This explains the reason why the algorithm performs so well on games with aversion to enemies preferences.

Finally, an important further question is what is the largest number of players for which the algorithm works. The preliminary experimental results presented in this paper show that, for the games with large number of players and aversion to enemies preferences, the memory consumption will likely not be the bottleneck for the Alg. 1 in practice.

6 Conclusions

We have investigated properties of hedonic coalitional games, and have proven that in a Nash stable coalition structure certain players can never join the same coalition. Based on this result, we have presented a new algorithm for the problem of generating all Nash stable coalition structures in hedonic games.

We have done an extensive experimental analysis of hedonic game benchmark classes, thus demonstrating that empirical studies improve our understanding of Nash stability in hedonic games. Indeed, it turned out that a typical instance of an additively separable hedonic game has only a few, if any, Nash stable outcomes.

We have also presented extensive experimental results on several benchmark game classes, which demonstrate the efficiency of the presented algorithm in practice. Our experiments indicate that, usually, in order to generate the set of Nash stable coalition structures our algorithm needs to check only a very small portion of all possible coalition structures. For example, in case of aversion to enemies hedonic games our algorithm is able to generate the set of all Nash stable coalition structures by checking only a 0.0024% portion of all possible coalition structures of the games.

Finally, the results in this paper give a baseline for even improved algorithmics which rely on stronger heuristics, which allow one to avoid considering even fever coalition structures. It would be interesting to consider extensions of Theorem 2 to more generic game classes in coalition formation.

References

1. Conitzer, V., Sandholm, T.: Complexity of constructing solutions in the core based on synergies among coalitions. Artificial Intelligence 170, 607–619 (2006)
2. Sandholm, T., Lesser, V.R.: Coalition formation among bounded rational agents. Artificial Intelligence 94, 99–137 (1997)
3. Dang, V.D., Dash, R.K., Rogers, A., Jennings, N.R.: Overlapping coalition formation for efficient data fusion in multi-sensor networks. In: AAAI 2006: Proceedings of the 21st national conference on Artificial intelligence, pp. 635–640. AAAI Press, Menlo Park (2006)
4. Drèze, J.H., Greenberg, J.: Hedonic optimality and stability. Econometrica 4, 987–1003 (1980)
5. Bogomolnaia, A., Jackson, M.O.: The stability of hedonic coalition structures. Games and Economic Behavior 38, 201–230 (2002)
6. Burani, N., Zwicker, W.S.: Coalition formation games with separable preferences. Mathematical Social Sciences 45, 27–52 (2003)
7. Elkind, E., Wooldridge, M.: Hedonic coalition nets. In: AAMAS 2009: Proceedings of The 8th International Conference on Autonomous Agents and Multiagent Systems, Richland, SC, International Foundation for Autonomous Agents and Multiagent Systems, pp. 417–424 (2009)
8. Banerjee, S., Konishi, H., Sonmez, T.: Core in a simple coalition formation game. Social Choice and Welfare 18, 135–153 (2001)
9. Dimitrov, D., Borm, P., Hendrickx, R., Sung, S.C.: Simple priorities and core stability in hedonic games. Social Choice and Welfare 26, 421–433 (2006)
10. Sung, S.C., Dimitrov, D.: On core membership testing for hedonic coalition formation games. Operations Research Letters 35, 155–158 (2007)
11. Ballester, C.: NP-completeness in hedonic games. Games and Economic Behavior 49, 1–30 (2004)
12. Olsen, M.: Nash stability in additively separable hedonic games and community structures. Theory of Computing Systems (2009) (forthcoming)
13. Sandholm, T., Larson, K., Andesson, M., Shehory, O., Tohmé, F.: Coalition structure generation with worst case guarantees. Artificial Intelligence 111, 209–238 (1999)
14. Larson, K.S., Sandholm, T.W.: Anytime coalition structure generation: an average case study. J. Expt. Theor. Artif. Intell. 12, 23–42 (2000)
15. Kernighan, B.W., Ritchie, D.M.: The C Programming Language. Prentice Hall, Englewood Cliffs (1988)

Conjunctive Queries with Constraints: Homomorphism, Containment and Rewriting

Ali Kiani and Nematollaah Shiri

Dept. of Computer Science and Software Engineering,
Concordia University,
Montreal, Quebec, Canada
{ali_kian,shiri}@cse.concordia.ca

Abstract. While the complexity of containment of standard conjunctive queries is NP-complete, adding constraints changes the complexity to Π_2^P-complete. This is because in such cases, we need to consider a team of containment mappings to establish containment as opposed to considering a single containment mapping in the standard case. The situation is different when *homomorphism property* holds, where considering a single mapping is sufficient to establish containment. We identify new classes of conjunctive queries with constraints that satisfy homomorphism property. For each class, we introduce a set of polynomial membership tests. Based on this, we propose an algorithm for rewriting of conjunctive queries with linear arithmetic constraints for which homomorphism property holds.

1 Introduction

It has been shown that the complexity of containment of conjunctive queries with constraints is Π_2^P-complete [9,13]. The reason is that for such queries, two or more containment mappings may team up and satisfy the containment requirement. In other words, to reject the containment of Q_1 in Q_2, we should consider all possible combinations of containment mappings. On the other hand, there are conjunctive queries with constraints that are exceptions and do not require such an expensive test. The class of such queries satisfy the *homomorphism property*. Containment for queries with homomorphism property can be established using a single mapping. Previous studies have investigated classes of conjunctive queries with constraints that satisfy homomorphism property [9,4,2]. In this work, we identify new classes of such queries that satisfy homomorphism property, which extends the results in [2].

In order to keep the complexity of the problem limited to query containment and not directly "dictated" by constraint solving, we restricted the constraints in our study to *Linear Arithmetic Constraints* which can be solved in polynomial time. We called such conjunctive queries as Conjunctive queries with Linear Arithmetic Constraints, or CLAC for short [8,6]. CLAC queries are frequent in real-life database applications. For instance, consider the following query Q on relation schemas $r(A, B, C)$ and $s(D, E, F)$.

S. Link and H. Prade (Eds.): FoIKS 2010, LNCS 5956, pp. 40–57, 2010.

Example 1. [CLAC Query]

$Q_1: \quad h(X) \text{:-} r(X,Y,Z), s(Z,U,W), X < 2Y,\ U \leq W$

The next example shows how the presence of constraints would affect the problems of query containment and rewriting.

Example 2. [Contained rewriting] Consider the same relations r and s and query Q_1 in Example 1, and views V_1 and V_2, defined as follows.

$V_1: \quad v_1(A,Y,Z) \text{:-}\ r(X,Y,Z),\ A = X + Y$

$V_2: \quad v_2(T) \text{:-}\ s(T,U,W), s(T,W,U)$

The following query Q_2 is a contained rewriting for Q_1 using v_1 and v_2.

$Q_2: \quad h(X) \text{:-}\ v_1(A,Y,Z), v_2(Z),\ X = A - Y,\ A < 3Y$

This can be easily shown through unfolding Q_2 using the views [12].

Note that the constraint $U <= W$ in Q_1 does not explicitly appear in the body of Q_2, but implicitly enforced through repeated occurrences of some subgoals and/or their arguments. Identifying such implicit constraints poses challenges in containment test and rewriting of conjunctive queries with constraints. Also, note that the constraint $X = A - Y$ is important as it makes Q_2 a safe query.

The class of CLAC queries extend *conjunctive queries with arithmetic comparisons* which were studied in previous work on query containment and rewriting [9,4,2,1]. Conjunctive queries with arithmetic comparison (AC queries) include constraints in the form $A\theta B$ and $A\theta c$, where A and B are variables, c is a constant, and $\theta \in \{<, \leq, >, \geq, =, \neq\}$.

The organization of the rest of this paper is as follows. Next, we study the containment of queries in standard form and queries with constraints. In section 3, we introduce new classes of conjunctive queries that satisfy homomorphism property and in section 4, we provide a rewriting algorithm for CLAC queries with homomorphism property. Section 5 considers a class of queries without homomorphism property, and Section 6 provides concluding remarks and future work.

2 Containment of Queries with Constraints

As shown in previous studies, the presence of constraints affects the requirements for query containment. To illustrate this point, we compare query containment for conjunctive queries with no constraints (standard case) [3] with query containment for conjunctive queries with arithmetic comparisons [9,5], and show that the results from [9] carry over to CLAC queries. For that, we first review the definition of *containment mapping.*

Consider a partial mapping μ from variables of a query Q_1 to variables/constant of a query Q_2 and extend μ to subgoals so that it maps the subgoals with same predicate name. Such a mapping is a containment mapping if it makes the subgoals in the body of Q_1 a subset of the subgoals in the body of Q_2, and the heads identical.

The following theorems characterize containment of conjunctive queries for the standard case [3], and for conjunctive queries with arithmetic comparisons (*AC queries*)[9].

Theorem 1. *[Containment of Standard Conjunctive Queries] [3]: Let Q_1 and Q_2 be conjunctive queries defined as follows:*

$$Q_1: \quad h(\bar{X}) \text{ :- } g_1(\bar{X_1}), \cdots, g_k(\bar{X_k})$$
$$Q_2: \quad h(\bar{Y}) \text{ :- } p_1(\bar{Y_1}), \cdots, p_l(\bar{Y_l})$$

Q_2 is contained in Q_1 (denoted $Q_2 \sqsubseteq Q_1$) if and only if there exists a containment mapping from Q_1 to Q_2.

Theorem 2. *[Containment of AC Queries] [9]: Let Q_1 and Q_2 be conjunctive queries defined as follows:*

$$Q_1: \quad h(\bar{X}) \text{ :- } g_1(\bar{X_1}), \cdots, g_k(\bar{X_k}), \alpha_1, \cdots, \alpha_n$$
$$Q_2: \quad h(\bar{Y}) \text{ :- } p_1(\bar{Y_1}), \cdots, p_l(\bar{Y_l}), \beta_1, \cdots, \beta_m$$

where α_is and β_js are of the form $A\theta B$ (A is a single variable and B is either a single variable or a constant). Let $C = \{\alpha_i | 1 \leq i \leq n\}$ and $D = \{\beta_j | 1 \leq j \leq m\}$ be the set of constraints in Q_1 and Q_2, respectively. Then, $Q_2 \sqsubseteq Q_1$ if and only if $D \Rightarrow \mu_1(C) \vee \cdots \vee \mu_q(C)$, where μ_i is a containment mapping from Q_1 to Q_2.

Note that the added complexity in containment test for AC queries compared to the standard case is due to the disjunction in the implication test defined in Theorem 2.

2.1 Containment of CLAC Queries

We next show that Theorem 2 is also applicable for testing the containment for CLAC queries. In Lemma 1 below, we show that every CLAC query can be transformed to an equivalent AC query with a set of auxiliary views. Based on this, in Theorem 3, we show that the requirements for containment of CLAC queries are the same as AC queries.

Lemma 1. *Every CLAC query can be transformed to a conjunctive query with arithmetic comparison and a set of auxiliary views.*

Proof: *Let Q_1 be a CLAC query, defined as follows.*

$$Q_1: \quad h(\bar{X}) \text{ :- } g_1(\bar{X_1}), \cdots, g_k(\bar{X_k}), \alpha_1, \cdots, \alpha_n$$

where $\alpha_i = L_i\theta_i R_i$, in which L_i is the left hand side, R_i is the right hand side, and θ_i is a comparison operator.

For each constraint α_i in Q_1 and based on the subgoals in Q_1 that share some variable with L_i or R_i in α_i, we define an auxiliary view v_i, as follows.

$$v_i(N_i, M_i, \bar{X_{i_1}}, \cdots, \bar{X_{i_n}}) \text{ :- } g_{i_1}(\bar{X_{i_1}}), \cdots, g_{i_n}(\bar{X_{i_n}}),\ N_i = L_i,\ M_i = R_i$$

where the N_i and M_i in v_i are two new variables defined based on L_i and R_i (appeared in α_i), and the rest of variables in v_i are the variables/constants of the subgoals contributed to the definition of v_i, in the order in which the subgoals *appear. Note that some variables might appear multiple times in the head of view v_i, however, this is for ease of presentation of the proof and with a minor modification we can remove repeated variables from the head.*

Also, note that not all the subgoals in the query body contribute to the definition of a constraint. For such subgoals, we define a view v_0. The difference between v_0 and other views is that in the head of v_0, we introduce no new variable.

$$v_0(\bar{X_{0_1}}, \cdots, \bar{X_{0_n}}) \text{:-} g_{0_1}(\bar{X_{0_1}}), \cdots, g_{0_n}(\bar{X_{0_n}})$$

Now, we define Q_2 using the set of views as follows.

$$\begin{aligned} Q_2 : \quad h(\bar{X}) \text{:-}\, & v_0(\bar{X_{0_1}}, \cdots, \bar{X_{0_p}}), \\ & v_1(N_1, M_1, \bar{X_{1_1}}, \cdots, \bar{X_{1_p}}),\ N_1\theta_1 M_1, \\ & \cdots, \\ & v_n(N_n, M_n, \bar{X_{n_1}}, \cdots, \bar{X_{n_p}}),\ N_n\theta_n M_n \end{aligned}$$

where θ_i is the comparison operator in α_i. Also, note that for each α_i in the original query, we considered a constraint $N_i\theta_i M_i$ in Q_2 which is in the form of arithmetic comparison.

It can be shown that $Q_2 \equiv Q_1$. For this, we unfold Q_2 based on the definition of v_is [12]. After expansion, we will get the same set of subgoals (some of which might be repeated) and the same set of arithmetic constraints (α_is), because originally we defined every $N_i\theta_i M_i$ based on a constraint α_i. This means, we can transform every CLAC query to an AC query and a set of auxiliary views. ■

The following example illustrates the steps in Lemma 1.

Example 3. We transform the following CLAC query Q_1 to an AC query Q'_1 and a set of auxiliary views.

$$Q_1 : \quad h(X) \text{:-} r(X, Y, Z), s(Z, U, W), X + 2Y > Z,\ U \geq W$$

Here we have two constraints, for each we define an auxiliary view.

[1] $(X + 2Y > Z)$ defines:

$$v_1^1(N_1, M_1, X, Y, Z, Z, U, W) \text{:-} r(X, Y, Z), s(Z, U, W), N_1 = X + 2Y, M_1 = Z$$

[2] $(U \leq W)$ defines:

$$v_2^1(N_2, M_2, Z, U, W) \text{:-} s(Z, U, W), N_2 = U, M_2 = W$$

Note that since the variables in the first constraint come from both subgoals r and s, both subgoals will contribute to the definition of v_1^1, (i.e., subgoals of $(X + 2Y) = \{r(X, Y, Z), s(Z, U, W)\}$). However, since all the variables in the second constraint appear only in subgoal s, only s and its variables appear in the definition of v_2^1. Based on these two views, we define below an AC query Q'_1 that is equivalent to Q_1.

$$\begin{aligned} Q'_1 : \quad h(X) \text{:-}\, & v_1^1(N_1, M_1, X, Y, Z, Z, U, W),\ N_1 > M_1, \\ & v_2^1(N_2, M_2, Z, U, W),\ N_2 \geq M_2 \end{aligned}$$

Unfolding the views will yield the same original query. Note that since the auxiliary views are conjunctive queries, we regard and treat each v_j^i in the body of Q'_1 as a base relation.

Next, we define the containment for CLAC queries.

Theorem 3. *[Containment of CLAC Queries]:*
Let Q_1 and Q_2 be CLAC queries. Then $Q_2 \sqsubseteq Q_1$ iff $D \Rightarrow \mu_1(C) \vee \cdots \vee \mu_q(C)$, where C and D are linear arithmetic constraints in Q_1 and Q_2, respectively, and μ_i is a containment mapping from Q_1 to Q_2.

Proof: *Based on Lemma 1, we create the following AC queries Q_1' and Q_2' which are equivalent to Q_1 and Q_2, respectively.*

$$\begin{aligned} Q_1': \quad & h(\bar{X}) \text{ :- } v_0^1(\bar{X_{0_1}}, \cdots, \bar{X_{0_p}}),\ v_1^1(N_1^1, M_1^1, \bar{X_{1_1}}, \cdots, \bar{X_{1_p}}),\ \cdots, \\ & v_n^1(N_n^1, M_n^1, \bar{X_{n_1}}, \cdots, \bar{X_{n_p}}),\ N_1^1\theta_1^1 M_1^1, \cdots, N_n^1\theta_n^1 M_n^1 \end{aligned}$$

$$\begin{aligned} Q_2': \quad & h(\bar{X}) \text{ :- } v_0^2(\bar{X_{0_1}}, \cdots, \bar{X_{0_q}}),\ v_1^2(N_1^2, M_1^2, \bar{X_{1_1}}, \cdots, \bar{X_{1_q}}),\ \cdots, \\ & v_m^2(N_m^2, M_m^2, \bar{X_{m_1}}, \cdots, \bar{X_{m_q}}),\ N_1^2\theta_1^2 M_1^2, \cdots, N_m^2\theta_m^2 M_m^2 \end{aligned}$$

It is easy to see that there is no containment mapping from Q_1' to Q_2', because their subgoals are distinct. In order to get the same set of containment mappings that exist from Q_1 to Q_2, we modify Q_2' while maintaining its equivalence to Q_2.

For this, we find all the containment mappings μ_j from Q_1 to Q_2. If there is no containment mapping, then the containment test fails. Otherwise, for every containment mapping μ_j, we apply μ_j on every subgoal of Q_1' (i.e., v_i^1) and append the new subgoal $\mu_j(v_i^1)$ to the body of Q_2'. We call the resulting query Q_2''. Note that $Q_2'' \equiv Q_2'$ because every newly added subgoal originates from a containment mapping from Q_1 to Q_2 whose target subgoals have already been in Q_2, and hence in Q_2'. Therefore, only some repeated subgoals are added to Q_2'', and $Q_2'' \equiv Q_2'$.

As an example, assume that u is the only containment mapping from Q_1 to Q_2. It maps subgoals in the body of Q_1 to a subset of subgoals in Q_2. Corresponding to this mapping, we append $v_i'' = \mu(v_i')$ to Q_2' and define Q_2''. If we unfold both Q_2' and Q_2'', we can see that unfolding Q_2'' might generate more subgoals all of which already appeared in the unfold of Q_2'.

Now, we consider the containment of Q_2'' in Q_1'. As these two are conjunctive queries with arithmetic comparisons, we can apply Theorem 2. Note that we already have all the containment mappings from Q_1' to Q_2'', since we actually built Q_2'' based on such mappings, i.e., those from Q_1 to Q_2.

So, $Q_2'' \sqsubseteq Q_1'$ iff $D' \Rightarrow \mu_1(C') \vee \cdots \vee \mu_q(C')$, where D' is the set of constraints $N_i^2\theta_i^2 M_i^2$ of $Q"_2$, and C' contains the constraints $N_j^1\theta_j^2 M_j^2$ in Q_1'.

Now, if we replace $N_i^2\theta_i^2 M_i^2$s and $N_j^2\theta_j^2 M_j^2$s with their original definitions in the body of v_i^1s and v_j^2s, we get a new implication to be tested which is nothing but $D \Rightarrow \mu_1(C) \vee \cdots \vee \mu_q(C)$, where D (C) is the set of constraints in Q_2 (Q_1). ■

Figure 1 illustrates the steps of the above proof. Theorem 3 characterizes containment of CLAC queries. It confirms that containment of CLAC queries and AC queries have the same requirement and complexity. Note that to test the containment of CLAC queries we do not need to transform them to AC queries.

$$\begin{array}{ccc} Q_1 & \rightsquigarrow & Q_1' \\ \downarrow & & \downarrow \\ Q_2 & \rightsquigarrow & Q_2'' \end{array}$$

Fig. 1. Transforming CLAC queries to AC queries while maintaining the containment mapping. To test $Q_2 \sqsubseteq Q_1$ (CLAC queries), we can build Q_1' and Q_2'' (AC queries) that are equivalent to Q_1 and Q_2 respectively and test containment of Q_2'' in Q_1'.

The following example illustrates the importance of disjunction in the implication test in Theorems 2 and 3.

Example 4. [Complexity of Containment Test]

Q_1: $h(X)$:- $p(X), r(Y,Z), Y \leq Z$
Q_2: $h(X)$:- $p(X), r(Y,Z), r(Z,Y)$

Here, to test if $Q_2 \sqsubseteq Q_1$ we apply Theorem 2. For that, we find all the containment mappings from Q_1 to Q_2 ($\mu_1 = \{X/X, Y/Y, Z/Z\}$, $\mu_2 = \{X/X, Y/Z, Z/Y\}$), and accordingly, test the implication as follows.

$True \Rightarrow (Y \leq Z) \vee (Z \leq Y)$

It is easy to see that no single term on the right hand side can be derived from the left hand side, however, the disjunction is always true, making the whole implication true.

Since testing the implication for disjunction is expensive, this example raises a question whether there are classes of queries for which we do not need to consider the expensive test for disjunctions. Next, we review and introduce such queries and their syntactical characteristics.

2.2 Importance of Homomorphism Property

So far we know that the complexity of testing containment for standard case is in NP-complete, and for CLAC queries and AC queries is Π_2^P-complete. This difference is due to the following disjunction in the implication test in Theorem 2.
$D \Rightarrow \mu_1(C) \vee \cdots \vee \mu_q(C)$
Based on this, the containment for queries with homomorphism property can be expressed as follows:

Theorem 4. *[Containment of Queries with Homomorphism property] [9,4,2]: Let Q_1 and Q_2 below be conjunctive queries that satisfy homomorphism property.*

Q_1 : $h(\bar{X})$:- $g_1(\bar{X}_1), \cdots, g_k(\bar{X}_k), \alpha_1, \cdots, \alpha_n$
Q_2 : $h(\bar{Y})$:- $p_1(\bar{Y}_1), \cdots, p_l(\bar{Y}_l), \beta_1, \cdots, \beta_m$

where α_i's and β_j's are constraints. Let $C = \{\alpha_i | 1 \leq i \leq n\}$ and $D = \{\beta_j | 1 \leq j \leq m]\}$. Then, $Q_2 \sqsubseteq Q_1$ if and only if $D \Rightarrow \mu_i(C)$, where μ_i is a containment mapping from Q_1 to Q_2.

We should note that if homomorphism property holds, to reject the containment, all the containment mappings should be examined.

In the next section, we introduce two new classes of queries with homomorphism property and also extend the class of AC queries with homomorphism property defined in [2]. The desired characteristics of such a class H is that checking membership of a given pair of queries Q_1 and Q_2 in H is syntactically polynomial.

3 Classes of Queries with Homomorphism Property

In this section, we study the existence of homomorphism property for the following classes of queries.

1. Conjunctive Queries with Equality Constraints (CQEC Queries)
2. Conjunctive Queries with Arithmetic Comparison (AC Queries)
3. CQEC+AC Queries

Next, we define each of these classes.

3.1 Conjunctive Queries with Equality Constraints

We define Conjunctive Queries with linear Equality Constraints (CQEC) in which the constraints include only equality relation (=) [6]. A CQEC query Q which is a special type of CLAC queries is a statement of the form:

$Q: \quad h(\bar{X}) \text{:-}\, g_1(\bar{X_1}), \ldots, g_k(\bar{X_k}), \alpha_1, \ldots, \alpha_n$

where $g_i(\bar{X_i})$ is an ordinary subgoal and α_j is a linear equality constraint. Let A be the set of variables and constants in Q. Every predicate g_i in the rule body is a base relation. Every constraint α_j is of the form $E_l = E_r$, where E_l and E_r are linear expressions over A. Every variable in query head $\bar{X}$ is defined as a function over variables appearing in some regular subgoals in the query body. Examples of such queries are as follows.

1. $p_1(X,T) \text{:-}\, r_1(X,H),\ r_2(X,N,R), T = H - R + 100N$
   Note that in this query, the head variable T is defined as an expression over the variables of the subgoals in the body.
2. $p_2(X) \text{:-}\, s_1(X,V_1,C_1,Y),\ s_2(X,V_2,C_2,Y_1), V_1 + C_1 = V_2 - C_2,\ Y_1 = Y + 1$
   In this query, the constraint part filters out the tuples for which the conditions $V_1 + C_1 = V_2 - C_2$ and $Y_1 = Y + 1$ are not satisfied.

Such constraints play two major roles in CQEC. One is asserting a condition on the query which is the most common role and the other is defining a variable in the query head based on other variables. Next, we show that CQEC queries satisfy homomorphism property.

Theorem 5. *[Containment of CQEC Queries]:*
Let Q_1 be a CQEC query and Q_2 any CLAC query. Then the homomorphism property holds for containment of Q_2 in Q_1.

Proof: *Based on Theorem 3, we can transform Q_1 and Q_2 to conjunctive queries with arithmetic comparison in which Q_1 would have variable equality constraints (i.e., constraints in the form of $A = B$ where A and B are single variables). When testing implication, we can see that on the right hand side of the implication we have only equality constraints that cannot form a "coupling" (a disjunction of terms on the right hand side where none of them alone is implied by the left hand side of the implication, however, together they are implied). As a result, either there exists a single term that satisfies the implication or the implication fails. This is nothing but homomorphism property.* ■

3.2 Homomorphism Property and Conjunctive Queries with Arithmetic Comparison

Following the work by Klug [9], Afrati et al. [2] identified a number of classes of queries that satisfy homomorphism property. These classes are special cases of AC queries. Next, we review the conditions $\mathcal{L}_1$, $\mathcal{L}_2$, $\mathcal{L}_3$, $\mathcal{R}_1$, $\mathcal{R}_2$, and $\mathcal{R}_3$ that identify such classes [2].

- Left Semi-Interval Comparisons (LSI) [2]:
  Let Q_1 be an AC query with Left semi-interval arithmetic comparisons only ($X < c$ or $X \leq c$, where X is a variable and c is a constant), and Q_2 is any AC query. If Q_1 and Q_2 satisfy all the following conditions, then homomorphism property holds. First, we define the terms used in these conditions.
  - Close-LSI: A constraint in the form of $X \leq c$.
  - Open-LSI: A constraint in the form of $X < c$.
  - Core(Q_1): the set of ordinary subgoals in the body of Q_1.
  - $AC(Q_1)$: the set of AC constraints in the body of Q_1.

  Note that similar terms are defined for RSI queries.

  $\mathcal{L}_1$: *There are not subgoals as follows which all share the same constant: An open-LSI subgoal in $AC(Q_1)$, a closed-LSI subgoal in the closure of $AC(Q_2)$, and a subgoal in core(Q_1).* This basically prevents forming the following coupling.

  $$X \leq \alpha \Rightarrow (X < \alpha \ \vee \ X = \alpha)$$

  where α is the shared constant.

  $\mathcal{L}_2$: *Either core(Q_1) has no shared variables or there are not subgoals as follows which all share the same constant: An open-LSI subgoal in $AC(Q_1)$, a closed-LSI subgoal in the closure of $AC(Q_2)$ and, a subgoal in core(Q_2).* This basically prevents forming the following coupling.

  $$(X \leq \alpha \ \wedge \ Y = \alpha) \Rightarrow (X < \alpha \ \vee \ X = Y)$$

  where α is the shared constant.

  $\mathcal{L}_3$: *Either core(Q_1) has no shared variables or there are not subgoals as follows which all share the same constant: An open-LSI subgoal in $AC(Q_1)$*

and two closed-LSI subgoals in the closure of $AC(Q_2)$. This basically prevents forming the following coupling.

$$(X \leq \alpha \ \wedge \ Y \leq \alpha) \ \Rightarrow (X < \alpha \ \vee \ Y < \alpha \ \vee \ X = Y)$$

where α is the shared constant.

- Right Semi-Interval Comparisons (RSI) [2]:
  Three conditions, $\mathcal{R}_1$, $\mathcal{R}_2$, and $\mathcal{R}_3$ that are similar to those in LSI, $\mathcal{L}_1$, $\mathcal{L}_2$, and $\mathcal{L}_3$ but revised for RSI, identify when homomorphism property holds for a given pair of AC queries, Q_1 and Q_2.
- Please see [2] for characterization of homomorphism property for queries with both LSI and RSI constraints (Semi-Interval), and queries with Point-Inequality (constraints of the form $A \neq c$ for constant c).

Example 5. Consider the following queries Q_1 and Q_2 where Q_2 is contained in Q_1. Since Q_1 has only LSI constraints, we test the conditions $\mathcal{L}_1$, $\mathcal{L}_2$, and $\mathcal{L}_3$.

$Q_1 :\quad h(X) \text{:-}\, r(X, Y, 4),\ Y < 4$
$Q_2 :\quad h(X) \text{:-}\, r(X, A, 4), r(X, 3, A), A \leq 4$

Note that in this example, conditions $\mathcal{L}_1$ and $\mathcal{L}_2$ are applicable, however, neither one is satisfied. This is because the following three subgoals share constant 4: (1) $Y < 4$ (open-LSI in Q_1), (2) $A \leq 4$ (closed-LSI in Q_2), and (3) $r(X, Y, 4)$ (a subgoal in $Core(Q_1)$). Accordingly, we cannot conclude whether homomorphism property holds or not, and to check the containment, we need to apply Theorem 2. For that, we find all the mappings and test if the constraint implication holds as follows. For Q_1 and Q_2, there are two mappings from Q_1 to Q_2, $\mu_1 = \{X/X, Y/A\}$, $\mu_2 = \{X/X, Y/3, (N' = 4)/A\}$.

$$A \leq 4 \Rightarrow \mu_1(Y < 4) \vee \mu_2(Y < 4) \equiv$$
$$A \leq 4 \Rightarrow (A < 4) \vee (3 < 4 \wedge A = 4) \equiv$$
$$A \leq 4 \Rightarrow (A < 4) \vee (A = 4).$$

That is implication holds, but as we can see, it is based on a team up of the two mappings and not a single mapping.

The following example illustrates a case where homomorphism property holds, however, the conditions in [2] could not help. In section 3.3, we extend these conditions which help identifying more classes of AC queries that satisfy homomorphism property.

Example 6. Consider the following queries which are very similar to those in Example 5.

$Q_1 :\quad h() \text{:-}\, r(X, Y, Z),\ Y < 4$
$Q_2 :\quad h() \text{:-}\, r(X, A, 4), r(4, 3, A), A \leq 4$

Again, for the same reason, the LSI test rejects this case and we conclude that we do not know if the homomorphism property holds. However, if we form the implication test, we can see that actually homomorphism property holds. Containment mappings include $\mu_1 = \{X/X, Y/A, Z/4\}$, $\mu_2 = \{X/4, Y/3, Z/A\}$, for which we have:

$A \leq 4 \Rightarrow \mu_1(Y < 4) \vee \mu_2(Y < 4) \equiv$
$A \leq 4 \Rightarrow (A < 4) \vee (3 < 4).$

The implication holds since the constraint $3 < 4$ in the disjunction evaluates the whole right hand side to *True*. That is, the containment test is established using only a single containment mapping.

In the next section, we define new conditions to identify more AC queries that satisfy homomorphism property.

3.3 More AC Queries with Homomorphism Property

During our studies, we identified some other conditions that are necessary for forming a coupling in the test of implication. As a result, if they are violated, then the implication holds if and only if at least a single containment mapping satisfies the test, i.e., homomorphism property holds. The important point is that testing these conditions is polynomial. Moreover, these conditions are applicable on all subclasses of AC queries defined in section 3.2. This extends the classes of AC queries with homomorphism property identified in [2]. Before defining the conditions, we define the notion of Join-Closure as follows.

Definition 1. *Consider a variable A and the set of subgoals that A appears in at least one of them where each subgoal shares at least one variable with another subgoal. We refer to this set of subgoals as the Join-Closure of A, denoted* $\mathbb{J}^*_A$.

In the rest of this paper, we use "repeated subgoal" to refer to occurrences of subgoals with the same predicate name.

Let Q_1 be an AC query with Left Semi-Interval (LSI) constraints only ($X < c$ or $X \leq c$, where X is a variable and c is a constant), and Q_2 is any AC query. If Q_1 and Q_2 satisfy at least one of the following conditions, then the homomorphism property holds [2].

$\mathcal{L}_4$: *The variable in every closed-LSI in the closure of $AC(Q_2)$ has appeared in less than two repeated subgoals in $Core(Q_2)$.*
This basically prevents from forming the following couplings.

1. $X \leq \alpha \Rightarrow (X < \alpha \ \vee \ X = \alpha)$
2. $(X \leq \alpha \ \wedge \ Y = \alpha) \Rightarrow (X < \alpha \ \vee \ X = Y)$
3. $(X \leq \alpha \ \wedge \ Y \leq \alpha) \Rightarrow (X < \alpha \ \vee \ Y < \alpha \ \vee \ X = Y)$

where α is the shared constant.
It is easy to verify $\mathcal{L}_4$ by looking at the right hand side of these couplings. That is, variable X has appeared in at least two subgoals in $Core(Q_2)$, and in different positions; otherwise we would not have X in different disjunctions.

$\mathcal{L}_5$: *There is no repeated subgoal in $Core(Q_2)$, or the variable in the subject closed-LSI in the closure of $AC(Q_2)$ has not appeared in the repeated subgoals or it has appeared in the same positions in the repeated subgoals.*
We can see that $\mathcal{L}_5$ does not allow forming any of the three types of coupling. It can be explained as follows. If there is no repeated subgoals then we do

not have multiple containment mappings. If we have repeated subgoals and variable X in the closed-LSI has not appeared in the repeated subgoals or it has appeared in the same position, then it cannot appear in two different constraints on the right hand side of disjunctions in the implication test, hence coupling cannot happen.

The following example illustrates the details of $\mathcal{L}_5$. That is, if the variable in closed-LSI is not connecting the repeated subgoals then the homomorphism property holds.

Example 7. Consider the following queries Q_1 and Q_2 where repeated subgoals in Q_2 do not share a variable. Condition $\mathcal{L}_1$ does not recognize these queries as queries with homomorphism property, however, since variable A and constant 4 that appeared in the first subgoal do not appear in the second subgoal, the homomorphism property holds.

$$\begin{array}{ll} Q_1: & h() \text{:-}\, r(X,4),\ X < 4 \\ Q_2: & h() \text{:-}\, r(A,4), r(3,D), A \leq 4 \end{array}$$

$\mathcal{L}_6$: *The variable A in the open-LSI appears in the head of Q_1.*
If $\mathcal{L}_6$ holds, then A cannot be mapped to different variables of Q_2. The reason is that in such cases the heads do not match and we do not have multiple containment mappings.

Also, we note that to form a coupling, it is vital that variable A in the open-LSI is mapped to different variables; otherwise, it would form the same constraint in the disjunctions.

Example 8. Consider the following queries that are similar to those in Example 5. Note that variable Y in open-LSI is a head variable and cannot be mapped to two variables.

$$\begin{array}{ll} Q_1: & h(Y) \text{:-}\, r(X,Y,Z),\ Y < 4 \\ Q_2: & h(A) \text{:-}\, r(X,A,4), r(3,4,A), A \leq 4 \end{array}$$

The LSI conditions in [2] cannot determine homomorphism property, however, our condition $\mathcal{L}_6$ implies that the homomorphism property hold.

$\mathcal{L}_7$: Recall the notion of Join-Closure. Assume that A is the variable in the open-LSI in Q_1, and that there exists a variable B in the head of Q_1 where $\mathbb{J}^*_A = \mathbb{J}^*_B$. Let S be the set of join variables in the subgoals in $\mathbb{J}^*_B$. If there exists a variable $X \in S$ such that no two partial mappings map X to the same variable, then the homomorphism property holds.

The reason is that to form a coupling we need at least two containment mappings. Since B is a head variable it has to be mapped to the same target in every containment mapping so does every variable X that is a join variable in the subgoal of B and some other subgoal. If there exists X for which no two partial mappings map it to the same variable we are sure that coupling cannot happen. We note that $\mathcal{L}_7$ requires to have partial containment mappings which in general, is exponential but since during the containment test, in the worst case, we need to find all mappings testing $\mathcal{L}_7$ seems practical.

The following example illustrates a case where $\mathcal{L}_7$ is applicable.

Example 9. Consider the following AC queries.

$$Q_1: \quad h(M) \text{:-} r(X,Y,Z), s(Z,M),\ Y < 4$$
$$Q_2: \quad h(N) \text{:-} r(X,A,4), r(4,3,A), s(A,N),\ A \leq 4$$

LSI conditions defined in [2] cannot determine homomorphism property, however, based on our condition $\mathcal{L}_7$, we can conclude that the homomorphism property holds. The reason is that M is a head variable in Q_1 which is chained with Y (open-LSI variable), and there is only a single subgoal with a potential target for M, i.e., variable N. To form the coupling, we need some containment mappings that map Y to different variables, and map Z to the same variable which is not possible.

Similar to conditions, $\mathcal{L}_4$, $\mathcal{L}_5$, $\mathcal{L}_6$, and $\mathcal{L}_7$ for LSI, we can define conditions $\mathcal{R}_4$, $\mathcal{R}_5$, $\mathcal{R}_6$, and $\mathcal{R}_7$, for RSI queries. Using these conditions we can identify more AC queries that enjoy homomorphism property.

Next, we introduce another subclass of CLAC queries for which we identify some necessary conditions of homomorphism property.

3.4 Homomorphism Property and Conjunctive Queries with Equality Constraint and Arithmetic Comparison

In this section, we introduce Conjunctive Queries With Equality Expression and Arithmetic Comparison (CQEC+AC) which contains both equality constraints (as defined in section 3.1), and LSI or RSI constraints (as defined in section 3.2), and show the conditions under which CQEC+AC queries satisfy homomorphism property. The general form of CQEC+AC queries is defined as follows.

Definition 2. *A CQEC+AC query Q is a statement of the form:*

$$Q: \quad h(\bar{X}) \text{:-} g_1(\bar{X}_1), \cdots, g_k(\bar{X}_k), \alpha_1, \cdots, \alpha_n$$

where $g_i(\bar{X}_i)$ are subgoals and α_j are arithmetic comparison or linear equality constraints. Let S be the set of variables and constants in Q. Each predicate g_i in the body of Q is a base relation, and every argument in g_i is an expression over S. Every constraint α_j is either of the form (1) $E_l = E_r$ or (2) $A\theta B$, where E_l and E_r are linear arithmetic expressions over $\mathbb{R}$, A is a variable, B is either a variable or a constant, and θ is a comparison operator in $\{=, \leq, <, >, \geq\}$.

We realized that adding equality expression to AC queries changes the way that subgoals and variables are related. Intuitively, in CQEC+AC queries, a variable might not be used in a subgoal r but could be related to r indirectly through other variables. For this matter, we define the notion of Variable-Subgoal Relationship as follows.

Definition 3. *[Variable-Subgoal Relationship] If variable A can be defined as a function of the variables in a subgoal $s(\bar{X})$, then we say that A has appeared in $s(\bar{X})$, directly or indirectly. If the function is the identity function, then A has appeared directly; otherwise, A has appeared indirectly.*

Based on this, we extend conditions $\mathcal{L}_4$, $\mathcal{L}_5$, $\mathcal{L}_6$ and $\mathcal{L}_7$ for testing homomorphism property for CQEC+AC queries considering the cases where variable appear in some subgoals indirectly.

Let Q_1 be a CQEC+AC query with Left semi-interval arithmetic comparisons and Q_2 is any CQEC+AC query. If Q_1 and Q_2 satisfy at least one of the following conditions, then the homomorphism property holds.

$\mathcal{L}'_4$: *The variable in every closed-LSI in the closure of $AC(Q_2)$ has appeared in less than two repeated subgoals in $Core(Q_2)$, directly or indirectly.*

$\mathcal{L}'_5$: *There is no repeated subgoal in $Core(Q_2)$, or the variable in the subject closed-LSI in the closure of $AC(Q_2)$ has not appeared in the repeated subgoals (directly or indirectly) or it has appeared in the same positions in the repeated subgoals (directly or indirectly).*

$\mathcal{L}'_6$: *The variable A in the open-LSI appears in the head of Q_1, directly or indirectly.*

Example 10. Consider the following queries that are similar to those in Example 5. Note that variable W in the head can be defined based on variable Y which is in the open-LSI hence, Y cannot be mapped to two variables.

$$Q_1 \quad h(W) :\!- r(X,Y,Z),\ Y < 4, W = 2Y$$
$$Q_2 \quad h(A) :\!- r(X,A,4), r(3,4,A), A \leq 4$$

The LSI conditions cannot determine homomorphism property, however, based on condition $\mathcal{L}'_6$, we can conclude that the homomorphism property holds.

$\mathcal{L}'_7$: Assume that A is the variable in the open-LSI in Q_1. Moreover, assume that there exists a variable B appeared in the head of Q_1, directly or indirectly, where $\mathbb{J}^*_A = \mathbb{J}^*_B$. Let S be the set of join variables in the subgoals in $\mathbb{J}^*_B$. If there exists a variable $X \in S$ such that no two partial mappings map X to the same variable, then the homomorphism property holds.

The following example in which the homomorphism property does not hold, shows that conditions for identifying homomorphism property in AC queries are not sufficient for CQAC+EC queries.

Example 11. Consider the following queries Q_1 and Q_2.

$$Q_1 \quad h() :\!- r(X,Y),\ Y < 5, X + 1 = Y$$
$$Q_2 \quad h() :\!- r(A,B), r(3,A), A \leq 4, A + 1 = B$$

Here, since OLSI ($Y < 5$) and $core(Q_2)$ ($\{r(A,B), r(3,A)\}$) in this example do not share a constant, the conditions defined for AC queries in [2] confirm that homomorphism property holds, however, this is not the case. That is, unlike AC queries, in order to show that homomorphism property does not hold in CQAC+EC queries, the terms OLSI, CLSI, and a subgoal in $core(Q_2)$ do not need to share the same constant. The implication test based on the containment mappings $\mu_1 = \{X/A, Y/B\}$ and $\mu_2 = \{X/3, Y/A\}$ is as follows.

$$(A \leq 4 \wedge A + 1 = B) \Rightarrow (B < 5 \wedge A + 1 = B) \vee (A < 5 \wedge 3 + 1 = A) \equiv$$
$$(A \leq 4 \wedge A + 1 = B) \Rightarrow (A < 4 \wedge A + 1 = B) \vee (A = 4) \equiv$$
$$(A \leq 4 \wedge A + 1 = B) \Rightarrow (A < 4) \vee (A = 4) \equiv True.$$

That means $Q_2 \sqsubseteq Q_1$.

Similar to conditions, $\mathcal{L}'_4$, $\mathcal{L}'_5$, $\mathcal{L}'_6$, and $\mathcal{L}'_7$ for CQEC+LSI queries, we can define conditions $\mathcal{R}'_4$, $\mathcal{R}'_5$, $\mathcal{R}'_6$, and $\mathcal{R}'_7$ for CQEC+RSI queries.

4 Query Rewriting for CLAC Queries with Homomorphism Property

In this section, we extend the rewriting algorithm introduced in [7] to support CLAC queries with Homomorphism Property.

A rewriting R is a set of rules (conjunctive queries), each of which is contained in Q. We get the union of contained rules to generate the *Maximally Contained Rewriting.* An MCR is a rewriting that is contained in Q and contains every other contained rewriting of Q. To show the containment of each rule in Q, we can use unfolding and then use Theorem 4. In general, in order to generate a rewriting, one should consider different combinations of view heads (head specialization) to generate rules, and ensure that each rule is contained in the query, and get the union of all such contained rules. Different algorithms use similar concept to create head specialization. For example, Bucket algorithm [10] uses *Bucket* and Minicon uses *MCD.* We refer to all of these as *Coverage.* We elaborate on this, using the concept of Coverage, as follows.

Definition 4. *[Coverage][7]*
A coverage C is a data structure of the form C=<S, ϕ, h, δ> where S is a subset of the subgoals in query Q, ϕ is a mapping from subgoals in S to a subset T of subgoals in view v_i, h is the head of v_i, i.e., $v_i(\bar{X}_i)$, and δ is a variable substitution that unifies every group of variables in ϕ that are mapped to the same variable.

Intuitively, every coverage is based on a view and can represent a set of subgoals. When generating rewriting, we can remove subgoals of S from Q, add a specialization of v_i head as well as the required constraints, and apply δ on Q so that the resulting query becomes contained in Q. The specialization of v_i is generated by applying the δ on $\phi^{-1}(v_i(\bar{X}_i))$, where ϕ^{-1} is the inverse of the mapping ϕ.

Example 12. Let $r(A, B)$ and $s(C, D, E)$ be relations. Consider the following query Q and views v_1, v_2, and v_3.

Q : $h(X, Y)$:- $r(X, Y)$, $s(Y, Z, W), X \leq 3, Z = 2$
V_1 : $v_1(A, B)$:- $r(A, B)$
V_2 : $v_2(B, D)$:- $s(B, D, D)$
V_3 : $v_3(A, C)$:- $r(A, D), s(D, C, D)$

Rule R below is a contained rewriting for Q. The reason is that unfolding every rule in R using views v_1, v_2 and v_3 would generate a query that is contained in Q, hence R is contained in Q.

$$R: \quad h(X,Y) \text{ :- } v_1(X,Y),\ v_2(Y,Z), X \leq 3, Z = 2$$
$$\quad\quad h(X,Y) \text{ :- } v_3(X,Z), X \leq 3, Z = 2$$

In example 12, there are coverages C_1=<$\{r(X,Y)\}, \{X/A, Y/B\}, v_1(A,B), \{\}$>, C_2=<$\{s(Y,Z,W)\}, \{Y/B, Z/D, W/D\}, v_2(B,D), \{W/Z\}$>, and C_3=<$\{r(X,Y), s(Y,Z,W)\}, \{X/A, Y/D, Z/C, W/D\}, v_3(A,C), \{\}$>, where C_1, C_2 and C_3 generate specializations $v_1(X,Y)$, $v_2(Y,Z)$, and $v_3(X,Z)$, respectively. These specializations contribute to defining the rules in the rewriting.

Next, we will explain how to find coverages and combine them to form a contained rewriting.

4.1 Rewriting Steps

Before going into the rewriting steps, we first test if homomorphism property holds for the input, i.e., query and every view should satisfy homomorphism property. If we could not confirm homomorphism property then we might not be able to return maximally contained rewriting. Otherwise, we proceed with (1) finding coverages, (2) combining coverages to generate rules, and (3) checking every generated rule to see if it can satisfy the constraints of the query.

Finding Coverages: To find coverages, we consider a set S with a single subgoal g_i in the query Q and its join variables J_S. For every view v_j that includes g_i, we ensure that all the variables A in J_S are accessible through v_j, that is A is distinguished in v_j. If this is the case, we create a new coverage C_{ji} based on v_j and assign S to C_{ji}. Otherwise, we add more subgoals to S, update J_S accordingly, and inquire if v_j can be useful in forming a coverage.

Adding subgoals to S is based on the join variable(s) that is not distinguished. If $A \in J_S$ is not accessible, we include all the subgoals in Q in which A has appeared. After adding the new subgoals to S, we update J_S and repeat the test. We continue this process until we find a coverage for g_i (and possibly some other subgoals), otherwise we fail.

Combining Coverages: To combine the coverages, we assign a bucket to each subgoal in the query and place coverages in the corresponding buckets. Note that a coverage might appear in several buckets as it may cover several subgoals. It has been shown that overlapping coverages (coverages with common subgoals) should not be considered in the same combination since otherwise they introduce unnecessary restrictions [11].

In general, to find coverages that do not overlap and generate contained rules there are three possible approaches defined as follows.

1. Detection and Recovery using Simple Cartesian Product:
   In this approach which is used in Bucket algorithm, to find all the rules, we perform the cartesian product of the buckets, and eliminate rows that contain overlapping coverages.

2. Avoidance using Optimized Cartesian Product:
This approach improves the performance of a simple cartesian product by discarding redundant or useless combinations during the cartesian product, explained as follows. At each iteration, before adding to R a coverage C_j from bucket B_i, we check all the subgoals already covered by coverages in R, and ensure there is no overlap. Minicon uses this approach to perform the Cartesian product.
3. Prevention using Pattern-based Cartesian Product:
Proposed in [7], this approach is based on the idea of identifying non-overlapping patterns to be used to break a original Cartesian product into a set of smaller Cartesian Products. It assigns to query the number $2^n - 1$, where n is the number of subgoals in the query Q, and assigns an identifier to every coverage C_i that is based on the subgoals it contains and their positions in the query body. For instance, in Example 12, identifiers assigned to C_1, C_2 and C_3 would be 2 ($= 10_{binary}$), 1 ($= 01_{binary}$), and 3 ($= 11_{binary}$) , respectively. The identifiers that do not have overlap define the patters. For example, here, we have two patters $\{\{3\}, \{1, 2\}\}$.

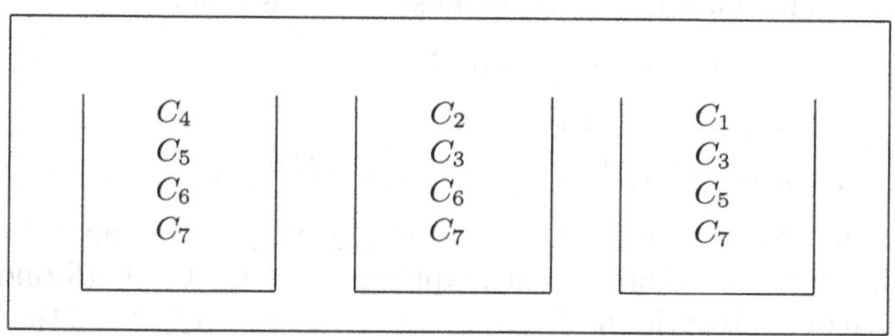

Fig. 2. Possible identifiers for coverages for a query with 3 subgoals. C_i represents the class of coverages with i as identifier. Subgoals that every coverage include identify its id, e.g., $C_5 = 101_{binary}$ covers subgoal at position 0 and 2.

For each pattern $\mathbb{P}$, we create buckets and perform a simple Cartesian product. For example, for the coverages listed in Figure 2, we break down the original bucket structure into smaller buckets and perform 5 different Cartesian products (since there are 5 different pattern combinations) listed as follows.

(a) $\mathbb{P}_1 = \{\{7\}\}$: No need for Cartesian product
(b) $\mathbb{P}_2 = \{\{3\}, \{4\}\}$: Cartesian product of two buckets: $[C_3] \times [C_4]$
(c) $\mathbb{P}_3 = \{\{6\}, \{1\}\}$:Cartesian product of two buckets: $[C_6] \times [C_1]$
(d) $\mathbb{P}_4 = \{\{5\}, \{2\}\}$: Cartesian product of two buckets:$[C_5] \times [C_2]$
(e) $\mathbb{P}_5 = \{\{4\}, \{2\}, \{1\}\}$: Cartesian product of three buckets: $[C_4] \times [C_2] \times [C_1]$

Note that the bitwise AND of the numbers in each pattern is 0 and their sum is $7{=}2^3 - 1$, e.g., for $\mathbb{P}_5 = \{\{4\}, \{2\}, \{1\}\}$, 7=1+2+4 and 1, 2, and 4 are bitwise disjoint. This makes sure that the coverages in the same pattern are non-overlapping and they cover all the subgoals in the query.

Handling the Constraints: For every combination generated in the previous step, we create a rule R_i, and check R_i to see if the constraints in query Q can be implied by the constraints in the views used in R_i. In that case, we add R_i to the rewriting R. Recall that the rewriting R is the union of contained conjunctive queries. For every rule R_i, we may have three possibilities, as follows.

1. The constraints C in Q are implied by the constraints in the views used in R_i. In this case, we simply add R_i to the result R.
2. A set of constraints C' can be added to R_i so that C' and the constraints in the views of R_i imply C. In this case, we append C' to R_i, and add R_i to R.
3. None of the above. The rule R_i should be discarded.

Finally, we return R, the union of R_is, as the MCR for Q.

5 Discussion

So far, we have investigated classes of conjunctive queries that satisfy homomorphism property. We believe identifying classes of queries for which homomorphism property does not hold but containment test can be done efficiently is also important. The following example illustrates this point.

Example 13. [Queries without homomorphism property]

Q_1: $h(X) \text{:-} p(X), r(U,Y,Z), Y \leq Z$
Q_2: $h(X) \text{:-} p(X), r(M,A,B), r(N,B,C), r(K,C,A)$

Intuitively, when testing the containment of Q_2 in Q_1, the constraint $Y \leq Z$ in Q_1 together with the containment mappings from Q_1 to Q_2 (formed by the special pattern of the repetition of variables in subgoals of Q_2) yield the following implication, and hence confirm the containment of Q_2 in Q_1.

$$I: \quad True \Rightarrow (A \leq B) \vee (B \leq C) \vee (C \leq A)$$

Implication I includes True (on the left hand side) and a disjunction of constraints (on the right hand side, where each constraint uses $\leq$ or $\geq$ as the comparison operator). Such kind of implications can be verified in an efficient way.

To see this, consider the variables of constraint $Y \leq Z$ together with the subgoal $r(U,Y,Z)$ in Q_1 in which these variables are used. Variables Y and Z are in positions 2 and 3 in $r(U,Y,Z)$, respectively. Let S be the set of all subgoals r in Q_2 (all subgoals with the same name as $r(U,Y,Z)$), and create a graph G whose nodes are variables in positions 2 or 3 in subgoals in S. That is, the nodes in graph G include variables A, B, and C, taken from positions 2 and 3 in subgoals $r(M,A,B)$, $r(N,B,C)$ and $r(K,C,A)$. For every subgoal $r(p_1,p_2,p_3)$ in S, define an edge from the node p_2 (position 2) to the node p_3 (position 3). Then, implication I holds if there is a cycle in G.

6 Conclusion and Future Work

In this paper, we studied the problems of containment and rewriting for conjunctive queries with linear arithmetic constraints. We identified new classes of queries that satisfy homomorphism property. These include conjunctive queries with equality constraints, conjunctive queries with arithmetic comparison, and conjunctive queries with linear equality constraint and arithmetic comparison. We also extended the query rewriting algorithm we proposed in [7] to support conjunctive queries with linear arithmetic constraints that enjoy homomorphism property.

Current studies have focused on identifying classes of queries with homomorphism property. Example 13 shows a case of queries for which homomorphism property does not hold, however, containment could be efficiently verified. We believe identifying classes of such queries is as important as identifying classes of queries with homomorphism property.

Acknowledgments. This work was supported in part by Natural Sciences and Engineering Research Council of Canada (NSERC), and by Concordia University.

References

1. Afrati, F., Li, C., Prasenjit, M.: Answering queries using views with arithmetic comparisons. In: Proceedings of the PODS 2002, pp. 209–220 (2002)
2. Afrati, F.N., Li, C., Mitra, P.: On containment of conjunctive queries with arithmetic comparisons. In: Bertino, E., Christodoulakis, S., Plexousakis, D., Christophides, V., Koubarakis, M., Böhm, K., Ferrari, E. (eds.) EDBT 2004. LNCS, vol. 2992, pp. 459–476. Springer, Heidelberg (2004)
3. Chandra, A.K., Merlin, P.M.: Optimal implementation of conjunctive queries in relational databases. In: Proceeding of the 9th Annual ACM Symp. on the Theory of Computing, pp. 77–90 (1977)
4. Gupta, A., Sagiv, Y., Ullman, J.D., Widom, J.: Constraint checking with partial information. In: Proceedings of the PODS 1994, pp. 45–55 (1994)
5. Ibarra, O.H., Su, J.: A technique for proving decidability of containment and equivalence of linear constraint queries. Journal of Computer and System Sciences 59(1), 1–28 (1999)
6. Kiani, A., Shiri, N.: Containment of conjunctive queries with arithmetic expressions. In: Proceedings of the 13th Int'l Conf. on Cooperative Information Systems (CoopIS), Agia Napa, Cyprus, pp. 439–452. Springer, Heidelberg (2005)
7. Kiani, A., Shiri, N.: Using patterns for faster and scalable rewriting of conjunctive queries. In: Proceedings of the 3rd Alberto Mendelzon International Workshop on Foundations of Data Management (2009)
8. Kiani, A., Shiri, N.: Answering queries in heterogenuous information systems. In: Proceedings of the ACM Workshop on Interoperability of Heterogeneous Information Systems, Bremen, Germany, November 4 (2005)
9. Klug, A.: On conjunctive queries containing inequalities. Journal of the ACM 35(1), 146–160 (1988)
10. Levy, A.Y., Rajaraman, A., Ordille, J.J.: Querying heterogeneous information sources using source descriptions. In: Proceedings of the 22nd VLDB Conf., pp. 251–262 (1996)
11. Pottinger, R., Levy, A.Y.: A scalable algorithm for answering queries using views. The VLDB Journal, 484–495 (2000)
12. Ullman Jeffrey, D.: Information integration using logical views. In: Afrati, F.N., Kolaitis, P.G. (eds.) ICDT 1997. LNCS, vol. 1186. Springer, Heidelberg (1996)
13. van der Meyden, R.: The complexity of querying indefinite data about linearly ordered domains. In: Proceedings of the PODS 1992, San Diego, CA, pp. 331–345 (1992)

Enhancing Dung's Preferred Semantics

Zhihu Zhang and Zuoquan Lin

School of Mathematical Sciences, Peking University,
100871 Beijing, China
{zhzhang,lzq}@is.pku.edu.cn

Abstract. Conflict resolution is an important issue. Dung's preferred semantics is a promising approach to resolving conflicts. However, such semantics is not capable of dealing with conflicts satisfactorily in the argumentation frameworks wherein there exists only empty admissible set. To enhance Dung's preferred semantics, we propose a novel semantics which follows the philosophy of Dung's preferred semantics, while satisfactorily resolving conflicts among arguments. In order to define our semantics, we first redefine Dung's basic notion *acceptability* by using pairs of sets of arguments and then propose the admissible semantics based on such notion. Relationships with Dung's preferred semantics, ideal semantics and semi-stable semantics are analyzed, and comparisons with other approaches such as *CF2* semantics are also discussed.

Keywords: argumentation, extensions, preferred semantics, conflict resolution.

1 Introduction

Conflict resolution is an important issue in constructing knowledge-based systems since conflicts always exist in knowledge-based systems. Over the past few years, formal argumentation systems [1,2,3,4,5,6,7] have been developed to deal with conflicts. Among them, Dung's argumentation system [1] is one of the most popular approaches, as it provides a powerful framework for managing conflicts.

Dung's argumentation system (also called argumentation framework), based on a set of arguments and a binary attack relation representing the attack relationship between arguments, is captured by four extensional argumentation semantics: *preferred semantics*, *stable semantics*, *grounded semantics* and *complete semantics*. Among these semantics, preferred semantics has been considered as one of the most popular semantics in resolving conflicts, in that it is sometimes able to discriminate some arguments that are left undecided by ground semantics [8], and resolve conflicts in such situations that there exists no stable extension. While in comparison with complete semantics, preferred semantics gains maximal information.

However, many individuals [8,9,10,11,12,13] have pointed out some disadvantages of Dung's preferred semantics. The most notable disadvantage of preferred semantics is the emptiness problem that results in the unsatisfactory way of resolving conflicts. For instance, provided that Dung's argumentation frameworks are considered as attack graphs, then in the presence of odd-length cycles in the attack relation between arguments, the empty set may be the unique preferred extension. An example is given in

S. Link and H. Prade (Eds.): FoIKS 2010, LNCS 5956, pp. 58–75, 2010.

[13]: an argument A attacks an argument B; B attacks an argument C; C attacks A; A, B and C all attack an argument D, and D attacks an argument E. In such an argumentation framework, Dung's preferred semantics adopts the way of assigning the status of *rejected* to all the arguments by deriving a unique preferred extension: $\emptyset$. That is, all the arguments should not be accepted. However, as pointed out in [13], E should be acceptable: although E is attacked by D, it is reinstated by all of A, B and C, since all of these arguments attack D. Thus, the authors of the paper [13] pointed out that the peculiar way of coping with such conflicts is one of the main unsolved problems in Dung's preferred semantics. To resolve the conflicts more satisfactorily in the presence of empty preferred extension, a new argumentation semantics is needed.

This paper presents a new approach to resolving conflicts among arguments. Our approach is based on the idea of ignoring some arguments alternately to make the left argumentation framework without unresolvable conflicts appearing in odd-length cycles. We redefine Dung's basic notion *acceptability* by using a pair of sets of arguments where the second element represents a set of arguments to be ignored. On the basis of this notion, we then define two other notions, *admissible pair* and *minimal admissible pair* that captures some kind of minimal change. By pursuing maximal information, we finally propose a new semantics named *enhanced preferred semantics*. Our semantics follows the nice philosophy of Dung's preferred semantics, while is more powerful in resolving conflicts among arguments. The relationships with Dung's preferred semantics, ideal semantics [14], and semi-stable semantics [15] are analyzed and comparisons with $CF2$ semantics [8], $preferred^+$ semantics [12] and preference-based approach [16] are briefly shown.

The remainder of this paper is organized as follows: Section 2 recapitulates some concepts of Dung's argumentation system. In Section 3, we first relax Dung's notion of acceptability by incorporating pairs of sets of argument instead of only sets in the definition, then propose the admissible semantics and enhanced preferred semantics. The intuition of enhanced preferred semantics is finally shown through some practical examples in this section. Section 4 presents the relations between Dung's preferred semantics and our semantics. We briefly discuss the related work in Section 5. Conclusions are drawn in the finial section. For the sake of the space limitation, proofs are omitted.[1]

2 Preliminaries

One of basic notions within Dung's abstract argumentation system [1] is the notion of argumentation framework defined as follows:

Definition 1. *(argumentation framework). An argumentation framework is a pair AF=<AR,attacks>, where AR is a set, and* $attacks \subseteq AR \times AR$ *is a binary relation on AR, called attack relation.*

In the above argumentation framework, $(A, B) \in attacks$ means that the argument A attacks the argument B (or B is attacked by A). In other words, A is in conflict

[1] The proofs can be found over `http://www.is.pku.edu.cn/~zzh/EPS_Tech.pdf`

with B. In the following part of this paper, an argumentation framework always means a finite argumentation framework such that it contains finitely many arguments. An argumentation framework $AF = < AR, attacks >$ is usually represented as a directed graph, called *attack graph*, where nodes are arguments and edges correspond to the elements of the attack relation.

Example 1. *Let* $AF = < \{A, B, C\}, \{(A, B), (B, C)\} >$. *AF can be depicted as in Fig. 1.*

$$A \longrightarrow B \longrightarrow C$$

Fig. 1. Graph representation of $AF = < \{A, B, C\}, \{(A, B), (B, C)\} >$

Another basic concept "conflict-free" in Dung's argumentation framework is defined as follows:

Definition 2. *(conflict-free set). Given an argumentation framework* $AF = <AR, attacks>$, *a set* $S \subseteq AR$ *is conflict-free iff* $\nexists A, B \in S$ *such that* $(A, B) \in attacks$.

The concept "acceptability" in Dung's argumentation framework is defined as follows:

Definition 3. *Let* $AF = < AR, attacks >$ *be an argumentation framework. An argument* $A \in AR$ *is acceptable with respect to (shortly, w.r.t) a set* $S \subseteq AR$ *of arguments iff for each argument* $B \in AR$*: if B attacks A then B is attacked by* S.[2]

We can also say that S defends A if A is acceptable with respect to S. On the basis of the above notion, Dung defined the notion of admissible set as follows:

Definition 4. *(admissible set). Let* $AF = < AR, attacks >$ *be an argumentation framework. A conflict-free set of arguments* S *is admissible iff each argument in* S *is acceptable with respect to* S.

Preferred semantics in Dung's system [1] is defined by the notion of preferred extension:

Definition 5. *(preferred semantics). A* preferred extension *of an argumentation framework AF is a maximal (w.r.t set inclusion) admissible set of AF.*

Complete semantics in Dung's system [1] is defined by the notion of complete extension:

Definition 6. *(complete extension). A set of arguments S is called a* complete extension *iff S is admissible, and each argument, which is acceptable w.r.t S, belongs to S.*

[2] B is attacked by S if B is attacked by an argument in S.

We use $\mathcal{PS}$ to denote Dung's preferred semantics. Note that when we say Dung's preferred semantics, we mean preferred semantics defined in Definition 5 or preferred semantics defined in [1]. The notion of preferred extension is popular because it represents the arguments in AF, which are defensible against all attacks and which can not be further extended without introducing a conflict. For convenience, the set of extensions prescribed by $\mathcal{PS}$ for a given argumentation framework AF is denoted as $\mathcal{E}_{\mathcal{PS}}(AF)$. For instance, the preferred extension of the argumentation framework in Example 1 is $\{A, C\}$, then $\mathcal{E}_{\mathcal{PS}}(AF) = \{\{A, C\}\}$.

As we mainly discuss preferred semantics in this paper, we do not review the other two traditional semantics including stable semantics and grounded semantics here. For understanding these semantics, please refer to the paper [1].

The notion of status of arguments is an important element of an argumentation framework. Its definition reveals which arguments should be accepted, which arguments should be rejected, and which arguments can not be decided whether accepted or rejected. The arguments in an argumentation framework are usually classified into three sets which represent different status of arguments.

Definition 7. *Let AF=<AR,attacks> be an argumentation framework, the set of arguments AR can be partitioned into three sets with respect to preferred semantics $\mathcal{PS}$:*

- *the set of* acceptable arguments $Acc_{\mathcal{PS}}(AF) = \{\alpha \in AR \mid \forall E \in \mathcal{E}_{\mathcal{PS}}(AF)\ \alpha \in E\}$
- *the set of* rejected arguments $Rej_{\mathcal{PS}}(AF) = \{\alpha \in AR \mid \forall E \in \mathcal{E}_{\mathcal{PS}}(AF)\ \alpha \notin E\}$
- *the set of* arguments in abeyance $Ab_{\mathcal{PS}}(AF) = \{\alpha \in AR \mid \exists E_1, E_2 \in \mathcal{E}_{\mathcal{PS}}(AF) : \alpha \in E_1 \wedge \alpha \notin E_2\}$.

In Example 1, $Acc_{\mathcal{PS}}(AF) = \{A, C\}$, that is, the arguments A and C are acceptable. $Rej_{\mathcal{PS}}(AF) = \{B\}$, that is, the argument B is rejected. $Ab_{\mathcal{PS}}(AF) = \emptyset$, that is, no arguments are in abeyance.

Dung [1] identified some special kinds of argumentation frameworks based on the following concepts.

Definition 8. *Let $AF =< AR, attacks >$ be an argumentation framework.*

- *An argument B* indirectly attacks *A if there exists a finite sequence $A_0, ..., A_{2n+1}$ such that* (1) *$A = A_0$ and $B = A_{2n+1}$, and* (2) *for each i, $0 \leq i \leq 2n$, A_{i+1} attacks A_i.*
- *An argument B* indirectly defends *A if there exists a finite sequence $A_0, ..., A_{2n}$ such that* (1) *$A = A_0$ and $B = A_{2n}$, and* (2) *for each i, $0 \leq i < 2n$, A_{i+1} attacks A_i.*
- *An argument B is said to be* controversial *w.r.t A if B indirectly attacks A and indirectly defends A.*
- *An argument is controversial if it is controversial w.r.t some argument A.*

Then, two special kinds, uncontroversial argumentation framework and limited controversial argumentation framework, are defined.

Definition 9. *Let* $AF =< AR, attacks >$ *be an argumentation framework.*

- *An argumentation framework is uncontroversial if none of its arguments is controversial.*
- *An argumentation framework is limited controversial if there exists no infinite sequence of arguments* $A_0, ..., A_n, ...$ *such that* A_{i+1} *is controversial w.r.t* A_i.

Since an argumentation framework is actually a graph, the following property characterizes the relations between preferred semantics and the cycles in the attack graph.

Proposition 1 [17,18]**.** *Let* $\mathcal{G}$ *denote the attack graph associated with an argumentation framework* $AF =< AR, attacks >$.

- *If* $\mathcal{G}$ *contains no cycle, AF has a unique preferred extension.*
- *If* $\mathcal{G}$ *contains no even-length cycle, AF has a unique preferred extension.*
- *If* $\emptyset$ *is the unique preferred extension of AF,* $\mathcal{G}$ *contains an odd-length cycle.*

3 Enhancing Dung's Preferred Semantics

In this section, we propose a new extensional semantics for Dung's argumentation frameworks based on a new notion of acceptability proposed below.

3.1 Defining the Acceptability of Arguments

We first informally show our basic idea by continuing the discussion about the argumentation framework mentioned in the introduction section. Such argumentation framework is depicted in Fig. 2.

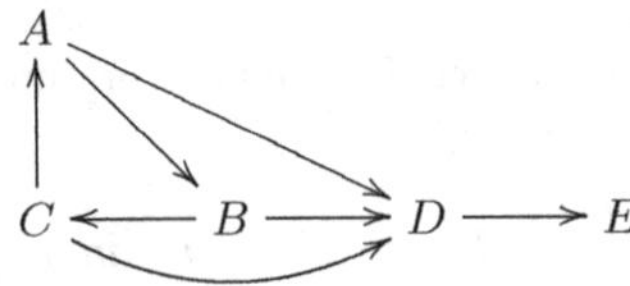

Fig. 2. Graph representation of the argumentation framework $< \{A, B, C, D, E\}, \{(A, B), (B, C), (C, A), (A, D), (B, D), (C, D), (D, E)\} >$

By the definition of preferred extension, we can know that $\emptyset$ is the unique preferred extension because of the conflicts among A, B and C. The reason for such conflicts in this argumentation framework is that there exists some "problematic"[3] argument among A, B and C, and an agent can not decide which one is "problematic" because of the limited information in his or her knowledge base. In order to cope with the conflicts without resorting to new information outside the current knowledge base and continue the reasoning task, the agent may naturally assume that one of arguments among A, B and C is "problematic". Once an argument is presumed to be "problematic", such attack

[3] Here, we use "problematic" argument to represent the argument that should not be acceptable.

relations orienting from it cannot hold. Furthermore, those attack relation that have this argument being attacked is of no use in that they do not affect the reasoning process of deriving the acceptable arguments. Then, for the argumentation framework depicted in Fig. 2, the agent may first presume that A is "problematic". Hence, A, (A, B), (C, A) and (A, D) should be ignored. Now, the current argumentation framework can be seen in Fig. 3. Based on this framework, the agent can easily derive a nonempty extension $\{B, E\}$.

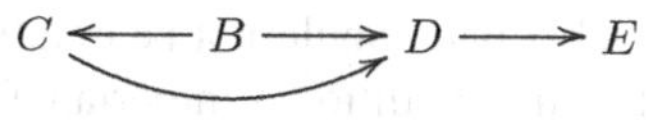

$$C \longleftarrow B \longrightarrow D \longrightarrow E$$

Fig. 3. Graph representation of the argumentation framework $< \{B, C, D, E\}, \{(B, C), (B, D), (C, D), (D, E)\} >$

The question is: should $\{B, E\}$ be the unique result accepted by the agent? Of course, the answer is no, since $\{B, E\}$ is inferred based on the hypothesis that A is ignored. To be fair, the agent can ignore the argument B. Then, C is not attacked. Similarly, the agent can ignore the argument C, then, the argument A is not attacked. Finally, the agent would get three results: $\{A, E\}$, $\{B, E\}$ and $\{C, E\}$. From hereon, we can learn that the argument E is acceptable, A, B, C are in abeyance, and D is rejected, which coincide with what we expect.

We now formally redefine Dung's basic notion of acceptability by incorporating the above idea.

Definition 10. *Let $AF =< AR, attacks >$ be an argumentation framework and $S, H \subseteq AR$. An argument $A \in AR$ is acceptable with respect to a pair of sets of arguments (S, H) iff:*

1. *$A \notin H$,*
2. *$S \cap H = \emptyset$,*
3. *$\forall B \in AR \setminus H$: if $(B, A) \in attacks$, then $\exists C \in S$ $(C, B) \in attacks$.*

In the above definition, H contains the arguments that are presumed to be ignored. The intuition behind the first condition is that an argument to be accepted by a rational agent should not be assumed to be ignored. The second condition demands that an argument to be ignored should not be considered to be acceptable. The third condition means that an argument A is accepted by an agent if A can be defended by the agent against all attacks on A except those attacks caused by arguments in H.

Example 2. *In the argumentation framework (see Fig. 2), A is acceptable w.r.t $(\emptyset, \{C\})$, and A is also acceptable w.r.t $(\{A\}, \{C\})$.*

Note that Dung's acceptability of arguments (see Definition 3) is a special case of the above definition.

Proposition 2. *Let $AF =< AR, attacks >$ be an argumentation framework. An argument $A \in AR$ is acceptable w.r.t a set S of arguments iff A is acceptable w.r.t a pair $(S, \emptyset)$.*

3.2 Defining Admissible Semantics

We are ready to give the new notion of admissible pair:

Definition 11. *Let $AF =< AR, attacks >$ be an argumentation framework, $S \subseteq AR$ be a conflict-free set and $H \subseteq AR$. A pair (S, H) is admissible iff:*

1. *$S \neq \emptyset$ or $H = AR$,*
2. *each argument in S is acceptable w.r.t (S, H).*

Note that the first condition in the above definition imposes the constraint that the first element of each admissible pair should not be empty unless the second element is the whole set of arguments. This condition is important for guaranteeing the non-emptiness of our semantics, which will be defined in the following part of the paper. The intuition behind this definition is that an agent tries to obtain nonempty results from an argumentation framework until he or she can not find any nonempty result after ignoring all the arguments.

By the definition, it is not difficult to see that the pair $(\emptyset, AR)$ is always an admissible pair of any argumentation framework $< AR, attacks >$.

Example 3. *Reconsider the argumentation framework in Example 1. $(\{A, C\}, \emptyset)$, $(\{B\}, \{A\})$ and $(\{A, C\}, \{B\})$ are admissible. Other admissible pairs are not listed here.*

Example 4. *Reconsider the argumentation framework in Example 2 . $(\{A\}, \{C\})$, $(\{A, E\}, \{C\})$, $(\{A\}, \{B, C\})$ and $(\{B\}, \{A\})$ are admissible. Other admissible pairs are not listed here.*

For any argumentation framework, there usually exist too many admissible pairs. The problem is how to make choices between these admissible pairs for a rational agent? A reasonable way to solve such problem is to fulfill the principle of minimal change: as few arguments as possible are ignored. So, an agent may prefer to choose the former one when making choice between the two admissible pairs $(\{A\}, \{C\})$ and $(\{A\}, \{B, C\})$ in Example 2. In Example 1, $(\{A, C\}, \emptyset)$ will be more likely to be selected rather than $(\{B\}, \{A\})$. Thus, we introduce the notion of minimal admissibility. We use $|H|$ to denote the number of arguments in the set H. In particular, $|\emptyset| = 0$.

Definition 12. *Given an argumentation framework $AF =< AR, attacks >$, let $S \subseteq AR$ be a conflict-free set and $H \subseteq AR$. (S, H) is a minimal admissible pair if and only if it is admissible and its second element H is minimal (w.r.t $|\ |$) among the admissible pairs of AF.*

By the above definition, if (S, H) and (S', H') are two minimal admissible pairs of the same argumentation framework, then $|H| = |H'|$.

In Example 1, the argumentation framework has two minimal admissible pairs: $(\{A, C\}, \emptyset)$ and $(\{A\}, \emptyset)$. In Example 2, the argumentation framework possesses seven minimal admissible pairs: $(\{A\}, \{C\})$,$(\{A, E\}, \{C\})$, $(\{B\}, \{A\})$,$(\{B, E\}, \{A\})$, $(\{C\}, \{A\})$, $(\{C, E\}, \{A\})$ and $(\{E\}, \{D\})$.

There exists a close relationship between an admissible set (see Definition 4) and a minimal admissible pair. Recall that an admissible set S is a set whose contained arguments are acceptable w.r.t S. Then, one can find that the following property holds.

Proposition 3. *For each nonempty set $S \subseteq AR$ of an argumentation framework $< AR, attacks >$, if S is an admissible set, then $(S, \emptyset)$ is a minimal admissible pair. The converse also holds.*

In the above proposition, the condition that S is nonempty is important, since $(\emptyset, \emptyset)$ is not a minimal admissible pair when the argumentation framework has at least one argument.

Reconsider Example 1, $\{A\}$ and $\{A, C\}$ are admissible, and the corresponding pair $(\{A\}, \emptyset)$ and $(\{A, C\}, \emptyset)$ are minimal admissible pairs. However, for an arbitrary minimal admissible pair, its first element is not an admissible set in general. For instance, in Example 2, $(\{A\}, \{C\})$ is a minimal admissible pair, but $\{A\}$ is not admissible in Dung's semantics.

One might argue that why not define minimal admissible pair w.r.t set inclusion rather than $|\ |$. Thus, Definition 12 can be modified as follows: (S, H) is a minimal admissible pair if and only if it is admissible and its second element H is minimal (w.r.t set inclusion) among the admissible pairs of AF. This approach seems also reasonable. However, in this paper, we suppose that a clever agent is usually lazy.[4] That is to say, assuming that $S \cap S' = \emptyset$ and $|H'|$ is much greater than $|H|$, we can ignore either $|H|$ arguments or $|H'|$ arguments. Obviously, ignoring $|H'|$ arguments pays more cost.

3.3 Defining Enhanced Preferred Semantics

Enhanced preferred semantics is defined by the notion of enhanced preferred extension:

Definition 13. *Let $AF =< AR, attacks >$ be an argumentation framework, $S, H \subseteq AR$. A pair $E = (S, H)$ is an* enhanced preferred extension *of AF if and only if E is a minimal admissible pair and its first element S is maximal (w.r.t set inclusion) among the minimal admissible pairs of AF. We define the first element of an enhanced preferred extension of AF as the* proper *enhanced preferred extension of AF.*

Given an enhance preferred extension E of an argumentation framework AF, we use $prop(E)$ to denote the proper enhanced preferred extension, $\mathcal{EPS}$ to denote enhanced preferred semantics, and $\mathcal{E}_{\mathcal{EPS}}(AF)$ to denote the set of all enhanced preferred extensions of AF.

Enhanced preferred semantics actually follows the philosophy of Dung's preferred semantics by pursuing maximal information. For instance, in Example 1, $(\{A, C\}, \emptyset)$ is an enhanced preferred extension while $(\{A\}, \emptyset)$ is not.

The status of arguments under enhanced preferred semantics is defined in the same way as in Definition 7. That is: $Acc_{\mathcal{EPS}}(AF) = \{\alpha \in AR \mid \forall E \in \mathcal{E}_{\mathcal{EPS}}(AF)\ \alpha \in prop(E)\}$; $Rej_{\mathcal{EPS}}(AF) = \{\alpha \in AR \mid \forall E \in \mathcal{E}_{\mathcal{EPS}}(AF)\ \alpha \notin prop(E)\}$; $Ab_{\mathcal{EPS}}(AF) = \{\alpha \in AR \mid \exists E_1, E_2 \in \mathcal{E}_{\mathcal{EPS}}(AF) : \alpha \in prop(E_1) \wedge \alpha \notin prop(E_2)\}$.

Example 5. *Reconsider the argumentation framework in Example 1. Here $(\{A, C\}, \emptyset)$ is the unique enhanced preferred extension. $\mathcal{E}_{\mathcal{EPS}}(AF) = \{(\{A, C\}, \emptyset)\}$. Therefore,*

[4] Here, we mean that a clever agent usually likes to learn more and do less.

$Acc_{\mathcal{EPS}}(AF) = \{A, C\}$, $Rej_{\mathcal{EPS}}(AF) = \{B\}$ and $Ab_{\mathcal{EPS}}(AF) = \emptyset$. Then, A and C are acceptable. B is rejected. This status of these arguments coincides with that in Dung's preferred semantics.

Example 6. *Reconsider the argumentation framework in Example 2.*
$\mathcal{E}_{\mathcal{EPS}}(AF) = \{(\{A, E\}, \{C\}), (\{B, E\}, \{A\}), (\{C, E\}, \{B\})\}$. Therefore, $Acc_{\mathcal{EPS}}(AF) = \{E\}$, $Rej_{\mathcal{EPS}}(AF) = \{D\}$ and $Ab_{\mathcal{EPS}}(AF) = \{A, B, C\}$. Then, E is acceptable, D is rejected, A, B, and C are in abeyance. This result is in accord with the floating acceptance [13] *and our intuition: since A, B, C attacks each other in an odd-length cycle, it is difficult to decide which one is acceptable. That is, they are in abeyance. But A, B and C all attack D, then D should be rejected, which instead makes E become acceptable.*

The fundamental lemma similar to Dung's [1] holds in our semantics.

Proposition 4. *Let $AF =< AR, attacks >$ be an argumentation framework, (S, H) be a minimal admissible pair where $S, H \subseteq AR$, and let A and $A^{'}$ be arguments which are acceptable w.r.t (S, H). Then:*

1. *$(S \cup \{A\}, H)$ is a minimal admissible pair.*
2. *$A^{'}$ is acceptable w.r.t $(S \cup \{A\}, H)$.*

Proposition 5 follows directly from Proposition 4.

Proposition 5. *Let $AF =< AR, attacks >$ be an argumentation framework. For each minimal admissible pair (S, H) of AF, there exists an enhanced preferred extension $(S^{'}, H)$ of AF such that $S \subseteq S^{'}$.*

Since $(\emptyset, AR)$ is always an admissible pair of every argumentation framework $AF =< AR, attacks >$, Proposition 5 implies the following property,

Corollary 1. *Every argumentation framework possesses at least one enhanced preferred extension.*

3.4 The Intuition of Enhanced Preferred Semantics

Let us consider two practical examples showing the intuition of our semantics. The first one is a practical version of the argumentation framework in Fig. 2.

Example 7. *Five witnesses (Smith, Jones, Robertson, Jack and Susan) discuss the reliability of each other in the following way:*

A: *"Smith says that Jones and Jack are unreliable, hence Jones and Jack are unreliable";*
B: *"Jones says that Robertson and Jack are unreliable, hence Robertson and Jack are unreliable";*
C: *"Robertson says that Smith and Jack are unreliable, hence Smith and Jack are unreliable";*
D: *"Jack says that Susan is unreliable, hence Susan is unreliable";*
E: *"Susan says that it is raining, hence it is raining".*

The graph representation of the argumentation framework in the above discussion can be seen in Fig. 2. Resorting to Dung's preferred semantics, the only preferred extension is $\emptyset$, as a result, all the arguments should be rejected, which is not intuitive. In contrast, in our semantics, as analyzed in Example 6, E is acceptable, D is rejected, A, B, and C are in abeyance, as intuitively should be.

The other example is about an agent who wants to buy a given pair of shoes.

Example 8. *Suppose that an agent is wondering whether to buy a pair of shoes or not. A person from China says that the pair of shoes sold for \$99.99 is produced by Adidas, that's why it is expensive. At the same time, another people called Tom says that the shoes are not produced by Adidas because the tag of those shoes is a counterfeit. Furthermore, a person from England tells the agent that the shoes are not expensive, so there is no necessary to forge the tag, thus, the tag of the shoes is genuine. The agent therefore gets the following knowledge: formally, A, B and C representing what the person from China says, what Tom says and what the person from England says respectively. Note that C, B and A provide reasons against the premises of B, A and C respectively. Then, the current knowledge base obtained by the agent can be described as an argumentation framework in Fig. 4. On the basis of such an argumentation framework, the arbitrary acceptance of one of these three arguments is more reasonable than the rejection of all three arguments since each of these arguments could have been accepted depending on what the agent itself thinks. For instance, if the agent think that the price is not high, then the shoes are genuine, and maybe the agent will buy them.*

Resorting to Dung's preferred semantics, we can obtain that only empty preferred extension, which means that all the three arguments should be rejected. However, in our semantics, three enhanced preferred semantics can be obtained, namely, $(\{A\},\{B\})$, $(\{B\},\{C\}),(\{C\},\{A\})$. Then, A, B, and C are all in abeyance, which coincides with what we analyzed above.

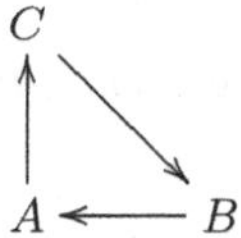

Fig. 4. Graph representation of the argumentation framework $<\{A,B,C\},\{(A,C),(C,B),(B,A)\}>$

4 Relationships with Dung's Preferred Semantics

For convenience, we classify Dung's argumentation frameworks into two basic classes according to different kinds of conflicts: inharmonious argumentation frameworks and harmonious argumentation frameworks. Before delivering such classification, we introduce the following intuitive definition.

Definition 14. *An argumentation framework AF is said to be nonempty if AF has at least one argument, empty otherwise.*

Definition 15. *A nonempty argumentation framework is said to be inharmonious if it has no nonempty admissible set, harmonious otherwise.*

The intuition behind an inharmonious argumentation framework is that, when people can not arrive at a nonempty decision or conclusion during a conversation or a dialog, it means that people can not persuade each other or reach an agreement, then we may say that the conversation or the dialog is inharmonious. By the definition, an inharmonious argumentation framework possesses only empty admissible set. Instead, we can see that a harmonious argumentation framework that is nonempty possesses at least one nonempty preferred extension.

Proposition 6. *Every nonempty harmonious argumentation framework possesses at least one nonempty preferred extension.*

There exists a close relation between limited controversial argumentation frameworks and harmonious argumentation frameworks. Dung proved that every nonempty limited controversial argumentation framework has at least a nonempty admissible set [1], then we get the following property.

Proposition 7. *Every limited controversial argumentation framework is harmonious, but not vice versa.*

The converse of the above proposition is not always true, as we can see the counter-example below:

Example 9. *Let $AF =< AR, attacks >$ be an argumentation framework with $AR = \{A, B, C\}$ and $attacks = \{(A,C), (B,C), (B,B)\}$. Here $\{A\}$ is a nonempty admissible set. Hence, AF is harmonious. Since there exists an infinite sequence $B, ..., B, ...$ such that B is controversial w.r.t B, AF is not limited controversial.*

As Dung pointed out that every uncontroversial argumentation framework is limited controversial [1], Proposition 7 implies that:

Corollary 2. *Every uncontroversial argumentation framework is harmonious, but not vice versa.*

On the basis of Proposition 6, for any harmonious argumentation framework, the following theory holds.

Theorem 1. *Let $AF =< AR, attacks >$ be a harmonious argumentation framework. Then each preferred extension of AF is a proper enhanced preferred extension of AF and vice versa.*

The above theorem shows that Dung's preferred semantics coincides with our semantics for harmonious argumentation frameworks. On the basis of Proposition 7, Theorem 1 implies that Dung's semantics coincides with our semantics for limited controversial argumentation frameworks.

Corollary 3. *Let $AF =< AR, attacks >$ be a limited controversial argumentation framework. Then each preferred extension of AF is a proper enhanced preferred extension of AF and vice versa.*

Directly from Corollary 2, Dung's preferred semantics coincides with enhanced preferred semantics for uncontroversial argumentation frameworks.

Corollary 4. *Let $AF =< AR, attacks >$ be an uncontroversial argumentation framework. Each preferred extension of AF is a proper enhanced preferred extension of AF and vice versa.*

With the consideration of any general argumentation framework, enhanced preferred semantics is an extension of Dung's preferred semantics.

Theorem 2. *Let $AF =< AR, attacks >$ be an argumentation framework. If dE is a preferred extension of AF, then there is an enhanced preferred extension E such that $dE \subseteq prop(E)$.*

By the above discussion, we conclude that enhanced preferred semantics not only preserves the advantages of Dung's preferred semantics, but also solves the emptiness problem of Dung's preferred semantics in many situations. However, there exist such argumentation frameworks that nonempty extensions can not be guaranteed by our semantics. We identify such argumentation frameworks as totally inharmonious argumentation frameworks.

Definition 16. *A nonempty argumentation framework is totally inharmonious iff $\forall A \in AR, \exists (A, A) \in attacks$.*

Example 10. *Let $AF =< AR, attacks >$ be an argumentation framework with $AR = \{A\}$ and $attacks = \{(A, A)\}$. Then this argumentation framework is totally inharmonious. Note that $\emptyset$ is the unique proper enhanced preferred extension of AF.*

We state an obvious property of totally inharmonious argumentation frameworks as follows:

Proposition 8. *A totally inharmonious argumentation framework is an inharmonious argumentation framework, but not vice versa.*

As inharmonious argumentation frameworks possess no nonempty preferred extension, then Proposition 8 implies that:

Corollary 5. *Let $AF =< AR, attacks >$ be a totally inharmonious argumentation framework. Then AF has a unique preferred extension: $\emptyset$.*

The proper enhanced preferred extension of such kind of argumentation framework is also empty.

Theorem 3. *Let $AF =< AR, attacks >$ be a nonempty argumentation framework. If AF is totally inharmonious, then AF has exactly one enhanced preferred extension $(\emptyset, AR)$ and vice verse.*

We argue that for this kind of argumentation frameworks, the behavior that our semantics results in empty proper enhanced preferred extension is not problematic. Since all the arguments in totally inharmonious argumentation frameworks are self-attacked[5],

[5] All the arguments attack themselves.

all the arguments should be rejected. Thus, it is intuitive to take $\emptyset$ as the result. Reconsider the argumentation frameworks in Example 10, since the argumentation framework contains only one self-attacked argument, $(\emptyset, \{A\})$ is the only enhanced preferred extension. However, for any argumentation frameworks that is not totally inharmonious, our semantics can guarantee non-emptiness.

Theorem 4. *For any nonempty argumentation framework AF, if AF is not totally inharmonious, then there exists at least one nonempty proper enhanced preferred extension.*

Corresponding to Proposition 1, the following property holds in our semantics.

Theorem 5. *Let $\mathcal{G}$ denote the attack graph associated with an argumentation framework $AF = < AR, attacks >$.*

- *If $\mathcal{G}$ contains no cycle, AF possesses a unique enhanced preferred extension whose second element is $\emptyset$.*
- *If $(\emptyset, AR)$ is the unique enhanced preferred extension of AF, $\mathcal{G}$ contains at least $\mid AR \mid$ odd-length cycles, wherein $\mid AR \mid$ denotes the number of the arguments in AF.*

To conclude this section, we have shown that our semantics guarantees nonempty extensions for argumentation frameworks except totally inharmonious argumentation frameworks. In addition, our semantics coincides with Dung's preferred semantics for harmonious argumentation frameworks. Thus, we showed that our semantics indeed enhances Dung's preferred semantics in resolving conflicts in argumentation frameworks.

5 Comparisons with Other Approaches

As we mentioned in the introduction section, there exist some disadvantages for the traditional semantics including stable, grounded and preferred semantics [1]. Therefore, some novel approaches such as *ideal semantics* [14], *semi-stable semantics* [15] and *CF2 semantics* [8] have been proposed to overcome the limits of these traditional semantics. It is interesting to discuss the relationships between those approaches and our approach.

Ideal semantics [14] provides a unique-status approach[6]. Ideal semantics is defined by the notion of ideal extension:

Definition 17. *Let $AF = < AR, attacks >$ be an argumentation framework, $S \subseteq AR$. S is* ideal *iff S is admissible and $\forall E \in \mathcal{E}_{PS}(AF)$ $S \subseteq E$. The* ideal extension *is the maximal (w.r.t set-inclusion) ideal set.*

Ideal semantics is less sceptical than grounded semantics which is overly sceptical. Meanwhile, it is more sceptical than just taking the intersection of all preferred extensions. From here, we can learn that the philosophy of ideal semantics is quite different from ours: ideal semantics is more sceptical than preferred semantics, while our

[6] The unique-status approach here means that only a unique extension exists for any argumentation framework under such an approach.

semantics is more credulous than preferred semantics. Consequently, multi-enhanced preferred extensions can not be avoided in our semantics while the unique ideal extension is obtained for any argumentation framework under ideal semantics. Nothing can be stated that which one, between the credulous approach and the sceptical approach, is better. However, there is a connection between these two semantics.

Theorem 6. *Let $AF =< AR, attacks >$ be an argumentation framework. If IE is the ideal extension of AF, then $\forall E \in \mathcal{E}_{\mathcal{EPS}}(AF)$ $IE \subseteq prop(E)$.*

Example 11. *Consider the argumentation framework in Fig. 5. It is not difficult to see that $\{A\}$ is the ideal extension, and $(\{A, C\}, \emptyset)$ and $(\{A, D\}, \emptyset)$ are the enhanced preferred extensions. The ideal extension is contained in each proper enhanced preferred extension.*

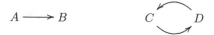

Fig. 5. Graph representation of the argumentation framework $< \{A, B, C, D\}, \{(A, B), (C, D), (D, C)\} >$

Semi-stable semantics [15] aiming at resolving the problem of non-existent extension in Dung's stable semantics is defined by the notion of semi-stable extension:

Definition 18 [15]. *Let $AF =< AR, attacks >$ be an argumentation framework, $S \subseteq AR$ and S^+ denote the set $\{B \mid (A, B) \in attacks\ for\ some\ A \in S\}$. S is a semi-stable extension iff S is a complete extension (see Definition 6) where $S \cup S^+$ is maximal (w.r.t set-inclusion).*

Semi-stable semantics adopts an ingenious approach to weakening the definition of stable semantics, which guaranteeing the existence of extensions while coinciding with stable semantics when stable extensions exist. As shown in [15], semi-semantics also has such a nice property that every semi-stable extension is a preferred extension. Therefore, we can get the following connection between semi-stable semantics and our semantics:

Theorem 7. *Let $AF =< AR, attacks >$ be an argumentation framework. If SE is a semi-stable extension of AF, then there is an enhanced preferred extension E such that $SE \subseteq prop(E)$.*

In particular, for the case of harmonious argumentation frameworks, the following property holds.

Theorem 8. *Let $AF =< AR, attacks >$ be a harmonious argumentation framework. Then each semi-stable extension of AF is a proper enhanced preferred extension of AF, but not vice versa.*

The converse of the above theorem is not always true. A counter-example can be shown as follows:

Example 12. *Consider the argumentation framework in Fig. 6. In this argumentation framework, $\{B, D\}$ is a semi-stable extension. Meanwhile $(\{B, D\}, \emptyset)$ is an enhanced preferred extension. $(\{A\}, \emptyset)$ is the other enhanced preferred extension, however, $\{A\}$ is not semi-stable.*

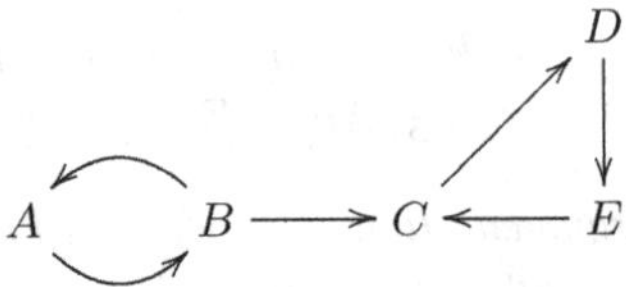

Fig. 6. Graph representation of the argumentation framework $< \{A, B, C, D, E\}, \{(A, B), (B, A), (B, C), (C, D), (D, E), (E, C)\} >$

The novel semantics discussed above are based on Dung's admissible semantics [1] while other novel approaches are not admissible. $CF2$ semantics [8,9] is a typical one which does not rely on Dung's admissible semantics.[7] In contrast, $CF2$ semantics relies on the following notions of *maximal conflict-free set* and *strongly connected components*:

Definition 19 [9]**.** *Given an argumentation framework $AF =< AR, attacks >$, a set $S \subseteq AR$ is* maximal conflict-free *iff it is maximal (w.r.t set inclusion) among the conflict-free sets of AF. The set made up of all the maximal conflict-free sets of AF is denoted as $\mathcal{MCF}(AF)$.*

Definition 20 [8]**.** *Given an argumentation framework $AF =< AR, attacks >$, two nodes $A, B \in AR$ are path-equivalent iff either $A = B$ or there is a path from A to B and a path from B to A. The* strongly connected components *of AF are the equivalence classes of nodes under the relation of path-equivalence. The set of strongly connected components of AF is denoted as $SCC(AF)$.*

Another basic notion related to a strong connected component is the parents of a strong connected component.

Definition 21 [8]**.** *Given an argumentation framework $AF =< AR, attacks >$, and a strongly connected component $S \in SCC(AF)$, $parents(S) = \{P \in SCC(AF) \mid P \neq S, and\ \exists A \in P,\ B \in S : (A, B) \in attacks\}$, and $parents^*(S) = \{A \in AR \mid A \notin S, and\ \exists B \in S : (A, B) \in attacks\}$.*

$CF2$ semantics is defined by the set of extensions, denoted as $\mathcal{CF}(AF)$:

Definition 22 [8]**.** *Given an argumentation framework $AF =< AR, attacks >$ and a set $E \subseteq AR$, $E \in \mathcal{CF}(AF)$ iff $\forall S \in SCC(AF)$*

1. $S^D(E) \cap E = \emptyset$; *and*
2. $S^U(E) \cap E \begin{cases} \in \mathcal{MCF}(AF\downarrow_{S^U(E)}), & \text{if} \mid SCC(AF\downarrow_{S^U(E)}) \mid = 1; \\ \in \mathcal{CF}(AF\downarrow_{S^U(E)}), & \text{otherwise.} \end{cases}$

[7] Actually, our approach is not admissible in the sense of [1].

wherein $S^D(E) = \{A \in S \mid \exists B \in parent^*(S) : B \in E, \text{ and } (B, A) \in attacks\}$, $S^U(E) = S \backslash S^D(E)$, *and* $AF \downarrow_{S^U(E)}$ *denotes the argumenation framework* $< S^U(E), attacks \cap (S^U(E) \times S^U(E)) >$.

The motivation of $CF2$ semantics which is quite similar to ours is to solve semantical problems with odd-length cycles in argumentation frameworks. However, we argue that, although $CF2$ semantics is able to deal with odd-length cycles properly, such semantics based on the notion of maximal conflict-free set is a little weak to preserve all the properties of Dung's preferred semantics. For instance, as Dung's preferred semantics shows its utility, if an argumentation framework has nonempty preferred extensions, it is reasonable to consider these extensions as the final results. However, $CF2$ semantics behaves differently in this case. Let us consider the following example.

Example 13. *Consider the argumentation framework AF in Fig. 7. In the argumentation framework, we have that* $SCC(AF) = \{S\}$, *where* $S = \{A, B, C, D\}$. *By Definition 22,* $parent(S) = \emptyset$. *Thus for any E* $S^D(E) = \emptyset$ *and* $S^U(E) = S$. *Also by Definition 22,* $S^U(E) \cap E \in \mathcal{MCF}(AF \downarrow_{S^U(E)})$, *i.e.* $(S \cap E) = \{\{D, B\}, \{D, A\}, \{C\}\}$, *thus,* $\mathcal{CF}(AF) = \{\{D, B\}, \{D, A\}, \{C\}\}$. *Since D defends A, then D indirectly attacks B. Thus, it is not reasonable for* $CF2$ *semantics to consider* $\{B, D\}$ *as one of its extensions. However, in Dung's preferred semantics and our semantics,* $\mathcal{E}_{\mathcal{EPS}}(AF) = \mathcal{E}_{\mathcal{PS}}(AF) = \{\{D, A\}\}$, *namely,* $\{D, A\}$ *is the unique extension, which is intuitive.*

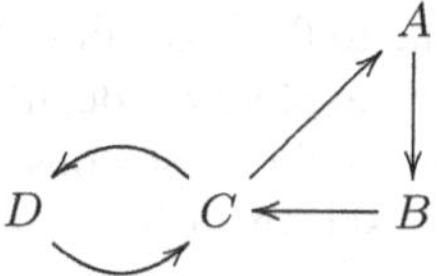

Fig. 7. Graph representation of the argumentation framework $AF =< \{A, B, C, D\}, \{(A, B), (B, C), (C, A), (C, D), (D, C)\} >$

The $preferred^+$ semantics [12] generalizes Dung's preferred semantics by resorting to the minimal models of logic programs and follows the philosophy of Dung's preferred semantics. However, there are such situations that the minimal models of logic programs do not exist. Then the $preferred^+$ semantics can not cope with these situations. For instance, recall the argumentation framework in Example 2, the minimal models of the program translated from the argumentation framework do not exist. However, our semantics can deal with these situations properly.

In some sense, our approach is similar to the preference-based approach proposed in [16]. In fact, ignoring some arguments means making these arguments less preferred to other arguments. But, in this paper, we do not rely on the preference explicitly. In addition, the semantics [16] relies on Dung's grounded semantics, while we refine Dung's preferred semantics.

6 Conclusions and Future Works

The work reported here concerns resolving conflicts in Dung's argumentation system [1]. Because of the unsatisfactory way of resolving conflicts in Dung's preferred semantics, we proposed a new approach to enhancing preferred semantics. The main innovations of our paper can be summarized as follows:

- We defined two new kinds of argumentation frameworks: harmonious argumentation frameworks and inharmonious argumentation frameworks. The relationships between these kinds of argumentation frameworks and other kinds defined in [1] are discussed. We also identified a particular argumentation frameworks namely, totally inharmonious argumentation frameworks in inharmonious argumentation frameworks.
- We redefined Dung's basic notion of acceptability by incorporating a new type of sets of arguments, which represents the set of arguments that are presumed to be ignored.
- Based on the notion of acceptability, we also defined the admissible semantics.
- Following the philosophy of Dung's preferred semantics, we defined a new semantics, enhanced preferred semantics, to resolve conflicts in Dung's argumentation system. Some good properties of the enhanced preferred semantics were also given.

We plan to extend our work in several directions:

- As one referee suggested, the implementation or the algorithm for computing enhanced preferred extensions should be investigated. As proved in this paper, enhanced preferred semantics coincides with preferred semantics if a given argumentation system has at least one non-empty preferred extension, a tree-based algorithm could be developed by resorting to the algorithms for computing preferred semantics [19,20,21,22], which will be explored in details in our future work. In addition, the complexity of our semantics will be also considered.
- Though some brief results are shown in Section 5, the deep comparison of our semantics to $CF2$ semantics and other semantics, such as the resolution-based semantics [23] is another direction of research.
- Some nice approaches proposed to refine Dung's preferred semantics such as prudent semantics [24] proposed to handle controversial arguments better can also be introduced to improve our semantics.
- One more direction of our work is to refine Dung's other extensional semantics such as the stable semantics based on the minimal admissible semantics. Furthermore, defining the minimal admissible semantics with respect to set inclusion is also worthy to be considered in the future.

Acknowledgments. We are grateful to Guilin Qi for his useful discussions. We would like to thank the referees for their helpful comments. This research was supported by the National Natural Science Foundation of China under number 60973003.

References

1. Dung, P.M.: On the acceptability of arguments and its fundamental role in nonmonotonic reasoning, logic programming and n-person games. Artificial Intelligence 77(2), 321–358 (1995)
2. Elvang-Gøransson, M., Hunter, A.: Argumentative logics: Reasoning with classically inconsistent information. Data Knowledge Engineering 16(2), 125–145 (1995)
3. Modgil, S.: Reasoning about preferences in argumentation frameworks. Artificial Intelligence 173(9-10), 901–934 (2009)

4. Pollock, J.L.: How to reason defeasibly. Artificial Intelligence 57(1), 1–42 (1992)
5. Prakken, H., Sartor, G.: Argument-based extended logic programming with defeasible priorities. Journal of Applied Non-Classical Logics 7(1), 25–75 (1997)
6. Vreeswijk, G.: Abstract argumentation systems. Artificial Intelligence 90(1-2), 225–279 (1997)
7. Amgoud, L., Cayrol, C.: Inferring from inconsistency in preference-based argumentation frameworks. Journal of Automated Reasoning 29(2), 125–169 (2002)
8. Baroni, P., Giacomin, M.: Solving semantic problems with odd-length cycles in argumentation. In: Nielsen, T.D., Zhang, N.L. (eds.) ECSQARU 2003. LNCS (LNAI), vol. 2711, pp. 440–451. Springer, Heidelberg (2003)
9. Baroni, P., Giacomin, M., Guida, G.: Scc-recursiveness: a general schema for argumentation semantics. Artificial Intelligence 168(1-2), 162–210 (2005)
10. Bench-Capon, T.J.M., Dunne, P.E.: Argumentation in artificial intelligence. Artificial Intelligence 171(10-15), 619–641 (2007)
11. Caminada, M.: Contamination in formal argumentation systems. In: Verbeeck, K., Tuyls, K., Nowé, A., Manderick, B., Kuijpers, B. (eds.) Proceedings of the 17th Belgium-Netherlands Conference on Artificial Intelligence 2005, pp. 59–65. Koninklijke Vlaamse Academie van Belie voor Wetenschappen en Kunsten, Brussels (2005)
12. Nieves, J.C., Cortés, U., Osorio, M., Olmos, I., Gonzalez, J.A.: Defining new argumentation-based semantics by minimal models. In: 7th Mexican International Conference on Computer Science, pp. 210–220. IEEE Computer Society, Los Alamitos (2006)
13. Prakken, H., Vreeswijk, G.: Logics for defeasible argumentation. In: Handbook of Philosophical Logic, 2nd edn. Kluwer Academic, Dordrecht (2002)
14. Dung, P.M., Mancarella, P., Toni, F.: A dialectic procedure for sceptical, assumption-based argumentation. In: Dunne, P.E., Bench-Capon, T.J.M. (eds.) COMMA 2006. Frontiers in Artificial Intelligence and Applications, vol. 144, pp. 145–156. IOS Press, Amsterdam (2006)
15. Caminada, M.: Semi-stable semantics. In: Dunne, P.E., Bench-Capon, T.J.M. (eds.) COMMA 2006. Frontiers in Artificial Intelligence and Applications, vol. 144, pp. 121–130. IOS Press, Amsterdam (2006)
16. Amgoud, L., Cayrol, C.: A reasoning model based on the production of acceptable arguments. Annals of Mathematics and Artificial Intelligence 34(1-3), 197–215 (2002)
17. Dunne, P.E., Bench-Capon, T.J.M.: Complexity and Combinatorial Properties of Argument Systems. Technical report, University of Liverpool, Department of Computer Science (2001)
18. Dunne, P.E., Bench-Capon, T.J.M.: Coherence in finite argument systems. Artificial Intelligence 141(1/2), 187–203 (2002)
19. Doutre, S., Mengin, J.: Preferred extensions of argumentation frameworks: Query answering and computation. In: Goré, R.P., Leitsch, A., Nipkow, T. (eds.) IJCAR 2001. LNCS (LNAI), vol. 2083, pp. 272–288. Springer, Heidelberg (2001)
20. Cayrol, C., Doutre, S., Mengin, J.: On decision problems related to the preferred semantics for argumentation frameworks. Journal of Logic and Computation 13(3), 377–403 (2003)
21. Dung, P.M., Mancarella, P., Toni, F.: Computing ideal sceptical argumentation. Artificial Intelligence 171(10-15), 642–674 (2007)
22. Nieves, J.C., Cortés, U., Osorio, M.: Preferred extensions as stable models. Theory and Practice of Logic Programming 8(4), 527–543 (2008)
23. Baroni, P., Giacomin, M.: Resolution-based argumentation semantics. In: Besnard, P., Doutre, S., Hunter, A. (eds.) COMMA 2008. Frontiers in Artificial Intelligence and Applications, vol. 172, pp. 25–36. IOS Press, Amsterdam (2008)
24. Marquis, S.C., Devred, C., Marquis, P.: Prudent semantics for argumentation frameworks. In: 17th International Conference on Tools with Artificial Intelligence, pp. 568–572. IEEE Computer Society, Los Alamitos (2005)

On the Distance of Databases

Gyula O.H. Katona[1], Anita Keszler[2], and Attila Sali[1]

[1] Alfréd Rényi Institute of Mathematics, Hungarian Academy of Sciences, Budapest, P.O.B. 127, H-1364 Hungary
{ohkatona,sali}@renyi.hu

[2] Distributed Events Analysis Research Group, Computer and Automation Research Institute, Hungarian Academy of Sciences, 1111 Budapest, Kende u. 13-17, Hungary
keszler@sztaki.hu

Abstract. In the present paper a distance concept of databases is investigated. Two database instances are of distance 0, if they have the same number of attributes and satisfy exactly the same set of functional dependencies. This naturally leads to the poset of closures as a model of changing database. The distance of two databases (closures) is defined to be the distance of the two closures in the Hasse diagram of that poset. We determine the diameter of the poset and show that the distance of two closures is equal to the natural lower bound, that is to the size of the symmetric difference of the collections of closed sets. We also investigate the diameter of the set of databases with a given system of keys. Sharp upper bounds are given in the case when the minimal keys are 2 (or r)-element sets.

Keywords: distance of databases, keys, antikeys, closures, poset, Hasse diagram.

1 Introduction

We are trying to investigate questions like "when two databases are the same?" or "are two databases similar?". For instance if we add or delete a row of the database we say that it changes only the instance of the given database. But in a strict sense this change may change the dependency structure, so we might say that a new database has been obtained. Another example is when a new attribute is added to the database, all rows are completed with the value of the new attribute. Is this another database?

If two databases should be considered different, how different they are? One needs a notion of *distance* between databases. If such a notion is introduced one can ask interesting questions like the following one. Knowing the distance between two databases what will be the distance between the merged database and one of the original ones?

Papers of Müller et.al. [9,10] treat this question from the point of view of conflicting copies of scientific databases. Today, many scientific databases overlap in

S. Link and H. Prade (Eds.): FoIKS 2010, LNCS 5956, pp. 76–93, 2010.

their sets of represented objects due to redundant data generation or data replication. For instance, in life science research it is common practice to distribute the same set of samples, such as clones, proteins, or patient's blood, to different laboratories to enhance the reliability of analysis results. Whenever overlapping data is generated or administered at different sites, there is a high probability of differences in results. These differences do not need to be accidental, but could be the result of different data production and processing workflows. For example, the three protein structure databases OpenMMS [2], MSD [3], and Columba [11] are all copies of the Protein Data Bank PDB [1]. However, due to different cleansing strategies, these copies vary substantially. Thus, a biologist is often faced with conflicting copies of the same set of real world objects and with the problem of solving these conflicts to produce a consistent view of the data. Thus, Müller et.al propose a distance concept that is similar to the edit distance of strings. They study the problem of efficiently computing the update distance for a pair of relational databases. In analogy to the edit distance of strings, the update distance of two databases is defined as the minimal number of set-oriented insert, delete and modification operations necessary to transform one database into the other.

In the present paper we take a data-mining oriented approach. We are mostly interested in the apparent dependency structure of a database, that is all functional dependencies that are satisfied by the given instance. We answer the questions posed above only for a very modest special case. It will be supposed that two database are the same if they have the same number of attributes and the system of functional dependencies are identical. The distance is introduced only between two databases having the same number of attributes.

Functional dependencies lead naturally to closure operations on the set of attributes $\mathcal{R}$. A subset $A \subseteq \mathcal{R}$ is *closed* if $A \rightarrow B$ implies $B \subseteq A$. Thus we take the approach of Burosch et.al. [4] by considering the *poset of closures* as a model of changing database. We will define the distance of two instances as the distance of the closures they generate, in that poset. We will see that this distance of two database instances is equal to the minimum number of tuples that are needed to be added to or removed from one instance to obtain the other instance of the database schema. In Section 2 we show that the largest distance possible is attained between the minimal and maximal elements of this poset. It will turn out that the distance of two closures is in fact *equal to* the obvious lower bound, that is to the size of the symmetric difference of the sets of closed sets. In Section 3 we consider the question that if something is known about the database instance, what is the diameter of the space database instances, that is the set of closures that satisfies the given information. This could be interpreted as a datamining question, we have some information given, and we wonder what is the size of the search space of databases based on that information. Our particular interest is in the case when the number of minimal keys is given. We give an upper bound in the general case. It is also interesting if the minimal keys all have the same cardinality r. We give a sharp upper bound for the diameter in the case $r = 2$. Finally, Section 4 contains conclusions and future research directions.

Some combinatorics notation we use that may not be well known for database people are as follows. $[n]$ denotes the set of first n positive integers, that is $[n] = \{1, 2, \ldots, n\}$. If Z is a set, then 2^Z denotes the collection of all subsets of Z, while $\binom{Z}{r}$ denotes the set of all r-element subsets of Z. These latter notations are mostly used in case of $Z = [n]$.

2 Poset of Closures

In what follows a schema $\mathcal{R}$ is considered fixed and every instance $\mathbf{r}$ is considered together with *all* functional dependencies $A \to B$ such that $\mathbf{r} \models A \to B$. For a set of attributes $A \subseteq \mathcal{R}$ the *closure* of A is given by $\ell(\mathbf{r})(A) = \{a \in \mathcal{R} : \mathbf{r} \models A \to a\}$. It is well known that the function $\ell(\mathbf{r}) \colon 2^{\mathcal{R}} \longrightarrow 2^{\mathcal{R}}$ is a *closure* that is it satisfies the properties

$$\begin{aligned} &A \subseteq \ell(A) \\ &A \subset B \Longrightarrow \ell(A) \subseteq \ell(B) \\ &\ell(\ell(A)) = \ell(A). \end{aligned} \tag{1}$$

Since constant columns are not really interesting, we assume that $\ell(\emptyset) = \emptyset$. Attribute set A is *closed* if $A = \ell(A)$. It is known that the family of closed attribute sets forms an intersection-semilattice. It was observed in [4], and is also well known consequence of the fact that functional dependencies can be expressed in First Order Logic by universal sentences, that if $\mathbf{r} \models A \to B$, then $\mathbf{r}' \models A \to B$ holds for any $\mathbf{r}' \subset \mathbf{r}$. That is a valid functional dependency stays valid if a record is removed from the database. This suggested the investigation of the poset of closures as a model of changing databases. Closure ℓ_1 is said to be *richer* than or equal to ℓ_2, $\ell_1 \geq \ell_2$ in notation, iff $\ell_1(A) \subseteq \ell_2(A)$ for all attribute sets. Let $\mathcal{F}(\ell)$ denote the collection of closed attribute sets for closure ℓ. It was proved in [4] that

Proposition 2.1. *$\ell_1 \leq \ell_2$ iff $\mathcal{F}(\ell_1) \subseteq \mathcal{F}(\ell_2)$.*

If $\mathbf{r_1}$ and $\mathbf{r_2}$ are two instances of schema $\mathcal{R}$, then $\ell(\mathbf{r_2})$ is richer than $\ell(\mathbf{r_1})$ can be interpreted as follows. In $\mathbf{r_2}$ there are more subsets of attributes that only determine attributes inside them, that is in $\mathbf{r_2}$ we need more attributes to determine some attribute, so $\mathbf{r_2}$ conveys 'more information" in the sense that the values of tuples are more abitrary.

The covering relation was also characterized. ℓ_2 *covers* ℓ_1, if $\ell_1 \leq \ell_2$ and for all ℓ' such that $\ell_1 \leq \ell' \leq \ell_2$ either $\ell' = \ell_1$ or $\ell' = \ell_2$.

Proposition 2.2 ([4])**.** *ℓ_2 covers ℓ_1 iff $\mathcal{F}(\ell_1) \subseteq \mathcal{F}(\ell_2)$ and $|\mathcal{F}(\ell_2) \setminus \mathcal{F}(\ell_1)| = 1$.*

Proposition 2.2 shows that the poset of all closures over a given schema $\mathcal{R}$, $\mathbf{P}(\mathcal{R})$, is *ranked*: its elements are distributed in *levels* and if ℓ_2 covers ℓ_1, then ℓ_2 is in the next level above ℓ_1's one. Let $|\mathcal{R}| = n$. Since $\emptyset$ and $\mathcal{R}$ are both closed for any closure considered, we obtain [4] that the *height* of $\mathbf{P}(\mathcal{R})$ is $2^n - 2$.

Let $\ell(\mathbf{r})$ denote the closure obtained from using all functional dependencies satisfied by instance $\mathbf{r}$. Since removing a record or adding a record to an instance changes the rank (level) of $\ell(\mathbf{r})$ by at most one, the distance of two instances are defined as follows.

Definition 2.1. *Let* $\mathbf{r}$ *and* $\mathbf{r}'$ *be two instances of schema* $\mathcal{R}$. *Their distance* $d(\mathbf{r},\mathbf{r}')$ *is defined to be the graph theoretic distance of* $\ell(\mathbf{r})$ *and* $\ell(\mathbf{r}')$ *in the Hasse diagram of* $\mathbf{P}(\mathcal{R})$. *That is, the length of the shortest path between points* $\ell(\mathbf{r})$ *and* $\ell(\mathbf{r}')$ *using only covering edges.*

It is easy to check that $d(\mathbf{r},\mathbf{r}')$ satisfies the triangle condition. The following proposition shows that the height of $\mathbf{P}(\mathcal{R})$ is an upper bound of the largest possible distance between two instances of the schema $\mathcal{R}$. Since the distance of instances is defined using the distance of closures, we allow the ambiguity of speaking of distance of an instance and a closure, as well.

Proposition 2.3. *Let* $|\mathcal{R}| = n$. *Then* $d(\mathbf{r},\mathbf{r}') \leq 2^n - 2$ *for any two instances of the schema* $\mathcal{R}$.

Proof. Let ℓ_m be the minimum element of $\mathbf{P}(\mathcal{R})$ furthermore let ℓ^M be the maximum element. Then

$$d(\mathbf{r},\mathbf{r}') \leq d(\mathbf{r},\ell_m) + d(\ell_m,\mathbf{r}') \tag{2}$$

and

$$d(\mathbf{r},\mathbf{r}') \leq d(\mathbf{r},\ell^M) + d(\ell^M,\mathbf{r}') \tag{3}$$

Thus,

$$2\,d(\mathbf{r},\mathbf{r}') \leq d(\mathbf{r},\ell_m) + d(\mathbf{r},\ell^M) + d(\ell_m,\mathbf{r}') + d(\ell^M,\mathbf{r}') = 2^n - 2 + 2^n - 2. \tag{4}$$

□

For any instance $\mathbf{r}$ the rank of $\ell(\mathbf{r})$ in $\mathbf{P}(\mathcal{R})$ is $d(\ell_m,\mathbf{r}) = |\mathcal{F}(\ell(\mathbf{r}))| - 2$.

The following is the main result of this section.

Theorem 2.1. *For any two instances* $\mathbf{r}$ *and* $\mathbf{r}'$ *of the schema* $\mathcal{R}$ *we have*

$$d(\mathbf{r},\mathbf{r}') = |\mathcal{F}(\ell(\mathbf{r}))\,\Delta\,\mathcal{F}(\ell(\mathbf{r}'))|, \tag{5}$$

where $A\,\Delta\,B$ *denotes the symmetric difference of the two sets, i.e.,* $A\,\Delta\,B = A \setminus B \cup B \setminus A$.

In order to prove Theorem 2.1 we need to recall the following result.

Theorem 2.2 ([5]). *A collection* $\mathcal{F}$ *of subsets of* $\mathcal{R}$ *is the collection of closed sets of some closure* $\ell(\mathbf{r})$ *for an appropriate instance* $\mathbf{r}$ *of* $\mathcal{R}$ *iff* $\emptyset, \mathcal{R} \in \mathcal{F}$ *and* $\mathcal{F}$ *is closed under intersection.*

Proof (of Theorem 2.1). According to Proposition 2.2 the size of $\mathcal{F}(\ell(\mathbf{r}))$ changes by one when we traverse along a covering edge in the Hasse diagram of $\mathbf{P}(\mathcal{R})$.

This immediately gives that the left hand side of (5) is at least as large as the right hand side, that is the distance of two instances is lower bounded by the size of the symmetric difference of the respective families of closed sets.

In order to prove the inequality in the other direction we have to find a way to move from $\ell(\mathbf{r})$ to $\ell(\mathbf{r}')$ using $|\mathcal{F}(\ell(\mathbf{r}))\,\Delta\,\mathcal{F}(\ell(\mathbf{r}'))|$ covering edges. That is, according to Theorem 2.2 we have to move from $\mathcal{F}(\ell(\mathbf{r}))$ to $\mathcal{F}(\ell(\mathbf{r}'))$ by successively removing a set of $\mathcal{F}(\ell(\mathbf{r})) \setminus \mathcal{F}(\ell(\mathbf{r}'))$ or adding a set of $\mathcal{F}(\ell(\mathbf{r}')) \setminus \mathcal{F}(\ell(\mathbf{r}))$ so that the property of being closed under intersection is preserved in each step. Note, that $\emptyset, \mathcal{R}$ are members of both closed sets system and the operations done do not change this.

First, we "peel off" sets of $\mathcal{F}(\ell(\mathbf{r})) \setminus \mathcal{F}(\ell(\mathbf{r}'))$ one by one. Let $F \in \mathcal{F}(\ell(\mathbf{r})) \setminus \mathcal{F}(\ell(\mathbf{r}'))$ such that $F \neq \mathcal{R}$, but there is no $F' \in \mathcal{F}(\ell(\mathbf{r})) \setminus \mathcal{F}(\ell(\mathbf{r}'))$ such that $F \subsetneqq F' \subsetneqq \mathcal{R}$. F can be removed from $\mathcal{F}(\ell(\mathbf{r}))$ if it is not an intersection of two closed sets both different from F. However, if $F = F' \cap F''$, then by the maximality property of F, both $F', F'' \in \mathcal{F}(\ell(\mathbf{r}'))$ yielding the contradiction that $F \in \mathcal{F}(\ell(\mathbf{r}'))$, as well.

Repeating the step above as long as $\mathcal{F}(\ell(\mathbf{r}))\setminus\mathcal{F}(\ell(\mathbf{r}'))$ is nonempty we arrive at $\mathcal{F}(\ell(\mathbf{r}))\cap\mathcal{F}(\ell(\mathbf{r}'))$. Now, we have to add the sets of $\mathcal{F}(\ell(\mathbf{r}'))\setminus\mathcal{F}(\ell(\mathbf{r}))$ one-by-one. In order to do so, let G be a minimal element of $\mathcal{F}(\ell(\mathbf{r}')) \setminus \mathcal{F}(\ell(\mathbf{r}))$ with respect to set containment. If $F \in \mathcal{F}(\ell(\mathbf{r})) \cap \mathcal{F}(\ell(\mathbf{r}'))$ then $F \cap G$ is a proper subset of G, or G itself. Since $\mathcal{F}(\ell(\mathbf{r}'))$ closed under intersection, $F \cap G \in \mathcal{F}(\ell(\mathbf{r}'))$. If it is a proper subset of G, then by the minimality of G, $F \cap G \in \mathcal{F}(\ell(\mathbf{r}))$, holds as well. In either case, adding G to $\mathcal{F}(\ell(\mathbf{r}))\cap\mathcal{F}(\ell(\mathbf{r}'))$, the collection remains closed under intersection. □

3 Diameter of Collection of Databases with the Same Set of Minimal Keys

It was investigated in [4] when does the system of keys determine uniquely the closure, that is the system of functional dependencies. A subset $K \subseteq \mathcal{R}$ is a *key* if $K \to \mathcal{R}$, and it is *minimal* if no proper subset of K has this property. In other words, K is a minimal key for instance $\mathbf{r}$ if there are no two rows of $\mathbf{r}$ that agree on K, but K is minimal with respect to this property. Let $\mathcal{K}(\mathbf{r})$ denote the system of minimal keys of instance $\mathbf{r}$. It is clear that $\ell(\mathbf{r})$ uniquely determines $\mathcal{K}(\mathbf{r})$, since $A \to B$ holds iff $B \subseteq \ell(A)$. On the other hand, $\mathcal{K}(\mathbf{r})$ does not determine $\ell(\mathbf{r})$. A simple example is the following. Let $\mathcal{R} = \{a, b, c, d\}$, and let the family of keys be $\mathcal{K} = \{\{a, c\}, \{a, d\}, \{b, c\}, \{b, d\}\}$. Then closure ℓ_1 has $\mathcal{K}$ as system of keys, where ℓ_1-closed sets are $\emptyset, \{a, b\}, \{c, d\}, \{a, b, c, d\}$. On the other hand, $\ell_2 > \ell_1$ has the same key system, where additionally the one-element subsets are ℓ_2-closed, too.

In order to determine the space of closures with a given minimal key system we need to introduce the concept of *maximal antikeys*. A subset $A \subset \mathcal{R}$ is a maximal antikey if it does not contain any key, and maximal with respect to this property. The collection of antikeys for a minimal key system $\mathcal{K}$ is usually denoted by $\mathcal{K}^{-1}$. This is justified by the fact that minimal keys and maximal antikeys determine each other, respectively [5]. In fact maximal antikeys are maximal sets that do not contain any key and keys are minimal sets that are not contained in any antikey. Clearly, both minimal key systems and maximal antikey systems form inclusion-free families of subsets of $\mathcal{R}$, that is no minimal key/antikey can contain another minimal key/antikey. For a set system $\mathcal{A}$ of subsets of $\mathcal{R}$ let $\mathcal{A}{\downarrow} = \{B \subseteq \mathcal{R}\colon \exists A \in \mathcal{A} \text{ with } B \subseteq A\} \cup \{\mathcal{R}\}$. Furthermore, let $\mathcal{A}_\cap = \{B \subseteq \mathcal{R}\colon \exists i \geq 1, A_1, A_2, \ldots A_i \in \mathcal{A} \text{ with } B = A_1 \cap A_2 \cap \ldots \cap A_i\} \cup \{\mathcal{R}\}$. That is, $\mathcal{A}{\downarrow}$ is the down-set generated by $\mathcal{A}$ appended with $\mathcal{R}$ and $\mathcal{A}^\cap$ is the set system closed under intersection generated by $\mathcal{A}$.

Theorem 3.1. *Let $\mathcal{K}$ be an inclusion-free family of subsets of $\mathcal{R}$. Then the closures whose minimal key system is $\mathcal{K}$ form an interval in the poset of closures $\mathbf{P}(\mathcal{R})$ whose smallest element is the closure with closed sets $\mathcal{K}_\cap^{-1}$ and largest element is the closure with closed sets $\mathcal{K}^{-1}{\downarrow}$.*

Proof. Let us suppose that $\mathbf{r}$ is an Armstrong-instance of $\mathcal{K}$ and let A be an antikey. For any $b \in \mathcal{R} \setminus A$, $A \cup \{b\}$ is a key, thus $A \cup \{b\} \to \mathcal{R}$ holds. If $A \to b$ held, then by the transitivity rule $A \to \mathcal{R}$ would hold, contradicting to the antikey property of A. Thus $\ell(\mathbf{r})(A) = A$ for every antikey $A \in \mathcal{K}^{-1}$. Since $\mathcal{F}(\ell(\mathbf{r}))$ is closed under intersection, $\mathcal{K}_\cap^{-1} \subseteq \mathcal{F}(\ell(\mathbf{r}))$ follows. On the other hand, if $\ell(\mathbf{r})(X) = X$ holds for some $X \subsetneq \mathcal{R}$, then X cannot contain any key. Thus, there exists a containment-wise maximal set $A \supset X$, that still does not contain any key. This implies that A is an antikey, hence $X \in \mathcal{K}^{-1}{\downarrow}$. It is easy to check, that both $\mathcal{K}_\cap^{-1}$ and $\mathcal{K}^{-1}{\downarrow}$ are closed under intersection and contain both, $\emptyset$ and $\mathcal{R}$. □

Corollary 3.1. *The diameter, that is the largest distance between any two elements of the collection of closures with given key system $\mathcal{K}$ is $|\mathcal{K}^{-1}{\downarrow}| - |\mathcal{K}_\cap^{-1}|$.* □

Corollary 3.1 determined the diameter if S was the set of databases (closures) with a given set of minimal keys. But the result, in this generality cannot be more than algorithmic.

In what follows we try to give a more precise, numerical answer in several special cases.

3.1 Unique Minimal Key

In this subsection we treat the case when the database has a unique minimal key. This is indeed very frequent case in practice.

Theorem 3.2. *The diameter of the set of closures having exactly one minimal key A where $0 < |A| = r < n$ is $2^n - 2^r - 2^{n-r}$.*

Proof. Let $A \subseteq B \subsetneq \mathcal{R}$. Then $A \to \mathcal{R}$ implies $B \to \mathcal{R}$, therefore B cannot be a closed set. The members F of a family of closed sets $\mathcal{F}$ with unique minimal key A satisfy either $A \not\subseteq F$ or $F = R$.

Let $a \in A$ be an attribute and $B = \mathcal{R} - \{a\}$. If B is not a closed set, then $B \to \mathcal{R}$, that is, B is a key, which does not contain A, hence there should exist another minimal key. This contradiction gives that $B = \mathcal{R} \setminus \{a\}$ is a closed set, $B = \mathcal{R} \setminus \{a\} \in \mathcal{F}$. Since $\mathcal{F}$ is closed under intersection, every set satisfying $F \supseteq \mathcal{R} \setminus A$ must be closed.

We can conclude that the family of closed sets $\mathcal{F}$ satisfies the following conditions.

$$\text{If } F \supseteq \mathcal{R} \setminus A \text{ then } F \in \mathcal{F}, \tag{6}$$

$$\text{if } F \supseteq A, F \neq R \text{ then } F \notin \mathcal{F}, \tag{7}$$

$$\emptyset \in \mathcal{F}. \tag{8}$$

$\mathcal{F}$ is also closed under intersection, but it is not needed for the proof of the upper estimation. On the other hand it is easy to see that if $\mathcal{F}$ satisfies these conditions then there is a closure in which the family of closed sets is $\mathcal{F}$ and A is the only minimal key.

We have to find the maximum size of the symmetric difference of two families $\mathcal{F}_1$ and $\mathcal{F}_2$ satisfying (6)–(8). The symmetric difference does not contain the sets under (6) (7) and (8). The number of subsets of $\mathcal{R}$ satisfying (6) is 2^r (they are in one-to-one correspondance with subsets of A). The number of subsets of $\mathcal{R}$ satisfying (7) is $2^{n-r} - 1$ (they are in one-to-one correspondance with subsets of $\mathcal{R} \setminus A$ except $\mathcal{R} \setminus A$ itself). Therefore we have

$$|\mathcal{F}_1 \,\Delta\, \mathcal{F}_2| \leq 2^n - 2^r - 2^{n-r} + 1 - 1 = 2^n - 2^r - 2^{n-r}. \tag{9}$$

Choose $\mathcal{F}_1$ containing all sets satisfying (6) (7) and (8), while let $\mathcal{F}_2$ have all the sets satisfying (6) and (8). It is easy to see that these families are closed under intersection and satisfy (9) with equality. □

Remark 3.1. The diameter will be the smallest if $r = 1$, then it becomes about half of the diameter of the space without restriction given by Proposition 2.3. This coincides with our expectation: the "smaller" key is "stronger" in the sense that it determines more, the diameter of the space of possible databases becomes the smallest.

Remark 3.2. It is interesting to compare this result with the case when it is only known that A is a minimal key, but we do not know if there are other keys or not. That means we only have (7) and (8) as restrictions. Then the diameter is, of course, larger: $2^n - 2^{n-r} + 1$.

3.2 Upper Bound for Non-uniform Minimal Key System

In this subsection we assume that $\mathcal{R} = \{1, 2, \dots n\} = [n]$, for the sake of convenience. Let $\binom{[n]}{r}$ denote the collection of r-subsets of $[n]$. Let $\mathcal{M}$ be a non-empty, inclusion-free family. Define

$$\mathcal{D}(\mathcal{M}) = \{H : \exists M \in \mathcal{M} \text{ such that } H \subseteq M\}, \tag{10}$$

$$\mathcal{U}(\mathcal{M}) = \{H :\ \exists M \in \mathcal{M} \text{ such that } H \supseteq M\}. \tag{11}$$

The *characteristic vector* $v(A)$ of the set $A \subseteq [n]$ is a 0,1 vector in which the i-th coordinate is 1 iff $i \in A$. If $A \in \binom{[n]}{r}$ then $v(A)$ contains exactly r 1's. $b(A)$ is the integer obtained by reading $v(A)$ as a binary number. Now an ordering "$<$" is introduced among the elements of $\binom{[n]}{r}$. Let $A, B \in \binom{[n]}{r}$ then $A < B$ iff $b(A) < b(B)$. This ordering is called *lexicographic.*

Define the (r,ℓ)-*shadow* of a family of r-element sets $\mathcal{A} \subseteq \binom{[n]}{r}$ for $\ell < r$:

$$\sigma_{r,\ell}(\mathcal{A}) = \{H :\ |H| = \ell, \exists A \in \mathcal{A} \text{ such that } H \subset A\}. \tag{12}$$

Proposition 3.1 ([7,8]). *If $\mathcal{A}$ consists of some lexicographically first members of $\binom{[n]}{r}$, $\ell < r$ then $\sigma_{r,\ell}(\mathcal{A})$ is a family of some lexicographically first members of $\binom{[n]}{\ell}$.*

Theorem 3.3 (Shadow Theorem, [7,8]). *If $\mathcal{A} \subseteq \binom{[n]}{r}$, $|\mathcal{A}| = m$ then $|\sigma_{r,\ell}(\mathcal{A})|$ is at least as large as the (r,ℓ)-shadow of the family of the lexicographically first m members of $\binom{[n]}{r}$, that is, the size of the (r,ℓ)-shadow attains its minimum for the lexicographically first r-element sets.*

If $\mathcal{A} \subseteq 2^{[n]}$ is a family then $\mathcal{A}_r$ denotes the subfamily consisting of all r-element members:

$$\mathcal{A}_r = \mathcal{A} \cap \binom{[n]}{r}. \tag{13}$$

The *profile vector* of the family $\mathcal{A} \subseteq 2^{[n]}$ is $p = (p_0, p_1, \dots p_n)$ where $p_r = p_r(\mathcal{A}) = |\mathcal{A}_r|$.

Lemma 3.1. *Let $\mathcal{M}$ be a non-empty inclusion-free family of subsets of $[n]$ with fixed $|\mathcal{M}| \geq n$. Then $|\mathcal{D}(\mathcal{M})|$ attains its minimum for a family satisfying the following conditions with some $2 \leq r \leq n$.*

$$p_n = \dots = p_{r+1} = p_{r-2} = \dots = p_1 = p_0 = 0, \tag{14}$$

$$\mathcal{M}_r \text{ consists of the lexicographically first } r-\text{element subsets,} \tag{15}$$

$$\mathcal{M}_{r-1} = \binom{[n]}{r-1} \setminus \sigma_{r,r-1}(\mathcal{M}_r). \tag{16}$$

We do not claim that this is the only optimal solution. On the other hand, if $|\mathcal{M}| \leq n$ then the best construction consists of $|\mathcal{M}|$ pieces of 1-element sets.

Proof. Suppose that $p_n = \dots = p_{r+1} = 0, p_r > 0$. Consider $\mathcal{D}(\mathcal{M})_{r-1}$. Its size is at least $|\sigma_{r,r-1}(\mathcal{M}_r)|$ what is minimum, by the Shadow Theorem, if $\mathcal{M}_r$ consists of the lexicographically first r-element subsets. By the proposition, $\sigma_{r,r-1}(\mathcal{M}_r)$ is a family of some first $r-1$-element subsets. $\mathcal{M}_{r-1}$ is disjoint from $\sigma_{r,r-1}(\mathcal{M}_r)$, since $\mathcal{M}$ is inclusion-free. $|\mathcal{D}(\mathcal{M})_{r-2}|$ is at least $|\sigma_{r-1,r-2}(\sigma_{r,r-1}(\mathcal{M}_r) \cup \mathcal{M}_{r-1})|$ by the Shadow Theorem with equality if $\sigma_{r,r-1}(\mathcal{M}_r) \cup \mathcal{M}_{r-1}$ is "lexicographically first", that is, if $\mathcal{M}_{r-1}$ is the "continuation" of $\sigma_{r,r-1}(\mathcal{M}_r)$ in the lexicographic ordering. Continuing in this way, we can see that $\mathcal{D}(\mathcal{M})_\ell$ will be minimum for a fixed profile for the following construction. Choose the lexicographically first p_r r-element sets, the lexicographically first $r-1$-element sets following $\sigma_{r,r-1}(\mathcal{M}_r)$, the lexicographically first $r-2$-element sets following $\sigma_{r-1,r-2}(\sigma_{r,r-1}(\mathcal{M}_r) \cup \mathcal{M}_{r-1})$, and so on. Since this construction does not depend on ℓ, this construction minimizes also $|\mathcal{D}(\mathcal{M})|$. Now, that we know what structure minimizes $|\mathcal{D}(\mathcal{M})|$ for a fixed profile of $\mathcal{M}$, we need to show that the best profile is when only two set sizes occure.

Suppose that

$$|\mathcal{D}(\mathcal{M})_s| < \binom{n}{s}, |\mathcal{D}(\mathcal{M})_{s-1}| = \binom{n}{s-1}, \ldots, |\mathcal{D}(\mathcal{M})_1| = \binom{n}{1}, |\mathcal{D}(\mathcal{M})_0| = 1 \tag{17}$$

holds for some integer $1 \leq s \leq r$.

If $s = r$, we are done, $\mathcal{M}$ has only r and $r-1$-element sets, ordered according to the statement of the Lemma. Otherwise suppose that $|\mathcal{D}(\mathcal{M})|$ is minimum for the given size $|\mathcal{M}|$ and $r-s$ is the smallest possible. Let A and B be the lexicographically last member of $\mathcal{M}_r$ and the lexicographically first non-member of $\mathcal{M}_s$. Replace A with B. All proper subsets of B are in $\mathcal{D}(\mathcal{M})$, therefore this operation cannot increase $|\mathcal{D}(\mathcal{M})|$. Repeat this step until either $\mathcal{M}_r$ becomes empty, or $\mathcal{M}_s$ "full": $|\mathcal{D}(\mathcal{M})_s| = \binom{n}{s}$. In both cases, the difference $r-s$ becomes smaller. This contradiction finishes the proof. □

Theorem 3.4. *Let $\mathcal{K}$ be a non-empty inclusion-free family of subsets of $[n]$, where $|\mathcal{K}| \geq n$ is fixed. Furthermore, let $S(\mathcal{K})$ denote the set of all closures in which the family of minimal keys is exactly $\mathcal{K}$. Then*

$$\mathsf{diam}(S(\mathcal{K})) \leq 2^n - |\mathcal{U}(\mathcal{K}^*)|, \tag{18}$$

where $\mathcal{K}^*$ consists of some lexicographically last sets of size s and all the $s+1$-element sets not containing the selected s-element ones, for some $0 \leq s \leq n-2$ and $|\mathcal{K}^*| = |\mathcal{K}|$.

Proof. It is obvious that the members of $\mathcal{U}(\mathcal{K}) \setminus \{[n]\}$ are not closed sets in a closure belonging to $S(\mathcal{K})$.

Define $X = \bigcap_{K \in \mathcal{K}} K$ and let x be an element of X. If $[n] \setminus \{x\}$ is not closed then it is a key. It must contain a minimal key as a subset. This contradiction shows that $[n] \setminus \{x\}$ is closed. The intersection of closed sets is closed [5], therefore all sets containing $[n] \setminus X$ are closed. Denote the family of these sets by $2^X + ([n] \setminus X)$. (This notation may sound peculiar, but reflects the idea, that any set containing $[n] \setminus X$ consists of the union of a subset of X and the set $[n] \setminus X$.)

Concluding, if $\mathcal{F}$ is a family of closed sets in a closure from $S(\mathcal{K})$, then the followings are true.

$$\mathcal{F} \cap (\mathcal{U}(\mathcal{K}) \setminus \{[n]\}) = \emptyset, \tag{19}$$

$$\mathcal{F} \supseteq 2^X + ([n] \setminus X). \tag{20}$$

Therefore, if $\mathcal{F}_1$ and $\mathcal{F}_2$ are two families of closed sets of two closures from $S(\mathcal{K})$, then $\mathcal{F}_1 \Delta \mathcal{F}_2$ cannot contain $\emptyset$, $[n]$ and members of $\mathcal{U}(\mathcal{K}) \setminus \{[n]\}$ and $2^X + ([n] \setminus X)$:

$$\mathcal{F}_1 \Delta \mathcal{F}_2 \subseteq 2^{[n]} \setminus \mathcal{U}(\mathcal{K}) \setminus (2^X + ([n] \setminus X)) \setminus \{\emptyset\}. \tag{21}$$

Hence we have, considering that the only common element of $\mathcal{U}(\mathcal{K})$ and $2^X + ([n] \setminus X)$ is $[n]$, that

$$|\mathcal{F}_1 \Delta \mathcal{F}_2| \leq 2^n - |\mathcal{U}(\mathcal{K})| - 2^{|X|}. \tag{22}$$

By Theorem 2.1 the left hand side is the distance of the two closures, (22) gives an upper bound on the diameter. Thus, to find a valid estimate, we need to minimize

$$|\mathcal{U}(\mathcal{K})| + 2^{|X|} \tag{23}$$

for fixed $|\mathcal{K}|$.

Let $\mathcal{K}^- = \{\overline{K} : K \in \mathcal{K}\}$, where $\overline{K}$ denotes the complement of K. It is easy to see that $\mathcal{U}(\mathcal{K})^- = \mathcal{D}(\mathcal{K}^-)$ and $X = [n] \setminus \bigcup_{M \in \mathcal{K}^-} M$. Therefore the minimum of (23) can be found minimizing

$$|\mathcal{D}(\mathcal{M})| + 2^{n - |\bigcup_{M \in \mathcal{M}} M|} \tag{24}$$

for fixed $|\mathcal{M}|$. If $\bigcup_{M \in \mathcal{M}} M \neq [n]$ then replace a member of $\mathcal{M}$ that is covered by the union of the other members,by a 1-element set in $[n] \setminus \bigcup_{M \in \mathcal{M}} M$. If no member of $\mathcal{M}$ is covered by other members, then each one has an "own" element that is not contained in any other member of $\mathcal{M}$. Since $|\mathcal{M}| \geq n$, that is impossible, so a covered member must exist. This operation obviously does not increase (24). Repeated application of this step shows that $\bigcup_{M \in \mathcal{M}} M = [n]$ can be supposed. Lemma3.1 determines the minimum of $\mathcal{D}(\mathcal{M})$, the substitution $\mathcal{M}^- = \mathcal{K}$ gives the construction showing the desired upper bound. □

Remark 3.3. If $|\mathcal{K}| = \binom{n}{s}$ then (18) becomes

$$diam(S(\mathcal{K})) \leq 2^n - \sum_{i=s}^{n} \binom{n}{i} = \sum_{i=0}^{s-1} \binom{n}{i}. \tag{25}$$

Indeed, $\binom{n}{s} = |\mathcal{K}| = |\mathcal{K}^*|$ by Theorem 3.4. There is only one possibility to have $\mathcal{K}^*$ of the structure given by Theorem 3.4 of this size, namely if $\mathcal{K}^* = \binom{[n]}{s}$.

3.3 Uniform Minimal Key Systems

If all keys are one-element sets, then $\mathcal{K}^{-1}$ consists of a single set A, thus $\mathcal{K}^{-1}\downarrow$ consists of all subsets of A and $\mathcal{R}$, while $\mathcal{K}_{\cap}^{-1}$ consists of two sets, A and $\mathcal{R}$, i.e. the diameter is $2^{|A|} - 1$. Hence, we start with the special case when the minimal keys have size 2.

Let D be a closure whose minimal keys have exactly two elements. $G = ([n], E)$ be the graph where $[n] = \{1, 2, \ldots, n\}$ stands for the set of attributes of D and $\{i, j\} \in E (i \neq j)$ is an edge of the graph iff $\{i, j\}$ is not a minimal key in D. That is, $\mathcal{K}$ is equal to $\binom{[n]}{2} - E$. Let $|E| = e$. The set of closures having $\binom{[n]}{2} - E$ as the set of minimal keys will be denoted by $S_2(G)$. We want to give an upper estimate on $\mathsf{diam} S_2(G)$ depending only on e, that is, actually we give upper bound for

$$s_2(e) = \max_{\{G=([n],E):\ |E|=e\}} \mathsf{diam} S_2(G). \tag{26}$$

First we consider the case when G has one non-trivial connected component.

Theorem 3.5. *If $e = \binom{t}{2} + r$, where $0 < r \leq t$, then*

$$\mathsf{diam} S_2(G) \leq \begin{cases} 2^t + 2^r - 4 & \textit{if } r < t \\ 2^{t+1} - 2 & \textit{if } r = t \end{cases} \tag{27}$$

for a graph G whose connected components are isolated vertices except for one component. Furthermore, this bound is sharp.

Proof. Let $D \in S_2(G)$ where $G = ([n], E)$ is a graph with $|E| = e$ edges. Since the family of minimal keys is $\mathcal{K} = \binom{[n]}{2} - E$, the family of maximal antikeys $\mathcal{K}^{-1}$ consists of sets containing no key (that is, no edge in $\binom{[n]}{2} - E$ and maximal for this property). Then the members of $\mathcal{K}^{-1}$ are maximal complete subgraphs in G. These are called the *cliques* of G. $\mathcal{K}^{-1}\!\downarrow$ consists of all complete subgraphs of G, while $\mathcal{K}_{\cap}^{-1}$ consists of those complete subgraphs that are intersections of cliques. We will show that

$$|\mathcal{K}^{-1}\!\downarrow| \leq 2^t + 2^r + n - t - 1. \tag{28}$$

We apply the following theorem of Erdős [6].

Theorem 3.6 (Erdős, 1962). *Let $G = (V, E)$ be a connected graph of e edges. Assume, that $e = \binom{t}{2} + r$, where $0 < r \leq t$. Then the number of complete k-subgraphs $C_k(G)$ of G is at most*

$$C_k(G) \leq \binom{t}{k} + \binom{r}{k-1}. \tag{29}$$

This estimate is sharp.

Note that Theorem 3.6 is valid for all k, since if $k > \max(t, r+1)$, then both binomial coefficients are 0, and no complete subgraph of that size could exist.

Since $\mathcal{K}^{-1}\!\downarrow$ consists of all complete subgraphs of G, so we just have to sum up (29) for all k in the non-trivial component of G, and add the number of isolated vertices. That is

$$|\mathcal{K}^{-1}\!\downarrow| = \sum_{k\geq 0} \binom{t}{k} + \sum_{k\geq 0} \binom{r}{k-1} + n - t - 1, \tag{30}$$

that results in (28). The optimum construction takes a complete graph on t vertices and add an extra vertex connected to r vertices of the complete graph. If $r < t$, then the nontrivial component of G cannot be a complete graph, so there are at least two maximal cliques. Then $\emptyset$, the two maximal cliques, and their intersection is in $\mathcal{K}_{\cap}^{-1}$. Furthermore, the isolated vertices are maximal cliques themselves, so they are contained in $\mathcal{K}_{\cap}^{-1}$, that is $|\mathcal{K}_{\cap}^{-1}| \geq 4+n-t-1$. Applying (28) and Corollary3.1 the upper bound in (27) follows. □

If we have more than one non-trivial component of G, then Erdős' theorem does not apply. In fact, for small e, we can have better construction. For example, if $e = 2$, then the best graph consists of two independent edges. In the Appendix we give a proof of a general upper bound. With the notations of Theorem 3.5 it says that $s_2(e) \leq 2^{t+1} - 2$.

We can have some upper bound in the case of r-uniform minimal key system. Let D be a closure whose minimal keys have exactly $r(\geq 2)$ elements. $H = ([n], \mathcal{E})$ be the hypergraph where $[n] = \{1, 2, \ldots, n\}$ stands for the set of attributes of D and the r-element set $R \in \binom{[n]}{r}$ is a hyperedge of the hypergraph H, that is a member of the family $\mathcal{E}$ iff R is not a minimal key in D. That is, $\mathcal{K}$ is equal to $\binom{[n]}{r} \setminus \mathcal{E}$. We also suppose that $|\mathcal{E}| = e$. The set of closures having $\binom{[n]}{r} \setminus \mathcal{E}$ as the set of minimal keys will be denoted by $S_r(H)$. We want to give an upper estimate on $\mathsf{diam} S_r(H)$ depending only on e, that is, actually we will give an upper estimate on

$$\max_{\{H=([n],\mathcal{E}):\ |\mathcal{E}|=e\}} \mathsf{diam} S_r(H). \tag{31}$$

Theorem 3.7. *If* $e \leq \binom{a}{r}$ *then* $\mathsf{diam}(S_r(H)) \leq 2^a + e2^r$.

Proof. Let $D \in S_r(H)$ where $H = ([n], \mathcal{E})$ is a graph with $|\mathcal{E}| = e$ hyperedges. Since the family of minimal keys is $\mathcal{K} = \binom{[n]}{r} \setminus \mathcal{E}$, the family of antikeys $\mathcal{K}^{-1}$ consists of sets containing no key (that is, no edge in $\binom{[n]}{r} \setminus \mathcal{E}$ and maximal for this property). Then the members of $\mathcal{K}^{-1}$ are vertex sets of maximal complete subhypergraphs in H. That is sets $B \subset [n]$ such that $\binom{B}{r} \subset \mathcal{E}$ but for all $B' \supsetneq B$ $\binom{B'}{r} \setminus \mathcal{E} \neq \emptyset$. These are called the *(hyper)cliques* of H.

We have to prove that the number of sets of the vertices of H which are subsets of at least one hyperclique and are not intersections of those is at most $2^a + e2^r$. That is, $|\mathcal{K}^{-1}\downarrow| - |\mathcal{K}_{\cap}^{-1}| \leq 2^a + e2^r$. We will actually prove something stronger, namely we will show that $|\mathcal{K}^{-1}\downarrow| \leq 2^a + e2^r$.

Suppose first $0 < i \leq r$ and consider the number of i-element subsets of the hypercliques. Such a set must be a subset of an r-element set which is either $\in \mathcal{E}$ or is a subset of a larger maximal non-key. Therefore, in the worst case their number is at most $e\binom{r}{i}$.

Suppose now $r < i$. Let $A_1, \ldots, A_m$ be the family of i-element subsets, whose all r-element subsets are in $\mathcal{E}$ (They are not necessarily cliques!) If $m > \binom{a}{i}$ then by the Shadow Theorem the number of r-element subsets (hyperedges) is $> \binom{a}{r} \geq e$. Indeed, the we are considering here (i, r)-shadows, whixh is minimalized by the lexicographically first m i-sets. If $m > \binom{a}{i}$ then these lexicographically first sets contain all i-subsets of $\{1, 2, \ldots, a\}$, so their (i, r)-shadows contain all r subsets of $\{1, 2, \ldots, a\}$. By the strict inequality $m > \binom{a}{i}$, some i-sets containing $a + 1$ are also in the lexicographically first m, so some r-sets containing $a + 1$ are in the shados. This contradiction shows that $m \leq \binom{a}{i}$.

Add up these maximums:

$$e \sum_{i=1}^{r} \binom{r}{i} + \sum_{i=r+1}^{a} \binom{a}{i} \leq 2^a + e2^r. \tag{32}$$

□

4 Conclusions, Further Research

In the present paper we have introduced a distance concept of databases. It is data-mining based, that is we start with the collection of functional dependencies that a given instance **r** of schema $\mathcal{R}$ satisfies. Two databases are considered to be the same, if their numbers of attributes agree and they satisfy exactly the same collection of functional dependencies. It has turned out that this concept fits nicely with the poset of closures as a model of changing databases, which was introduced sometimes ago.

Our research concentrates on how much different two databases are. On the other hand, Müller et. al. discussed distance of databases from the point of view how much work one needs to do in order to synchronize them. Their approach is algorithmic, our approach is more theoretical.

We have done the first steps by determining the largest possible distance between two databases of the same number of attributes. Next, we determined the distance of any two databases by showing that it is the size of the symmetric difference of the collections of closed sets. Then we investigated the diameter of the set of databases with a given system of (minimal) keys. This led to interesting discrete mathematics problems. Namely, given a hypergraph $H = (V, \mathcal{E})$, what is the number of complete subhypergraphs that are *not* intersections of *maximal subhypergraphs*? We have given good upper bounds in the case of ordinary graphs and k-uniform hypergraphs, when the number of hyperedges is fixed. Further research topic is to improve these bounds and find those key systems that achieve the extremal values. We in fact *conjecture* that if minimal keys are 2-element sets, and the number of keys is of form $\binom{n}{2} - \binom{a}{2}$, then the maximum diameter is $2^a - 2$ and it is attained when the two-element sets that are non-keys form a complete graph on a vertices.

Of course, our distance concept is restricted in the sense that it takes into account only the system of functional dependencies. This model can basically be extended in two directions. On the one hand it could be refined in the way distinguishing two databases when their system of functional dependencies are identical, but they are different in some other sense, for instance some other dependencies are satisfied in one of them while they are not satisfied in the other one. The other direction of extension is the "generalization". Then the distance is introduced for databases with different numbers of attributes, as well.

This can be imagined in the following way. The "space" of all possible databases is given with a hypothetical distance in this space. In our model we forgot about the distance between two databases with the same system of functional dependencies (and, of course the same number of attributes) considering them the same and setting their distance 0, that is they are considered to be one "point" in the space. We introduce the distance between the databases within the set of the databases with the same number of attributes. We do not define here the distance between databases with different numbers of attributes, that is databases in distinct sets. Then the "real" or "hypothetical" distance could be a combination of the distance between databases in different sets and the distance between "refined elements" within one "point".

On the other hand, one may try to find a distance concept that takes into account that a database can have several different statuses during its lifetime. A first and very strict identity definition could be to declare instances $\mathbf{r}$ and $\mathbf{r}'$ of schemata $\mathcal{R}$ and $\mathcal{R}'$, respectively, being the same if there exist one-to-one mappings $\Phi\colon \mathbf{Attr} \to \mathbf{Attr}$ and $\Psi\colon \mathbf{Dom} \to \mathbf{Dom}$ such that

$$\Phi(\mathcal{R}) = \mathcal{R}' \text{ and } \Psi(\mathbf{r}) = \mathbf{r}'. \tag{33}$$

However, this does not allow adding or deleting records, or modifying the schema. These latter two could be incorporated by saying that $\mathbf{r}$ and $\mathbf{r}'$ of schemata $\mathcal{R}$ and $\mathcal{R}'$ are instances of the same database if instead of (33), only

$$\Phi(\mathcal{R}) \subseteq \mathcal{R}' \text{ and } \Psi(\mathbf{r}) \subseteq \mathbf{r}' \tag{34}$$

or

$$\Phi(\mathcal{R}) \subseteq \mathcal{R}' \text{ and } \Psi(\mathbf{r}) \supseteq \mathbf{r}' \tag{35}$$

is required. However, while (33) is overly restrictive, (34) and (35) are too loose. They would allow the empty database to be the same with any other database. Thus, another future research topic is finding the proper balance between (33), and (34) and (35).

References

1. Berman, H.M., Westbrook, J., Feng, Z., Gilliland, G., Bhat, T.N., Weissig, H., Shindyalov, I.N., Bourne, P.E.: The Protein Data Bank. Nucleic Acids Research 28(1), 235–242 (2000)
2. Bhat, T.N., et al.: The PDB data uniformity project. Nucleic Acid Research 29(1), 214–218 (2001)
3. Boutselakis, H., et al.: E-MSD: the European Bioinformatics Institute Macromolecular Structure Database. Nucleic Acid Research 31(1), 458–462 (2003)
4. Burosch, G., Demetrovics, J., Katona, G.O.H.: The Poset of Closures as a Model of Changing Databases. Order 4, 127–142 (1987)
5. Demetrovics, J., Katona, G.O.H.: Extremal combinatorial problems in relational data base. In: Gecseg, F. (ed.) FCT 1981. LNCS, vol. 117, pp. 110–119. Springer, Heidelberg (1981)

6. Erdős, P.: On the number of complete subgraphs contained in certain graphs. Publ. Math. Inst. Hung. Acad. Sci. VII, Ser. A3, 459–464 (1962), http://www.math-inst.hu/~p_erdos/1962-14.pdf
7. Katona, G.: A theorem on finite sets. In: Theory of Graphs, Proc. Coll. held at Tihany, 1966, Akadémiai Kiadó, pp. 187–207 (1968)
8. Kruskal, J.B.: The number of simplices in a complex. In: Mathematical Optimization Techniques, pp. 251–278. University of California Press, Berkeley (1963)
9. Müller, H., Freytag, J.-C., Leser, U.: On the Distance of Databases, Technical Report, HUB-IB-199 (March 2006)
10. Müller, H., Freytag, J.-C., Leser, U.: Describing differences between databases. In: CIKM 2006: Proceedings of the 15th ACM international conference on Information and knowledge management, Arlington, Virginia, USA, pp. 612–621 (2006)
11. Rother, K., Müller, H., Trissl, S., Koch, I., Steinke, T., Preissner, R., Frömmel, C., Leser, U.: COLUMBA: Multidimensional Data Integration of Protein Annotations. In: Rahm, E. (ed.) DILS 2004. LNCS (LNBI), vol. 2994. Springer, Heidelberg (2004)

A Appendix

Here we give a general upper bound for $s_2(e)$, through a series of lemmata. We suppose that $G = ([n], E)$ is a graph with e edges, where $e \leq \binom{a}{2}$. The number of subsets of $[n]$ which span a complete graph, not equal to $\emptyset$ and a clique is $c(G)$. Our final aim is to prove $c(G) \leq 2^a - 2$.

Theorem A.1. *If* $9 < a$ *and the number of edges* e *in* G *is at most* $\binom{a}{2}$ *then the number of subsets spanning a complete graph in* G*, not counting the empty set and the cliques is at most* $2^a - 2$*, that is* $s_2(e) \leq 2^a - 2$.

Remark A.1. If $e = \binom{a}{2}$ then the complete graph shows that our estimate is sharp. The statement of the theorem is also true when $a \leq 9$. This can be shown with ugly case analysis.

Remark A.2. With the notations of Theorem 3.5, $a = t + 1$.

Lemma A.1. *If* G *contains a* K_{a-1} *then* $c(G) \leq 2^a - 2$.

Proof. K_{a-1} contains $\binom{a-1}{2}$ edges, at most $a - 1$ are left. Denote the vertex set of K_{a-1} by A.

Suppose first that the graph M determined by the edges not in the K_{a-1} is connected and is not a tree. Then the number of vertices covered by them, that is, the number of vertices of M is at most $a-1$. The number of subsets of the vertices of M is at most 2^{a-1}. Since a the vertex set of a complete subgraph in G is either a subset of A or of the vertex set of M, we obtained $c(G) \leq 2^{a-1}-2+2^{a-1} = 2^a-2$.

If M is connected, but is a tree, the previous consideration does not work only when the number of edges of M is $a - 1$. Denote the set of vertices of M not in A by B, the subgraph of M spanned by B will be denoted by M_B. If $|B| = 1$ then G is a complete graph on a vertices, the statement is obvious. Suppose that M_B is connected and $|B| = b \geq 2$. Denote the set of vertices in

A adjacent to $i \in B$ by A_i. These sets $A_i (1 \le i \le b)$ are disjoint, because M is a tree. The number of edges of M_B is $b-1$. The sum of the sizes of A_i is $a-1-(b-1)=a-b$. The complete subgraphs not in A are either 2-element subsets of B (edges of M_B) or a vertex $i \in B$ plus a subset of A_i. Their total number is $\sum_{i=1} 2^{|A_i|} + b - 1 \le 2^{a-b} + b - 1 + b - 1$. The maximum of the right hand side in the interval $1 \le b \le a-1$ is 2^{a-1}, this case is also settled.

Suppose now that M_B consists of $k \ge 2$ components: $N_1, \ldots, N_k$. The number of edges of N_i is denoted by f_i. On the other hand, let the number of edges having at least one end in N_i be $a_i - 1$. Here $\sum_{i=1}^{k}(a_i - 1) = a - 1$. Every new complete graph must contain one vertex from B but cannot contain vertices from distinct components N_i. Consider those new complete graphs containing at least one vertex from N_i. The statement of the previous paragraph can be repeated replacing $a-1$ by $a_i - 1$ and b by f_i. Therefore the number of these complete graphs is at most $2^{a_1 - 1}$. The total number of complete graphs is at most

$$\sum_{i=1}^{k} 2^{a_i - 1}. \tag{36}$$

Here $a_i = 1$ is impossible because then N_i would be an isolated point. Then 2^{a-1} is an upper estimate on (36), like in the previous cases.

Suppose now that M consists of $k \ge 2$ components: $M_1, \ldots, M_k$. The number of edges of M_i is denoted by m_i. Here $\sum_{i=1}^{k} m_i \le a-1$. The result of the previous sections can be used: the number of new complete graphs in the ith component is at most 2^{m_i}. Altogether: $\sum_{i=1}^{k} 2^{m_i}$. Since every m_i is at least 1, therefore they cannot exceed $a-k$. This case can be finished exactly like the previous one. □

The *difference* of two complete graphs is the $|V_1 - V_2|$ where V_1 and V_2 are the two vertex sets.

Lemma A.2. *Suppose that G contains no K_{a-1} but it contains (at least) two K_{a-2} with the vertex sets V_1 and V_2, respectively. Then one of the followings holds:*

$$|V_1 - V_2| = 1, \tag{37}$$

$$|V_1 - V_2| = 2, \tag{38}$$

$$|V_1 - V_2| \ge 3 \text{ and } a \le 9. \tag{39}$$

Proof. Let $|V_1 - V_2|$ be denoted by i. Then the total number of edges in the two K_{a-2} is

$$2\binom{a-2}{2} - \binom{a-2-i}{2} \le \binom{a}{2}. \tag{40}$$

Easy algebra leads to the inequality

$$0 \le i^2 - i(2a-5) + 4a - 6. \tag{41}$$

This holds iff i is not in the interval determined by the solutions of the quadratic equation:

$$\frac{2a-5\pm\sqrt{(2a-5)^2-16a+24}}{2}=\frac{2a-5\pm\sqrt{4a^2-36a+49}}{2}. \tag{42}$$

Here $2a-11<\sqrt{4a^2-36a+49}$ holds when $9<a$. Using this inequality we obtain strict upper and a lower estimate on the "smaller" and the "larger" roots, respectively:

$$3=\frac{2a-5-(2a-11)}{2}, \quad 2a-8=\frac{2a-5+(2a-11)}{2}. \tag{43}$$

This proves that $9<a$ implies $i<3$. (The other estimate becomes $a+1<i$ when $9<a$ what is impossible.) □

Lemma A.3. *Suppose that* $5\le a$ *and* G *contains no copy of* K_{a-1}, *but contains a pair of* K_{a-2}*'s with difference* 2. *Then* $c(G)<2^a-2$.

Proof. It is easy to see that there might be at most 4 extra edges which can be added to the two K_{a-2}'s. The number of subsets spanning complete graphs in them, not counting themselves and the empty set, is $2^{a-2}-3+32^{a-4}$.

First suppose that there is no edge between V_1-V_2 and V_2-V_1, where V_1 and V_2 are the vertex sets of the K_{a-2}'s. Then all of the 4 new edges have one vertex not in $V_1\cup V_2$. It is easy to see by case analysis, that the maximum number of sets spanning a complete graph is 16, using the extra 4 edges. The total number of sets in question is at most $2^{a-2}-3+32^{a-4}+16\le 2^a-2$ when, say $5\le a$.

Suppose now that there is exactly one edge between V_1-V_2 and V_2-V_1. This addition creates a new K_{a-2}. It also adds $2^{a-4}-1$ new complete graphs. the remaining 3 edges may add at most 8. Altogether: $2^{a-2}-3+32^{a-4}+2^{a-4}-1+8=2^{a-1}+4$ what cannot be more than 2^a-2 when $4\le a$.

If there are two adjacent edges between V_1-V_2 and V_2-V_1 then they form a K_{a-1} contradicting our assumptions. Therefore if we suppose that there are at least two edges between V_1-V_2 and V_2-V_1 then it is possible only when there are exactly two of them and they have no common vertex. The so obtained graph contains $2^{a-1}+2^{a-4}-5$ proper complete graphs. The remaining two edges may add 4 more complete graphs: $2^{a-1}+2^{a-4}-1\le 2^a-2$. □

Lemma A.4. *Suppose that* $4\le a$, G *contains no copy of* K_{a-1}, *contains no pair of* K_{a-2}*s with difference 2, but contains 3 copies of* K_{a-2} *with pairwise difference* 1. *Then* $c(G)\le 2^a-2$.

Proof. Take two of the complete graphs. Let K denote the intersection of their vertex sets. ($|K|=a-3\ge 2$.) The vertex sets have the respective forms $K\cup u$ and $K\cup v (u\ne v)$. It is easy to see that the vertex set of the third K_{a-2} cannot contain either of u and v, therefore it also has the form $K\cup w$ where $(w\ne u),(w\ne v)$. Denote the graph obtained as their union by $K(3)$.

The number of edges of $K(3)$ is $\binom{a-3}{2}+3(a-3)=\binom{a}{2}-3$, that is, only 3 edges remained.

The number of proper complete subgraphs is $2^{a-3} + 32^{a-3} - 4 = 2^{a-1} - 4$. The addition of 3 edges may create at most 8 complete graphs. $2^{a-1} + 4 \leq 2^a - 2$ holds. □

Proof (of Theorem A.1). Start like in the proof of Theorem 3.5 The number of 1-element sets in question is at most $2e$, the number of 2-element subsets is e. The number of i-element subsets (≥ 3) is at most $\binom{a}{i}$. The only novelty here is the treatment of the large sets. We saw in Lemmata A.1-A.4 that the statement of the theorem holds when there is a K_{a-1} or at least three K_{a-2} in G. If we suppose the contrary then the number of $a-2$ and $a-1$-element sets in question is 0, the number of such $a-3$-element subsets is at most $2(a-2)$ (subsets of the two possible K_{a-2}). The total number of sets is at most

$$3e + \sum_{i=3}^{a-4} \binom{a}{i} + 2(a-2) = 3e + 2^a - 1 - a - \binom{a}{2} - \binom{a}{a-3} + 2(a-2) -$$

$$- \binom{a}{a-2} - \binom{a}{a-1} - 1$$

$$= 2^a - 6 + \left(2e - 2\binom{a}{2}\right) + \left(e - \binom{a}{3}\right). \tag{44}$$

Here $e \leq \binom{a}{2}$ by definition, $e \leq \binom{a}{3}$ is an easy consequence when $5 \leq a$. (44) can be upper estimated by $2^a - 6$. □

On the Existence of Armstrong Data Trees for XML Functional Dependencies

Sven Hartmann[1], Henning Köhler[2], and Thu Trinh[1]

[1] Clausthal University of Technology, Germany
[2] The University of Queensland, Brisbane, Australia

Abstract. Armstrong databases are a popular tool in example-based database design. An Armstrong database for a given constraint set Σ from a fixed constraint class $\mathcal{Z}$ satisfies precisely those constraints from $\mathcal{Z}$ that are logically implied by Σ. In this paper we study Armstrong data trees for functional dependencies for XML-encoded data in the context of a simple XML tree model reflecting the permitted parent-child relationships together with their frequencies.

1 Introduction

Armstrong databases have attracted a good deal of attention ever since the introduction of the relational data model in the 1970s. Given a class of integrity constraints $\mathcal{Z}$ an Armstrong database for a set $\Sigma \subseteq \mathcal{Z}$ satisfies a constraint from $\mathcal{Z}$ if and only if it is logically implied by Σ. Armstrong databases have proven useful both in database theory as well as in practical database design. The story of Armstrong databases goes back to Armstrong [3] who proved their existence for sets of functional dependencies defined on a relation schema. Later this result was extended to other constraint classes, e.g., implicational dependencies [20], multi-valued dependencies [5,7,61], and strong dependencies [17,52]. For other constraint classes Armstrong databases only exist under certain conditions, e.g., inclusion dependencies [21], boolean dependencies [16], branching dependencies [13,35], partition dependencies [25], domain constraints [48], and participation constraints [28,29]. Further works investigated the existence of Armstrong databases in the context of more advanced data models such as incomplete relations [38,39] and nested attribute models [46,47], or studied properties of Armstrong databases such as their size and structure, cf. [6,11,12,14,18,26,34,36,43,53,54].

The usefulness of Armstrong databases in database design practice was first noted by Silva and Melkanoff [49]. Mannila and Räihä's seminal work on design-by-example [44,45] was probably most influential in promoting Armstrong databases as a design aid. It is widely accepted now that sample databases are a valuable tool for data architects, e.g., to communicate with domain experts, to study consequences of particular design decisions, and to foresee potential anomalies during database operation. Eventually the generation, analysis and evolution of good sample databases at design time may help data architects to

S. Link and H. Prade (Eds.): FoIKS 2010, LNCS 5956, pp. 94–113, 2010.

specify a constraint set that best reflects the business rules for the application under development, thus preventing expensive corrections at run time. Armstrong databases are often regarded as perfect sample databases as they satisfy precisely all logical consequences of the specified constraint set Σ within some constraint class $\mathcal{Z}$. Thus they do double duty as an example for all constraints from $\mathcal{Z}$ that are logically implied by Σ, but also as a counterexample for all those that are not logically implied. Various aspects of applying Armstrong databases in database design were addressed, e.g., in [8,9,10]. Algorithms for generating Armstrong databases were presented, e.g., in [15,19,42], and dependency discovery in Armstrong databases was discussed, e.g., in [27].

In this paper we study Armstrong databases for functional dependencies in the context of XML-encoded data. XML (the Extensible Markup Language) has recently gained popularity as a standard for sharing and integrating data on the Web and elsewhere. There is wide consensus among researchers and practitioners that data architects require commensurate tool support for designing XML-encoded data. As for relational databases integrity constraints should be used to reflect the business rules of the application domain, cf. [22]. Specifying integrity constraints and enforcing them at run time helps to keep the database meaningful for the application of interest. There is evidence that the thorough use of integrity constraints is beneficial for a range of tasks such as schema design, query optimisation, efficient access and updating, privacy assurance, and data cleaning. Unfortunately, however, the inherent syntactic flexibility and hierarchical structure of XML make it a challenge to precisely describe desirable business rules in terms of integrity constraints, and to provide methods for efficiently validating and enforcing such constraints, cf. [22,23,24,50,56].

For relational databases, functional dependencies have been vital in the investigation of how to design "good" database schemas to avoid or minimise problems related to data redundancy and inconsistency. The same problems can be shown to exist in poorly designed XML databases. Consequently, functional dependencies for XML (often referred to as XFDs) have recently gained considerable interest by the database community. In the relational data model a functional dependency states that some set of attributes X determines some other set of attributes Y. A relation satisfies such a functional dependency if any two tuples in the relation that agree on the attributes in X also agree on the attributes in Y. In order to extend the concept of functional dependency from relational databases to XML one needs to decide how to interpret the notions of tuple, attribute and agreement in the context of XML. The difficulty with XML data is that its nested structure is more complex than the rigid structure of relational data, and thus may well observe a larger variety of functional dependencies. In the literature [2,30,31,37,41,57,58,59,60,62,64], several generalisations of functional dependencies to XML have been proposed, and these proposals differ quite considerably in their expressiveness.

In this paper we use the approach of [31,32] that defines XFDs on the basis of a schema tree that serves as a structural summary of the XML-encoded data and reflects the permitted parent-child relationships. Satisfaction of an XFD in a data tree is defined based on pairwise comparison of almost-copies of the schema tree that can be found in that data tree. XML schema trees capture information on the frequency of parent-child relationships, that is, they show whether a child is optional or required, and whether it is unique or may occur multiple times. Our approach comes close to the approach taken in [2] where XFDs are defined on the basis of paths evolving from the root element in an XML data tree. Both approaches are motivated by earlier studies of functional dependencies in semantic and object-oriented data models [51,63]. It should be noted that XFDs may well interact in a non-trivial way with chosen structural summaries as demonstrated, e.g., in [1,2].

This paper is organised as follows. In Section 2, we assemble preliminary notions like XML schema trees and data trees that form the context for our investigation. In Section 3, we provide the definition of XFDs. In Section 4, we recall the axiomatisation of XFDs presented in [32]. In Section 5, we define and characterise Armstrong data trees for XFDs. Section 6 studies the construction of Armstrong data trees for XFDs as far as they exist. Section 7 concludes the paper.

2 Preliminaries

It is common to represent XML-encoded data by node-labelled trees, cf. the representations used in DOM, XPath, XQuery, XSL, and XML Schema. Fig. 1 shows an example of an XML tree in which nodes are labelled by their names (chosen from a suitable set $Names$ of element and attribute names) and annotated by their kind: E for element, A for attribute, and S for text (PCDATA).

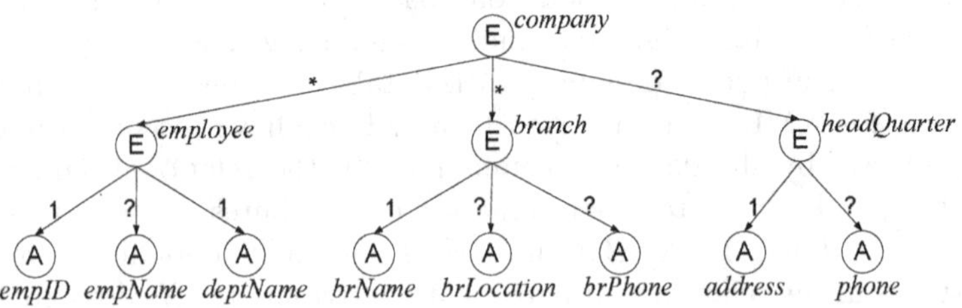

Fig. 1. The XML schema tree PD

We recall the tree model used in [32]. For that we assume familiarity with basic concepts from graph theory, cf. [33]. An *XML tree* T is a rooted tree with node set V_T, arc set A_T, a unique and distinguished root node r_T, and mappings $name : V_T \to Names$ and $kind : V_T \to \{E, A, S\}$ that assign every node its name and kind, respectively. T is said to be *finite* if V_T is finite, and is said to be *empty* if V_T consists of the root node only. All trees considered in this paper are finite. In an XML tree each node $v \in V_T$ can be reached from the root by a unique path from r_T to v. For every (directed) path from a node u to a node v in T

we call u an ancestor of v and v a descendant of u. For an arc (u, v) of T, in particular, u is the parent of v while v is a child of u. A node without children is a *leaf*. We further suppose that, in an XML tree, nodes of kind A and S are always leaves. Let L_T denote the set of all leaves of kind A or S in T.

We call every path from the root to some leaf in L_T a *walk* of T. Given a set of walks W of T the graph union of the walks in W is said to be an r_T-*subgraph* of T. For convenience we will usually not distinguish between the set W and the r_T-subgraph formed by it.

An *XML data tree* is an XML tree T' together with an evaluation $val : L_{T'} \to STRING$ assigning every leaf $v \in L_{T'}$ a string $val(v)$.

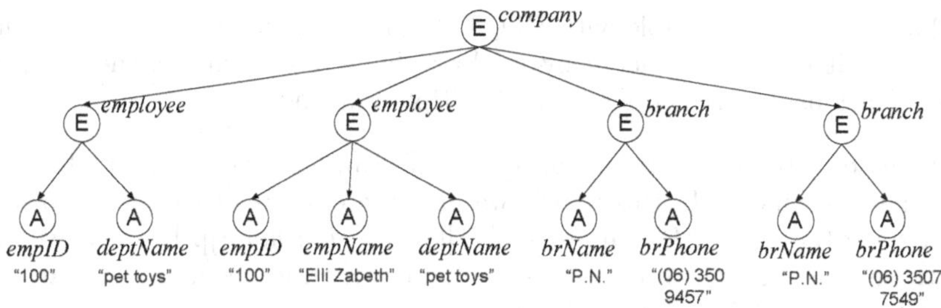

Fig. 2. An XML data tree PD′ compatible with PD

Given two XML trees T and T', a mapping $\phi : V_{T'} \to V_T$ is called a *homomorphism* between T' and T if ϕ is

i) root-preserving, that is, $\phi(r_{T'}) = r_T$,
ii) arc-preserving, that is, $(\phi(u'), \phi(v')) \in A_T$ for all $(u', v') \in A_{T'}$,
iii) name-preserving, that is, $name(\phi(u')) = name(u')$ for all $u' \in V_{T'}$, and
iv) kind-preserving, that is, $kind(\phi(u')) = kind(u')$ for all $u' \in V_{T'}$.

A homomorphism ϕ is an *isomorphism* if ϕ is bijective and ϕ^{-1} is again a homomorphism. When an isomorphism exists between two XML trees T' and T then T' is said to be *isomorphic* to T or a *copy* of T, denoted by $T' \cong T$. Two isomorphic XML data trees T' and T are said to be *value-equal*, denoted by $T' =_v T$, if an isomorphism between them is evaluation-preserving, that is, $val(\phi(v')) = val(v')$ for all leaves $v' \in L_{T'}$.

An *XML schema tree* is an XML tree T together with a mapping $freq : A_T \to \{?, 1, +, *\}$ that assigns every arc its frequency. We suppose that no node in V_T has two successors with the same name and the same kind. Further, we suppose that an arc $a = (u, v)$ has frequency $freq(a) = ?$ or 1 whenever v is of kind A, and frequency $freq(a) = 1$ whenever v is of kind S. Note that this convention is a legacy of XML and can be skipped if desired. In particular it is not essential for the investigations undertaken in this paper. For convenience, arcs of frequency f will be called f-arcs, and arcs of frequency f or g will be called f/g-arcs. For example, a ?-arc refers to an arc of frequency ?, while a */+-arc refers to an arc of frequency * or +.

An XML data tree T' is *compatible* with an XML schema tree T, denoted by $T' \rhd T$, if there is a homomorphism ϕ between T' and T such that for each node v' of T' and each arc $a = (\phi(v'), w)$ of T, the number of arcs $a' = (v', w'_i)$ mapped to a is at most 1 if $freq(a) =?$, exactly 1 if $freq(a) = 1$, at least 1 if $freq(a) = +$, and arbitrarily many if $freq(a) = *$. Due to the definition of a schema tree, this homomorphism is unique if it exists.

Example 1. To illustrate the concepts introduced above we provide a small example that captures the information needs of a product development company and its employees. Fig. 1 shows the schema tree PD used in our example, and Fig. 2 a data tree compatible with PD. The company may have one head quarter office that can be contacted by mail or phone. Furthermore, the company operates various departments and has multiple branches. □

For the ease of presentation, we have chosen example schema trees where the leaf names are unique. In this paper, we may therefore refer to a walk to some leaf carrying the name "B" simply as $[[\,B\,]]$, e.g., $[[\,brPhone\,]]$. Further, we may refer to an r_T-subgraph X by listing the names of all leaves in X separated by white spaces, e.g., $[[\,brName\ brPhone\,]]$.

An XML schema tree may be developed by a data architect similar to a database schema, or it can be derived from other specifications (such as DTDs or XSDs). Alternatively, an XML schema tree may also be derived from an XML document itself, cf. [31]. Note, however, there is usually more than just a single schema tree T for a given data tree T'. For example, T may well be extended by adjoining new nodes and arcs.

Note also that in this paper we are not concerned with the question of whether some information are better modelled as an attribute or text element. In our example, we do not use text elements, purely for keeping the schema graph as compact as possible. In principle, however, each attribute in a schema tree T with name "B" may alternatively be modelled by a node u of kind E with name "B" and a child v of kind S.

To conclude this section we briefly recall operators for constructing new trees from given ones. Let X, Y be subgraphs of an XML tree T. The *union* of X and Y, denoted by $X \cup Y$, is the restriction of the graph union of X and Y to the maximal r_T-subgraph of T contained in it. For convenience, we sometimes omit the union symbol and write XY instead of $X \cup Y$. The *intersection* of X and Y, denoted by $X \cap Y$, is the union of all walks that belong to both X and Y. The *difference* between X and Y, denoted by $X - Y$, is the union of all walks that belong to X but not to Y. In particular, $X \cup Y$, $X \cap Y$ and $X - Y$ are r_T-subgraphs of T. In the absence of parentheses, we suppose that the union and intersection operators bind tighter than the difference operator. For example, by $X \cup Y - Z$ we mean $(X \cup Y) - Z$.

3 Functional Dependencies for XML

For the rest of this paper let T be an XML schema tree, T' an XML data tree compatible with T, and ϕ the unique homomorphism between T' and T. In this section we outline the concept of a functional dependency for XML as discussed in [31,32]. Recall that a subgraph H' of T' is a copy of a subgraph H of T if the restriction of ϕ to H' and H is an isomorphism between H' and H. A subgraph H' of T' is a *subcopy of* T if it is a copy of some r_T-subgraph H of T. A subcopy of T is maximal if it is not properly contained in any other subcopy of T. We call each non-empty maximal subcopy of T an *almost-copy* of T. In the context of XML, almost-copies will play the role of tuples in relations. In the literature [2] the term "tree tuple" has been suggested for generalisations of relational tuples to XML, however, we will keep using the term "almost-copy" to avoid confusion with other definitions.

Given an r_T-subgraph H of T, the *projection* of T' to H, denoted by $T'|_H$, is the union of all subcopies of H in T'.

Example 2. The XML data tree PD$'$ in Fig. 2 contains four almost-copies of PD, two of which are shown in Fig. 3. The other two almost-copies can be obtained as follows:

$$P_3' = P_1'|_{[[\ empID\ empName\ deptName\ ]]} \cup P_2'|_{[[\ brName\ brPhone\ ]]}$$
$$P_4' = P_2'|_{[[\ empID\ empName\ deptName\ ]]} \cup P_1'|_{[[\ brName\ brPhone\ ]]}$$

□

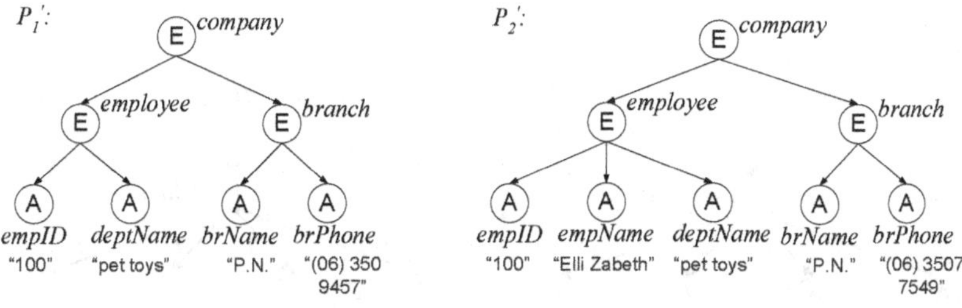

Fig. 3. Two of the four almost-copies of PD contained in PD$'$

An *XML functional dependency* (or XFD for short) on T is an expression $X \to Y$ where X and Y are non-empty r_T-subgraphs of T. Two almost-copies T_1', T_2' of T in T' are *value-equal* on some r_T-subgraph X of T if $T_1'|_X =_v T_2'|_X$, they *agree* on X if $T_1'|_X =_v T_2'|_X \cong X$, and they *differ* on X if $T_1'|_X \neq_v T_2'|_X$. That is, they agree on X if they are copies of X and value-equal to one another, while they differ on X if they are not value-equal. An XML data tree T' *satisfies* the XFD $X \to Y$, written as $\models_{T'} X \to Y$, if and only if for every pair of distinct almost-copies T_1', T_2' of T in T' it is true that T_1', T_2' are value-equal on Y whenever T_1', T_2' agree on X.

Example 3. Suppose each department of our product development company is located at a single branch. Branches are located in unique locations. Employees are assigned unique employee IDs. We use the following XFDs to model the product development company information:

$$\begin{array}{lll} (F1) & [[\,brName\,]] & \to [[\,brLocation\,]] \\ (F2) & [[\,brName\,]] & \to [[\,brPhone\,]] \\ (F3) & [[\,brLocation\,]] & \to [[\,brName\,]] \\ (F4) & [[\,deptName\,]] & \to [[\,brName\,]] \\ (F5) & [[\,empID\,]] & \to [[\,empName\,]] \end{array}$$

The almost-copies P_1', P_2' agree on $[[\,brName\,]]$, and both do not contain a copy of $[[\,brLocation\,]]$. As P_3', P_4' can be obtained from P_1', P_2', it is easy to see that any two almost-copies of PD in PD$'$ agree on $[[\,brName\,]]$, but contain no copy of $[[\,brLocation\,]]$. Hence, PD$'$ satisfies (F1). While the almost-copies P_1', P_2' agree on $[[\,brName\,]]$ they differ on $[[\,brPhone\,]]$. Therefore PD$'$ does not satisfy (F2). On the other hand, (F3) is trivially satisfied because PD$'$ contains no copy of $[[\,brLocation\,]]$. Moreover, any two almost-copies agree on $[[\,deptName\,]]$, but are also value-equal on $[[\,brName\,]]$. Thus PD$'$ satisfies (F4). Finally, P_1', P_2' agree on $[[\,empID\,]]$, but differ on $[[\,empName\,]]$. Thus, PD$'$ does not satisfy (F5). □

Note that the concept of XFDs presented here is a proper generalisation of the notion of functional dependency as defined in the relational data model and studied under the assumption of complete information or (partly) incomplete information, cf. [4,40]. Fig. 4 shows how to translate a relation schema into a corresponding XML schema tree.

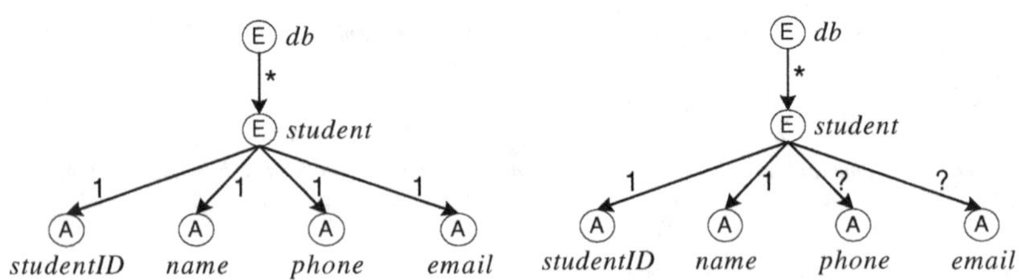

Fig. 4. XML schema trees corresponding to a relation schema Student = {studentID, name, phone, email} assuming complete or incomplete information

4 Reasoning about XFDs

An XML data tree T' satisfies a given set Σ of XFDs, denoted by $\models_{T'} \Sigma$, if T' satisfies each individual XFD in Σ. Satisfaction of some given set of XFDs by an XML data tree often implies the satisfaction of further XFDs. The notions of implication and derivability (with respect to a rule system $\mathcal{R}$) can be defined similar to the analogous notions in the relational context, cf. [45,51].

Let Σ be a set of XFDs and $X \to Y$ a single XFD. If $X \to Y$ is satisfied in every XML data tree which satisfies Σ, then Σ *implies* $X \to Y$, written as $\Sigma \models X \to Y$. The *semantic closure of* Σ, denoted by Σ^*, is the set of all XFDs that are implied by Σ, that is, $\Sigma^* = \{X \to Y \mid \Sigma \models X \to Y\}$.

Given a system of inference rules $\mathcal{R}$, we call an XFD $X \to Y$ *derivable* from Σ by $\mathcal{R}$, denoted by $\Sigma \vdash_{\mathcal{R}} X \to Y$, if there is a finite sequence of XFDs, whose last element is $X \to Y$ such that each XFD in the sequence is in Σ or can be obtained from Σ by applying one of the inference rules in the rule system $\mathcal{R}$ to a finite number of previous XFDs in the sequence. The *syntactic closure of* Σ *with respect to the rule system* $\mathcal{R}$, denoted $\Sigma^+_{\mathcal{R}}$, is the set of all XFDs which are derivable from Σ by means of inference rules in $\mathcal{R}$, that is, $\Sigma^+_{\mathcal{R}} = \{X \to Y \mid \Sigma \vdash_{\mathcal{R}} X \to Y\}$.

$$\frac{}{X \to Y}\; Y \subseteq X \qquad \frac{X \to Y}{X \to Z}\; Z \subseteq Y \qquad \frac{W \to Y}{X \to Y}\; W \subseteq X$$

(reflexivity axiom) (subtree rule) (supertree rule)

$$\frac{X \to Y, X \to Z}{X \to Y \cup Z} \qquad \frac{X \to Y, Y \to Z}{X \to Z}\; Y \text{ is } X, Z\text{-compliant}$$

(union rule) (restricted-transitivity rule)

$$\frac{}{X \to R_T} \qquad \frac{\big((X \cup B) \cup T_{\geq 1} - U_B\big) \cup X \to B}{X \to B}$$

(root axiom) (noname rule)

Fig. 5. A sound and complete rule system for the implication of XFDs

An inference rule is called *sound* if for any given set Σ of XFDs, every XFD which may be derived from Σ due to that rule is also implied by Σ. A rule system $\mathcal{R}$ is *sound* if all inference rules in $\mathcal{R}$ are sound. In other words, $\mathcal{R}$ is sound if every XFD which is derivable from Σ by $\mathcal{R}$ is also implied by Σ (i.e. $\Sigma^+_{\mathcal{R}} \subseteq \Sigma^*$). A rule system is said to be *complete* if it is possible to derive every XFD which is implied by Σ (i.e. $\Sigma^* \subseteq \Sigma^+_{\mathcal{R}}$). A sound and complete rule system is also called an *axiomatisation.*

In [32] we presented a sound and complete rule system for the implication of XFDs, cf. Fig. 5. Let $T_{\leq 1}$ be the graph union of all ?/1-arcs of T, $T_{\geq 1}$ the graph union of all 1/+-arcs of T, and R_T the maximal r_T-subgraph of $T_{\leq 1}$. Note that $T_{\leq 1}$ and $T_{\geq 1}$ might not be r_T-subgraphs of T. It should be emphasised that this axiomatisation is considerably different from the well-known Armstrong system [3] for functional dependencies in the relation model.

Most noteworthy, Armstrong's transitivity rule no longer holds for XFDs. Given r_T-subgraphs X, Y, Z of T we say that Y *is* X, Z*-compliant* if $Y \subseteq (X \cup C) \cup T_{\geq 1}$ for each walk C in Z. The restricted-transitivity rule is somewhat reminiscent of a similar rule in the axiomatisation of functional dependencies in the context of incomplete relations with null values proposed by Lien [40] and by Atzeni and Morfuni [4]. Note, however, that the axiomatisation of XFDs contains further rules that do not have natural counterparts in the systems of Lien, Atzeni and Morfuni.

We call a maximal non-empty r_T-subgraph consisting of walks of T that share a common $+/*$-arc a *unit* of T. Given a walk B of T let the *unit of* B, denoted by U_B, be the union of all walks sharing some $*/+$-arc with B. Note that we have $U_C = U_B$ for each walk $C \in U_B$. Walks without any $*/+$-arc do not have a unit.

Example 4. The XML schema tree PD in Fig. 1 has two units: $[[empID\ empName\ deptName]]$ and $[[brName\ brLocation\ brPhone]]$. Furthermore there are two $r_{\rm PD}$-walks in $\mathrm{PD}_{\leq 1}$ which do not have a unit by definition. This gives us $R_{\rm PD} = [[address\ phone]]$. □

Every XML schema tree T possesses a unique partition of its walks into its units and R_T. Units play a core role in constructions of XML data trees that are compatible with a given schema tree. For more details we refer to the paragraph before Lemma 12 below. The noname rule in Fig. 5 reflects the nested nature of XML-encoded data induced by the units, while the root axiom arises from the choice of frequencies.

Example 5. Recall the XFDs (F1), (F4) specified in Example 3 on the schema tree PD. Using the *root axiom* we can derive for example the XFDs $[[empName\ brName]] \to [[address\ phone]]$ and $[[empID]] \to [[address\ phone]]$.

It is easy to see that $[[brName]] \subseteq ([[deptName]] \cup [[brLocation]]) \cup \mathrm{PD}_{\geq 1}$ holds so that $[[brName]]$ is $[[deptName]], [[brLocation]]$-compliant. Applying the *restricted-transitivity rule* to (F4) and (F1) gives us the XFD $[[deptName]] \to [[brLocation]]$.

The *noname rule* gives the new XFD $[[empName]]{\to}[[brName]]$. To see this, first determine $([[empName]] \cup [[brName]]) \cup \mathrm{PD}_{\geq 1}$ as the $r_{\rm PD}$-subgraph $[[empID\ empName\ deptName\ brName]]$ of PD, and the unit $U_{[[brName]]}$ as the $r_{\rm PD}$-subgraph $[[brName\ brLocation\ brPhone]]$. This amounts to the premise of the noname rule being $[[empID\ empName\ deptName]]{\to}[[brName]]$, which can be derived from (F4) using the supertree rule. □

5 Characterising Armstrong Data Trees

We will now focus on Armstrong data trees. In the sequel let Σ be a set of XFDs on T. The XML data tree $T' \rhd T$ is said to be *Armstrong* for Σ if T' satisfies every XFD $X \to Y$ in Σ and violates every XFD not in Σ^*.

Agree sets and difference sets play a core role in dependency discovery in relational databases, cf. [44]. We can easily extend these notions to XML data trees. Given two almost-copies T'_1, T'_2 of T in T' let $ag_T(T'_1, T'_2)$ be the union of all walks of T on which T'_1, T'_2 agree, and let $diff_T(T'_1, T'_2)$ be the union of all walks of T on which T'_1, T'_2 differ. We call $ag_T(T'_1, T'_2)$ and $diff_T(T'_1, T'_2)$ the *agree set* and the *difference set* of T'_1, T'_2, respectively. Note, however, that there may exist walks in T that belong neither to the agree set nor to the difference set of two almost-copies T'_1, T'_2 - these are just those walks in T of which neither T'_1

nor T'_2 contains a copy. Moreover, let $ag_T(T')$ and $\mathit{diff}_T(T')$ be the families of agree sets and difference sets of T', respectively, where T'_1, T'_2 runs through all pairs of distinct almost-copies of T in T'.

Example 6. Recall the schema tree PD in Fig. 1 and the data tree in Fig. 2. Consider the almost-copies P'_1, P'_2 of PD shown in Fig. 3. Their agree set is $ag_{\mathrm{PD}}(P'_1, P'_2) = [[\mathit{empID}\ \mathit{deptName}\ \mathit{brName}]]$ and their difference set is $\mathit{diff}_{\mathrm{PD}}(P'_1, P'_2) = [[\mathit{empName}\ \mathit{brPhone}]]$. □

Next we extend the notion of a closure of an attribute set from the relational model to XML. Given an r_T-subgraph X let X^*_Σ be the union of all r_T-subgraphs Y of T for which $X \to Y$ is implied by Σ, that is, $X^*_\Sigma = \bigcup\{Y \mid X \to Y \in \Sigma^*\}$. For a study into properties and computational aspects of the operator $(.)^*_\Sigma$ we refer the interested reader to [55].

Our interest in agree sets and closed sets was triggered by a nice characterisation of Armstrong relations for functional dependencies that is due to Beeri et al. [6]. They proved that a relation r' over a relation schema r is Armstrong for a set of functional dependencies Σ defined on r if and only if $gen_r(\Sigma) \subseteq ag_r(r') \subseteq cl_r(\Sigma)$ holds. Herein, $cl_r(\Sigma)$ denotes the family of all closed sets in r, and $gen_r(\Sigma)$ the family of intersection generators of $cl_r(\Sigma)$. The difficulty with generalising this characterisation to XFDs is due to the fact that the operator $(.)^*_\Sigma$ is not idempotent for sets of XFDs, and thus no longer a closure operator (though still extensive and monotone). In consequence we are losing the concept of closed sets that is central to this characterisation.

To overcome the aforementioned difficulty we define the *divergence* $\delta(X)$ of an r_T-subgraph X of T as the union of the difference sets $\mathit{diff}_T(T'_1, T'_2)$ where T'_1, T'_2 runs through all pairs of distinct almost-copies of T in T' with agree set $ag_T(T'_1, T'_2) = X$.

Lemma 7. *Let T be an XML schema tree, Σ be a set of XFDs defined on T, and T' be an XML data tree compatible with T. Then T' satisfies Σ if and only if $X^*_\Sigma \cap \delta(X) = \emptyset$ holds for every $X \in ag_T(T')$.*

Proof. Suppose that T' satisfies Σ, and let $X \in ag_T(T')$. If $\delta(X)$ is empty the claim follows trivially. Otherwise let B be a walk in $\delta(X)$. There exists two distinct almost-copies T'_1, T'_2 of T in T' with $X = ag_T(T'_1, T'_2)$ and $B \in \mathit{diff}_T(T'_1, T'_2)$. That is, T'_1, T'_2 agree on X but differ on B. Hence, T' violates the XFD $X \to B$ and therefore $B \notin X^*_\Sigma$. This proves $X^*_\Sigma \cap \delta(X) = \emptyset$.

Conversely, suppose that T' satisfies the condition $X^*_\Sigma \cap \delta(X) = \emptyset$ for every $X \in ag_T(T')$. Let $X \to B$ be an XFD in Σ. Now assume there are two distinct almost-copies T'_1, T'_2 that agree on X. Then $X \subseteq Y := ag_T(T'_1, T'_2)$. Since $X \to B \in \Sigma$ we have $B \in X^*_\Sigma$ and thus $B \in Y^*_\Sigma$. By the condition of the lemma, B is not in $\delta(Y)$. In particular, B is not in $\mathit{diff}_T(T'_1, T'_2)$ and therefore T'_1, T'_2 are at least value-equal on B. This proves that T' satisfies Σ. □

For a walk B of T let $max_T(\Sigma, B)$ be the family of all non-empty r_T-subgraphs X of T that are maximal with the property that Σ does not imply $X \to B$, and let $max_T(\Sigma)$ be the union of all these families. Note that for relational databases Mannila and Rähiä [44] showed the equivalence $gen_r(\Sigma) = max_r(\Sigma)$ and emphasised that the definition of $max_r(\Sigma)$ is constructive, thus guiding a way for computing $max_r(\Sigma)$ from Σ. By further strengthening this observation to condition *ii)* in Theorem 9 below and by applying the previous lemma we obtain a characterisation of Armstrong data trees for XFDs.

Example 8. Recall the schema tree PD in Fig. 1 and let Σ consist of the XFDs (F1), (F4) specified in Example 3. Then $max_T(\Sigma, [[brLocation]])$ contains only the r_{PD}-subgraph $[[brPhone\ address\ phone]]$. □

Theorem 9. *Let T be an XML schema tree, Σ be a set of XFDs defined on T, and T' be an XML data tree compatible with T. Then T' is an Armstrong data tree for Σ if and only if the following conditions hold:*

i) $\forall X \in ag_T(T').X^*_\Sigma \cap \delta(X) = \emptyset$, *and*
ii) $\forall$ *walk* B *of* $T.\forall X \in max_T(\Sigma, B).B \in \delta(X)$.

Proof. From Lemma 7 we already know that T' satisfies Σ if and only if condition *i)* holds. Now suppose T' satisfies conditions *i)* and *ii)*. It remains to show that T' violates every XFD not implied by Σ. Assume there exists some XFD $X \to B$ that is not implied by Σ. Then there is some $Y \in max_T(\Sigma, B)$ that contains X. By condition *ii)* we have $B \in \delta(Y)$. By definition, there are two distinct almost-copies T_1', T_2' with $Y = ag_T(T_1', T_2')$ and $B \in diff_T(T_1', T_2')$. Thus, T_1', T_2' agree on X but differ on B. That is, T' violates $X \to B$. This proves that T' is Armstrong for Σ.

Conversely, suppose T' is Armstrong for Σ. It remains to validate conditions *i)* and *ii)*. Assume there exists a walk B of T, and some $X \in max_T(\Sigma, B)$. That is, $X \to B$ is not implied by Σ while Σ implies $W \to B$ for every r_T-subgraph W that properly contains X. Since T' is Armstrong for Σ there exist two distinct almost-copies T_1', T_2' of T in T' that agree on X but differ on B. Let $Y := ag_T(T_1', T_2')$ and note that $B \in diff_T(T_1', T_2') \subseteq \delta(Y)$ by definition. Hence, $Y \to B$ is not implied by Σ either, and from $X \subseteq Y$ we conclude $X = Y$ due to $X \in max_T(\Sigma, B)$. This proves $X \in ag_T(T')$ and $B \in \delta(X)$. That is, T' satisfies condition *ii)*. Recall that condition *i)* holds, too, in consequence of Lemma 7. □

A constraint class $\mathcal{Z}$ is said *to enjoy Armstrong data trees* if for every XML schema tree T and for each constraint set $\Sigma \subseteq \mathcal{Z}$ defined on T there exists an XML data tree $T' \rhd T$ such that T' satisfies Σ but violates all constraints from $\mathcal{Z}$ that are not implied by Σ. The question is now whether it is always possible to find a data tree T' that meets the characterisation of Theorem 9. Unfortunately it is not hard to see that the answer to this question is negative.

Theorem 10. *XFDs do not enjoy Armstrong data trees.*

Proof. It is sufficient to give an example of an XML schema tree and a set of XFDs Σ on it for which no Armstrong data tree exists. Consider the schema tree S in Fig. 6, and $\Sigma = \{[[accountNo]] \to [[empID]], [[empID]] \to$

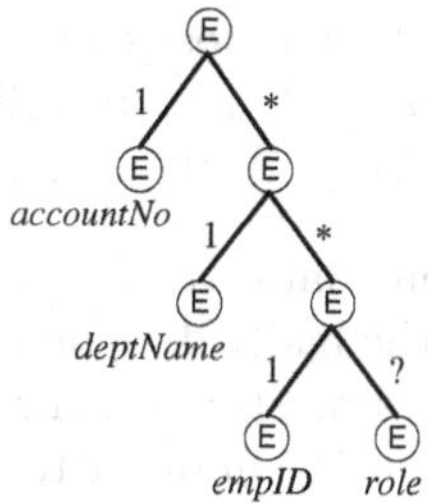

Fig. 6. An XML schema tree S with constraint set $\Sigma = \{[[accountNo]] \rightarrow [[empID]], [[empID]] \rightarrow [[deptName]]\}$ defined on S

$[[deptName]]\}$. A schema tree like this one might be used as a view for data exchange within a company for auditing purposes. The data trees in Fig. 7(a) and Fig. 7(b) show, respectively, that the XFDs $[[accountNo]] \rightarrow [[deptName]]$ and $[[accountNo]] \rightarrow [[role]]$ are not implied by Σ. We proceed by demonstrating that no XML data tree $T' \rhd S$ can satisfy Σ but violate both $[[accountNo]] \rightarrow [[role]]$ and $[[accountNo]] \rightarrow [[deptName]]$.

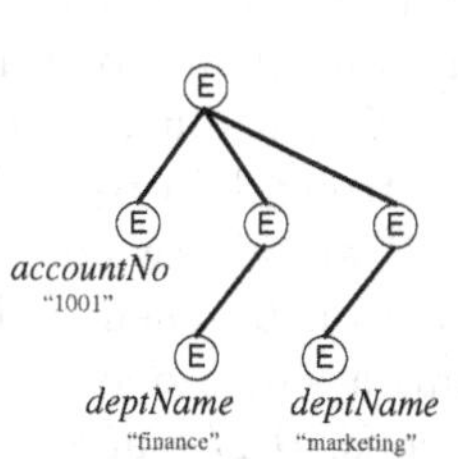

(a) XML data tree S' violates $[[accountNo]] \rightarrow [[deptName]]$ but satisfies $[[accountNo]] \rightarrow [[role]]$.

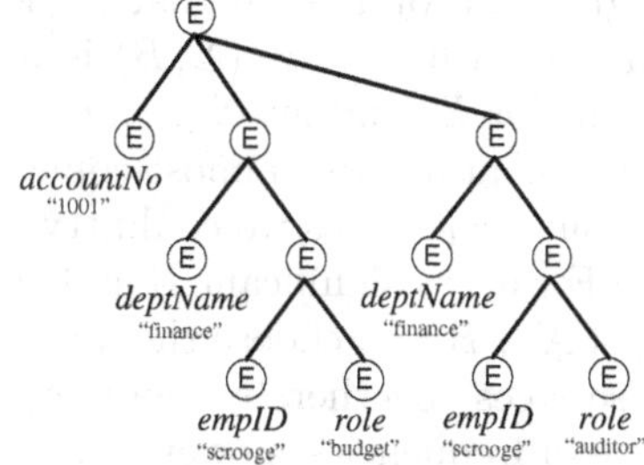

(b) XML data tree S'' violates $[[accountNo]] \rightarrow [[role]]$ but satisfies $[[accountNo]] \rightarrow [[deptName]]$.

Fig. 7. Two XML data trees S' and S'' that are compatible with the XML schema tree S and satisfy Σ. They illustrate that the occurrence (or non-occurrence) of certain walks in a data tree may cause the satisfaction of non-implied XFDs.

Assume, on the contrary, there exists some data tree $T' \rhd S$ which satisfies Σ and violates the XFDs $[[accountNo]] \rightarrow [[deptName]]$ and $[[accountNo]] \rightarrow [[role]]$ simultaneously. Since T' violates the XFD $[[accountNo]] \rightarrow [[role]]$ there are two almost-copies T_1', T_2' of S in T' that agree on $[[accountNo]]$ and differ on $[[role]]$. At least one of them, say T_1' contains a copy of the walk $[[role]]$. Consequently T_1' also contains copies of $[[empID]]$ and $[[deptName]]$. Since T' satisfies the XFD $[[accountNo]] \rightarrow [[empID]]$ in Σ all almost-copies of T in T' contain mutually value-equal copies of $[[empID]]$. Further, since T' satisfies the

XFD $[[empID]] \to [[deptName]]$ all almost-copies of T in T' contain mutually value-equal copies of $[[deptName]]$. This contradicts our assumption that T' violates $[[accountNo]] \to [[deptName]]$. Hence our proof is complete. □

The non-existence of an Armstrong data tree for the example in Theorem 10 can be regarded as a consequence of the badly designed schema tree S. Obviously for a data tree compatible with S all almost-copies have to agree on $[[accountNo]]$, and must be value-equal on $[[empID]]$ in order to satisfy Σ. In the schema tree in Fig. 6, it would probably be more appropriate to place the $empID$-node as a sibling of the $accountNo$-node and to assign its incoming arc the frequency ? to account for almost-copies that do not contain a copy of $[[empID]]$, as for example in Fig. 7(a). We hasten to point out that the suggested transformation is not lossless, but a thorough discussion of quality aspects in schema tree design is out of the scope of this paper. In the next section, however, we will investigate whether Armstrong data trees can be guaranteed when badly designed schema trees like the one here are excluded.

6 Existence of Armstrong Data Trees

A walk B of T is *significant* for a given constraint set Σ if there is at least one r_T-subgraph X of T such that Σ does not imply $X \to B$. Note that this is just the case when $max_T(\Sigma, B)$ is non-empty. Now consider a walk B that is significant for Σ, and let $X \in max_T(\Sigma, B)$. By Theorem 9 an Armstrong data tree for Σ contains two almost-copies of T with agree set X that differ on B.

Without frequencies we could try to construct an Armstrong data tree T' as follows: For every significant B and every $X \in max_T(\Sigma, B)$ we take two disjoint copies of $X \cup B$ and choose them value-equal on X but not on B. Then we link all these trees together by merging their root nodes into a single one. In the presence of frequencies, however, we face some complications: *i)* 1/+-arcs could necessitate further arcs or even whole walks to exist in T', and *ii)* ?/1-arcs could necessitate the merger of further arcs when forming T' from the ingredient trees. Such merger is not always possible, in particular if leave nodes carrying different string values need to be merged. In both situations the necessary changes have to be inspected for their impact on the criteria of Theorem 9. We therefore propose a more systematic approach towards almost-copies in data trees.

For later use we record the next simple observations on almost-copies. Recall that ϕ denotes the unique homomorphism between T' and T, and that R_T is the maximal r_T-subgraph of $T_{\leq 1}$. A data tree T' compatible with T contains at most one copy of every walk in R_T, and none of the walks in R_T is significant.

Lemma 11. *Let T be an XML schema tree, T' be an XML data tree compatible with T, and X be an r_T-subgraph of T. For every almost-copy X' of X in T' the following two statements hold:*

i) $X'|_{R_T} = T'|_{R_T \cap X}$, *and*
ii) $\phi(X')$ *contains* $(\phi(X') \cup T_{\geq 1}) \cap X$.

Let U be a unit of T, that is, a maximal non-empty r_T-subgraph consisting of walks of T that share a common $+/*$-arc. Observe that any almost-copy of $T - U$ in T' together with any almost-copy U' of U form an almost-copy of T in T'. In particular, for two almost-copies T'_1, T'_2 of T in T', it holds that $T'_1|_{T-U} \cup T'_2|_U$ and $T'_2|_{T-U} \cup T'_1|_U$ are almost-copies of T in T' again. The mix-and-match approach is possible since $T'_2|_U$ shares with $T'_1|_{T-U}$ exactly those arcs that $T'_2|_U$ shares with $T'_2|_{T-U}$, and likewise $T'_1|_U$ shares with $T'_2|_{T-U}$ exactly those arcs that $T'_1|_U$ shares with $T'_1|_{T-U}$.

Lemma 12. *Let T be an XML schema tree, T' be an XML data tree compatible with T, and U a unit of T. Then the following statements hold:*

i) For every almost-copy T'_a of T in T' we have that $T'_a|_U$ is an almost-copy of U in T', and
ii) for every almost-copy T'_a of T in T' and for every almost-copy U' of U in T' we have that $T'_a - T'_a|_U \cup U'$ is an almost-copy of T in T'.

For an r_T-subgraph X of T let $\mathcal{A}_{T'}(X)$ denote the set of all almost-copies of X in T'. Lemma 12 gives us a one-to-one correspondence between the members of $\mathcal{A}_{T'}(T)$ and the members of the cross-product over the sets $\mathcal{A}_{T'}(U)$ where U runs through the units of T for which $\mathcal{A}_{T'}(U)$ is non-empty.

Next we introduce an operation that allows us to extend XML data trees by new almost-copies of units. Let T' be an XML data tree compatible with T, and U' an XML data tree compatible with a unit U of T. Let (v, w) be the top-most $*/+$-arc of U. To *adjoin* U' to T' we take the disjoint union of T' and U', and then for each ancestor u of v (including r_T and u itself) we merge the pre-images of u in T' and U' if they exist. Due to the frequencies defined on T every such ancestor u has at most one pre-image in each of T' and U'. Recall that arc (v, w) has frequency $+/*$ in T. The resultant data tree satisfies the frequencies defined on T, and thus is again compatible with T.

Now we are in the position to present a first result on the existence of Armstrong data trees. We assume $(R_T)^*_\Sigma = R_T$ to exclude situations as the one encountered in Theorem 10. Note that this assumption prohibits non-trivial XFDs $X \to Y$ with $X \subseteq R_T$ to be specified in Σ, but still permits XFDs $X \to Y$ where X has non-empty intersection with both R_T and $T - R_T$.

Theorem 13. *Let T be an XML schema tree, and Σ a set of XFDs defined on T. Assume that $(R_T)^*_\Sigma = R_T$ and that T has only a single unit U. Then there exists an Armstrong data tree T' for Σ if and only if all members of $max_T(\Sigma)$ contain R_T.*

Proof. Due to space limitations we do not give a complete proof here, but only sketch the major ideas. Suppose T' Armstrong for Σ. Then the condition on $max_T(\Sigma)$ follows from part *ii)* of Theorem 9 and part *i)* of Lemma 11.

Conversely, suppose the condition on $max_T(\Sigma)$ holds. We outline the construction of a suitable data tree $T' \rhd T$ that turns out to be Armstrong for Σ. Let $\mathcal{P}$ denote the set of all pairs (X, B) with B a walk of T and $X \in max_T(\Sigma, B)$, and let $s_0, s_{(X,B)}, s'_{(X,B)}$, with $(X,B) \in \mathcal{P}$, be mutually distinct strings from *STRING*. If T has no significant walk for Σ we choose T' as a copy of T with string value s_0 assigned to all its leaves of kind A or S. Then T' is clearly Armstrong for Σ. Otherwise, if T has at least one significant walk for Σ, we start with a copy R' of R_T and assign each of its leaves the string value s_0. It is easy to see that by the root axiom $R_T \cap T_{\geq 1}$ is contained in every $X \in max_T(\Sigma)$. Next we consider one by one the pairs $(X,B) \in \mathcal{P}$. For every pair $(X,B) \in \mathcal{P}$ we take a copy of $((X \cup B) \cup T_{\geq 1}) \cap U$ and assign each of its leaves of type A or S the value $s_{(X,B)}$. Let $U'_{(X,B),a}$ denote the resultant data tree. Then we take another copy of $((X \cup B) \cup T_{\geq 1}) \cap U$ and assign each of its leaves v of type A or S the string value $s_{(X,B)}$ if the walk to $\phi(v)$ in T belongs to X, and $s'_{(X,B)}$ otherwise. Let $U'_{(X,B),b}$ denote the resultant data tree. We adjoin both $U'_{(X,B),a}$ and $U'_{(X,B),b}$ to T'.

Observe that the almost-copies of T in T' are just the subgraphs $R' \cup U'_{(X,B),i}$ with $(X,B) \in \mathcal{P}$ and $i = a, b$. By construction we have

$$ag_T(R' \cup U'_{(X,B),a}, R' \cup U'_{(X,B),b}) = X,$$

while for any pair of almost-copies that stem from two different pairs $(X,B) \in \mathcal{P}$ the agree set is R_T. For the same reason we have

$$\mathit{diff}_T(R' \cup U'_{(X,B),a}, R' \cup U'_{(X,B),b}) = ((X \cup B) \cup T_{\geq 1}) \cap U - X.$$

It remains to verify that T' is indeed Armstrong for Σ. By construction the almost-copies $R' \cup U'_{(X,B),a}$ and $R' \cup U'_{(X,B),b}$ differ on B, hence $B \in \delta(X)$ holds. This gives us condition *ii)* of Theorem 9. To verify also condition *i)* we inspect the agree sets of T'. These are just R_T and the members of $max_T(\Sigma)$. Firstly, $(R_T)^*_\Sigma \cap \delta(R_T) = \emptyset$ holds since $(R_T)^*_\Sigma = R_T$ while $\delta(R_T) \subseteq U$. For $X \in max_T(\Sigma)$ the divergence $\delta(X)$ is the union of the subgraphs $((X \cup B) \cup T_{\geq 1}) \cap U - X$ with B running through all walks of T with $(X,B) \in \mathcal{P}$. This can be used to show $X^*_\Sigma \cap \delta(X) = \emptyset$, and thus gives us condition *i)* of Theorem 9. Applying Theorem 9 then shows T' to be Armstrong for Σ. □

In the case of multiple units we face some additional complications that are the consequence of the composability of almost-copies of units when forming almost-copies of T as described in Lemma 12. For every walk B of T and any r_T-subgraph Y of T let $\mathcal{M}_Y(\Sigma, B)$ denote the collection of r_T-subgraphs $X \cap Y$ where X runs through $max_T(\Sigma, B)$. We call an almost-copy W' of U in T' *big* if $\phi(W')$ is maximal in the set $\{\phi(U') : U' \in \mathcal{A}_{T'}(U)\}$.

Lemma 14. *Let T be an XML schema tree, Σ be a set of XFDs defined on T, T' be an Armstrong data tree for Σ that is compatible with T, U a unit of T, and B a walk of T that is significant for Σ and not in U. Then for every $W \in \mathcal{M}_U(\Sigma, B)$ there exists a big almost-copy W' of U in T' such that $\phi(W') = W$, and vice versa.*

Lemma 14 imposes some non-trivial preconditions for an Armstrong data tree to exist under the assumptions stated in the lemma. Firstly, no member of

$\mathcal{M}_U(\Sigma, B)$ may be properly contained in any other member of $\mathcal{M}_U(\Sigma, B)$. In particular, the members W of $\mathcal{M}_U(\Sigma, B)$ satisfy the identity $W = (W \cup T_{\geq 1}) \cap U$. Secondly, $\mathcal{M}_U(\Sigma, B)$ is independent from the particular choice of the walk $B \notin U$. Given an r_T-subgraph Y of T we use $\mathcal{M}_Y(\Sigma)$ to denote the union of the sets $\mathcal{M}_Y(\Sigma, B)$ where B runs through all walks B of T, while we use $\mathcal{M}_Y^f(\Sigma)$ to denote the union of the sets $\mathcal{M}_Y(\Sigma, B)$ where B only runs through all walks $B \notin Y$. The superscript f stands for "foreign" and indicates the restriction to walks not in Y.

Corollary 15. *Let T be an XML schema tree, Σ be a set of XFDs defined on T, $T' \rhd T$ be an Armstrong data tree for Σ, U a unit of T, and B a walk of T that is significant for Σ and not in U. Then the following statements hold:*

i) The members of $\mathcal{M}_U(\Sigma, B)$ form a Sperner family over the walks in U, and
ii) $\mathcal{M}_U(\Sigma, B)$ equals $\mathcal{M}_U^f(\Sigma)$.

The arguments that we used for proving Lemma 14 may be further refined to conclude the following slightly stronger observation.

Lemma 16. *Let T be an XML schema tree, Σ be a set of XFDs defined on T, $T' \rhd T$ be an Armstrong data tree for Σ, and B a walk of T that is significant for Σ. Then for every $X \in max_T(\Sigma, B)$ there exists a collection of big almost-copies W'_U of U in T' with U running through all units of T other than U_B for which $\mathcal{A}_{T'}(U)$ is non-empty such that $\phi(W'_U) = X \cap U$, and vice versa.*

The previous lemma states that $\mathcal{M}_{T-U-R_T}(\Sigma, B)$ corresponds to the cross-union $\mathcal{C}_U(\Sigma, B)$ of the sets $\mathcal{M}_Q(\Sigma, B)$ where Q runs through all units of T other than U for which $\mathcal{M}_Q(\Sigma, B)$ is non-empty. This gives us a third non-trivial precondition for an Armstrong data tree to exist under the stated assumptions.

Eventually, we are in the position to present our major result on the existence of Armstrong data trees. To exclude situations as the one encountered in Theorem 10 we assume $(T-U)^*_\Sigma = T-U$ for every unit U. If T has a single unit only, this corresponds to our earlier assumption $(R_T)^*_\Sigma = R_T$ used in Theorem 13. It states that no walk in unit U is completely determined by the exterior $T-U$ of the unit. Again we argue that this assumption reflects a certain level of design quality that schema trees should meet in practice.

Theorem 17. *Let T be an XML schema tree, and Σ a set of XFDs defined on T. Assume that $(T-U)^*_\Sigma = T-U$ for every unit U of T. Then there exists an Armstrong data tree T' for Σ if and only if all members of $max_T(\Sigma)$ contain R_T, and the following conditions hold for every unit U of T and every walk $B \notin U$ that is significant for Σ:*

i) The members of $\mathcal{M}_U(\Sigma, B)$ form a Sperner family over the walks in U,
ii) $\mathcal{M}_U(\Sigma, B)$ equals $\mathcal{M}_U^f(\Sigma)$, and
iii) $\mathcal{M}_{T-U-R_T}(\Sigma, B)$ equals $\mathcal{C}_U(\Sigma, B)$.

7 Conclusion

In this paper we have started to investigate Armstrong data trees for functional dependencies in the context of XML. For relational databases Armstrong databases have been investigated for a variety of constraint classes. Some of these classes enjoy Armstrong databases, while for some others necessary and/or sufficient conditions on the constraint sets are known under which Armstrong databases can be generated. As pointed out by several authors, Armstrong databases are a powerful design aid for data architects and a useful tool for proving results on the satisfaction or implication of constraints. To the best of our knowledge Armstrong databases have not been investigated so far in the context of XML. However with the growing use of XML-encoded data there is an interest in developing a well-founded design theory for XML upon which design practice can be built. Armstrong data trees can play an important role in design theory and practice similar to their counterparts in the relational context.

Above we proved a characterisation of Armstrong data trees for constraints sets chosen from the class of XFDs studied in this paper. We motivated this result by comparing it to the known characterisation of Armstrong relations for functional dependencies, but also pointed out essential differences that are due to the semi-structured nature of XML-encoded data. We further proved that the class of XFDs does not enjoy Armstrong data trees. On the positive side we derived necessary and sufficient conditions for the existence of Armstrong data trees under a reasonable assumption that excludes certain badly designed schema trees. If a schema tree has only single unit our conditions are fairly simple to validate while they appear rather complex in the case of multiple units.

Clearly this paper does not answer all questions related to Armstrong data trees for XFDs. In future research we want to investigate computational aspects of generating Armstrong data trees. In particular the computation of the sets $max_T(\Sigma, B)$ requires further discussion in order to turn the constructions used in proving Theorems 13 and 17 into algorithms. In principle this can be done similar to the approach proposed in [45] for functional dependencies in the relational context. Note that the number of maximal sets in $max_T(\Sigma, B)$ can be exponential, so that we cannot expect a polynomial time method for constructing Armstrong data trees but rather an algorithm that is exponential in the number of walks of the schema tree and the size of Σ. Another question that is suggested for future research is that of how data architects can make best use of Armstrong data trees when communicating with fellow designers or domain experts. For relational databases this question was investigated in more detail, e.g., in [6,8,9,10,44,45]. Similar investigations are certainly required for XML-encoded data, too.

In the definition of XFDs used in this paper, walks and almost-copies play the role of attributes and tuples from the relational data model, while agreement is based on value-equality. In the literature, several other different extensions of functional dependencies to XML and semi-structured data have been suggested, e.g. [30,47,41,57,58,62,64]. Due to the disparate underlying semantics, these proposals allow data architects to capture different data dependencies and have different axiomatisations. Therefore we do not expect that our results here readily carry over to these proposals.

References

1. Arenas, M., Fan, W., Libkin, L.: What's hard about XML schema constraints? In: Hameurlain, A., Cicchetti, R., Traunmüller, R. (eds.) DEXA 2002. LNCS, vol. 2453, pp. 269–278. Springer, Heidelberg (2002)
2. Arenas, M., Libkin, L.: A normal form for XML documents. ACM ToDS 29, 195–232 (2004)
3. Armstrong, W.W.: Dependency structures of data base relationships. In: IFIP, pp. 580–583 (1974)
4. Atzeni, P., Morfuni, N.M.: Functional dependencies and constraints on null values in database relations. Inform. and Control 70, 1–31 (1986)
5. Baixeries, J., Balcázar, J.L.: Characterization and Armstrong relations for degenerate multivalued dependencies using formal concept analysis. In: Ganter, B., Godin, R. (eds.) ICFCA 2005. LNCS (LNAI), vol. 3403, pp. 162–175. Springer, Heidelberg (2005)
6. Beeri, C., Dowd, M., Fagin, R., Statman, R.: On the structure of Armstrong relations for functional dependencies. J. ACM 31, 30–46 (1984)
7. Beeri, C., Fagin, R., Howard, J.H.: A complete axiomatization for functional and multivalued dependencies in database relations. In: SIGMOD, pp. 47–61. ACM, New York (1977)
8. De Marchi, F., Lopes, S., Petit, J.-M.: Informative Armstrong relations: Application to database analysis. In: BDA (2001)
9. De Marchi, F., Lopes, S., Petit, J.-M., Toumani, F.: Analysis of existing databases at the logical level: the DBA companion project. SIGMOD Rec. 32, 47–52 (2003)
10. De Marchi, F., Petit, J.-M.: Semantic sampling of existing databases through informative Armstrong databases. Inf. Syst. 32, 446–457 (2007)
11. Demetrovics, J., Gyepesi, G.: A note on minimal matrix representation of closure operations. Combinatorica 3, 177–179 (1983)
12. Demetrovics, J., Katona, G.O.H.: A survey of some combinatorial results concerning functional dependencies. Ann. Math. Artificial Intelligence 7, 63–82 (1993)
13. Demetrovics, J., Katona, G.O.H., Sali, A.: The characterization of branching dependencies. Discrete Appl. Math. 40, 139–153 (1992)
14. Demetrovics, J., Katona, G.O.H., Sali, A.: Design type problems motivated by database theory. J. Statist. Plann. Inference 72, 149–164 (1998)
15. Demetrovics, J., Libkin, L., Muchnik, I.B.: Functional dependencies in relational databases: A lattice point of view. Discrete Appl. Math. 40, 155–185 (1992)
16. Demetrovics, J., Rónyai, L., Nam Son, H.: On the representation of dependencies by propositional logic. In: Thalheim, B., Gerhardt, H.-D., Demetrovics, J. (eds.) MFDBS 1991. LNCS, vol. 495, pp. 230–242. Springer, Heidelberg (1991)
17. Demetrovics, J., Thi, V.D.: Armstrong relations, functional dependencies and strong dependencies. Comp. Artif. Intell. 14 (1995)
18. Demetrovics, J., Thi, V.D.: Some observations on the minimal Armstrong relations for normalised relation schemes. Comp. Artif. Intell. 14 (1995)
19. Demetrovics, J., Thi, V.D.: Some remarks on generating Armstrong and inferring functional dependencies relation. Acta Cybernet. 12, 167–180 (1995)
20. Fagin, R.: Horn clauses and database dependencies. J. ACM 29, 952–985 (1982)

21. Fagin, R., Vardi, M.Y.: Armstrong databases for functional and inclusion dependencies. Inf. Process. Lett. 16, 13–19 (1983)
22. Fan, W.: XML constraints. In: DEXA Workshops, pp. 805–809. IEEE, Los Alamitos (2005)
23. Fan, W., Libkin, L.: On XML integrity constraints in the presence of DTDs. J. ACM 49(3), 368–406 (2002)
24. Fan, W., Siméon, J.: Integrity constraints for XML. J. Comput. Syst. Sci. 66(1), 254–291 (2003)
25. Felea, V.: Armstrong-like relations for functional partition dependencies. Sci. Ann. Cuza Univ. 1, 69–76 (1992)
26. Gottlob, G., Libkin, L.: Investigation on Armstrong relations, dependency inference, and excluded functional dependencies. Acta Cybernet. 9, 385–402 (1990)
27. Gunopulos, D., Khardon, R., Mannila, H., Saluja, S., Toivonen, H., Sharma, R.S.: Discovering all most specific sentences. ACM ToDS 28, 140–174 (2003)
28. Hartmann, S.: On the implication problem for cardinality constraints and functional dependencies. Ann. Math. Artificial Intelligence 33, 253–307 (2001)
29. Hartmann, S., Leck, U., Link, S.: On matrix representations of participation constraints. In: Heuser, C.A., Pernul, G. (eds.) ER 2009 Workshops. LNCS, vol. 5833, pp. 75–84. Springer, Heidelberg (2009)
30. Hartmann, S., Link, S.: More functional dependencies for XML. In: Kalinichenko, L.A., Manthey, R., Thalheim, B., Wloka, U. (eds.) ADBIS 2003. LNCS, vol. 2798, pp. 355–369. Springer, Heidelberg (2003)
31. Hartmann, S., Link, S., Kirchberg, M.: A subgraph-based approach towards functional dependencies for XML. In: SCI, IIIS, pp. 200–205 (2003)
32. Hartmann, S., Trinh, T.: Axiomatising functional dependencies for XML with frequencies. In: Dix, J., Hegner, S.J. (eds.) FoIKS 2006. LNCS, vol. 3861, pp. 159–178. Springer, Heidelberg (2006)
33. Jungnickel, D.: Graphs, Networks and Algorithms. Springer, Heidelberg (1999)
34. Katona, G.O.H.: Combinatorial and algebraic results for database relations. In: Hull, R., Biskup, J. (eds.) ICDT 1992. LNCS, vol. 646, pp. 1–20. Springer, Heidelberg (1992)
35. Katona, G.O.H., Sali, A.: New type of coding problem motivated by database theory. Discrete Appl. Math. 144, 140–148 (2004)
36. Katona, G.O.H., Tichler, K.: Some contributions to the minimum representation problem of key systems. In: Dix, J., Hegner, S.J. (eds.) FoIKS 2006. LNCS, vol. 3861, pp. 240–257. Springer, Heidelberg (2006)
37. Lee, M.-L., Ling, T.W., Low, W.L.: Designing functional dependencies for XML. In: Jensen, C.S., Jeffery, K., Pokorný, J., Šaltenis, S., Bertino, E., Böhm, K., Jarke, M. (eds.) EDBT 2002. LNCS, vol. 2287, pp. 124–141. Springer, Heidelberg (2002)
38. Levene, M., Loizou, G.: Axiomatisation of functional dependencies in incomplete relations. Theoret. Comput. Sci. 206, 283–300 (1998)
39. Levene, M., Loizou, G.: Database design for incomplete relations. ACM ToDS 24, 80–125 (1999)
40. Lien, Y.E.: On the equivalence of database models. J. ACM 29, 333–362 (1982)
41. Liu, J., Vincent, M., Liu, C.: Functional dependencies, from relational to XML. In: PSI, pp. 531–538 (2003)
42. Lopes, S., Petit, J.-M., Lakhal, L.: Efficient discovery of functional dependencies and Armstrong relations. In: Zaniolo, C., Grust, T., Scholl, M.H., Lockemann, P.C. (eds.) EDBT 2000. LNCS, vol. 1777, pp. 350–364. Springer, Heidelberg (2000)
43. Mannila, H., Räihä, K.-J.: Small Armstrong relations for database design. In: PODS, pp. 245–250. ACM, New York (1985)

44. Mannila, H., Räihä, K.-J.: Design by example: An application of Armstrong relations. J. Comput. Syst. Sci. 33, 126–141 (1986)
45. Mannila, H., Räihä, K.-J.: Design of Relational Databases. Addison-Wesley, Reading (1992)
46. Sali, A.: Minimal keys in higher-order datamodels. In: Seipel, D., Turull-Torres, J.M.a. (eds.) FoIKS 2004. LNCS, vol. 2942, pp. 242–251. Springer, Heidelberg (2004)
47. Sali, A., Schewe, K.-D.: Keys and Armstrong databases in trees with restructuring. Acta Cybernet. 18, 529–556 (2008)
48. Sali, A., Székely, L.A.: On the existence of Armstrong instances with bounded domains. In: Hartmann, S., Kern-Isberner, G. (eds.) FoIKS 2008. LNCS, vol. 4932, pp. 151–157. Springer, Heidelberg (2008)
49. Silva, A.M., Melkanoff, M.A.: A method for helping discover the dependencies of a relation. In: Advances in Data Base Theory, pp. 115–133 (1979)
50. Suciu, D.: On database theory and XML. SIGMOD Rec. 30, 39–45 (2001)
51. Thalheim, B.: Entity-Relationship Modeling. Springer, Heidelberg (2000)
52. Thi, V.D., Son, N.H.: On Armstrong relations for strong dependencies. Acta Cybernet. 17 (2006)
53. Tichler, K.: Minimum matrix representation of some key system. Discrete Appl. Math. 117, 267–277 (2002)
54. Tichler, K.: Extremal theorems for databases. Ann. Math. Artificial Intelligence 40, 165–182 (2004)
55. Trinh, T.: Functional dependencies for XML: Axiomatisation and normal form in the presence of frequencies and identifiers. MSc thesis, Massey University (2004)
56. Vianu, V.: A web odyssey: from Codd to XML. SIGMOD Rec. 32, 68–77 (2003)
57. Vincent, M.W., Liu, J.: Functional dependencies for XML. In: Zhou, X., Zhang, Y., Orlowska, M.E. (eds.) APWeb 2003. LNCS, vol. 2642, pp. 22–34. Springer, Heidelberg (2003)
58. Vincent, M., Liu, J.: Strong functional dependencies and a redundancy free normal form for XML. In: SCI, IIIS, pp. 218–223 (2003)
59. Vincent, M., Liu, J., Liu, C.: Redundancy free mappings from relations to XML. In: WAIM, pp. 55–67 (2003)
60. Vincent, M., Liu, J., Liu, C.: Strong functional dependencies and their application to normal forms in XML. ACM ToDS 29, 445–462 (2004)
61. Vincent, M.W., Srinivasan, B.: Armstrong relations for functional and multivalued dependencies in relational databases. In: ADC, pp. 317–328 (1993)
62. Wang, J., Topor, R.W.: Removing XML data redundancies using functional and equality-generating dependencies. In: ADC, pp. 65–74 (2005)
63. Weddell, G.E.: Reasoning about functional dependencies generalized for semantic data models. ACM ToDS 17, 32–64 (1992)
64. Yu, C., Jagadish, H.V.: XML schema refinement through redundancy detection and normalization. VLDB J. 17, 203–223 (2008)

Polymorphism in Datalog and Inheritance in a Metamodel

Paolo Atzeni, Giorgio Gianforme, and Daniele Toti

Dipartimento di Informatica e Automazione
Roma Tre University, Rome, Italy
atzeni@dia.uniroma3.it, giorgio.gianforme@gmail.com,
daniele@final-fantasy.it

Abstract. We discuss the restructuring of a metamodel designed for representing several data models in a uniform way. This metamodel is currently used within MIDST, our Model Management proposal for performing translations of schemas and databases from a model to another. Such a restructuring is carried out by introducing hierarchies and, consequently, extending Datalog by providing it with inheritance and polymorphism in order to take advantage of them. In comparable scenarios, where predicates of the metamodel share structural elements and rules are syntactically and semantically similar, the use of hierarchies and a particular form of polymorphism provide significant advantages. These advantages range from simplifying the specification of elementary and complete translations (i.e. Datalog rules and programs, respectively) to ensuring a higher level of reuse for them, thus further improving the development of such rule-based systems.

1 Introduction

Datalog-based languages have been recently used in various experimental and research projects related to the database field [3,6,9,13,18,15], with relevant benefits thanks to the simplicity of the language, its declarativeness, and the possibility of separating the "rules" that describe the problem of interest and its solution from the engine that implements the solution itself.

In this paper, we discuss a restructuring of a metamodel for model management purposes by providing Datalog with inheritance features and a particular kind of polymorphism[1], proving this way the possibility of increasing the effectiveness of the language and the chance of reuse for its individual rules.

The case study is represented by our model-independent schema and data translation platform (MIDST) [4], where data models and schemas are represented in a uniform way by means of a set of constructs, and translations are coded in a Datalog variant featuring OID invention. The usefulness of the MIDST proposal lies in the expressive power of its supermodel, a special model that is

[1] This topic had been previously introduced as a preliminary study in short paper form (see [5]).

S. Link and H. Prade (Eds.): FoIKS 2010, LNCS 5956, pp. 114–132, 2010.

able to represent all the single models by using a limited number of constructs, which are stored in an appropriate data dictionary. In order to improve the supermodel's expressive power, it has been necessary to introduce new constructs, often just variants of preexisting ones: such an introduction inevitably resulted in a growth in the number of constructs.

A key observation is the following: many constructs, despite small differences in their syntactical structures, are semantically similar or almost identical. Our purpose is to introduce inheritance in our data model by relying upon structural and semantical similarities among its constructs, and to subsequently extend Datalog in order to take advantage of such hierarchies.

This extension is meant to provide Datalog with polymorphic features similar to those typical of the Object-Oriented Paradigm, so that the hierarchies introduced in our data model can be successfully handled by the translation rules. This way, in fact, it will be no longer necessary to write a specific rule for each variant of a single construct, but it will instead be possible to write just one polymorphic rule for each root construct of a generalization: the rule engine will be the one responsible for compiling the Datalog rules by performing the necesessary replacements for the specified polymorphic variables, thus producing specific rules accordingly.

The paper is structured as follows. In Section 2 we discuss related work. In Section 3 we illustrate our MIDST project and highlight the problem tackled in the paper. In Section 4 we explain the restructuring of our data model by introducing PolyDatalog, an extension of Datalog with concepts of polymorphism and inheritance. In Section 5 we show our experimental results. And finally, in Section 6 we draw our conclusions.

2 Related Work

The idea of extending logics and rule-based systems with concepts like polymorphism, typing, and inheritance goes back to the beginning of 80's [19]. Recent approaches [11,12,1,2,14,17,16] adapt theories and methodologies of object-oriented programming and systems, proposing several techniques to deal with methods, typing, overriding, and multiple inheritance.

Gulog [11,12] is a deductive object-oriented logic (alternatively, according to its creators, a deductive object-oriented database programming language) with hierarchies, inheritance, overriding, and late binding; every Gulog program can be translated into an equivalent Datalog program with negation ($Datalog^{neg}$), where negated predicates are used to discern applicability of a rule to a class or subclass. Many works proposed extensions of Datalog and provided algorithms to translate their custom Datalog programs into "classic" Datalog with negation. In $Datalog^{meth}$ [1], a deductive object-oriented database query language, Datalog is extended with classes and methods; its programs can be translated into Datalog with negation as well. Selflog is a modular logic programming with non-monotonic inheritance. In [2], moving from SelfLog and $Datalog^{meth}$, Datalog is extended with inheritance (there are explicit precedence rules among classes) with or without overriding; programs can be rewritten in Datalog with an extra-predicate to mark rules and make them applicable only for a certain class or subclass; they propose also a fine-grained form of inheritance for Datalog systems, where specialization of method definitions in subclasses is allowed and, when a local definition is not applicable, a class hierarchy is traversed bottom-up (from subclass to superclass) until a class with an applicable method is reached. *Datalog*$^{++}$ [14] is an extension of Datalog with classes, objects, signatures, is-a relationships, methods, and inheritance with overriding; *Datalog*$^{++}$ programs can be rewritten in Datalog with negation. A language with encapsulation of rule-based methods in classes and non-monotonic behavioral inheritance with overriding, conflict resolution, and blocking (two features missing in other languages, according to the authors) is presented in [17]. In f-logic [16], by limiting to topics of interest, there are polymorphic types, classes, and subclasses; it is possible to distinguish between two kinds of inheritance: structural, where subclasses inherit attributes of super-classes, and behavioral, where subclasses inherit methods of super-classes; three methodologies (pointwise, global-method, user-controlled) to manage the overriding with behavioral inheritance are provided.

Our approach differs from the aforementioned proposals. They introduce concepts of object-oriented programming and, in particular, propose overriding of methods for sub-classes, where needed. We have a different goal: neither do we need overriding nor define anything for sub-classes (sub-predicates, in our case). Instead, using object-oriented programming terminology, we define a method (the rule) for the super-class (the polymorphic construct) and, consequently, generate specific methods (other rules) for the sub-classes (children constructs). From this point of view, our work has something in common with [10] where reusing and modification of rules is allowed by defining "ad hoc" rules for replacing predicate names featured in other rules.

3 MIDST Framework

The MIDST proposal is an implementation of the ModelGen operator, one of the Model Management operators [7,8]. Recalling the Introduction, the goal of ModelGen is to translate schemas from a data model to another: given a (source) data model M_1, a (source) schema S_1 (in data model M_1) and a (target) data model M_2, it produces a schema S_2 in M_2 that suitably corresponds to S_1; the data level extension, given also a database D_1 over schema S_1, generates a corresponding database D_2.

We use the notion of construct to represent and manage different models in a uniform way. Constructs with the "same" meaning in different models are defined in terms of the same generic construct; for example, entity in an ER model and class in an object-oriented model both correspond to the Abstract construct. We assume the availability of a universe of constructs. Each construct has a set of references (which relate its occurrences to other constructs) and boolean properties. Constructs also have names and possibly types (this is the case of lexical elements with a value like attributes of entities in the ER model). Let us comment on the various aspects with respect to the ER model. References are rather intuitive and two examples are enough to explain them: each attribute of entity (a specific construct) has a reference to the entity (another construct) it belongs to; each relationship (binary here, for the sake of simplicity) has references to the two entities involved. Properties require some explanation: for each attribute of an entity, by using two properties we can specify whether it belongs to the primary key and whether it allows for null values; for each relationship, by using properties we can describe its cardinality and whether it contributes to the identification of an entity or not: two properties tell us whether the participation of the first and second entity is optional or compulsory (that is, whether its minimum cardinality is 0 or 1), two properties tell whether maximum cardinality of the first and second entity is 1 or is unbounded (N, as we usually write), and another property tells us whether the first entity has an external identifier which this relationship contributes to (i.e. whether it is a weak entity).

In this framework, given a set of constructs, the references are always required to build schemas for meaningful models (for example, a relationship without references to entities makes no sense), whereas properties could be restricted in some way (for example, we can think of models where all cardinalities for relationships are allowed and models where many-to-many relationships are not allowed). Therefore, we can consider models to be defined by means of their constructs, each with a condition on its properties. The supermodel is a model that is able to include all the constructs of the universe in the most general form (i.e. with no restrictions).

In MIDST, translations are specified in a Datalog variant, where the predicate names are names of constructs and argument names may be OIDs, names, types, names of references and properties. A specific, important feature is the OID-invention, obtained by means of Skolem functors. Without loss of generality, we assume that our rules satisfy the standard safety requirements: all construct fields in their head have to be defined with a constant, a variable, that cannot be left undefined (i.e. it must appear somewhere in the body of the rule) or a Skolem term. It is the same for arguments of Skolem terms. Besides, our Datalog programs are assumed to be coherent with respect to referential constraints. More precisely, if there is a rule that produces a construct N that refers to a construct N', then there is another rule that generates a suitable N' which guarantees the satisfaction of the constraint.

4 Refactoring the Data Model

In this section we will now thoroughly discuss an extension to the Datalog language that is meant to introduce typical Object-Oriented Paradigm (OOP) features like polymorphism and inheritance, in order to take advantage of a possible restructuring of our data dictionary which must be laid out beforehand.

4.1 Increasing Complexity of the Data Model

The necessity of properly representing a large number of heterogeneous models, and therefore being able to define even more complex schemas, led to the introduction of a significant quantity of new constructs within our data dictionary, as we have briefly underlined.

A key observation is the following: many constructs, despite differences in their syntactical structures, are semantically similar or almost identical. For example, attributes of entities and of relationships in the ER model and columns in the relational model show some similarity: they all represent a lexical value. In these cases, two or more constructs can be collapsed into a single construct retaining their common semantics and possessing a structure obtained by the union of the structures of the constructs involved. Clearly, constructs obtained this way will have some optional references, together with some mandatory ones. This observation leads to the definition of a more compact (i.e. with a smaller number of constructs) and cohesive (i.e. where a single construct represents all the concepts sharing the same semantics) supermodel.

The increasing structural complexity of the data model that comes as a result has brought to our attention a certain amount of issues concerning the language used for defining the translation rules (i.e. Datalog), as well as the current structure of the dictionary itself: on one hand, the necessary Datalog rules have kept increasing, while on the other hand their scalability and reusability have consequently dropped. Therefore, it was necessary to take into account a possible refactoring for the dictionary and its constructs, in terms of establishing appropriate hierarchies for those constructs which allow for mutually exclusive references.

Actually, a generalization for each of the aforementioned constructs is introduced, whose parent is the generic construct without any references, and whose children are the different variants for the parent construct. Each of these variants features a set of mandatory references (corresponding to one of the original mutually exclusive subsets of references). Datalog with OID invention, though, is extremely limited when it comes to handling such generalizations. Being a logic programming language, in fact, it does not include a proper set of features to effectively represent and handle generalizations or hierarchies. That is why we have come to the definition of a new Datalog extension, which we have labelled PolyDatalog: by taking advantage of the introduced generalizations via an appropriate use of polymorphism and inheritance, we will show how it is possible to greatly increase scalability, maintainability and reusability for the translation rules and thus the whole translation processes.

4.2 Birth and Definition of PolyDatalog

The main idea behind PolyDatalog is born out of a simple observation. Despite the rising number of Datalog rules necessary for performing more and more complex translations, in fact, a large quantity of such rules feature rather apparent similarities from several points of view. From the syntactical point of view, a huge amount of rules actually share the same syntax, even though it is applied on different constructs, depending on the specific case. And from the semantical point of view, at least as many rules as the aforementioned set, despite being syntactically non-homogenous, share the same purpose, regardless of the constructs involved.

In order to make the following concepts clearer, let us consider a simplified version of the Object-Relational (OR) model (depicted in Figure 1) that involves the following constructs of the supermodel:

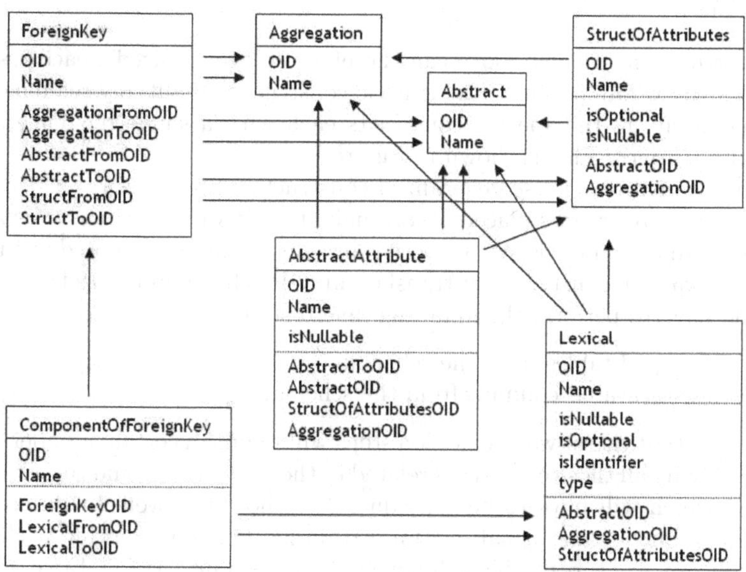

Fig. 1. A simplified Object-Relational model

- Abstract, representing typed tables;
- Aggregation, representing simple tables;
- StructOfAttributes, representing structured columns;
- AbstractAttribute, representing reference columns from tables (typed or not) or from structured columns, both pointing towards a typed table;
- Lexical, representing columns belonging to tables (typed or not) or to structured columns;
- ForeignKey, representing foreign keys from/to tables (typed or not) or structured columns;
- ComponentOfForeignKey, representing columns involved in a foreign key.

In accordance with the issues concerning mutually exclusive references that we have previously discussed, four constructs within this very model actually allow for this kind of references. Specifically, these constructs are the following:

- Lexical, allowing for three references in mutual exclusion: towards Abstract, Aggregation, or StructOfAttributes;
- StructOfAttributes, allowing for two references in mutual exclusion: towards Abstract or Aggregation;
- AbstractAttribute, allowing for three references in mutual exclusion: towards Abstract, Aggregation or StructOfAttributes;
- ForeignKey, allowing for three references in mutual exclusion: towards Abstract, Aggregation or StructOfAttributes; the presence of a couple of such references (as pointing - from - and pointed - to - construct involved in the foreign key) is mandatory, thus every possible combination of them is permitted.

The constructs mentioned above can therefore be "generalized", each resulting in a hierarchy whose parent is the generic construct without any references and whose children are the specific constructs each with a single reference among their optional ones. This is shown in Figure 2.

The rules involving these generalized constructs currently vary just in terms of the different references placed accordingly. Let us stress this via an example, taken from an actual translation whose source model is the proposed OR model. Should we want, for instance, to translate an OR schema into a Relational one, we would have to perform the following macro-steps:

- Remove typed tables from the schema;
- Remove structured columns from the schema.

It is obvious that these two translation steps will require a certain number of substeps, involving all those constructs related to the typed tables and the structured columns, which will have to be consequently removed as well. In this context, a removal is but a transformation of a certain kind of construct into a different one: for instance, a typed table will have to be turned into a plain table, in order to fit within the destination schema (a Relational one). According to our representation, all the Abstract constructs (representing typed tables) will be turned into Aggregation constructs (representing plain tables). Therefore, all those constructs, whose references point towards a typed table within the source schema, will have to be modified as well, by "changing" their respective references. This is obtained via a Datalog rule for each of these constructs.

For example, as we have already said, a Lexical (i.e. a column) allows for three references in mutual exclusion, towards an Aggregation (i.e. a table), an Abstract (i.e. a typed table), and a StructOfAttributes (i.e. a structured column). In the translation process mentioned above, as far as the typed tables removal is concerned, the rules involving Lexical will thus have the following semantics:

1. Copy all those Lexicals pointing towards Aggregation (i.e. leave the columns of plain tables as they are, therefore a simple copy is performed);
2. Turn the Lexicals pointing towards Abstract into Lexicals pointing towards Aggregation (i.e. transform all the columns belonging to a typed table into columns belonging to the plain table resulting from the translation of the original typed table);

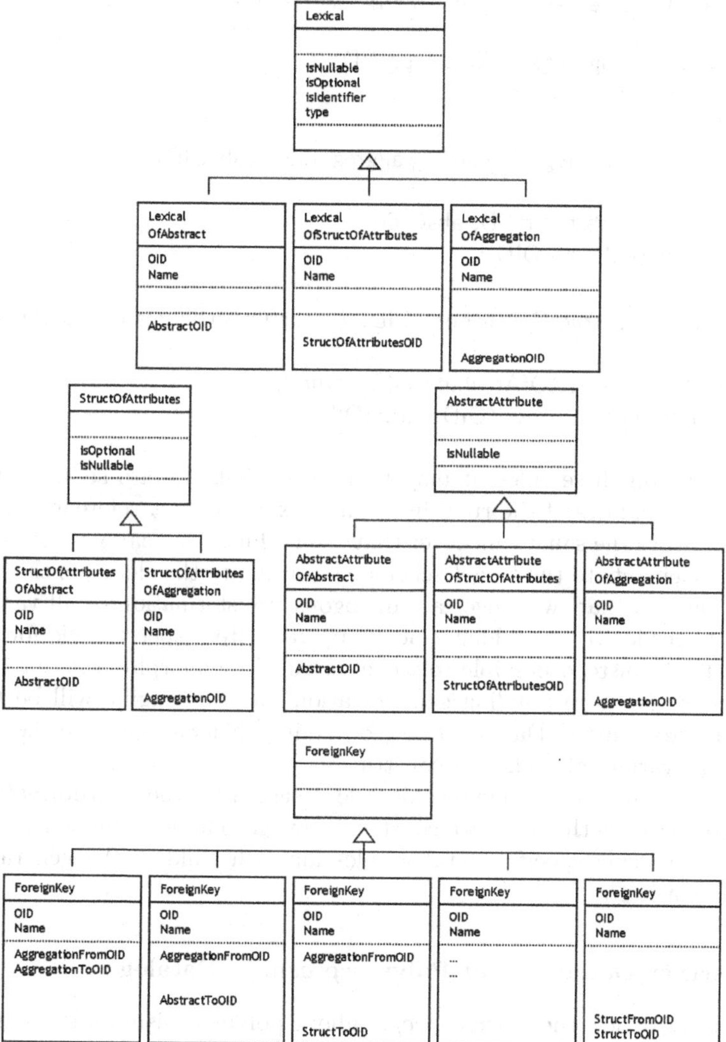

Fig. 2. Generalizations of Object-Relational model

3. Copy all those Lexicals pointing towards StructOfAttributes (as in step 1, i.e. leave the columns of StructOfAttributes exactly as they are).

Let us have a quick glance at the rules involved (where the # symbol denotes Skolem functors and we omit non-relevant fields) and double-check the correctness of our point:

```
LEXICAL (..., aggregationOID: #aggregation_0(aggOID))
←
LEXICAL (..., aggregationOID: aggOID),
AGGREGATION (OID: aggOID);

LEXICAL (..., aggregationOID: #aggregation_1(absOID))
←
LEXICAL (..., abstractOID: absOID),
ABSTRACT (OID: absOID);

LEXICAL (..., structOfAttributesOID: #structOfAttributes_0(structOID))
←
LEXICAL (..., structOfAttributesOID: structOID),
STRUCTOFATTRIBUTES (OID: structOID);
```

By reasoning on these rules, it may be noticed that, despite some syntactical differences, semantics of the rules involving Lexicals of (i.e. referencing) "something" is always the same whichever that "something" is: "carry over the values of various elements to the target schema, according to the elements they belong to". The idea is that, whenever an analogous transformation of all variants of Lexicals is needed, it is no longer necessary to write a specific rule for each of them, but it is instead possible to write just one polymorphic rule for the root construct of the corresponding generalization; the rule engine will be the one responsible to compile Datalog rules, obtaining this way one specific rule for every single variant of each root construct.

In the general case, a polymorphic rule designed for the transformation of a root construct C, with n children constructs, when compiled, will be instantiated n times, producing specific Datalog rules for each child of the generalization whose parent is C.

4.3 Further Details about Polymorphism in Datalog

Let's quickly review the basic concepts about polymorphism and then see how they fit in our Datalog extension. According to the object-oriented programming paradigm, polymorphism is the ability of objects belonging to different types to respond to method calls of the same name, each one according to an appropriate type-specific behavior. Specifically:

- an "object", in our approach, is an instantiated supermodel construct referenced by a construct that is parent of a generalization;
- a "method" is a reference within the Datalog rule specifying the translation for such a construct;
- a "type" is the specific reference (among those in mutual exclusion) possessed by a certain instance of the considered construct (e.g. StructOfAttributes, that is the reference featured by Lexical, being parent of a generalization, when instantiated as Lexical Of StructOfAttributes);

– the "behavior" is the outcome of the specific translation step (e.g.: "translate every Lexical by correctly placing its proper referenced construct, according to the way the latter has been previously translated by the other Datalog rules of the whole translation"), for each of the references featured by the children of a generalization.

4.4 Critical Aspects of Our Solution

There are a certain amount of issues we have to deal with when considering the semantics of a PolyDatalog rule, both at a theoretical and at a practical point of view.

Identifying the translated construct. First and foremost, the major critical aspect is strictly bound to the PolyDatalog semantics. So far, we have been considering the semantics of a classic Datalog rule as the following: "generate the construct expressed by the literal within the rule's head, beginning from the constructs featured as literals within the rule's body". While this certainly holds true for the instances of a polymorphic rule, when instantiating such a rule a critical issue arises: according to the PolyDatalog semantics, we need to know how the constructs referenced by a polymorphic construct have been translated within previous rules in the translation process. Let us go back to our previous example concerning Lexical. Its three referenced constructs in the considered model, i.e. Abstract, Aggregation and StructOfAttributes, have been respectively turned into Aggregation, Aggregation and StructOfAttributes throughout the whole translation process. But how do we know this? Clearly, by considering those rules where such (three) translations/copies occur. But then again, how do we identify such rules when scanning them? It might be perhaps enough for us to give just a quick glimpse at the various Datalog rules in the simplest cases. However, when dealing with more complex rules, with many constructs in their bodies and a wider range of syntax elements involved, things might get far more complicated. In general, in a Datalog rule, we therefore need to identify the "main", or "translated", construct within the rule's body. This means establishing which construct featured as a literal in its body is the actually translated construct into the construct expressed by the literal in its head.

Still referring to our example, as we have stated, Abstract becomes Aggregation. Therefore, there must be a rule, previously examined in the translation process, whose "main" construct is Abstract, and is right there translated into, namely, Aggregation. A rule like the following must thus exist:

```
AGGREGATION (OID: #AggregationOID_1(absOID), Name: name)
←
ABSTRACT (OID: absOID, Name: name);
```

Abstract is clearly the translated construct of this rule, becoming Aggregation in the destination schema. Abstract satisfies two paramount requirements: it is featured in the rule's body and its OID argument is found as the argument of the Skolem functor used to generate the head construct's OID.

The aforementioned requirements are necessary but not sufficient in order to correctly identify a main construct within a Datalog rule. In fact, a construct's OID (satisfying every previous requirement) must not be featured as an argument of a Skolem functor used to generate a value for a reference of the head construct. Furthermore, in our Datalog rules, we can refer to constructs belonging to the destination schema as well as to temporary views (used to store partial results): these constructs will never be the translated constructs of the corresponding rule.

By putting together what we have said so far, the algorithm for detecting whether a given construct is the one translated within a given rule will be the following:

Given a construct C and a Datalog rule R:

1. check that C is included within R's body;
2. check that C, found in R's body, is neither an explicit destination construct nor a temporary construct;
3. check that C's OID argument is found within the arguments of the Skolem functor used for generating the head literal's OID; if so, check that it does not also appear within the Skolem functor used for referencing some destination construct in the head literal.

If all these checks succeed, the considered construct is the translated construct of the given rule, and therefore its translation is represented by the construct expressed in the rule's head.

Multiple translations for a single construct. Strictly bound to the identification process for the main construct of a Datalog rule is the second, most relevant issue we are about to consider. As it results from our previous assertions, when we proceed to check whether a given construct is actually the translated construct of a rule, we are considering it in terms of its name. In other words, we check whether a construct, possessing a certain name, is the main construct within a set of scanned rules. At this high abstraction level, though, a seemingly critical scenario may occur: throughout a whole translation process, a given kind of construct bearing a specific name may have multiple translations, i.e. may result as the main construct of multiple rules. As an example, let us consider the elimination of many-to-many relationships within the ER family: many-to-many relationships have to be translated to entities, while other relationships have to be copied. When compiling a polymorphic rule meant to copy the attributes of relationships, we have to consider both the translations undergone by the relationships. The key point here is that multiple translations depend on the specific features of the constructs involved, expressed by different values of their respective parameters (some constraints on their properties, for instance). Therefore, those parameters, used to discriminate among the different instances of a construct, will have indeed to be included when instantiating the polymorphic rule involving the polymorphic construct which allows such a construct as one of its references.

Multiple polymorphic references within a single literal. The third critical aspect is related to a particular situation where a polymorphic construct allows for multiple mandatory references among its optional ones (e.g. ForeignKey). From a superficial view, we could think it is enough for these mandatory references to be handled separately. Actually, this is not the case: having their simultaneous presence as a constraint, we will have to generate, when instantiating the PolyDatalog rule, every possible combination for the allowed references, thus producing a minimum number of n^k rules, where n is the number of constructs featured as children of the considered generalization (whose parent is the polymorphic construct as previously stated), and k is the number of mandatory references to be simultaneously featured within the polymorphic construct. We say minimum under the assumption of a single translation for each of the referenced constructs: since there might be more than one translation, as we have shown in the previous paragraph, such a number is bound to exponentially increase. In fact, by defining m as the sum of the various m_i, i.e. the number of different translations for a construct n_i, the maximum number of rules produced turns out to be m^k. In our representation, we usually end up having $n \leq 3$, $m_i \leq 3$ and $k \leq 2$, therefore producing an amount of rules ranging from 1 and 9^2. It is rather apparent that these particular cases allow for the most effective use of PolyDatalog rules, whereas a single polymorphic rule succeeds in replacing several dozens of classic Datalog rules.

4.5 Syntax Template for PolyDatalog Rules and Processing Algorithm

By gathering together all the aforementioned considerations and statements, it is then possible to define a syntax template for a generic PolyDatalog rule, used within a translation process from a schema to another.

The syntax template for a polyDatalog rule would be as follows:

```
GENERICCONSTRUCT' (...,
        constructResult[k]OID: #skolemForConstruct[k](cOID[k]))
←
GENERICCONSTRUCT (...,
        constructOrigin[k]OID: cOID[k]),
CONSTRUCTORIGIN[K] (OID: cOID[k]);
```

Let us discuss the various elements. GenericConstruct is a root construct of a generalization that has to be turned into GenericConstruct′. The keyword ConstructOrigin[k] is used to indicate a generic construct referenced by a child of a generalization whose parent is GenericConstruct (the index k is indeed needed, for a GenericConstruct might possess multiple mandatory references); the value of its OID is denoted by cOID[k]. The keyword constructOrigin[k]OID is used as the name of a GenericConstruct's reference (whose value obviously is denoted by cOID[k]) towards the corresponding ConstructOrigin[k]. Assuming that a ConstructOrigin[k] has been translated to a certain ConstructResult[k], we use the keyword ConstructResult[k]OID as the name of the former's resulting reference within the head literal, and skolemForConstruct[k] as the name of the corresponding Skolem functor (that obviously has cOID[k] as an argument).

By taking into account all the various considerations and arguments we have discussed so far, we come up with the following algorithm to be used by a Poly-Datalog interpreter. It takes a translation process as input and refers to a given set of generalizations defined over the (meta)constructs of our data dictionary.

```
POLYDATALOGINTERPRETER(P)
 1  for each R of P
 2     if ISPOLYMORPHIC(R) then {
 3          rootC = GETPOLYMORPHICCONSTRUCT(R)
 4          refList = GETCONSTRUCTSREFERENCEDBYCHILDREN(rootC)
 5          for each ref_i in refList {
 6             ruleSet_i = FINDRULES(ref_i)
 7             IR_i = INSTANTIATERULES(ref_i, ruleSet_i)
 8             P = P − R + IR_i
 9          }
10     }
```

This algorithm analyzes every rule of the considered program (line 1). If a polymorphic variable is detected (line 2), it tracks down the corresponding polymorphic construct (line 3) and then obtains the list of its children (line 4), by analyzing the generalizations such a construct is involved in. For each child of a generalization C_i (line 5), it looks out for non polymorphic rules in **P** whose main construct is one of those referred by C_i (line 6). Now it can produce the needed non polymorphic rules (line 7) and replace the polymorphic rule in **P** with these generated rules. (line 8).

The PolyDatalog rule used to replace, for instance, the three rules listed in Section 4.2 will be as follows:

```
LEXICAL (...,
          constructResultOID: #skolemForConstruct(cOID))
←
LEXICAL (...,
          constructOriginOID: cOID),
CONSTRUCTORIGIN (OID: cOID);
```

When processing this rule, the interpreter will look at the generalization having Lexical as its parent, consequently finding out that Abstract, Aggregation and StructOfAttributes are the constructs referenced by its children (see Figure 2). By scanning the regular rules whose main construct is one of those three, the interpreter will therefore produce the aforementioned rules, having Abstract been turned into Aggregation, Aggregation into Aggregation, and StructOfAttributes into StructOfAttributes in the considered translation process from an OR schema to a Relational one.

5 Experimental Results

In this section we will proceed to show the relevant results we managed to achieve thanks to the restructuring of our data model and the consequent introduction of PolyDatalog rules within the schema translation processes. Results were encouraging: we detected far greater benefits in terms of reusability, maintainability and scalability of our solution than what was expected from the theoretical study.

Let us get back to considering the translation earlier proposed whose source was the Object-Relational from OR model to relational, which was meant to remove any typed tables within a given schema.

The outcome of PolyDatalog is rather straightforward: a translation originally made up of 27 rules gets stripped of nearly two-thirds of its rules, while obviously keeping safe both its correctness and completeness.

By extending this approach to the whole set of schema translations handled by our system, we have obtained significant advantages: an average 30% of Datalog rules (with peaks of 55%) have been removed by a handful of PolyDatalog rules. The summary of these results is shown in Figure 3.

	Number of Initial Rules	Number of Resulting Rules	Number of Replaced Rules	% of Replaced Rules
Entity-Relationship	21	18	3	14.3
Binary Entity-Relationship	83	79	4	4.8
Object (UML Class Diagram)	8	8	0	0
Object-Relational	480	325	155	32.3
Relational	7	7	0	0
XSD	138	88	50	36.2
Supermodel	29	13	16	55.2

Fig. 3. Experimental results

We must stress an important point: translations where more than 50% of their rules are removed by a very limited amount of PolyDatalog rules are not so rare at all. Actually, a significant number of them gets extraordinary results from the introduction of polymorphic rules: this is usually the case when dealing with source models featuring hierarchies of greater complexity, i.e. with more constructs involved and a larger number of children for each generalization. This happens therefore with models like XML-Schema and OR, the latter having been discussed with practical examples throughout this paper. On the other hand, it becomes crystal clear that models with limited and fewer hierarchies get less benefits from the use of polymorphism, as it is shown by our experimental results. Even so, PolyDatalog's inner parametricity in terms of the polymorphic references greatly enhances the schema translations within its range of applicability: first, the PolyDatalog rules would not change over time even though the construct referenced by the polymorphic one did indeed change; second, only a handful of them have succeeded in removing hundreds of original Datalog rules; and finally, we could anytime define more PolyDatalog rules should the need arise (for instance, when newer and more complex hierarchies are introduced in the data model), all the while getting even larger benefits from our Datalog extension.

6 Conclusion

In this paper we have proposed a restructuring of a data model carried out by the introduction of hierarchies among its constructs, along with their effective use by extending Datalog accordingly. Our polymorphic Datalog has therefore proven dramatically effective within our model management proposal. As we have found out during our experimentation of the PolyDatalog interpreter, the use of hierarchies has resulted in a major turning point as far as our schema translations are concerned. Thanks to the parametrical structure of the PolyDatalog rules, in fact, more than one-third of the previously necessary rules within our sytem's translation processes have been removed. Furthermore, reusability and scalability have been greatly enhanced, for PolyDatalog rules are highly reusable and can be successfully applied to future translations as well, whereas new models are conceptually defined and newer, more complex hierarchies are introduced in order to represent them correctly. Such a restructuring, combined with the polymorphic features introduced within Datalog, might prove just as successful in other scenarios as well, where predicates belonging to their data model feature a similar structure and syntactical and semantical similarities can be detected among their rules.

References

1. Abiteboul, S., Lausen, G., Uphoff, H., Waller, E.: Methods and rules. SIGMOD Rec. 22(2), 32–41 (1993)
2. Afrati, F.N., Karali, I., Mitakos, T.: On inheritance in object oriented datalog. In: IADT, pp. 280–289 (1998)
3. Atzeni, P., Cappellari, P., Bernstein, P.A.: Model-independent schema and data translation. In: Ioannidis, Y., Scholl, M.H., Schmidt, J.W., Matthes, F., Hatzopoulos, M., Böhm, K., Kemper, A., Grust, T., Böhm, C. (eds.) EDBT 2006. LNCS, vol. 3896, pp. 368–385. Springer, Heidelberg (2006)
4. Atzeni, P., Cappellari, P., Torlone, R., Bernstein, P.A., Gianforme, G.: Model-independent schema translation. VLDB J. 17(6), 1347–1370 (2008)
5. Atzeni, P., Gianforme, G.: Inheritance and polymorphism in datalog: an experience in model management. In: Information and Knowledge Bases XX, pp. 354–358 (2009)
6. Atzeni, P., Gianforme, G., Cappellari, P.: Reasoning on data models in schema translation. In: Hartmann, S., Kern-Isberner, G. (eds.) FoIKS 2008. LNCS, vol. 4932, pp. 158–177. Springer, Heidelberg (2008)

7. Bernstein, P.A.: Applying model management to classical meta data problems. In: CIDR, pp. 209–220 (2003)
8. Bernstein, P.A., Melnik, S.: Model management 2.0: manipulating richer mappings. In: SIGMOD Conference, pp. 1–12. ACM, New York (2007)
9. Bernstein, P.A., Melnik, S., Mork, P.: Interactive schema translation with instance-level mappings. In: VLDB, pp. 1283–1286 (2005)
10. Bonner, A.J., Imielinski, T.: Reusing and modifying rulebases by predicate substitution. J. Comput. Syst. Sci. 54(1), 136–166 (1997)
11. Dobbie, G., Topor, R.W.: A model for sets and multiple inheritance in deductive object-oriented systems. In: Ceri, S., Tsur, S., Tanaka, K. (eds.) DOOD 1993. LNCS, vol. 760, pp. 473–488. Springer, Heidelberg (1993)
12. Dobbie, G., Topor, R.W.: Representing inheritance and overriding in datalog. Computers and Artificial Intelligence 13, 133–158 (1994)
13. Gottlob, G., Koch, C., Baumgartner, R., Herzog, M., Flesca, S.: The lixto data extraction project - back and forth between theory and practice. In: PODS, pp. 1–12 (2004)
14. Jamil, H.M.: Implementing abstract objects with inheritance in datalog^{neg}. In: VLDB, pp. 56–65 (1997)
15. Kensche, D., Quix, C., Chatti, M.A., Jarke, M.: Gerome: A generic role based metamodel for model management. J. Data Semantics 8, 82–117 (2007)
16. Kifer, M., Lausen, G., Wu, J.: Logical foundations of object-oriented and frame-based languages. J. ACM 42(4), 741–843 (1995)
17. Liu, M., Dobbie, G., Ling, T.W.: A logical foundation for deductive object-oriented databases. ACM Trans. Database Syst. 27(1), 117–151 (2002)
18. Mork, P., Bernstein, P.A., Melnik, S.: Teaching a schema translator to produce O/R views. In: Parent, C., Schewe, K.-D., Storey, V.C., Thalheim, B. (eds.) ER 2007. LNCS, vol. 4801, pp. 102–119. Springer, Heidelberg (2007)
19. Mycroft, A., O'Keefe, R.A.: A polymorphic type system for PROLOG. Artif. Intell. 23(3), 295–307 (1984)

Appendix A

This appendix lists the actual rules used for the example mentioned throughout this paper (the translation from an Object-Relational schema to a Relational one), first as they appear before using PolyDatalog, and then as they result from the introduction of PolyDatalog rules. These PolyDatalog rules are explicitly shown at the end of the appendix.

OR-to-Relational Translation Rules without PolyDatalog

- copy Aggregations
- copy Lexicals of Aggregations;
- turn Abstracts into Aggregations;
- turn Lexicals of Abstracts into Lexicals of Aggregations;
- copy StructOfAttributes of Aggregations;
- turn StructOfAttributes of Abstracts into StructOfAttributes of Aggregations;
- copy Lexicals of StructOfAttributes;
- create key Lexicals of Aggregations for those generated from Abstracts;
- create Lexicals of Aggregations for those that have AbstractAttributes to define ForeignKeys;
- create Lexicals of Aggregations for those generated from Abstract that have AbstractAttributes to define Foreign Key;
- create Lexicals of StructOfAttributes for those that have AbstractAttributes to define ForeignKeys;
- create ForeignKeys for each AbstractAttribute of Abstract;
- create ForeignKeys for each AbstractAttribute of Aggregation;
- create ForeignKeys for each AbstractAttribute of StructOfAttributes;
- create ComponentsOfForeignKeys for each AbstractAttribute of Abstract;
- create ComponentsOfForeignKeys for each AbstractAttribute of Aggregation;
- create ComponentsOf ForeignKeys for each AbstractAttribute of StructOf-Attributes;
- copy ForeignKeys from Aggregation to Aggregation;
- copy ForeignKeys from StructOfAttributes to StructOfAttributes;
- copy ForeignKeys from Aggregation to StructOfAttributes;
- copy ForeignKeys from StructOfAttributes to Aggregation;
- turn ForeignKeys from Abstract to Abstract into ForeignKeys from Aggregation to Aggregation;
- turn ForeignKeys from Abstract to Aggregation into ForeignKeys from Aggregation to Aggregation;
- turn ForeignKeys from Aggregation to Abstract into ForeignKeys from Aggregation to Aggregation;
- turn ForeignKeys from Abstract to StructOfAttributes into ForeignKeys from Aggregation to StructofAttributes;
- turn ForeignKeys from StructOfAttributes into Abstract into ForeignKeys from StructOfAttributes to Aggregation;
- copy ComponentsOfForeignKeys.

OR-to-Relational Translation Rules with PolyDatalog

- copy Aggregations;
- turn Abstracts into Aggregations;
- transform StructOfAttributes;
- copy Lexicals;
- create key Lexicals of Aggregations for those generated from Abstracts;
- create Lexicals of Aggregations to define ForeignKeys;
- create ForeignKeys for each AbstractAttribute;
- create ComponentsOfForeignKeys for each AbstractAttribute;
- transform ForeignKeys;
- copy ComponentsOfForeignKeys.

PolyDatalog Rules

Copy/Transform Lexicals

```
LEXICAL (...,
        constructResultOID: #skolemForConstruct(cOID))
←
LEXICAL (...,
        constructOriginOID: cOID),
CONSTRUCTORIGIN (OID: cOID);
```

Copy/Transform StructOfAttributes

```
STRUCTOFATTRIBUTES (...,
        constructResultOID: #skolemForConstruct(cOID))
←
STRUCTOFATTRIBUTES (...,
        constructOriginOID: cOID),
CONSTRUCTORIGIN (OID: cOID);
```

Copy/Transform AbstractAttributes

```
ABSTRACTATTRIBUTE (...,
        constructResultOID: #skolemForConstruct(cOID),
        abstractToOID: #AbstractOID_0(absToOID))
←
ABSTRACTATTRIBUTE (...,
        constructOriginOID: cOID),
ABSTRACT (...,
        OID: absToOID),
CONSTRUCTORIGIN (OID: cOID);
```

Copy/Transform ForeignKeys

```
FOREIGNKEY (...,
        constructResultFromOID: #skolemForConstruct1(cOID1),
        constructResultToOID: #skolemForConstruct2(cOID2))
←
FOREIGNKEY (...,
        constructOriginFromOID: cOID1,
        constructOriginToOID: cOID2),
CONSTRUCTORIGIN1 (OID: cOID1),
CONSTRUCTORIGIN2 (OID: cOID2);
```

Possibilistic Semantics for Logic Programs with Ordered Disjunction

Roberto Confalonieri[1], Juan Carlos Nieves[1], Mauricio Osorio[2], and Javier Vázquez-Salceda[1]

[1] Universitat Politècnica de Catalunya
Dept. Llenguatges i Sistemes Informàtics
C/ Jordi Girona Salgado 1-3
E - 08034 Barcelona
{confalonieri,jcnieves,jvazquez}@lsi.upc.edu
[2] Universidad de las Américas, CENTIA
Sta. Catarina Mártir, Cholula
México - 72820 Puebla
osoriomauri@googlemail.com

Abstract. Logic programs with ordered disjunction (or LPODs) have shown to be a flexible specification language able to model and reason about preferences in a natural way. However, in some realistic applications which use user preferences in the reasoning, information can be pervaded with vagueness and a preference-aware reasoning process that can handle uncertainty is required. In this paper we address these issues, and we propose a framework which combines LPODs and possibilistic logic to be able to deal with a reasoning process that is preference-aware, non-monotonic, and uncertain. We define a possibilistic semantics for capturing logic programs with possibilistic ordered disjunction (or LPPODs) which is a generalization of the original semantics. Moreover, we present several transformation rules which can be used to optimize LPODs and LPPODs code and we show how the semantics of LPODs and the possibilistic semantics of LPPODs are invariant *w.r.t.* these transformations.

1 Introduction

With the increasing complexity of many type of decision-making contexts, such as online applications, system configuration and design, and real-word decision problems, new preference handling methods able to capture and reason efficiently about preferences are more and more required [6]. Classical preference modeling approaches assume in fact a complete specification of an utility function to map possible decisions' outcomes to numerical values and select the best choice. Although being a valid solution in quantitative approaches to knowledge representation, it is not fully adequate for supporting complex scenarios where the set of possible decisions tends to be either too large to be described explicitly or information is incomplete. Such concerns have fostered the study of compact preference representations and intelligent reasoning approaches [12, 17].

S. Link and H. Prade (Eds.): FoIKS 2010, LNCS 5956, pp. 133–152, 2010.

In this context nonmonotonic logics have shown to be a powerful knowledge formalism for preference representation and reasoning [12]. For instance, preference logics [42], default logic [41], and extended logic programs [26] can be considered logics with qualitative preferences which allow intelligent agents to prefer some belief sets to others. Generally speaking, several extensions of the basic formalism of Answer Set Programming (ASP) have been proposed to model preferences [17], showing how nonmonotonic logics constitute an effective way of resolving indeterminate solutions, reasoning in terms of preferred answer sets of a logic program. Unfortunately, nonmonotonic logics by themselves are not flexible enough and not well designed for modeling orderings on belief sets specified in terms of preferences on their elements [12].

Logic programs with ordered disjunction (or LPODs) offer one way to overcome this problem. LPODs are extended logic programs based on answer set semantics augmented by an *ordered disjunction* connector $\times$ which permits to explicitly represent preferences in the head of ordered disjunction rules [11]. Programs of this form can capture user qualitative preferences by means of ordered disjunction rules, represent choices among different alternatives and specify a preference order between the answer sets through predefined comparison criteria. In this way they are classes of logic programs that fit well in applications such as policy languages [5], planning [43], game theory [25] and user preference representation and reasoning [11]. Moreover, the system *psmodels* is an implemented computational engine for the semantics of LPODs [11]. Ordered disjunction has also been used to enhance an extended ASP formalism like CR-Prolog [2].

Although LPODs have shown to be a flexible ASP language able to model and reason about preferences in a natural way, they suffer from some issues when applied to some realistic scenarios [13, 14]. [14] proposes a service recommendation system which allows users to express preferences about service searches and which suggests the best results to the users *w.r.t.* their preferences. In this context the dynamicity and uncertainty of such an environment makes that the available information is pervaded with vagueness and the recommendation results keep on changing over the time. Hence an approach that considers and handles preferences in a static way is limited. A preference-aware reasoning method that can handle uncertainty can in fact drive the recommendation process to the selection of the best alternative and can help to adapt the preferences weights previously encoded according to the searches' results by means of uncertainty values [14]. Moreover, LPODs inherit the intrinsic difficulty of writing correct and efficient ASP programs due to the lack of ASP programming environments and friendly interfaces [29]. The coding of user preferences into LPODs would be much easier if users can be assisted by an LPOD preference editor that can makes the writing process transparent. For this reason the automatic generation of optimal LPOD code by means of a graphical tool is currently investigated in [13].

In this paper we will address part of these issues and we propose a theoretical framework which supports a reasoning that is at the same time preference-aware, non-monotonic, and uncertain. To be able to deal with preferences under uncertainty we combine possibilistic logic [20] and LPODs. We define a possibilistic semantics for capturing *logic programs with possibilistic ordered disjunction* (or LPPODs) which is a generalization of the original semantics presented in [11]. In particular, we extend the syntax and semantics of LPODs associating necessity degrees as uncertainty values to programs' clauses and we compute the possibilistic answer sets of an LPPOD using a direct definition of *possibilistic minimal model* of possibilistic definite logic programs following the approach of Nicolas *et al.* [34]. For achieving this we define a possibilistic reduction which reduce an LPPOD to a possibilistic definite logic program and we compute its possibilistic answer sets applying a fix-point operator ΠCn.

The main motivations for using possibilistic logic are essentially that possibilistic logic is a logic of uncertainty tailored for reasoning under incomplete evidence and partially inconsistent knowledge and that at mathematical level degrees of possibility and necessity are closely related to fuzzy sets [21]. Thus possibilistic logic is especially well adapted to automated reasoning when the available information is pervaded with vagueness as it happens in the scenarios we aim to deal with in [13, 14]. The use of possibilistic logic in ASP is not new in the literature and a good set of logic programming semantics has already been defined for capturing possibilistic logic programs [34, 36, 38]. In particular a first attempt to combine possibilistic logic and LPODs has been explored in [14].

On the other hand the idea of generating automatically LPOD code calls for optimization methods which remove redundancies. Assisting code generation tools usually tend to generate redundant code which in principle can increase the size of logic programs causing a fall off in the performance when computing possibilistic answer sets. For these purposes we study some *transformations rules* to transform LPODs into reduced ones. In most existing work, transformation rules have been defined especially in the context of rewriting systems and used to characterize logic programming semantics [7, 8, 19, 39]. In this paper we will follow an approach similar to [37] where equivalence and transformations between logic programs without preferences are studied. In our case we will generalize some program transformations rules of normal programs ($Contra$, RED^{+}, RED^{-}, SUC, $Failure$) [10, 40] for LPODs and we show that they preserve the semantics generalizing the notion of *weak equivalence* of logic programs [30] to LPODs. Based on these results we extend the transformation rules to LPPODs and we define a possibilistic preference relation which considers the necessity-value of ordered disjunction rules to decide the most preferred solution.

Our paper is organized as follows. After giving some background information of the concepts involved (Section 2), we introduce in Section 3 a set of transformation rules for LPODs. Our main result in this section is that the transformations reduce the size of LPODs without affecting their semantics. In Section 4 we describe our possibilistic extension for LPODs providing their syntax, semantics, the transformation rules and a preference relation. We conclude with Section 5. Throughout the paper we use a simple running example to explain our approach.

2 Background

In this section we introduce all the necessary terminology and relevant definitions in order to make this paper self-contained. We assume that the reader has familiarity with basic concepts of *classical logic*, *logic programming*, *answer set semantics*, and *lattices.*[1]

2.1 Extended Logic Programs

We consider extended logic programs which have two kinds of negation, strong negation $\neg$ and default negation *not*. A signature $\mathcal{L}$ is a finite set of elements that we call atoms, where atoms negated by $\neg$ are called extended atoms. Intuitively, *not* a is true whenever there is no reason to believe a, whereas $\neg a$ requires a proof of the negated atom. In the following we use the concept of atom without paying attention if it is an extended atom or not. A *literal* is either an atom a called *positive literal*, or the negation of an atom *not* a called *negative literal.* Given a set of atoms $\{a_1, ..., a_n\}$, we write *not* $\{a_1, ..., a_n\}$ to denote the set of atoms $\{not\ a_1, ..., not\ a_n\}$. An *extended normal rule* (rule, for short) r is a rule of the form

$$a \leftarrow b_1, \ldots, b_m,\ not\ b_{m+1}, \ldots,\ not\ b_{m+n}$$

where a and each of the b_i are atoms for $1 \leq i \leq m+n$. If $m+n=0$ the rule is an abbreviation of $a \leftarrow \top$ such that $\top$ is the proposition symbol that always evaluates to true; the rule is known as a *fact* and can be denoted just by a. If $n=0$ the rule is an extended definite rule. We denote a rule r by $a \leftarrow \mathcal{B}^+,\ not\ \mathcal{B}^-$ where the set $\{b_1, \ldots, b_m\}$ and the set $\{b_{m+1}, \ldots, b_{m+n}\}$ are denoted by $\mathcal{B}^+$ and $\mathcal{B}^-$ respectively. A *constraint* is a rule of the form $\leftarrow \mathcal{B}^+,\ not\ \mathcal{B}^-$. We denote by *head*($r$) the head a of rule r and by *body*(r) the $\mathcal{B}^+,\ not\ \mathcal{B}^-$ of the rule r. An *extended normal logic program* P is a finite set of extended normal rules and/or constraints. By $\mathcal{L}_P$ we denote the signature of P, *i.e.* the set of atoms that appear in the rules of P. If all the rules in P are extended definite rules we call the program P *extended positive logic program.* In our logic programs we will manage the strong negation $\neg$ as it is done in Answer Set Programming (ASP) [3]. Basically, each atom $\neg a$ is replaced by a new atom symbol a' which does not appear in the language of the program and we add the constraint $\leftarrow a, a'$ to the program. For managing the constraints in our logic programs, we will replace each rule of the form $\leftarrow \mathcal{B}^+\ not\ \mathcal{B}^-$ by a new rule of the from $f \leftarrow \mathcal{B}^+,\ not\ \mathcal{B}^-,\ not\ f$ such that f is a new atom symbol which does not appear in $\mathcal{L}_P$.

2.2 Logic Programs with Ordered Disjunction

Logic programs with ordered disjunction (LPODs) are extended logic programs augmented by an ordered disjunction connector $\times$ which allows to express qualitative preferences in the head of rules [11]. A LPOD is a finite collection of rules of the form

[1] For details the reader can refer to [3, 16, 31, 32].

$$r = c_1 \times \ldots \times c_k \leftarrow b_1, \ldots, b_m, \, not \; b_{m+1}, \ldots, \, not \; b_{m+n}$$

where c_i (for $1 \leq i \leq k$) and each of the b_j (for $1 \leq j \leq m+n$) are atoms. The rule r states that if the body is satisfied then some c_i must be in the answer set, if possible c_1, if not then c_2, and so on, and at least one of them must be true. Each of the c_i represents alternative, ranked options for problem solutions the user specifies according to a desired order. If $k = 1$ then the rule is an extended normal rule. The semantics of LPODs is based on the following reduction.

Definition 1 (×-reduction). [11] *Let $r = c_1 \times \ldots \times c_k \leftarrow b_1, \ldots, b_m, \, not \; b_{m+1}, \ldots, \, not \; b_{m+n}$ be an ordered disjunction rule and M be a set of atoms. The ×-reduct $r_\times^M$ is defined as*

$$r_\times^M := \{c_i \leftarrow b_1, \ldots, b_m | c_i \in M \text{ and } M \cap (\{c_1, \ldots, c_{i-1}\} \cup \{b_{m+1}, \ldots, b_{m+n}\}) = \emptyset\}$$

Let P be an LPOD and M be a set of atoms. The ×-reduct $P_\times^M$ is defined as

$$P_\times^M = \bigcup_{r \in P} r_\times^M$$

Definition 2 (SEM_{LPOD}). [11] *Let P be an LPOD and M a set of atoms. Then, M is an answer set of P if and only if M is a minimal model of $P_\times^M$. We denote by $SEM_{LPOD}(P)$ the mapping which assigns to P the set of all answer sets of P and by SEM_{LPOD} the semantics of LPODs.*

One interesting characteristic of LPODs is that they provide a mean to represent preferences among answer sets by considering the rule satisfaction degree [11].

Definition 3 (Rule Satisfaction Degree). [11] *Let M be an answer set of an ordered disjunction program P. Then M satisfies the rule $r = c_1 \times \ldots \times c_k \leftarrow b_1, \ldots, b_m, \, not \; b_{m+1} \ldots, \, not \; b_{m+n}$*

- *to degree 1 if $b_j \notin M$ for some j $(1 \leq j \leq m)$, or $b_i \in M$ for some i $(m+1 \leq i \leq m+n)$,*
- *to degree j $(1 \leq j \leq k)$ if all $b_l \in M$ $(1 \leq l \leq m)$, $b_i \notin M$ $(m+1 \leq i \leq m+n)$, and $j = min\{r \mid c_r \in M, 1 \leq r \leq k\}$.*

The degrees can be viewed as penalties: the higher the degree the less satisfied we are. If the body of a rule is not satisfied, then there is no reason to be dissatisfied and the best possible degree 1 is obtained [11]. The satisfaction degree of an answer set M *w.r.t.* a rule, denoted by $deg_M(r)$, provides a ranking of the answer sets of an LPOD, and a preference order on the answer sets can be obtained using some proposed combination strategies.[2]

[2] In [11], the authors have proposed some criteria for comparing answer sets namely *cardinality*, *inclusion* and *Pareto*.

2.3 Possibilistic Logic

A *necessity-valued formula* is a pair $(\varphi\ \alpha)$ where φ is a classical logic formula and $\alpha \in (0, 1]$ is a positive number. The pair $(\varphi\ \alpha)$ expresses that the formula φ is *certain* at least to the level α, *i.e.* $N(\varphi) \geq \alpha$, where N is a necessity measure modeling our possibly incomplete state knowledge [20]. α is not a probability (like it is in probability theory) but it induces a certainty (or confidence) scale. This value is determined by the expert providing the knowledge base. A necessity-valued knowledge base is then defined as a finite set (*i.e.* a conjunction) of necessity-valued formulae.

Dubois *et al.* [20] introduced a formal system for necessity-valued logic which is based in the following axiom schemata (propositional case):

(A1) $(\varphi \rightarrow (\psi \rightarrow \varphi)\ 1)$
(A2) $((\varphi \rightarrow (\psi \rightarrow \xi)) \rightarrow ((\varphi \rightarrow \psi) \rightarrow (\varphi \rightarrow \xi))\ 1)$
(A3) $((\neg\varphi \rightarrow \neg\psi) \rightarrow ((\neg\varphi \rightarrow \psi) \rightarrow \varphi)\ 1)$

The inference rules for the axioms above are:

(GMP) $(\varphi\ \alpha), (\varphi \rightarrow \psi\ \beta) \vdash (\psi\ GLB\{\alpha, \beta\})$
(S) $(\varphi\ \alpha) \vdash (\varphi\ \beta)$ if $\beta \leq \alpha$

According to Dubois *et al.* basically we need a complete lattice in order to express the levels of uncertainty in Possibilistic Logic. Dubois *et al.* extended the axioms schemata and the inference rules for considering partially ordered sets.

3 Transformation Rules for LPODs

Generally speaking a *transformation rule* is a syntactic rule which specifies the conditions under which a logic program P can be transformed to another program P'. In logic programming semantics literature, it has already been studied how transformation rules for classes of normal logic programs and disjunctive logic programs help to reduce the complexity in computing the semantics, *i.e.* to calculate the answer sets of a logic program [7, 8, 19, 39]. A common requirement is that the transformations can be used to reduce the size of a logic program provided that they do not affect its semantics. In this context several notions of equivalence between logic programs have been defined [22, 23, 30]. Here we generalize the notion of *weak equivalence* in [30], according to which we say that two LPODs P and P' are weak equivalent under SEM_{LPOD} (denoted $SEM_{LPOD}(P) \equiv SEM_{LPOD}(P')$) if both have exactly the same answer sets.

Starting from these results we generalize some basic transformations of normal programs ($Contra$, RED^+, RED^-, SUC, $Failure$) [10, 40] to LPODs and we show how these transformations reduce the structure of LPODs to a normal form without affecting their semantics. For achieving this, we first prove that the transformations preserve weak equivalence and we reuse the theory of rewriting system [18] to guarantee that the reduced LPOD is always unique. These results are especially important for the rest of the paper as they provide the necessary theoretical basis for defining transformation rules for LPPODs.

In the following we refer to a transformation rule as a program transformation as defined in [37]. A program transformation $\rightarrow$ is a binary relation on $Prog_{\mathcal{L}}$ where $Prog_{\mathcal{L}}$ is the set of all LPODs with atoms from the signature $\mathcal{L}$. $\rightarrow$ maps an LPOD P to another LPOD P'. We use $P \rightarrow_T P'$ for denoting that we get P' from P by applying a transformation rule T to P.

In a slight abuse of notation we denote an ordered disjunction rule r (see Section 2.2) by the formula $\mathcal{C}^{\times} \leftarrow \mathcal{B}^{+},\ not\ \mathcal{B}^{-}$ where the sets $\{c_1, \ldots, c_k\}$, $\{b_1, \ldots, b_m\}$, $\{b_{m+1}, \ldots, b_{m+n}\}$ will be denoted by $\mathcal{C}^{\times}$, $\mathcal{B}^{+}$, and $\mathcal{B}^{-}$ respectively. We write $HEAD(P)$ for the set of all atoms occurring in rule heads of an LPOD P. We will illustrate the transformations by means of the following running example:

Example 1 (Running Example P_0)

P_0:

$$
\begin{array}{ll}
r_1 = a \times b \leftarrow not\ c, not\ d. & r_5 = \quad \leftarrow a,\ b. \\
r_2 = c \times d \leftarrow e,\ not\ e. & r_6 = c \leftarrow not\ e. \\
r_3 = b \times a \leftarrow c. & r_7 = d \leftarrow not\ e. \\
r_4 = a \times b \leftarrow d. &
\end{array}
$$

Definition 4 (Contra). *Let P and P' be LPODs. P' results from P by elimination of contradictions ($P \rightarrow_C P'$) if P contains a rule $r = \mathcal{C}^{\times} \leftarrow \mathcal{B}^{+},\ not\ \mathcal{B}^{-}$ which has an atom b such that $b \in \mathcal{B}^{+}$ and $b \in \mathcal{B}^{-}$, and $P' = P \backslash \{r\}$.*

By applying the first transformation we get rid of rule r_2 and get P_1:

Example 2 ($P_0 \rightarrow_C P_1$)

P_1:

$$
\begin{array}{ll}
r_1 = a \times b \leftarrow not\ c, not\ d. & r_5 = \quad \leftarrow a,\ b. \\
r_3 = b \times a \leftarrow c. & r_6 = c \leftarrow not\ e. \\
r_4 = a \times b \leftarrow d. & r_7 = d \leftarrow not\ e.
\end{array}
$$

We would like also to be able to evaluate the negative literal *not* e to true, *i.e.* to delete *not* e from all rule bodies whenever e does not appear in the head of an LPOD. This can be guaranteed by *Positive Reduction.* Contrariwise if an LPOD contains $a \leftarrow \top$ then the atoms in the head must be true, so rules containing in their bodies negative literals *not* a are surely false and should be deleted. This can be guaranteed by *Negative Reduction.*

Definition 5 (Positive Reduction). *Let P and P' be LPODs. P' results from P by positive reduction RED^{+} ($P \rightarrow_{RED^{+}} P'$), if there is a rule $r = \mathcal{C}^{\times} \leftarrow \mathcal{B}^{+},\ not\ (\mathcal{B}^{-} \cup \{b\})$ in P such that $b \notin HEAD(P)$, and $P' = (P \backslash \{r\}) \cup \{\mathcal{C}^{\times} \leftarrow \mathcal{B}^{+},\ not\ \mathcal{B}^{-}\}$.*

Definition 6 (Negative Reduction). *Let P and P' be LPODs. P' results from P by negative reduction RED^{-} ($P \rightarrow_{RED^{-}} P'$), if P contains the rules: $r = a \leftarrow \top$ and $r' = \mathcal{C}^{\times} \leftarrow \mathcal{B}^{+},\ not\ (\mathcal{B}^{-} \cup \{a\})$, and $P' = (P \backslash \{r'\})$.*

An application of these reductions reduces the size of programs. In our example, we can apply $\rightarrow_{RED^{+}}$ (two times) to obtain P_2 and then $\rightarrow_{RED^{-}}$ to obtain P_3.

Example 3 ($P_1 \rightarrow_{RED^+} P_2$, $P_2 \rightarrow_{RED^-} P_3$)

P_2:

$r_1 = a \times b \leftarrow not\ c, not\ d.$
$r_3 = b \times a \leftarrow c.$
$r_4 = a \times b \leftarrow d.$
$r_5 = \quad \leftarrow a,\ b.$
$r_6 = c.$
$r_7 = d.$

P_3:

$r_3 = b \times a \leftarrow c.$
$r_4 = a \times b \leftarrow d.$
$r_5 = \quad \leftarrow a,\ b.$
$r_6 = c.$
$r_7 = d.$

The following two transformations are usually used to replace the Generalize Principle of Partial evaluation ($GPPE$) [7, 9].

Definition 7 (Success). *Let P and P' be LPODs. P' results from P by success ($P \rightarrow_S P'$), if P contains a fact $a \leftarrow \top$ and a rule $r = \mathcal{C}^\times \leftarrow \mathcal{B}^+$, not $\mathcal{B}^-$ such that $a \in \mathcal{B}^+$, and $P' = (P \backslash \{r\}) \cup \{\mathcal{C}^\times \leftarrow (\mathcal{B}^+ \backslash \{a\})$, not $\mathcal{B}^-\}$.*

Definition 8 (Failure). *Let P and P' be LPODs. P' results from P by failure ($P \rightarrow_F P'$), if P contains a rule $r = \mathcal{C}^\times \leftarrow \mathcal{B}^+$, not $\mathcal{B}^-$ such that $a \in \mathcal{B}^+$ and $a \notin HEAD(P)$, and $P' = (P \backslash \{r\})$.*

Example 4 ($P_3 \rightarrow_S P_4$)

P_4:

$r_3 = b \times a.$
$r_4 = a \times b.$
$r_5 = \quad \leftarrow a,\ b.$
$r_6 = c.$
$r_7 = d.$

Based on these transformations we can now define a rewriting system for LPODs.

Definition 9 (Rewriting System for LPODs). *Let $\mathcal{CS}_{LPOD}$ be the rewriting system based on the transformations $\{\rightarrow_C, \rightarrow_{RED^+}, \rightarrow_{RED^-}, \rightarrow_S, \rightarrow_F\}$ for LPOD. We denote a normal form of an LPOD program P w.r.t. $\mathcal{CS}_{LPOD}$ by $norm_{\mathcal{CS}_{LPOD}}(P)$.*[3]

As stated before, an essential requirement for program transformations is that they preserve the semantics of the programs to which they are applied. The following Lemma is an important result as it allows us to reduce LPODs without affecting their semantics.

Lemma 1 ($\mathcal{CS}_{LPOD}$ preserve SEM_{LPOD}). *Let P be an LPOD related to any transformation in $\mathcal{CS}_{LPOD}$. $SEM_{LPOD}(P) \equiv SEM_{LPOD}(norm_{\mathcal{CS}_{LPOD}}(P))$.*

Please note that $SEM_{LPOD}(P)$ is based on a syntactic reduction and stable semantics. Thus the proof of this Lemma consists in showing that the $\times$-reduction (see Section 2.2) is not affected by the transformation rules contained in $\mathcal{CS}_{LPOD}$ and that LPODs are weak equivalent under the stable semantics *w.r.t.* $\mathcal{CS}_{LPOD}$.[4]

It can be noticed that the program P_4 has the property that it cannot be further reduced because none of our transformations is applicable, thus $P_4 =$

[3] Soon we will show that the normal form is unique.

[4] In the paper we omit formal proofs of Lemmas, Propositions and Theorems due to space constraints.

$norm_{\mathcal{CS}_{LPOD}}(P_0)$. However, the normal form of P could have been obtained applying a different set of transformations. In particular the sequence $\rightarrow_{RED^+}$ (three times), $\rightarrow_{RED^-}$, $\rightarrow_F$ reduces the program P_0 to P_4. The following theorem is a strong result as it shows that a different application of our transformations (in a different ordering) leads to the same reduced program.

Theorem 1 (confluence and termination). *Let $\mathcal{CS}_{LPOD}$ be the rewriting system for LPODs, and P be an LPOD. $\mathcal{CS}_{LPOD}$ is confluent and noetherian and $norm_{\mathcal{CS}_{LPOD}}(P)$ is unique.*

The confluence of $\mathcal{CS}_{LPOD}$ can be proved using local confluence as stated by Newman's lemma [33]. According to [33] in fact a noetherian rewriting system is confluent if it is locally confluent. As we know that $\mathcal{CS}_{LPOD}$ terminates (all our program transformations decrease the clauses' size of an LPOD), it is a long but easy exercise to verify the local confluence of $\mathcal{CS}_{LPOD}$.

The results we have obtained are important as they have some theoretical and practical implications. In the case of the implementation of a solver which uses these kinds of transformations the confluence will guarantee that the order in which the transformations are applied does not matter (*e.g.* it allows to apply the less costly transformation) and that an LPOD can always be reduced to an unique normal form. The semantics of LPODs thus can be safely computed on their normal form.

4 Logic Programs with Possibilistic Ordered Disjunction

In order to reason in the presence of uncertainty we have extended the formalism of LPODs with possibilistic logic, a well-studied logic to reason about uncertain information [20, 21]. In this section we propose a syntax and semantics able to capture logic programs with possibilistic ordered disjunction.

4.1 Syntax

The syntax of a logic program with possibilistic ordered disjunction is based on the syntax of ordered disjunction rules (Section 2.2) and of possibilistic logic (Section 2.3). A *possibilistic atom* is a pair $p = (a, q) \in \mathcal{A} \times \mathcal{Q}$ where $\mathcal{A}$ is a set of atoms and $(\mathcal{Q},\leq)$ a finite lattice.[5] The projection $*$ for any possibilistic atom p is defined as: $p^* = a$. Given a set of possibilistic atoms M, the projection of $*$ over M is defined as: $M^* = \{p^* \mid p \in M\}$. Given $(\mathcal{Q},\leq)$, a possibilistic ordered disjunction rule r is of the form:

$$\alpha : c_1 \times \ldots \times c_k \leftarrow b_1, \ldots, b_m,\ not\ b_{m+1}, \ldots,\ not\ b_{m+n}$$

[5] In the paper we will only consider finite lattices. However, given that a program only will use a finite number of necessity values, it would be enough to require that such a lattice would be locally finite, *i.e.* the sublattice generated by a finite set of values is again finite (anonymous referee).

where $\alpha \in \mathcal{Q}$ and $c_1 \times \ldots \times c_k \leftarrow b_1, \ldots, b_m,\ not\ b_{m+1}, \ldots,\ not\ b_{m+n}$ is an ordered disjunction rule as defined in Section 2.2. We will denote a possibilistic ordered disjunction rule r by $\alpha : c_1 \times \ldots \times c_k \leftarrow \mathcal{B}^+,\ not\ \mathcal{B}^-$ where $\mathcal{B}^+ = \{b_1, \ldots, b_m\}$ and $\mathcal{B}^- = \{b_{m+1}, \ldots, b_{m+n}\}$.

The projection $*$ for a possibilistic ordered disjunction rule r, is $r^* = c_1 \times \ldots \times c_k \leftarrow \mathcal{B}^+,\ not\ \mathcal{B}^-$. $n(r) = \alpha$ is a necessity degree representing the certainty level of the information described by r. A possibilistic constraint c is of the form $\mathcal{TOP}_{\mathcal{Q}} :\leftarrow \mathcal{B}^+,\ not\ \mathcal{B}^-$, where $\mathcal{TOP}_{\mathcal{Q}}$ is the top of the lattice $(\mathcal{Q},\leq)$ and $\leftarrow \mathcal{B}^+,\ not\ \mathcal{B}^-$ is a constraint. Observe that any possibilistic constraint must have the top of the lattice $(\mathcal{Q},\leq)$. This restriction is motivated by the fact that, like constraints in standard ASP, the purpose of the possibilistic constraint is to eliminate possibilistic models. Hence, it is assumed that there is no uncertainty about the information captured by a possibilistic constraint. As in possibilistic ordered disjunction rules, the projection $*$ for a possibilistic constraint c is $c^* = \leftarrow \mathcal{B}^+,\ not\ \mathcal{B}^-$. For managing possibilistic constraints, we will replace each possibilistic rule of the form $\mathcal{TOP}_{\mathcal{Q}} :\leftarrow \mathcal{B}^+,\ not\ \mathcal{B}^-$ by a new rule of the from $\mathcal{TOP}_{\mathcal{Q}} : f \leftarrow \mathcal{B}^+,\ not\ \mathcal{B}^-,\ not\ f$ such that f is a new atom symbol which does not appear in the signature of the possibilistic program.

A *logic program with possibilistic ordered disjunction* (LPPOD) is a tuple of the form $P := \langle (Q, \leq), N \rangle$ such that N is a finite set of possibilistic ordered disjunction rules and/or possibilistic constraints. The projection of $*$ over P is defined as follows: $P^* := \{r^* \mid r \in N\}$. Notice that P^* is an LPOD.

Example 5. Let $P_0 = \langle (Q, \leq), N \rangle$ be an LPPOD such that $\mathcal{Q} = (\{0, 0.1, \ldots, 0.9, 1\}, \leq)$, $\leq$ be the standard relation between rational numbers, and N be the set of possibilistic rules of the program of Example 1 with possibilistic values $\alpha \in \mathcal{Q}$ associated to each rule:

P_0:
$r_1 = \mathbf{0.7} : a \times b \leftarrow not\ c, not\ d.$
$r_2 = \mathbf{0.2} : c \times d \leftarrow e,\ not\ e.$
$r_3 = \mathbf{0.5} : b \times a \leftarrow c.$
$r_4 = \mathbf{0.5} : a \times b \leftarrow d.$
$r_5 = \mathbf{1} : \quad \leftarrow a,\ b.$
$r_6 = \mathbf{0.6} : c \leftarrow not\ e.$
$r_7 = \mathbf{0.4} : d \leftarrow not\ e.$

4.2 Semantics

Our semantics is based on the definition of answer set semantics for extended normal programs [26]. For capturing LPPODs we are going to consider the basic idea of *possibilistic least model* [34]. Before defining the possibilistic semantics for LPOD we introduce some relevant concept. The first basic definition considers some basic operations between sets of possibilistic atoms and an order relation between them.

Definition 10. *Given $\mathcal{A}$ a finite set of atoms and $(\mathcal{Q},\leq)$ a lattice, we denote by $\mathcal{PS} = 2^{\mathcal{A} \times \mathcal{Q}}$ the finite set of all the possibilistic atom sets induced by $\mathcal{A}$ and $\mathcal{Q}$. Let $A, B \in \mathcal{PS}$, we define the operators $\sqcap$, $\sqcup$ and $\sqsubseteq$ as follows:*

$$
\begin{aligned}
A \sqcap B &= \{(x, GLB\{\alpha, \beta\}) | (x, \alpha) \in A \ \wedge (x, \beta) \in B\} \\
A \sqcup B &= \{(x, \alpha) | (x, \alpha) \in A \ \wedge \ x \notin B^*\} \cup \{(x, \beta) | x \notin A^* \ \wedge \ (x, \beta) \in B\} \cup \\
&\quad \{(x, LUB\{\alpha, \beta\})(x, \alpha) \in A \ \wedge \ (x, \beta) \in B\}. \\
A \sqsubseteq B &\Longleftrightarrow A^* \subseteq B^*, \ \textit{and} \ \forall x, \alpha, \beta, (x, \alpha) \in A \wedge (x, \beta) \in B \ \wedge \\
&\quad \beta = LUB\{\beta' | (x, \beta') \in B\} \ \textit{then} \ \alpha \leq \beta.
\end{aligned}
$$

Please notice that this definition suggests an extension of operators between sets in order to deal with uncertain values which belongs to a partially ordered set. The original definition considers in fact only totally ordered sets [34].

We will define a possibilistic semantics able to capture LPPODs based on a fix-point operator ΠCn which was introduced in [34] by considering possibilistic definite logic programs.[6] For this purpose our approach is based on a syntactic reduction. The following reduction is a possibilistic extension of the reduction defined in [11] able to consider the uncertainty values of the rules.

Definition 11 (Possibilistic Reduction $r_\times^M$). *Let $r = \alpha : c_1 \times \ldots \times c_k \leftarrow \mathcal{B}^+, \ not \ \mathcal{B}^-$ be a possibilistic ordered disjunction clause and M be a set of atoms. The $\times$-possibilistic reduct $r_\times^M$ is defined as follows:*

$$r_\times^M := \{\alpha : c_i \leftarrow \mathcal{B}^+ | c_i \in M \ \textit{and} \ M \cap (\{c_1, \ldots, c_{i-1}\} \cup \mathcal{B}^-) = \emptyset\}$$

Definition 12 (Possibilistic Reduction $P_\times^M$). *Let $P = \langle (Q, \leq), N \rangle$ be an LPPOD and M be a set of atoms. The $\times$-possibilistic reduct $P_\times^M$ is defined as follows:*

$$P_\times^M = \bigcup_{r \in N} r_\times^M$$

Example 6. Let P be the LPPOD in Example 5 and let M be the set of possibilistic atoms $M {=} \{(a, 0.5), (c, 0.6), (d, 0.4)\}$. We can see that:[7]

$P_\times^{M^*}$:

$r_2 = \mathbf{0.2} : c \leftarrow e.$ $\quad r_6 = \mathbf{0.6} : c.$

$r_3 = \mathbf{0.5} : a \leftarrow c.$ $\quad r_7 = \mathbf{0.4} : d.$

$r_4 = \mathbf{0.5} : a \leftarrow d.$

Observe that the program $P_\times^{M^*}$ is a possibilistic definite logic program. Once an LPPOD P has been reduced by a set of atoms M^*, it is possible to test whether M is a possibilistic answer set of the program P by considering ΠCn. Following this approach we will first define possibilistic answer sets for possibilistic definite logic programs and then generalize their definition to LPPODs. In order to define ΠCn, let us introduce some basic definitions. Given a possibilistic definite logic program P and $x \in \mathcal{L}_{P^*}$, $H(P, x) = \{r \in P | head(r^*) = x\}$.

Definition 13 (β-applicability). [34] *Let $P = \langle (Q, \leq), N \rangle$ be a possibilistic definite logic program, $r \in N$ such that r is of the form $\alpha : c \leftarrow b_1, \ldots, b_m$ and M be a set of possibilistic atoms,*

[6] A possibilistic definite logic program is a possibilistic positive normal logic program.

[7] Rule r_5 is deleted by our possibilistic reduction for the way we manage possibilistic constraints in our LPPODs (see Section 4.1).

- *r is β-applicable in M with $\beta = GLB\{\alpha, \alpha_1, \ldots, \alpha_n\}$ if $\{(b_1, \alpha_1), \ldots, (b_m, \alpha_n)\} \subseteq M$.*
- *r is $\bot_Q$-applicable otherwise.*

And then, for all atom $x \in \mathcal{L}_{P^*}$ we define:

$$App(P, M, x) = \{r \in H(P, x) | r \text{ is } \beta\text{-applicable in } M \text{ and } \beta > \bot_Q\}$$

where $\bot_Q$ denotes the bottom element of Q.

Observe that this definition is based on the inferences rules of possibilistic logic (Section 2.3). In order to illustrate this definition, let us consider the following example.

Example 7. Let P be the possibilistic definite logic program in Example 6 obtained by the possibilistic reduction $P_{\times}^{M^*}$. We can see that if we consider $M = \emptyset$ and the atoms c and d, then r_6 and r_7 are respectively 0.6-applicable and 0.4-applicable in M. In fact, we can see that $App(P, \{\emptyset\}, c) = \{r_6\}$ and $App(P, \{\emptyset\}, d) = \{r_7\}$. Also, we can see that if $M = \{(c, 0.6)\}$ and the atom a is considered, then r_3 is 0.5-applicable in M and if $M = \{(d, 0.4)\}$, r_4 is 0.4-applicable in M. Observe that if $M = \{(c, 0.6), (d, 0.4)\}$, then r_3 is 0.5-applicable in M and r_4 is 0.4-applicable in M and $App(P, \{(c, 0.6), (d, 0.4)\}, a) = \{r_3, r_4\}$.

Based on Definition 13 we can define a possibilistic consequence operator which can be used to compute possibilistic answer sets.

Definition 14 (Possibilistic consequence operator ΠT_P). [34] *Let P be a possibilistic definite logic program and M be a set of possibilistic atoms. The immediate possibilistic consequence operator ΠT_P maps a set of possibilistic atoms to another one by this way:*

$$\Pi T_P(M) = \{(x, \delta) | x \in HEAD(P^*), App(P, M, x) \neq \emptyset, \\ \delta = LUB_{r \in App(P,M,x)}\{\beta | r \text{ is } \beta\text{-applicable in } M\}\}$$

Then the iterated operator ΠT_P^k is defined by

$$\Pi T_P^k = \emptyset \text{ and } \Pi T_P^{n+1} = \Pi T_P(\Pi T_P^n), \forall n \geq 0$$

Observe that ΠT_P is a monotonic operator, *i.e.* let A and B be sets of possibilistic atoms, if $A \sqsubseteq B \Rightarrow \Pi T_P(A) \sqsubseteq \Pi T_P(B)$. Therefore, it can be guaranteed that ΠT_P always reaches a fix-point as stated by the following proposition.

Proposition 1. [34] *Let P be a possibilistic definite logic program, then ΠT_P has a least fix-point $\bigsqcup_{n \geq 0} \Pi T_P^n$ that we call the set of possibilistic consequences of P and we denote it by $\Pi Cn(P)$.*

By considering the fix-point operator $\Pi Cn(P)$ and the possibilistic reduction $P_{\times}^M$, we define the possibilistic semantics for LPPODs as follows:

Definition 15 (Possibilistic Answer Set for LPPODs). *Let $P = \langle (Q, \leq), N\rangle$ be an LPPOD, M be a set of possibilistic atoms such that M^* is an answer set of P^*. M is a possibilistic answer set of P if and only if $M = \Pi Cn(P_{\times}^{M^*})$. We denote by $SEM_{LPPOD}(P)$ the mapping which assigns to P the set of all possibilistic answer sets of P and by SEM_{LPPOD} the semantics of LPPODs.*

Example 8. Let $P = \langle (Q, \leq), N\rangle$ be the LPPOD introduced in Example 5 and M=$\{(a, 0.5), (c, 0.6), (d, 0.4)\}$. In order to infer the possibilistic answer sets of P, we have to infer the answer set of the LPOD P^*. First of all it can be proved that $M^* = \{a, c, d\}$ is an answer set of P^*. It is easy to see that $SEM_{LPOD}(P^*)$ is $M_1 = \{(a, c, d)\}$ and $M_2 = \{(b, c, d)\}$. As we saw in Example 6, $P_{\times}^{M_1}$ is a definite possibilistic logic program and

$$\Pi Cn(P_{\times}^{M^*}) = \{(a, 0.5), (c, 0.6), (d.0.4)\}$$

since

$$\begin{aligned}
\Pi T^0_{P_{\times}^M} &= \emptyset \\
\Pi T^1_{P_{\times}^M} &= \Pi T_{P_{\times}^M}(\emptyset) = \{(c, 0.6), (d.0.4)\} \\
\Pi T^2_{P_{\times}^M} &= \Pi T_{P_{\times}^M}(\{(c, 0.6), (d.0.4)\}) = \{(a, 0.5), (c, 0.6), (d.0.4)\} \\
\Pi T^3_{P_{\times}^M} &= \Pi T_{P_{\times}^M}(\{(a, 0.5), (c, 0.6), (d.0.4)\}) = \{(a, 0.5), (c, 0.6), (d.0.4)\} \\
\Pi T^{k+1}_{P_{\times}^M} &= \Pi T^k_{P_{\times}^M}, \forall k > 2
\end{aligned}$$

From Definition 15, we can observe that there is an important condition *w.r.t.* the definition of a *possibilistic answer set* of an LPPOD: a possibilistic set M cannot be a possibilistic answer set of an LPPOD P, if M^* is not an answer set of an LPOD P^*. Hence an important relation between the possibilistic semantics of LPPODs and the semantics of LPODs can be formalized by the following proposition.

Proposition 2 (Relation between Semantics). *Let $P = \langle (Q, \leq), N\rangle$ be an LPPOD and M be a set of possibilistic atoms. If M is a possibilistic answer set of P then M^* is an answer set of P^*.*

When all the possibilistic rules of an LPPOD P have as certainty level the top of the lattice $\mathcal{TOP}_{\mathcal{Q}}$, the answer sets of P^* can be directly generalized to the possibilistic answer sets of P.

Proposition 3 (Generalization). *Let $P = \langle (Q, \leq), N\rangle$ be an LPPOD and $\mathcal{TOP}_{\mathcal{Q}}$ be the top of the lattice $(Q, \leq)$. If $\forall r \in P$, $n(r) = \mathcal{TOP}_{\mathcal{Q}}$ and M' is an answer set of P^*, then $M := \{(a, \mathcal{TOP}_{\mathcal{Q}}) \mid a \in M'\}$ is a possibilistic answer set of P.*

4.3 Transformation Rules for LPPODs

In the following we provide the possibilistic version of the transformation rules we defined in Section 3. We denote a possibilistic ordered disjunction rule r (see

Section 4.1) by the formula $\alpha : \mathcal{C}^{\times} \leftarrow \mathcal{B}^{+}, \, not \, \mathcal{B}^{-}$. We write $HEAD(P)$ for the set of all atoms occurring in rule heads of an LPPOD P. Please notice that all the transformations include the possibilistic values we associated to each program's rules.

Definition 16 (Transformation rules for LPPODs). *Let P and P' be LPPODs. We define the following possibilistic transformation rules:*

Possibilistic Contra: *P' results from P by possibilistic elimination of contradictions ($P \rightarrow_{PC} P'$) if P contains a rule $r = \alpha : \mathcal{C}^{\times} \leftarrow \mathcal{B}^{+}, \, not \, \mathcal{B}^{-}$ which has an atom b such that $b \in \mathcal{B}^{+}$ and $b \in \mathcal{B}^{-}$, and $P' = P \backslash \{r\}$*

Possibilistic Positive Reduction: *P' results from P by possibilistic positive reduction $PRED^{+}$ ($P \rightarrow_{PRED^{+}} P'$), if there is a rule $r = \alpha : \mathcal{C}^{\times} \leftarrow \mathcal{B}^{+}, \, not \, (\mathcal{B}^{-} \cup \{b\})$ in P and such that $b \notin HEAD(P)$, and $P' = (P \backslash \{r\}) \cup \{\alpha : \mathcal{C}^{\times} \leftarrow \mathcal{B}^{+}, \, not \, \mathcal{B}^{-}\}$*

Possibilistic Negative Reduction: *P' results from P by possibilistic negative reduction $PRED^{-}$ ($P \rightarrow_{PRED^{-}} P'$), if P contains the rules $r = \alpha : a \leftarrow \top$, and $r' = \beta : \mathcal{C}^{\times} \leftarrow \mathcal{B}^{+}, \, not \, (\mathcal{B}^{-} \cup \{a\})$, and $P' = (P \backslash \{r'\})$*

Possibilistic Success: *P' results from P by possibilistic success ($P \rightarrow_{PS} P'$), if P contains a fact $\alpha : a \leftarrow \top$ and a rule $r = \beta : \mathcal{C}^{\times} \leftarrow \mathcal{B}^{+}, \, not \, \mathcal{B}^{-}$ such that $a \in \mathcal{B}^{+}$, and $P' = (P \backslash \{r\}) \cup \{GLB\{\alpha, \beta\} : \mathcal{C}^{\times} \leftarrow (\mathcal{B}^{+} \backslash \{a\}), \, not \, \mathcal{B}^{-}\}$*

Possibilistic Failure: *P' results from P by possibilistic failure ($P \rightarrow_{PF} P'$), if P contains a rule $r = \alpha : \mathcal{C}^{\times} \leftarrow \mathcal{B}^{+}, \, not \, \mathcal{B}^{-}$ such that $a \in \mathcal{B}^{+}$ and $a \notin HEAD(P)$, and $P' = (P \backslash \{r\})$.*

As done for the transformation rules for LPODs, we can define an abstract writing system that contains the possibilistic transformation rules we have introduced for LPPODs.

Definition 17 (Rewriting System for LPPODs). *Let P be an LPPOD and $\mathcal{CS}_{LPPOD}$ be the rewriting system based on the possibilistic transformation rules $\{\rightarrow_{PC}, \rightarrow_{PRED^{+}}, \rightarrow_{PRED^{-}}, \rightarrow_{PS}, \rightarrow_{PF}\}$. We denote a normal form of P w.r.t. $\mathcal{CS}_{LPPOD}$ by $norm_{\mathcal{CS}_{LPPOD}}(P)$.*

As by Proposition 2 and Proposition 3 we know that our possibilistic semantics is a generalization of the LPOD semantics and that our possibilistic $\times$-reduct (Definition 12) does not affect the $\times$-reduct (Definition 1), the following results are directly drawn from the results we obtained in Section 3. To keep the notation consistent we extend the concept of weak equivalence between LPODs to LPPODs. We say that two LPPODs P, and P' are weak equivalent under possibilistic semantics SEM_{LPPOD} (Definition 15), denoted by $P \equiv_p P'$ if both have exactly the same possibilistic answer sets.

Lemma 2 ($\mathcal{CS}_{LPPOD}$ preserve SEM_{LPPOD}). *Let P be an LPPOD related to any transformation defined in the rewriting system $\mathcal{CS}_{LPPOD}$. $SEM_{LPPOD}(P) \equiv_p SEM_{LPPOD}(norm_{\mathcal{CS}_{LPPOD}}(P))$.*

Theorem 2 (confluence and termination). *Let $\mathcal{CS}_{LPPOD}$ be the rewriting system for LPPODs, and P be an LPPOD. $\mathcal{CS}_{LPPOD}$ is confluent and noetherian and $norm_{\mathcal{CS}_{LPPOD}}(P)$ is unique.*

The *Possibilistic Success* is particularly important for LPPODs because it reflects the possibilistic modus ponens (GMP) (see Section 2.3). For instance, applying the program transformations $\rightarrow_{PC}$, $\rightarrow_{PRED^+}$, and $\rightarrow_{PRED^-}$ to the program in Example 5 we obtain the possibilistic program P_3 to which we can apply $\rightarrow_{PS}$ as shown:

Example 9 ($P_3 \rightarrow_{PS} P_4$)

	P_3:	P_4:
	$r_3 = \mathbf{0.5} : b \times a \leftarrow c.$	$r_3 = \mathbf{0.5} : b \times a.$
	$r_4 = \mathbf{0.5} : a \times b \leftarrow d.$	$r_4 = \mathbf{0.4} : a \times b.$
	$r_5 = \mathbf{1} : \quad \leftarrow a,\ b.$	$r_5 = \mathbf{1} : \quad \leftarrow a,\ b.$
	$r_6 = \mathbf{0.6} : c.$	$r_6 = \mathbf{0.6} : c.$
	$r_7 = \mathbf{0.4} : d.$	$r_7 = \mathbf{0.4} : d.$

In this way we can propagate the necessity-values of possibilistic facts without affecting the possibilistic semantics of LPPODs.

4.4 Possibilistic Preferred Answer Sets

In order to compare the possibilistic answer sets in LPPODs, we need to define a *possibilistic preference relation* that takes into account the necessity-values of the LPPOD rules. That is, we need a method to establish a partial order over the resulting possibilistic answer sets which is based on the rules' necessity values. In [14] an approach based on a necessity α-cut is presented. The preference relation depends just on the possibilistic values of the possibilistic answer sets, but does not consider rules' uncertainty degree directly. Although it is well suited for comparing possibilistic answer sets of some LPPODs, it cannot compare the ones presented in this paper. Here instead we define a *possibilistic preference relation* based on the normal form of a reduced LPPOD which compares possibilistic answer sets considering both rules' satisfaction degrees and rules' necessity values. In particular we are specially interested to define a relation where the necessity values of the possibilistic disjunction rules are influenced by the necessity values of the possibilistic facts (e.g. in program $P = \{r_3 = 0.5 : b \times a \leftarrow c.\ r_4 = 0.5 : a \times b \leftarrow d.\ r_6 = 0.6 : c.\ r_7 = 0.4 : d.\}$ we can achieve a ranking between r_3 and r_4 using the necessity values of facts c and d). For these purposes we reuse the definition and the notation of satisfaction degree of M *w.r.t.* a rule r as $deg_M(r)$ (see Section 2.2) and the normal form of an LPPOD.

Definition 18 (Possibilistic Preferred Relation). *Let P be an LPPOD, M_1 and M_2 be possibilistic answer sets of P, $norm_{CS_{LPPOD}}(P)$ be the normal form of P w.r.t. the rewriting system CS_{LPPOD}. M_1 is possibilistic preferred to M_2 ($M_1 >_{pp} M_2$) iff $\exists\ r \in norm_{CS_{LPPOD}}(P)$ such that $deg_{M_1}(r) < deg_{M_2}(r)$, and $\nexists r' \in norm_{CS_{LPPOD}}(P)$ such that $deg_{M_1}(r') > deg_{M_2}(r')$, and $n(r) < n(r')$.*[8]

This definition tells us that once we have reduced an LPPOD P applying the transformations seen before, we are able to compare possibilistic answer sets

[8] $n(r)$ denotes the uncertainty value of a rule r as explained in Section 4.1.

using directly the normal form of the program as the semantics is not affected by the transformations and the normal form is unique.

Example 10 ($M_2 >_{pp} M_1$).
$norm_{CS_{LPPOD}}(P_0) = P_4$:

$r_3 = \mathbf{0.5} : b \times a.$
$r_4 = \mathbf{0.4} : a \times b.$
$r_5 = \mathbf{1} : \quad \leftarrow a,\ b.$
$r_6 = \mathbf{0.6} : c.$
$r_7 = \mathbf{0.4} : d.$

First of all we can observe that P_4 is the normal form of the LPPOD program P_0 as it cannot be reduced by any possibilistic transformations. Secondly it can be proved that the program P_4 has two possibilistic answer sets $M_1 = \{(a, 0.5), (c, 0.6), (d, 0.4)\}$ and $M_2 = \{(b, 0.5), (c, 0.6), (d, 0.4)\}$ which are exactly the same possibilistic answer sets as for P_0. We can notice that in the program the rules r_3 and r_4 are satisfied by M_1 and M_2 with degree $(2, 1)$ and $(1, 2)$ respectively. Thus it is not possible to decide which possibilistic answer sets among M_1 and M_2 is the most preferred using the comparison criteria defined in [14]. Instead, if we apply Definition 18, we can conclude that $M_2 >_{pp} M_1$ as $n(r_3) > n(r_4)$ ($M_1 \not>_{pp} M_2$[9] follows by Definition 18 as well).

Moreover we can observe that there is an important property *w.r.t.* the possibilistic preference relation for possibilistic answer sets: a possibilistic answer set M is comparable if M^* is comparable in the LPOD P^*. Our extension maintains in fact the preference relation between answer sets of LPODs *w.r.t.* the Pareto criterium defined in [11].

Proposition 4. *Let M_1 and M_2 be possibilistic answer sets of an LPPOD P, $>_p$ be the Pareto criterium from [11], and $>_{pp}$ be the possibilistic preference relation. If $M_1^* >_p M_2^*$ then $M_1 >_{pp} M_2$.*

5 Conclusions

The main scope of this paper has been to define a possibilistic semantics for LPODs to deal with uncertainty degrees in the reasoning about preferences. In working towards this direction we have progressively built all the theoretical knowledge needed to define a framework which combines possibilistic logic and LPODs to support a reasoning which is preference-aware, non-monotonic, uncertain, and which could consider uncertainty values of program clauses. The use of possibilistic logic in LPODs is not new [14, 15]. In [14] a first attempt towards this direction is presented. However, in this paper we have followed a different approach in defining the LPPOD semantics as in [14] the semantics is specified in terms of possibilistic inference while here we have used a direct definition of possibilistic minimal model for possibilistic definite positive logic programs which has shown to be a more natural specification. The possibilistic comparison criteria used in [14] to compare possibilistic answer sets have shown to be efficient only in some cases as they do not consider the necessity-values of LPPOD rules.

[9] We denote by $M_1 \not>_{pp} M_2$ that M_1 is not possibilistic preferred to M_2.

In this paper instead we have specified a possibilistic preference relation which considers apart from the rule satisfaction degrees the uncertainty values of the program rules themselves (Definition 18). In [15] a semantics for capturing LP-PODs based on pstable semantics has been explored. Such semantics based on paraconsistent logic allows to treat inconsistent LPODs and LPPODs (under answer set semantics) which do not have any solution and it has been shown how it is a generalization of the original semantics of LPODs.

Concerning research on logic programming with uncertainty existing approaches in the literature employ different formalisms such as probabilistic logic [4, 27], fuzzy sets theory [35] and possibilistic logic [1, 34]. Basically these approaches differ in the underlying notion of uncertainty and how uncertainty values associated to clauses and facts are managed. As stated in Section 1 we are interested in modeling uncertainty related to preference representation and reasoning and especially when this information is vague and incomplete, *i.e.* the uncertainty we want to model is dynamic and it can not be related to previous observations. For this reason we manage the uncertainty by means of possibilistic logic and we have captured uncertain values considering lattices. The use of lattices is not new and one of the most important approaches is the one suggested by Fitting in [24]. Fitting showed that interlaced bilattices provide a simple and elegant setting for the consideration of logic programming extensions. At an abstract level he observed that all interlaced bilattices are quite natural, but not all are appropriate for computer implementation. It is not difficult to see that the semantics of an LPPOD is defined in the domain of the interlaced bilattice $\mathcal{B}(\{0,1\},\mathcal{Q})$. Since the possibilistic semantics defined in this paper is computable our approach is restricted to computable interlaced bilattices.

As far as code optimization is concerned we have first generalized some basic program transformations rules for normal programs (*Contra*, RED^+, RED^-, *SUC*, *Failure*) to LPODs and we have proved that they preserve weak equivalence between LPODs and always reduce an LPOD to a unique normal form. An important result of this section has been in fact that the rewriting system we defined based on these transformations is confluent and noetherian (Theorem 1). Concerning related works about program transformations and equivalence in logic programs, we have been inspired by the works presented in [23, 28, 30, 37].

As next step we have extended the syntax of LPODs associating necessity degrees as uncertainty values to programs' clauses and we have defined a semantics which is based on a fix-point operator. In this context, we have obtained some important results *w.r.t.* LPPODs. In particular we have shown how possibilistic answer sets of our extended version can be answer sets for no possibilistic programs as well (Proposition 2), and how our semantics is a generalization of the original semantics (Proposition 3). Based on these results we have generalized the transformations for LPODs providing their possibilistic extension for LP-PODs (Section 4.3). Such transformations are important for LPPODs. In fact apart from reducing the size of LPPODs while preserving their possibilistic semantics, they permit to define a possibilistic preference relation which considers both the satisfaction degree and the necessity values of possibilistic disjunction rules (Definition 18).

There are several interesting issues and challenges we want to pursue for future work. From a theoretical point of view we aim to explore a stronger notion of equivalence for LPODs such as the one studied in [23]. This notion is in fact stronger than our notion of weak equivalence because it generalizes the strong equivalence of normal programs to LPOD testing whether two LPODs have the same preferred answer sets in any context. At the same time, interesting challenges come into place whenever realistic applications are considered. The work in [14] describes a framework that aims to deal with user preferences about service searches by means of LPPODs. One of our objectives is to prototype the framework, to implement the possibilistic reasoner for LPPODs we have formalized in this paper and to allow users to express preferences by means of a preference editor. We believe that the transformations for LPPODs we have introduced can help in the code optimization and that the reasoning framework we have presented puts the necessary theoretical basis which will allow us to implement the system with a solid theoretical background.

Acknowledgements. This work has been funded by the European Commission Framework 7 funded project ALIVE (FP7-215890). Javier Vázquez-Salceda's work has been also partially funded by the Ramón y Cajal program of the Spanish Ministry of Education and Science. We thank anonymous referees for their valuable comments.

References

1. Alsinet, T., Godo, L.: Towards an Automated Deduction System for First-Order Possibilistic Logic Programming with Fuzzy Constants. International Journal of Intelligent Systems 17(9), 887–924 (2002)
2. Balduccini, M., Mellarkod, V.: CR-Prolog with Ordered Disjunction. In: Balduccini, M., Mellarkod, V. (eds.) Advances in Theory and Implementation. CEUR Workshop Proceedings, vol. 78, CEUR-WS.org, pp. 98–112 (2003)
3. Baral, C.: Knowledge Representation, Reasoning and Declarative Problem Solving. Cambridge University Press, Cambridge (2003)
4. Baral, C., Gelfond, M., Rushton, N.: Probabilistic Reasoning with Answer Sets. Theory and Practice of Logic Programming 9(1), 57–144 (2009)
5. Bertino, E., Mileo, A., Provetti, A.: PDL with Preferences. In: Proc. of the Sixth IEEE International Workshop on Policies for Distributed Systems and Networks, pp. 213–222. IEEE Computer Society, Washington (2005)
6. Brafman, R.I., Domshlak, C.: Preference Handling - An Introductory Tutorial. AI Magazine 30(1), 58–86 (2009)
7. Brass, S., Dix, J.: Characterizations of the Stable Semantics by Partial Evaluation. In: Marek, V.W., Nerode, A., Truszczyński, M. (eds.) LPNMR 1995. LNCS, vol. 928, pp. 85–98. Springer, Heidelberg (1995)
8. Brass, S., Dix, J.: Characterizations of the Disjunctive Well-Founded Semantics: Confluent Calculi and Iterated GCWA. Journal of Automated Reasoning 20(1-2), 143–165 (1998)

9. Brass, S., Dix, J.: Semantics of (disjunctive) Logic Programs Based on Partial Evaluation. Journal of Logic Programming 40(1), 1–46 (1999)
10. Brewka, G., Dix, J., Konolige, K.: Nonmonotonic Reasoning: An Overview. CSLI Lecture Notes, vol. 73. CSLI Publications, Stanford (1997)
11. Brewka, G., Niemelä, I., Syrjänen, T.: Logic Programs with Ordered Disjunction. Computational Intelligence 20(2), 333–357 (2004)
12. Brewka, G., Niemelä, I., Truszczyński, M.: Preferences and Nonmonotonic Reasoning. AI Magazine 29(4), 69–78 (2008)
13. Confalonieri, R., Nieves, J.C., Vázquez-Salceda, J.: A Preference Meta-Model for Logic Programs with Possibilistic Ordered Disjunction. In: Software Engineering for Answer Set Programming (SEA 2009) (September 2009), To appear in CEUR Workshop Proc. of SEA 2009, Co-located with LPNMR 2009, `http://sea09.cs.bath.ac.uk/downloads/sea09proceedings.pdf`
14. Confalonieri, R., Nieves, J.C., Vázquez-Salceda, J.: Logic Programs with Possibilistic Ordered Disjunction. Research Report LSI-09-19-R, UPC - LSI (2009), `http://www.lsi.upc.edu/~techreps/files/R09-19.zip`
15. Confalonieri, R., Nieves, J.C., Vázquez-Salceda, J.: Pstable Semantics for Logic Programs with Possibilistic Ordered Disjunction. In: Serra, R., Cucchiara, R. (eds.) AI*IA 2009. LNCS (LNAI), vol. 5883, pp. 52–61. Springer, Heidelberg (2009)
16. Davey, B.A., Priestly, H.A.: Introduction to Lattices and Order, 2nd edn. Cambridge University Press, Cambridge (2002)
17. Delgrande, J., Schaub, T., Tompits, H., Wang, K.: A classification and Survey of Preference Handling Approaches in Nonmonotonic Reasoning. Computational Intelligence 20(2), 308–334 (2004)
18. Dershowitz, N., Plaisted, D.A.: Rewriting. In: Handbook of Automated Reasoning, pp. 535–610. Elsevier/MIT Press (2001)
19. Dix, J., Osorio, M., Zepeda, C.: A General Theory of Confluent Rewriting Systems for Logic Programming and its Applications. Annals of Pure and Applied Logic 108(1-3), 153–188 (2001)
20. Dubois, D., Lang, J., Prade, H.: Possibilistic Logic. In: Handbook of Logic in Artificial Intelligence and Logic Programming, vol. 3, pp. 439–513. Oxford University Press, Oxford (1994)
21. Dubois, D., Prade, H.: Possibilistic logic: a retrospective and prospective view. Fuzzy Sets and Systems 144(1), 3–23 (2004)
22. Faber, W., Konczak, K.: Strong Equivalence for Logic Programs with Preferences. In: Proc. of the Nineteenth Int. Joint Conference on Artificial Intelligence, pp. 430–435. Professional Book Center (2005)
23. Faber, W., Tompits, H., Woltran, S.: Notions of Strong Equivalence for Logic Programs with Ordered Disjunction. In: Proc. of the 11th International Conference on Principles of Knowledge Representation and Reasoning, pp. 433–443. AAAI Press, Menlo Park (2008)
24. Fitting, S.: Bilattices and the Semantics of Logic Programming. Journal of Logic Programming 11(2), 91–116 (1991)
25. Foo, N.Y., Meyer, T., Brewka, G.: LPOD Answer Sets and Nash Equilibria. In: Maher, M.J. (ed.) ASIAN 2004. LNCS, vol. 3321, pp. 343–351. Springer, Heidelberg (2004)
26. Gelfond, M., Lifschitz, V.: Classical Negation in Logic Programs and Disjunctive Databases. New Generation Computing 9(3/4), 365–386 (1991)
27. Kern-Isberner, G., Lukasiewicz, T.: Combining probabilistic logic programming with the power of maximum entropy. Artificial Intelligence 157(1-2), 139–202 (2004)

28. Konczak, K.: Weak Order Equivalence for Logic Programs with Preferences. In: Workshop on Logic Programming, Technische Universität Wien, Austria. volume 1843-06-02 of INFSYS Research Report, pp. 154–163 (2006)
29. Leone, N.: Logic Programming and Nonmonotonic Reasoning: From Theory to Systems and Applications. In: Baral, C., Brewka, G., Schlipf, J. (eds.) LPNMR 2007. LNCS (LNAI), vol. 4483, p. 1. Springer, Heidelberg (2007)
30. Lifschitz, V., Pearce, D., Valverde, A.: Strongly equivalent logic programs. ACM Transaction on Computational Logic 2(4), 526–541 (2001)
31. Lloyd, J.W.: Foundations of logic programming. Springer, New York (1987)
32. Mendelson, E.: Introduction to Mathematical Logic. Chapman & Hall, Boca Raton (1997)
33. Newman, M.H.A.: On Theories with a Combinatorial Definition of Equivalence. The Annals of Mathematics 43(2), 223–243 (1942)
34. Nicolas, P., Garcia, L., Stéphan, I., Lefèvre, C.: Possibilistic Uncertainty Handling for Answer Set Programming. Annals of Mathematics and Artificial Intelligence 47(1-2), 139–181 (2006)
35. Nieuwenborgh, D., Cock, M., Vermeir, D.: An introduction to fuzzy answer set programming. Annals of Mathematics and Artificial Intelligence 50(3-4), 363–388 (2007)
36. Nieves, J.C., Osorio, M., Cortés, U.: Semantics for Possibilistic Disjunctive Programs. In: Baral, C., Brewka, G., Schlipf, J. (eds.) LPNMR 2007. LNCS (LNAI), vol. 4483, pp. 315–320. Springer, Heidelberg (2007)
37. Osorio, M., Navarro, J.A., Arrazola, J.: Equivalence in Answer Set Programming. In: Pettorossi, A. (ed.) LOPSTR 2001. LNCS, vol. 2372, pp. 57–75. Springer, Heidelberg (2002)
38. Osorio, M., Nieves, J.C.: PStable Semantics for Possibilistic Logic Programs. In: Gelbukh, A., Kuri Morales, Á.F. (eds.) MICAI 2007. LNCS (LNAI), vol. 4827, pp. 294–304. Springer, Heidelberg (2007)
39. Osorio, M., Nieves, J.C., Giannella, C.: Useful Transformations in Answer Set Programming. In: Provetti, A., Son, T.C. (eds.) AAAI 2001 Spring Symposium Series, Stanford, E.U., pp. 146–152. AAAI Press, Stanford (2001)
40. Pettorossi, A., Proietti, M.: Transformation of Logic Programs. In: Gabbay, D.M., Robinson, J.A., Hogger, C.J. (eds.) Handbook of Logic in Artificial Intelligence and Logic Programming, vol. 5, pp. 697–787. Oxford University Press, Oxford (1998)
41. Reiter, R.: Readings in Nonmonotonic Reasoning. In: Reiter, R. (ed.) Logic for Default Reasoning, ch. A, pp. 68–93. Morgan Kaufmann Publishers Inc., San Francisco (1987)
42. Shoham, Y.: A Semantical Approach to Nonmonotonic Logics. In: Ginsberg, M.L. (ed.) Readings in Nonmonotonic Reasoning, pp. 227–250. Morgan Kaufmann Publishers Inc., San Francisco (1987)
43. Zepeda, C., Osorio, M., Nieves, J.C., Solnon, C., Sol, D.: Applications of Preferences using Answer Set Programming. In: Zepeda, C., Osorio, M., Nieves, J.C., Solnon, C., Sol, D. (eds.) Proc. of the 3rd Intl. Workshop in Advances in Theory and Implementation. CEUR Workshop Proceedings, CEUR-WS.org, vol. 142 (2005)

Semantic Web Search Based on Ontological Conjunctive Queries

Bettina Fazzinga[1], Giorgio Gianforme[2], Georg Gottlob[3,4], and Thomas Lukasiewicz[3,5]

[1] Dipartimento di Elettronica, Informatica e Sistemistica, Università della Calabria, Italy
bfazzinga@deis.unical.it
[2] Dipartimento di Informatica e Automazione, Università Roma Tre, Italy
giorgio.gianforme@gmail.com
[3] Computing Laboratory, University of Oxford, UK
{georg.gottlob,thomas.lukasiewicz}@comlab.ox.ac.uk
[4] Oxford-Man Institute of Quantitative Finance, University of Oxford, UK
[5] Institut für Informationssysteme, TU Wien, Austria

Abstract. Many experts predict that the next huge step forward in Web information technology will be achieved by adding semantics to Web data, and will possibly consist of (some form of) the Semantic Web. In this paper, we present a novel approach to Semantic Web search, which is based on ontological conjunctive queries, and which combines standard Web search with ontological background knowledge, as it is, e.g., available in Semantic Web repositories. We show how standard Web search engines can be used as the main inference motor for processing ontology-based semantic search queries on the Web. We develop the formal model behind this approach and also provide an implementation in desktop search. Furthermore, we report on extensive experimental results.

1 Introduction

Web search is a key technology of the Web, since it is the primary way to access content in the ocean of Web data. Current Web search technologies are essentially based on a combination of textual keyword search with an importance ranking of documents via the link structure of the Web [6].

Web search, however, is about to change radically with the development of a more powerful future Web, called the *Semantic Web* [4, 5], which is a common framework that allows data to be shared and reused in different applications, enterprises, and communities. It is an extension of the current Web by standards and technologies that help machines to understand the information on the Web so that they can support richer discovery, data integration, navigation, and automation of tasks. It consists of several hierarchical layers, where the *Ontology layer*, in form of the *OWL Web Ontology Language* [13, 18, 3], is the highest layer that has currently reached a sufficient maturity. Some important layers below the Ontology layer are the *RDF* and *RDF Schema layers* along with the *SPARQL* query language. For the higher *Rules*, *Logic*, and *Proof layers* of the Semantic Web, one has especially developed languages integrating rules and ontologies, and languages supporting more sophisticated forms of knowledge. During

S. Link and H. Prade (Eds.): FoIKS 2010, LNCS 5956, pp. 153–172, 2010.

the recent decade, a huge amount of academic and commercial research activities has been spent towards realizing the Semantic Web. Hence, in addition to the traditional Web pages, future Web data are expected to be more and more organized in the new formalisms of the Semantic Web, and will thus also consist of RDF data along with ontological and rule-based knowledge.

The development of a new search technology for the Semantic Web, called *Semantic Web search*, is currently an extremely hot topic, both in Web-related companies and in academic research (see Section 8). In particular, there is a fast growing number of commercial and academic Semantic Web search engines. The research can be roughly divided into two main directions. The first (and most common) one is to develop a new form of search for searching the pieces of data and knowledge that are encoded in the new representation formalisms of the Semantic Web, while the second (and nearly unexplored) direction is to use the data and knowledge of the Semantic Web in order to add some semantics to Web search. The second direction is also a first step towards Web search queries that are formulated in natural language expressions.

In this paper, we follow the second line of research. The main ideas behind the present paper are to realize Semantic Web search by (i) using ontological conjunctive queries, (ii) combining standard Web search with ontological background knowledge (as, e.g., available in the Semantic Web), (iii) using the power of Semantic Web formalisms and technologies, and (iv) using standard Web search engines as the main inference motor of Semantic Web search. Our approach to Semantic Web search is realized on top of standard ontology reasoning technologies and Web search engines. As important advantages of this approach, (i) it can be applied to the whole existing Web (and not only to the future Semantic Web), (ii) it can be done immediately (and not only when the future Semantic Web is in place), and (iii) it can be done with existing Web search technology (and so does not require completely new technologies). This line of research aims at making current search engines for the existing Web more "semantic" (i.e., in a sense also more "intelligent") by combining the information on existing Web pages with background ontological knowledge. Intuitively, rather than being interpreted in a keyword-based syntactic fashion, the pieces of data on existing Web pages are interpreted by their connected ontology-based semantic background knowledge. That is, the pieces of data on Web pages are connected to a much more precise semantic and contextual meaning. This also allows for more complex ontology-based Web search queries, which are closer to complex natural language search expressions than current (Boolean) keyword-based Web search queries. Furthermore, it allows for answering Web search queries in a much more precise way, as the following simple examples show:

- As for complex queries, when searching for a movie, one may be interested in movies by a US company before 1999 and with a French director. Similarly, when buying a house in a town, one may be interested in large house selling companies within 50 miles of that town, existing for at least 15 years, and not known to be blacklisted by a consumer organization in the last 5 years. Such queries are answered by connecting the information on existing Web pages with available background knowledge.
- Suppose next one is searching for "laptop". Then, one is looking for laptops or synonyms/related concepts (e.g., "notebook"), but also for special kinds of laptops

that are not synonyms / related concepts, such as IBM ThinkPads. Semantic background knowledge now allows for obtaining a collection of contextually correct synonyms / related concepts and a collection of contextually correct special laptops.
- Similarly, a Web search for "president of the USA" should also return Web pages that contain "George W. Bush" (who is/was one of the presidents of the USA according to some background ontology). Also, a Web search for "the president of the USA on September 11, 2001" should return Web pages mentioning "George W. Bush" (who was the president of the USA on September 11, 2001, according to some background ontology). On the other hand, when searching for Web pages about the first president of the USA, "Washington", semantic annotations and background knowledge allow us to restrict our search to Web pages that are actually about Washington as the name of the president, and so to ignore, e.g., Web pages about the state or town.

The above are examples of very simple Web search queries, which can be handled in our Semantic Web search, but not appropriately in current Web search. The main contributions of this paper and the characteristic features of our approach are as follows:

- We present a novel approach to Semantic Web search, where standard Web search engines are combined with ontological background knowledge. We show how the approach can be implemented on top of standard Web search engines and ontological inference technologies, using lightweight user-site software clients.
- We develop the formal model behind this approach. More specifically, we introduce Semantic Web knowledge bases and Semantic Web search queries to them. We also generalize the PageRank technique to our approach to Semantic Web search.
- We provide a technique for processing Semantic Web search queries, which consists of an offline inference and an online reduction to a collection of standard Web search queries. We prove that this way of processing Semantic Web search queries is ontologically correct (and in many cases also ontologically complete).
- We report on a prototype implementation of our approach in the framework of desktop search. Experiments with half a million annotated facts show that the new methods are principally feasible and potentially scale to Web search (which is actually faster than desktop search, even with a dramatically larger search space).
- The offline inference step compiles terminological knowledge into so-called *completed* annotations. Since ontological hierarchies in practice (such as the GALEN ontology[1]) are generally not that deep (a concept has at most a dozen superconcepts), the generated completed annotations are generally only of a small size.
- Differently from conventional Boolean keyword-oriented Web search queries, the proposed Semantic Web search queries clearly empower the user to precisely describe her information need for certain kinds of queries, resulting in a very precise result set and a very high precision and recall for the query result.

The rest of this paper is organized as follows. In Section 2, we give an overview of our approach to Semantic Web search. In Section 3, we introduce Semantic Web knowledge bases and Semantic Web search queries, including a generalized PageRank technique.

[1] `http://www.co-ode.org/galen/`

Sections 4 to 6 describe how Semantic Web search queries are processed via offline inference and online reduction to standard Web search. In Section 7, we describe a first prototype implementation for semantic desktop search on top of Windows Desktop Search, along with experimental results. In Section 8, we discuss related work to Semantic Web search. Section 9 summarizes our main results and gives an outlook on future research. Detailed proofs of all results are given in the extended report [11].

2 System Overview

The overall architecture of our Semantic Web search system is shown in Fig. 1. It consists of the *Interface*, the *Query Evaluator*, and the *Inference Engine* (blue parts), where the Query Evaluator is implemented on top of standard Web *Search Engines*. Standard *Web* pages and their objects are enriched by *Annotation* pages, based on an *Ontology*.

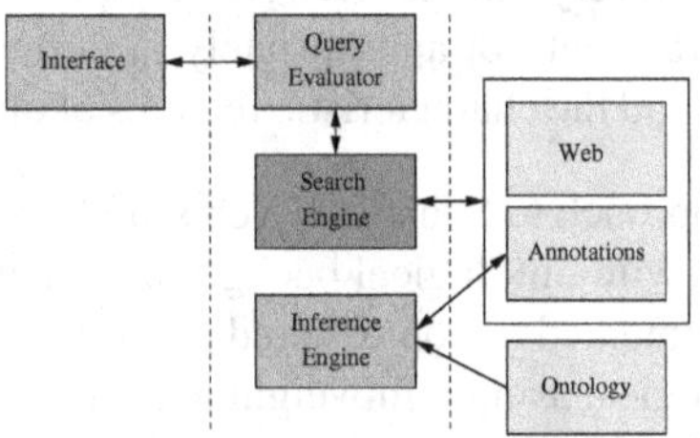

Fig. 1. System architecture

We thus make the standard assumption in the Semantic Web that there are semantic annotations to standard Web pages and to objects on standard Web pages. Such annotations are starting to be widely available for a large class of Web resources, especially user-defined annotations with the Web 2.0. They may also be automatically learned from the Web pages and the objects to be annotated (see, e.g., [8]), and/or they may be extracted from existing ontological knowledge bases on the Semantic Web.

For example, in a very simple scenario, a Web page i_1 may contain information about a Ph.D. student i_2, called Mary, and two of her papers: a conference paper i_3 with title

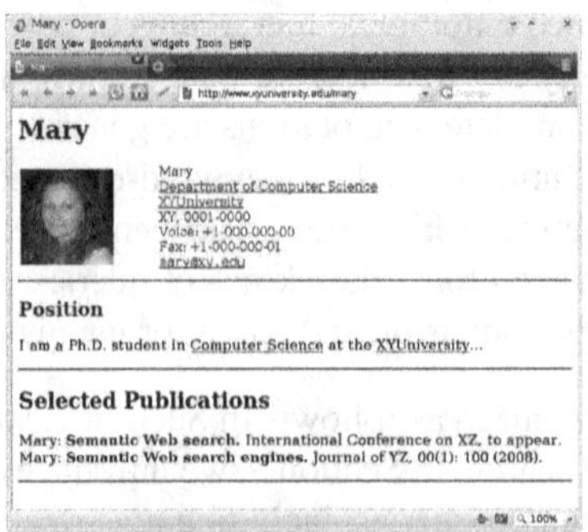

Fig. 2. HTML page

"*Semantic Web search*" and a journal paper i_4 entitled "*Semantic Web search engines*" and published in 2008. A simple HTML page representing this scenario is shown in Fig. 2. There may now exist one semantic annotation each for the Web page, the Ph.D. student Mary, the journal paper, and the conference paper. The annotation for the Web page may simply encode that it mentions Mary and the two papers, while the one for Mary may encode that she is a Ph.D. student with the name Mary and the author of the papers i_3 and i_4. The annotation for i_3 may encode that i_3 is a conference paper and has the title "*Semantic Web search*", while the one for i_4 may encode that i_4 is a journal paper, authored by Mary, has the title "*Semantic Web search engines*", was published in 2008, and has the keyword "RDF". The semantic annotations of i_1, i_2, i_3, and i_4 are formally expressed as the sets of axioms $\mathcal{A}_{i_1}$, $\mathcal{A}_{i_2}$, $\mathcal{A}_{i_3}$, and $\mathcal{A}_{i_4}$, respectively:

$$\begin{aligned}
\mathcal{A}_{i_1} &= \{\mathit{contains}(i_1, i_2), \mathit{contains}(i_1, i_3), \mathit{contains}(i_1, i_4)\},\\
\mathcal{A}_{i_2} &= \{\mathit{PhDStudent}(i_2),\ \mathit{name}(i_2, \text{"mary"}),\ \mathit{isAuthorOf}(i_2, i_3),\ \mathit{isAuthorOf}(i_2, i_4)\},\\
\mathcal{A}_{i_3} &= \{\mathit{ConferencePaper}(i_3), \mathit{title}(i_3, \text{"Semantic Web search"})\},\\
\mathcal{A}_{i_4} &= \{\mathit{JournalPaper}(i_4),\ \mathit{hasAuthor}(i_4, i_2),\ \mathit{title}(i_4, \text{"Semantic Web search engines"}),\\
&\qquad \mathit{yearOfPublication}(i_4, 2008), \mathit{keyword}(i_4, \text{"RDF"})\}.
\end{aligned} \tag{1}$$

Inference Engine. Using an ontology containing some background knowledge, these semantic annotations are then further enhanced in an offline ontology compilation step, where the *Inference Engine* adds all properties (i.e., ground atoms) that can be deduced from the semantic annotations and the ontology. The resulting (*completed*) semantic annotations are then published as Web pages, so that they can be searched by standard Web search engines. For example, an ontology may contain the knowledge that (i) conference and journal papers are articles, (ii) conference papers are not journal papers, (iii) *isAuthorOf* relates scientists and articles, (iv) *isAuthorOf* is the inverse of *hasAuthor*, and (v) *hasFirstAuthor* is a functional binary relationship, which is formalized by:

$$\begin{aligned}
&\mathit{ConferencePaper} \sqsubseteq \mathit{Article},\ \mathit{JournalPaper} \sqsubseteq \mathit{Article},\ \mathit{ConferencePaper} \sqsubseteq \neg\mathit{JournalPaper},\\
&\exists\mathit{isAuthorOf} \sqsubseteq \mathit{Scientist},\ \exists\mathit{isAuthorOf}^- \sqsubseteq \mathit{Article},\ \mathit{isAuthorOf}^- \sqsubseteq \mathit{hasAuthor},\\
&\mathit{hasAuthor}^- \sqsubseteq \mathit{isAuthorOf},\ (\mathsf{funct}\ \mathit{hasFirstAuthor}).
\end{aligned} \tag{2}$$

Using this ontological knowledge, we can derive from the above annotations that the two papers i_3 and i_4 are also articles, and both authored by Mary.

HTML Annotations. These resulting searchable (completed) semantic annotations of (objects on) standard Web pages are published as HTML Web pages with pointers to the respective object pages, so that they (in addition to the standard Web pages) can be

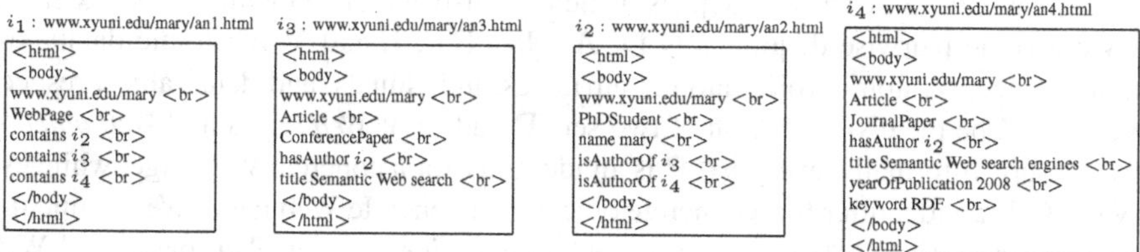

Fig. 3. Four HTML pages encoding the (completed) semantic annotations for the HTML page in Fig. 2 and the three objects on it

searched by standard search engines. For example, the HTML pages for the completed semantic annotations of the above $\mathcal{A}_{i_1}$, $\mathcal{A}_{i_2}$, $\mathcal{A}_{i_3}$, and $\mathcal{A}_{i_4}$ are shown in Fig. 3. We here use the HTML address of the Web page/object's annotation page as an identifier for that Web page/object. The plain textual representation of the completed semantic annotations allows their processing by existing standard search engines for the Web.

RDF Annotations. It is important to point out that this textual representation is simply a list of properties, each eventually along with an identifier or a data value as attribute value, and it can thus immediately be encoded as a list of RDF triples.

Query Evaluator. The *Query Evaluator* (see Fig. 1) reduces each Semantic Web search query of the user in an online query processing step to a sequence of standard Web search queries on standard Web and annotation pages, which are then processed by a standard Web *Search Engine*. The Query Evaluator also collects the results and re-transforms them into a single answer which is returned to the user. As an example of a Semantic Web search query, one may ask for all Ph.D. students who have published an article in 2008 with RDF as a keyword, which is formally expressed as follows:

$$Q(x) = \exists y\,(\mathit{PhDStudent}(x) \wedge \mathit{isAuthorOf}(x, y) \wedge \mathit{Article}(y) \wedge \\ \mathit{yearOfPublication}(y, 2008) \wedge \mathit{keyword}(y, \text{"}RDF\text{"}))\,.$$

This query is transformed into the two queries $Q_1 = \mathit{PhDStudent}$ AND *isAuthorOf* and $Q_2 = \mathit{Article}$ AND *"yearOfPublication* 2008" AND *"keyword* RDF", which can both be submitted to a standard Web search engine, such as Google. The result of the original query Q is then built from the results of the two queries Q_1 and Q_2. Note that a graphical user interface, such as the one of Google's advanced search, or even a natural language interface can help to hide the conceptual complexity of ontological queries to the user.

3 Semantic Web Search

We now introduce Semantic Web knowledge bases and the syntax and semantics of Semantic Web search queries to such knowledge bases. We then generalize the PageRank technique to our approach. We assume the reader is familiar with the syntax and the semantics of Description Logics (DLs) [1], used as underlying ontology languages.

Semantic Web Knowledge Bases. Intuitively, a Semantic Web knowledge base consists of a background TBox and a collection of ABoxes, one for every concrete Web page and for every object on a Web page. For example, the homepage of a scientist may be such a concrete Web page and be associated with an ABox, while the publications on the homepage may be such objects, which are also associated with one ABox each.

We assume pairwise disjoint sets $\mathbf{D}$, $\mathbf{A}$, $\mathbf{R}_A$, $\mathbf{R}_D$, $\mathbf{I}$, and $\mathbf{V}$ of atomic datatypes, atomic concepts, atomic roles, atomic attributes, individuals, and data values, respectively. Let $\mathbf{I}$ be the disjoint union of two sets $\mathbf{P}$ and $\mathbf{O}$ of *Web pages* and *Web objects*, respectively. Informally, every $p \in \mathbf{P}$ is an identifier for a concrete Web page, while every $o \in \mathbf{O}$ is an identifier for a concrete object on a concrete Web page. We assume the atomic roles *links_to* between Web pages and *contains* between Web pages and Web objects. The former represents the link structure between concrete Web pages, while the latter encodes the occurrences of concrete Web objects on concrete Web pages.

Definition 1. A *semantic annotation* $\mathcal{A}_a$ for a Web page or object $a \in \mathbf{P} \cup \mathbf{O}$ is a finite set of concept membership axioms $A(a)$, role membership axioms $P(a,b)$, and attribute membership axioms $U(a,v)$ (which all have the Web page or object a as first argument), where $A \in \mathbf{A}$, $P \in \mathbf{R}_A$, $U \in \mathbf{R}_D$, $b \in \mathbf{I}$, and $v \in \mathbf{V}$. A *Semantic Web knowledge base* $KB = (\mathcal{T}, (\mathcal{A}_a)_{a \in \mathbf{P} \cup \mathbf{O}})$ consists of a TBox $\mathcal{T}$ and one semantic annotation $\mathcal{A}_a$ for every Web page and object $a \in \mathbf{P} \cup \mathbf{O}$.

Informally, a Semantic Web knowledge base consists of some background terminological knowledge and some assertional knowledge for every concrete Web page and for every concrete object on a Web page. The background terminological knowledge may be an ontology from some global Semantic Web repository or an ontology defined locally by the user site. In contrast to the background terminological knowledge, the assertional knowledge will be directly stored on the Web (on annotation pages like the described standard Web pages) and is thus accessible via Web search engines.

Example 1. (Scientific Database). We use a knowledge base $KB = (\mathcal{T}, \mathcal{A})$ in *DL-Lite*$_A$ [17] to specify some simple information about scientists and their publications. The sets of atomic concepts, atomic roles, atomic attributes, and data values are:

$$\begin{aligned}
\mathbf{A} &= \{\mathit{Scientist}, \mathit{Article}, \mathit{ConferencePaper}, \mathit{JournalPaper}\},\\
\mathbf{R}_A &= \{\mathit{hasAuthor}, \mathit{hasFirstAuthor}, \mathit{isAuthorOf}, \mathit{contains}\},\\
\mathbf{R}_D &= \{\mathit{name}, \mathit{title}, \mathit{yearOfPublication}\},\\
\mathbf{V} &= \{\text{“}\mathit{mary}\text{”}, \text{“}\mathit{Semantic\ Web\ search}\text{”}, 2008, \text{“}\mathit{Semantic\ Web\ search\ engines}\text{”}\}.
\end{aligned}$$

Let $\mathbf{I} = \mathbf{P} \cup \mathbf{O}$ be the set of individuals, where $\mathbf{P} = \{i_1\}$ is the set of Web pages, and $\mathbf{O} = \{i_2, i_3, i_4\}$ is the set of Web objects on i_1. The TBox $\mathcal{T}$ contains the axioms in Eq. 2. Then, a Semantic Web knowledge base is given by $KB = (\mathcal{T}, (\mathcal{A}_a)_{a \in \mathbf{P} \cup \mathbf{O}})$, where the semantic annotations of the individuals in $\mathbf{P} \cup \mathbf{O}$ are the ones in Eq. 1.

Semantic Web Search Queries. We use unions of conjunctive queries with conjunctive and negated conjunctive subqueries as Semantic Web search queries to Semantic Web knowledge bases. We now first define the syntax of Semantic Web search queries and then the semantics of positive and general such queries.

Syntax. Let $\mathbf{X}$ be a finite set of variables. A *term* is either a Web page $p \in \mathbf{P}$, a Web object $o \in \mathbf{O}$, a data value $v \in \mathbf{V}$, or a variable $x \in \mathbf{X}$. An *atomic formula* (or *atom*) α is of one of the following forms: (i) $d(t)$, where d is an atomic datatype, and t is a term; (ii) $A(t)$, where A is an atomic concept, and t is a term; (iii) $P(t, t')$, where P is an atomic role, and t, t' are terms; and (iv) $U(t, t')$, where U is an atomic attribute, and t, t' are terms. An *equality* has the form $=(t, t')$, where t and t' are terms. A *conjunctive formula* $\exists \mathbf{y}\, \phi(\mathbf{x}, \mathbf{y})$ is an existentially quantified conjunction of atoms α and equalities $=(t, t')$, which have free variables among $\mathbf{x}$ and $\mathbf{y}$.

Definition 2. A *Semantic Web search query* $Q(\mathbf{x})$ is an expression $\bigvee_{i=1}^{n} \exists \mathbf{y}_i\, \phi_i(\mathbf{x}, \mathbf{y}_i)$, where each ϕ_i with $i \in \{1, \ldots, n\}$ is a conjunction of atoms α (also called *positive atoms*), conjunctive formulas ψ, negated conjunctive formulas $\mathit{not}\ \psi$, and equalities $=(t, t')$, which have free variables among $\mathbf{x}$ and $\mathbf{y}_i$.

Intuitively, using database and DL terminology, Semantic Web search queries are unions of conjunctive queries, which may contain conjunctive queries and negated conjunctive queries in addition to atoms and equalities as conjuncts.

Example 2. (Scientific Database cont'd). Two Semantic Web search queries are:

$$Q_1(x) = (\mathit{Scientist}(x) \wedge \mathit{not}\ \mathit{doctoralDegree}(x, \text{“oxford university”}) \wedge \mathit{worksFor}(x, \text{“oxford university”})) \vee (\mathit{Scientist}(x) \wedge \mathit{doctoralDegree}(x, \text{“oxford university”}) \wedge \mathit{not}\ \mathit{worksFor}(x, \text{“oxford university”}));$$

$$Q_2(x) = \exists y\, (\mathit{Scientist}(x) \wedge \mathit{worksFor}(x, \text{“oxford university”}) \wedge \mathit{isAuthorOf}(x, y) \wedge \mathit{not}\ \mathit{ConferencePaper}(y) \wedge \mathit{not}\ \exists z\, \mathit{yearOfPublication}(y, z)).$$

Informally, $Q_1(x)$ asks for scientists who are either working for *oxford university* and did not receive their Ph.D. from that university, or who received their Ph.D. from *oxford university* but do not work for it. Whereas $Q_2(x)$ asks for scientists of *oxford university* who are authors of at least one unpublished non-conference paper. Note that when searching for scientists, the system automatically searches for all subconcepts (known according to the background ontology), such as Ph.D. students or computer scientists.

Semantics of Positive Search Queries. We now define the semantics of positive Semantic Web search queries, which are free of negations, in terms of ground substitutions via the notion of logical consequence.

A search query $Q(\mathbf{x})$ is *positive* iff it contains no negated conjunctive subqueries. A *(variable) substitution* θ maps variables from $\mathbf{X}$ to terms. A substitution θ is *ground* iff it maps to Web pages $p \in \mathbf{P}$, Web objects $o \in \mathbf{O}$, and data values $v \in \mathbf{V}$. A closed first-order formula ϕ is a *logical consequence* of a knowledge base $KB = (\mathcal{T}, (\mathcal{A}_a)_{a \in \mathbf{P} \cup \mathbf{O}})$, denoted $KB \models \phi$, iff every first-order model $\mathcal{I}$ of $\mathcal{T} \cup \bigcup_{a \in \mathbf{P} \cup \mathbf{O}} \mathcal{A}_a$ also satisfies ϕ.

Definition 3. Given a Semantic Web knowledge base KB and a positive Semantic Web search query $Q(\mathbf{x})$, an *answer* for $Q(\mathbf{x})$ to KB is a ground substitution θ for the variables $\mathbf{x}$ with $KB \models Q(\mathbf{x}\theta)$.

Example 3. (Scientific Database cont'd). Consider the Semantic Web knowledge base KB of Example 1 and the following positive Semantic Web search query, asking for all scientists who author at least one published journal paper:

$$Q(x) = \exists y\, (\mathit{Scientist}(x) \wedge \mathit{isAuthorOf}(x, y) \wedge \mathit{JournalPaper}(y) \wedge \exists z\, \mathit{yearOfPublication}(y, z)).$$

An answer for $Q(x)$ to KB is $\theta = \{x/i_2\}$. Recall that i_2 represents the scientist Mary.

Semantics of General Search Queries. We next define the semantics of general search queries by reduction to the semantics of positive ones, interpreting negated conjunctive subqueries $\mathit{not}\ \psi$ as the lack of evidence about the truth of ψ. That is, negations are interpreted by a closed-world semantics on top of the open-world semantics of DLs.

Definition 4. Given a Semantic Web knowledge base KB and search query

$$Q(\mathbf{x}) = \bigvee_{i=1}^{n} \exists \mathbf{y}_i (\phi_{i,1}(\mathbf{x}, \mathbf{y}_i) \wedge \cdots \wedge \phi_{i,l_i}(\mathbf{x}, \mathbf{y}_i) \wedge \mathit{not}\ \phi_{i,l_i+1}(\mathbf{x}, \mathbf{y}_i) \wedge \cdots \wedge \mathit{not}\ \phi_{i,m_i}(\mathbf{x}, \mathbf{y}_i)),$$

an *answer* for $Q(\mathbf{x})$ to KB is a ground substitution θ for the variables $\mathbf{x}$ such that $KB \models Q^+(\mathbf{x}\theta)$ and $KB \not\models Q^-(\mathbf{x}\theta)$, where $Q^+(\mathbf{x})$ and $Q^-(\mathbf{x})$ are defined as follows:

$$Q^+(\mathbf{x}) = \bigvee_{i=1}^{n} \exists \mathbf{y}_i\,(\phi_{i,1}(\mathbf{x},\mathbf{y}_i) \wedge \cdots \wedge \phi_{i,l_i}(\mathbf{x},\mathbf{y}_i)) \text{ and}$$
$$Q^-(\mathbf{x}) = \bigvee_{i=1}^{n} \exists \mathbf{y}_i\,(\phi_{i,1}(\mathbf{x},\mathbf{y}_i) \wedge \cdots \wedge \phi_{i,l_i}(\mathbf{x},\mathbf{y}_i) \wedge (\phi_{i,l_i+1}(\mathbf{x},\mathbf{y}_i) \vee \cdots \vee \phi_{i,m_i}(\mathbf{x},\mathbf{y}_i)))\,.$$

Roughly, a ground substitution θ is an answer for $Q(\mathbf{x})$ to KB iff (i) θ is an answer for $Q^+(\mathbf{x})$ to KB, and (ii) θ is not an answer for $Q^-(\mathbf{x})$ to KB, where $Q^+(\mathbf{x})$ is the positive part of $Q(\mathbf{x})$, while $Q^-(\mathbf{x})$ is the positive part of $Q(\mathbf{x})$ combined with the complement of the negative one. Notice that both $Q^+(\mathbf{x})$ and $Q^-(\mathbf{x})$ are positive queries.

Example 4. (Scientific Database cont'd). Consider the Semantic Web knowledge base $KB = (\mathcal{T}, (\mathcal{A}_a)_{a \in \mathbf{P} \cup \mathbf{O}})$ of Example 1 and the following general Semantic Web search query, asking for Mary's unpublished non-journal papers:

$$Q(x) = \exists y\,(\mathit{Article}(x) \wedge \mathit{hasAuthor}(x,y) \wedge \mathit{name}(y, \text{“}\mathit{mary}\text{”}) \wedge \mathit{not}\ \mathit{JournalPaper}(x) \wedge \mathit{not}\ \exists z\, \mathit{yearOfPublication}(x,z)).$$

An answer for $Q(x)$ to KB is given by $\theta = \{\mathbf{x}/i_3\}$. Recall that i_3 represents an unpublished conference paper entitled “*Semantic Web search*”. Observe that the membership axioms $\mathit{Article}(i_3)$ and $\mathit{hasAuthor}(i_2, i_3)$ do not appear in the semantic annotations $\mathcal{A}_a$ with $a \in \mathbf{P} \cup \mathbf{O}$, but they can be inferred from them using the background ontology $\mathcal{T}$.

Ranking Answers. As for the ranking of all answers for a Semantic Web search query Q to a Semantic Web knowledge base KB (i.e., ground substitutions for all free variables in Q, which correspond to tuples of Web pages, Web objects, and data values), we use a generalization of the PageRank technique [6]: rather than considering only Web pages and the link structure between Web pages (expressed through the role *links_to* here), we also consider Web objects, which may occur on Web pages (expressed through the role *contains*), and which may also be related to other Web objects via other roles. More concretely, we define the *ObjectRank* of a Web page or a Web object a as follows:

$$R(a) = d \cdot \textstyle\sum_{b \in B_a} R(b)\,/\,N_b + (1-d) \cdot E(a)\,,$$

where (i) B_a is the set of all Web pages and Web objects o that relate to a (i.e., o relates to a via some role), (ii) N_b is the number of Web pages and Web objects o that relate from b (i.e., b relates to o via some role), (iii) d is a damping factor, and (iv) E associates with every Web page and every Web object an initial value, called *source of rank*. So, rather than depending only on the link structure between Web pages, the new ranking depends also on the relationships between Web pages and Web objects, and on the relationships between Web objects, where the user fixes the roles to be considered. Note that in some cases, only a subset of all relationships may be used for specifying ObjectRank. For example, in the running Scientific Database Example, the relationship “cites” alone between articles produces a very useful ranking on articles.

The ranking on Web pages and objects is then naturally extended to answers (i.e., tuples of Web pages, Web objects, and values) for Semantic Web search queries to Semantic Web knowledge bases. For example, the answers can be ordered lexicographically, or the rank of an answer can be defined as the minimum (or maximum) of the ranks of its Web pages and objects, and then ordered as usual.

4 Realizing Semantic Web Search

The main idea behind processing Semantic Web search queries Q to a knowledge base *KB* is to reduce them to standard Web search queries. To this end, the TBox $\mathcal{T}$ of *KB* must be considered when performing standard Web search. There are two main ways to do so. The first is to compile $\mathcal{T}$ into Q, yielding a new standard Web search query Q' on the ABox $\mathcal{A}$ of *KB*. The second, which we adopt here, is to compile $\mathcal{T}$ via offline ontology reasoning into the ABox $\mathcal{A}$ of *KB*, yielding a completed ABox $\mathcal{A}'$, which (being represented on the Web in addition to the standard Web pages) is then searched online by a collection of standard Web search queries depending on Q. So, processing Semantic Web search queries Q is divided into

- an offline ontology reasoning step, where roughly all semantic annotations of Web pages / objects are completed by entailed (from *KB*) membership axioms, and
- an online reduction to standard Web search, where Q is transformed into standard Web search queries whose answers are used to construct the answer for Q.

In the offline ontology reasoning step, we check whether the Semantic Web knowledge base is satisfiable, and we compute the completion of all semantic annotations, i.e., we augment the semantic annotations with all concept, role, and attribute membership axioms that can be deduced from the semantic annotations and the background ontology. We suggest to use only the so-called simple completion of all semantic annotations, which is sufficient for a large class of Semantic Web knowledge bases and search queries. It is important to point out that since ontology reasoning is done offline (like the construction of an index structure for Web search), its running time does not contribute to the running time of the actual online processing of Semantic Web search queries. Thus, the running time used for ontology reasoning can be fully neglected. Nonetheless, there are also tractable ontology languages, such as *DL-Lite*$_{\mathcal{A}}$, in which checking whether a Semantic Web knowledge base is satisfiable and computing its simple completion can both be done very efficiently (in LOGSPACE) in the data complexity, and one can use existing systems such as QuOnto [7].

In the online reduction to standard Web search, we decompose a given Semantic Web search query Q into a collection of standard Web search queries, of which the answers are then used to construct the answer for Q. The standard Web search queries are processed with existing search engines on the Web. It is important to point out that, publishing the completed semantic annotations as standard Web pages, as we propose here, this standard Web search can be done immediately with existing standard Web search engines (see Section 7 for an implementation of semantic desktop search on top of standard desktop search). Alternatively, we may also keep the completed semantic annotations in a virtual way only and use them for the construction of the index structure for Web search only. In that case, the offline ontology reasoning step can be combined with the construction of the index structure for Web search.

Note that the terms "online" and "offline" are here used in a computational sense. In the following, we describe the offline ontology reasoning step in Section 5 and the online reduction to standard Web search in Section 6. We finally describe the implementation of a semantic desktop search engine in Section 7.

5 Offline Ontology Compilation

The offline ontology reasoning step transforms the implicit terminological knowledge in the TBox of a Semantic Web knowledge base into explicit membership axioms in the ABox, i.e., in the semantic annotations of Web pages / objects, so that it (in addition to the standard Web pages) can be searched by standard Web search engines. In this section, we assume that Semantic Web knowledge bases are defined relative to *DL-Lite*$_{\mathcal{A}}$ [17] as underlying description logic. In the case of quantifier-free search queries under *DL-Lite*$_{\mathcal{A}}$, and when the TBox is equivalent to a Datalog program, it is sufficient to add all logically entailed membership axioms constructed from Web pages, Web objects, and data values.

Simple Completion. The compilation of TBox into ABox knowledge is as follows. Given a satisfiable Semantic Web knowledge base $KB = (\mathcal{T}, (\mathcal{A}_a)_{a \in \mathbf{P} \cup \mathbf{O}})$, the *simple completion* of KB is the Semantic Web knowledge base $KB' = (\emptyset, ({\mathcal{A}_a}')_{a \in \mathbf{P} \cup \mathbf{O}})$ such that every ${\mathcal{A}_a}'$ is the set of all concept membership axioms $A(a)$, role membership axioms $P(a, b)$, and attribute membership axioms $U(a, v)$ that logically follow from $\mathcal{T} \cup \bigcup_{a \in \mathbf{P} \cup \mathbf{O}} \mathcal{A}_a$, where $A \in \mathbf{A}$, $P \in \mathbf{R}_A$, $U \in \mathbf{R}_D$, $b \in \mathbf{I}$, and $v \in \mathbf{V}$. Informally, for every Web page and object, the simple completion collects all available and deducible facts (whose predicate symbols shall be usable in search queries) in a completed semantic annotation.

Example 5. Consider again the TBox $\mathcal{T}$ and the annotations $(\mathcal{A}_a)_{a \in \mathbf{P} \cup \mathbf{O}}$ of Example 1. The simple completion contains in particular the three new axioms *Article*(i_3), *hasAuthor*(i_3, i_2), and *Article*(i_4). The first two are added to A_{i_3} and the last one to A_{i_4}.

The following theorem shows that positive quantifier-free search queries to a Semantic Web knowledge base KB over *DL-Lite*$_{\mathcal{A}}$ can be evaluated on the simple completion of KB (which contains only compiled but no explicit TBox knowledge anymore).

Theorem 1. *Let KB be a satisfiable Semantic Web knowledge base over DL-Lite*$_{\mathcal{A}}$*, let* $Q(\mathbf{x})$ *be a positive Semantic Web search query without existential quantifiers, and let* θ *be a ground substitution for* $\mathbf{x}$*. Then,* θ *is an answer for* $Q(\mathbf{x})$ *to KB iff* θ *is an answer for* $Q(\mathbf{x})$ *to the simple completion of KB.*

As an immediate consequence, we obtain that general quantifier-free search queries to a Semantic Web knowledge base KB over *DL-Lite*$_{\mathcal{A}}$ can also be evaluated on the simple completion of KB, which is expressed by the next theorem.

Corollary 1. *Let KB be a satisfiable Semantic Web knowledge base over DL-Lite*$_{\mathcal{A}}$*,* $Q(\mathbf{x})$ *be a (general) Semantic Web search query without existential quantifiers, and* θ *be a ground substitution for* $\mathbf{x}$*. Then,* θ *is an answer for* $Q(\mathbf{x})$ *to KB iff* θ *is an answer for* $Q^+(\mathbf{x})$ *but not an answer for* $Q^-(\mathbf{x})$ *to the simple completion of KB.*

Similar results hold when the TBox of KB is equivalent to a Datalog program, and the query $Q(\mathbf{x})$ is fully general. Hence, the simple completion assures a complete query processing for a large class of search queries. For this reason, and since completeness of query processing is actually not that much an issue in an inherently incomplete environment like the Web, we propose to use the simple completion as the basis of our Semantic Web search. In general, for general Semantic Web knowledge bases over *DL-Lite*$_{\mathcal{A}}$, the completeness of query processing can be assured with the more general k-completion, instead of the simple one, which is described in detail in the extended report [11].

Semantic Web knowledge bases may be stored in a centralized way, and updated whenever a new Web page or a new object on a Web page is created (which does not have to be done simultaneously). Then, the completed semantic annotations of Web pages and their objects can be computed by their websites (using the centralized Semantic Web knowledge base), and also be published there in a distributed way; they are eventually updated when the centralized Semantic Web knowledge base is updated.

HTML Encoding. Once the completed semantic annotations are computed, we encode them as HTML pages, so that they are searchable via standard keyword search. We build one HTML page for the semantic annotation $\mathcal{A}_a$ of each individual $a \in \mathbf{P} \cup \mathbf{O}$. That is, for each individual a, we build a page p containing all the atomic concepts whose argument is a and all the atomic roles/attributes where the first argument is a.

After rewriting the annotations, also search queries are rewritten to deal with the new syntax of the annotations. Specifically, we remove all the variables and the brackets. For example, the query $Q(x) = \mathit{Article}(x) \wedge \mathit{yearOfPublication}(x, 2008) \wedge \mathit{keyword}(x, \text{“}\mathit{RDF}\text{”})$ is translated into *Article* AND “*yearOfPublication* 2008” AND “*keyword* RDF”. In this form, the query can be evaluated by Web search engines, since it is a simple query consisting of a conjunction of a keyword and a phrase (see Section 6).

We rely on the assumption that each Web page / object $a \in \mathbf{P} \cup \mathbf{O}$ is associated with an identifier, which uniquely characterizes the individual. Here, we use the HTML address of the Web page's / object's annotation page as identifier. We employ the identifiers to evaluate complex queries involving more than one atomic concept, thus involving several annotations. Consider the query $Q(x)$ of Section 2 and the standard queries $Q_1 = \mathit{PhDStudent}$ AND *isAuthorOf* and $Q_2 = \mathit{Article}$ AND “*yearOfPublication* 2008” obtained from it. To evaluate $Q(x)$, we submit Q_1 and Q_2 to a Web search engine, and we collect the results r_1 and r_2 of the two queries, which are the sets of annotation pages $\{i_2\}$ and $\{i_4\}$, respectively. We return the annotation page p belonging to r_1 if there exists an annotation page in r_2 that occurs beside *isAuthorOf* on p. Since i_4 occurs beside *isAuthorOf* on the annotation page i_2, we thus return i_2 as overall query result.

6 Online Query Processing

Each semantic annotation of a Web page or object $a \in \mathbf{P} \cup \mathbf{O}$ is stored on an HTML page in the Web. We now define safe search queries and describe how they are answered by reduction to collections of standard Web search queries.[2] We also show how the computation of ObjectRank can be reduced to the computation of the standard PageRank.

[2] Another important class of search queries are so-called *simple* search queries (they contain no equalities and only one free variable, being the first argument in every atom), which can be directly reduced to standard Boolean Web search queries; see the extended report [11].

Safe Search Queries. Search queries where all free variables in negated conjunctive formulas and in equalities also occur in positive atoms are safe queries. That is, we connect the use of negation to a safeness condition, as usual in databases. They are reduced to collections of standard atomic Web search queries, one collection for the positive part, and one for every negative subquery. Due to the safeness, we retain all results of the positive part that are not matching with any result of a negative subquery.

Definition 5. A Semantic Web search query $Q(\mathbf{x}) = \bigvee_{i=1}^{n} \exists \mathbf{y}_i\ \phi_i(\mathbf{x}, \mathbf{y}_i)$ is *safe* iff, for every $i \in \{1, \ldots, n\}$, each variable that occurs in an equality in ϕ_i and freely in a negated conjunctive formula also occurs in a positive atom in ϕ_i.

Example 6. (Scientific Database cont'd). The following Semantic Web search queries ask for all students who do not attend at least one existing course (resp., event):

$$Q_1(x) = \exists y\, (\mathit{Student}(x) \wedge \mathit{not\ attends}(x, y) \wedge \mathit{Course}(y)),$$
$$Q_2(x) = \exists y\, (\mathit{Student}(x) \wedge \mathit{not\ attends}(x, y)).$$

Observe that query $Q_1(x)$ is safe, whereas $Q_2(x)$ is not, since the variable y does not occur in any positive atom of $Q_2(x)$.

We now describe an algorithm for the online reduction of safe Semantic Web search queries Q to standard Web search queries. Since such Q's with equalities can easily be reduced to those without (via variable substitutions, if possible), we assume w.l.o.g. that Q is equality-free. So, the algorithm reduces fully general but safe (and equality-free) Semantic Web search queries to several standard Web search queries. Recall that the former are unions of conjunctive queries, which may contain negated conjunctive (and w.l.o.g. no conjunctive) subqueries in addition to atoms as conjuncts.

Algorithm **SWSearch** in Fig. 4 takes as input a Semantic Web knowledge base $KB = (\emptyset, (\mathcal{A}_a)_{a \in \mathbf{P} \cup \mathbf{O}})$ and a safe (equality-free) Semantic Web search query $Q(\mathbf{x})$, and it returns as output the set Θ of all answers θ for $Q(\mathbf{x})$ to KB. The main ideas behind it are informally described as follows. We first decompose the query $Q(\mathbf{x})$ into the positive subqueries $Q_{i,j}(\mathbf{x}, \mathbf{y}_i)$ with $i \in \{1, \ldots, n\}$ and $j \in \{0, \ldots, n_i\}$ whose free variables are among $\mathbf{x}$ and $\mathbf{y}_i$. Here, $Q_{i,0}(\mathbf{x}, \mathbf{y}_i)$ stands for the positive part of the i-th disjunct of $Q(\mathbf{x})$, while the $Q_{i,j}(\mathbf{x}, \mathbf{y}_i)$'s with $j > 0$ stand for the negative parts of the i-th disjunct of $Q(\mathbf{x})$. We then compute the answers for the positive subqueries $Q_{i,j}(\mathbf{x}, \mathbf{y}_i)$ via Algorithm **PositiveSWSearch** in Fig. 5 (lines 3 and 5). Thereafter, the result for the i-th disjunct of $Q(\mathbf{x})$ is computed by removing from the set of all answers for the positive part all tuples matching with an answer for one of the negative parts (line 6). Here, $t[R_{i,j}]$ denotes the restriction of the tuple t to the attributes of the tuples in $R_{i,j}$. Finally, the overall result is computed by projecting the results for all disjuncts onto the free variables $\mathbf{x}$ of $Q(\mathbf{x})$ and unifying the resulting answer sets (line 8).

Algorithm **PositiveSWSearch** in Fig. 5 computes the set of all answers for positive Semantic Web search queries. It takes as input a Semantic Web knowledge base $KB = (\emptyset, (\mathcal{A}_a)_{a \in \mathbf{P} \cup \mathbf{O}})$ and a positive (equality-free) Semantic Web search query $Q(\mathbf{x})$, and it returns as output the set Θ of all answers θ for $Q(\mathbf{x})$ to KB. We first decompose the query $Q(\mathbf{x})$ into the subqueries $Q_i(\mathbf{x}, \mathbf{y}) = \bigwedge_{j=1}^{n_i} \phi_{i,j}$, $i \in \{1, \ldots, n\}$, where $\phi_{i,j} = p_{i,j}(t_i)$ or $\phi_{i,j} = p_{i,j}(t_i, t_{i,j})$, whose free variables are among $\mathbf{x}$ and $\mathbf{y}$. Note

Algorithm SWSearch

Input: Semantic Web knowledge base $KB = (\emptyset, (\mathcal{A}_a)_{a \in \mathbf{P} \cup \mathbf{O}})$; safe (=-free) Semantic Web search query $Q(\mathbf{x}) = \bigvee_{i=1}^{n} \exists \mathbf{y}_i \, Q_i(\mathbf{x}, \mathbf{y}_i)$, where $Q_i(\mathbf{x}, \mathbf{y}_i) = Q_{i,0}(\mathbf{x}, \mathbf{y}_i) \wedge \bigwedge_{j=1}^{n_i} \mathit{not} \; Q_{i,j}(\mathbf{x}, \mathbf{y}_i)$ and the free variables of $Q_i(\mathbf{x}, \mathbf{y}_i)$ are among $\mathbf{x}, \mathbf{y}_i$.

Output: set Θ of all answers θ for $Q(\mathbf{x})$ to KB.

1. $R := \emptyset$;
2. **for** $i := 1$ **to** n **do begin**
3. $\quad R_{i,0} :=$ **PositiveSWSearch**$(KB, Q_{i,0}(\mathbf{x}, \mathbf{y}_i))$;
4. $\quad$ **for** $j := 1$ **to** n_i **do begin**
5. $\quad\quad R_{i,j} :=$ **PositiveSWSearch**$(KB, Q_{i,j}(\mathbf{x}, \mathbf{y}_i))$;
6. $\quad\quad R_{i,0} := \{t \in R_{i,0} \mid \forall t_{i,j} \in R_{i,j} : t[R_{i,j}] \neq t_{i,j}\}$
7. $\quad$ **end**;
8. $\quad R := R \cup \pi_{\mathbf{x}}(R_{i,0})$
9. **end**;
10. **return** R.

Fig. 4. Algorithm **SWSearch**

that all atoms in such queries have the same term t_i as first argument. We then collect the set I_i of all matching Web pages and objects in KB for t_i as follows. If t_i is already a Web page or object, then $I_i = \{t_i\}$ (line 2), and if t_i is a variable, then we collect in I_i all Web pages and objects a with a matching semantic annotation $\mathcal{A}_a$ in KB (line 3). These matching Web pages and objects in I_i are then used in a lookup step on KB to fill all the matching identifiers, identifier-value pairs, and identifier-identifier pairs from the semantic annotations $\mathcal{A}_a$ in KB for all atomic concepts $A(t_i)$, attributes $U(t_i, t_{i,j})$, and roles $P(t_i, t_{i,j})$ in $Q_i(\mathbf{x}, \mathbf{y})$ into collections of unary and/or binary relations $A[t_i]$, $U[t_i, t_{i,j}]$, and $P[t_i, t_{i,j}]$, respectively (line 6). These relations are then joined via common variables, individuals, and values in $Q_i(\mathbf{x}, \mathbf{y})$, and finally projected to all free variables $\mathbf{x}$ in $Q(\mathbf{x})$ (line 8).

The operations in lines 3 and 6 of Algorithm **PositiveSWSearch** are realized by a standard Web search and by a lookup on the Web, respectively. In detail, recall that the semantic annotation $\mathcal{A}_a$ for every Web page and object $a \in \mathbf{P} \cup \mathbf{O}$ is stored on the Web as an HTML annotation page. The annotation page for a contains a collection of URIs, namely, the HTML address of a's standard Web page, if a is a Web page, and all standard Web pages mentioning a, if a is a Web object. In addition, it contains all atomic concepts "A" such that $KB \models A(a)$, all atomic-attribute-value pairs "$U \; v$" such that $KB \models U(a, v)$, and all atomic-role-identifier pairs "$P \; b$" such that $KB \models P(a, b)$. Hence, the search in line 6 can be realized by searching for all the URIs whose pages contain all atomic concepts "A", attributes "U" (resp., "$U \; t_{i,j}$", if $t_{i,j}$ is a value), and roles "P" (resp., "$P \; t_{i,j}$", if $t_{i,j}$ is an identifier) such that $A(t_i)$, $U(t_i, t_{i,j})$, and $P(t_i, t_{i,j})$, respectively, occur in $Q_i(\mathbf{x}, \mathbf{y})$, while the operations in line 6 can be realized by lookups under the given URIs, collecting all the matching data. Note that the algebraic operations in Figs. 4 and 5 can be easily implemented and optimized using standard relational database operations and techniques.

Algorithm PositiveSWSearch

Input: Semantic Web knowledge base $KB = (\emptyset, (\mathcal{A}_a)_{a \in \mathbf{P} \cup \mathbf{O}})$; positive (=-free) Semantic Web search query $Q(\mathbf{x}) = \exists \mathbf{y} \bigwedge_{i=1}^{n} Q_i(\mathbf{x}, \mathbf{y})$, where $Q_i(\mathbf{x}, \mathbf{y}) = \bigwedge_{j=1}^{n_i} p_{i,j}(t_i, t_{i,j})$ and the free variables of $Q_i(\mathbf{x}, \mathbf{y})$ are among $\mathbf{x}, \mathbf{y}$.

Output: set Θ of all answers θ for $Q(\mathbf{x})$ to KB.

1. **for** $i := 1$ **to** n **do begin**
2.   **if** $t_i \in \mathbf{P} \cup \mathbf{O}$ **then** $I_i := \{t_i\}$
3.   **else** $I_i := \{a \in \mathbf{P} \cup \mathbf{O} \mid \exists \theta \forall j \colon p_{i,j}(a, t_{i,j}\theta) \in \mathcal{A}_a\}$;
4.   **for each** $a \in I_i$ **do**
5.   **for** $j := 1$ **to** n_i **do**
6. $R_{i,j}[t_i, t_{i,j}] := \{(a, t_{i,j}\theta) \mid p_{i,j}(a, t_{i,j}\theta) \in \mathcal{A}_a\}$
7. **end**;
8. **return** $\pi_{\mathbf{x}}(\bowtie_{i=1}^{n} \bowtie_{j=1}^{n_i} R_{i,j})$.

Fig. 5. Algorithm **PositiveSWSearch**

The following theorem shows that the two algorithms **SWSearch** and **PositiveSWSearch** in Figs. 4 and 5, respectively, are correct, i.e., they return the set of all answers for safe (and equality-free) general and positive Semantic Web search queries, respectively, to Semantic Web knowledge bases $KB = (\mathcal{T}, (\mathcal{A}_a)_{a \in \mathbf{P} \cup \mathbf{O}})$ with $\mathcal{T} = \emptyset$.

Theorem 2. *Let $KB = (\emptyset, (\mathcal{A}_a)_{a \in \mathbf{P} \cup \mathbf{O}})$ be a Semantic Web knowledge base, and let $Q(\mathbf{x})$ be a safe (and equality-free) Semantic Web search query. Then, Algorithm* **SWSearch** *on KB and $Q(\mathbf{x})$ returns the set of all answers for $Q(\mathbf{x})$ to KB.*

Ranking Answers. The following theorem shows that computing ObjectRank can be reduced to computing PageRank. That is, using the encoding of semantic annotations as HTML pages on the Web, the ObjectRank of all Web pages and objects is given by the PageRank of their HTML pages on the Web.

Theorem 3. *Let $KB = (\emptyset, (\mathcal{A}_a)_{a \in \mathbf{P} \cup \mathbf{O}})$ be a Semantic Web knowledge base, let E be a source of rank, and let d be a damping factor. Let the directed graph $G_{KB} = (V, E)$ be defined by $V = \mathbf{P} \cup \mathbf{O}$ and $(p, q) \in E$ iff $P(p, q) \in \mathcal{A}_p$. Then, for every $p \in \mathbf{P} \cup \mathbf{O}$, the ObjectRank of p relative to KB is given by the PageRank of p relative to G_{KB}.*

7 Implementation and Experiments

In this section, we describe our prototype implementation for a semantic desktop search engine. Furthermore, we report on experimental results on the size of completed annotations, the running time of the online query processing step, and the precision and the recall of our approach to Semantic Web search compared to Google.

Implementation. We have implemented a prototype for a semantic desktop search engine (in desktop search, it was possible to quickly index 500,000 facts at once). The implementation is based on the above offline inference technique and a (simplified)

Table 1. Size of completed annotations

Ontology	Average Size of a Completed Annotation (bytes)
FSM	202
SWM	173
NTN	229
SCIENCE	146
FINANCIAL	142

desktop version of the above online Semantic Web search (by reduction to standard Web search). The former uses the deductive database system DLV [16], while the latter is written in Java (nearly 2 000 lines of code) and uses Microsoft Windows Desktop Search 3.0 (WDS) as external desktop search engine; in detail, it uses the search index created by WDS, which is queried by a cmdlet script in Microsoft Powershell 1.0.

Size of Completed Annotations. Since ontological hierarchies in practice are generally not that deep (a concept has at most a dozen superconcepts), the generated completed semantic annotations are generally only of a small size. To prove this experimentally, we have measured the sizes of the generated completed annotations for the FINITE-STATE-MACHINE (FSM), the SURFACE-WATER-MODEL (SWM), the NEW-TESTAMENT-NAMES (NTN), and the SCIENCE ontologies from the Protégé Ontology Library[3], and for the FINANCIAL ontology[4]. Indeed, the experimental results in Table 1 show that the average size of a completed annotation is rather small.

Efficiency of Online Query Processing. First experiments with our implemented semantic desktop search engine show the principle feasibility of our approach, and that it scales quite well to very large collections of standard pages, annotation pages, and background ontologies. The results are summarized in Table 2, which shows in bold the net time (in ms) used by our system (without the WDS calls) for processing six different search queries ($Q_1, \ldots, Q_6$) on four different randomly generated knowledge bases (in the context of the running Scientific Database), consisting of up to 5 000 annotations with up to 590 027 facts; the processing times for further search queries are given in the extended report [11]. Notice that this net system time (for the decomposition of the query and the composition of the query results) is very small (at most 6 seconds in the worst case). Table 2 also shows the time used for calling WDS for processing all subqueries, as well as the different numbers of returned pages and objects. The biggest part of the total running time is used for these WDS calls, which is due to the fact that our current implementation is based on a file interface to WDS. This time can be dramatically reduced by using an API. Further dramatic reductions (even with much larger datasets) can be achieved by employing a (more efficient) Web search engine (such as Google) rather than a desktop search engine, and by using an efficient relational query engine for constructing the overall query result.

[3] http://protegewiki.stanford.edu/index.php/Protege_Ontology_Library

[4] http://www.cs.put.poznan.pl/alawrynowicz/financial.owl

Table 2. WDS and system time (in ms) used for processing $Q_1, \ldots, Q_6$ on four different KBs

	625	1250	2500	5000	No. Annotations
	69283	142565	292559	590270	No. Facts
	9498	10002	10888	12732	WDS Time
Q_1	**218**	**479**	**944**	**1860**	**System Time**
	73	162	304	529	No. URIs
	10914	12176	14748	19847	WDS Time
Q_2	**490**	**904**	**1632**	**3154**	**System Time**
	95	228	420	679	No. URIs
	10331	11096	12446	15532	WDS Time
Q_3	**138**	**191**	**359**	**732**	**System Time**
	23	48	95	204	No. URIs
	24735	26580	30352	37857	WDS Time
Q_4	**748**	**1523**	**2996**	**5990**	**System Time**
	112	235	431	687	No. URIs
	4777	4882	4878	4920	WDS Time
Q_5	**9**	**30**	**45**	**59**	**System Time**
	1	8	14	20	No. URIs
	16892	19524	24798	34218	WDS Time
Q_6	**593**	**1179**	**2297**	**4753**	**System Time**
	53	225	431	687	No. URIs

The six search queries $Q_1, \ldots, Q_6$ are more concretely given as follows; they ask for all the following individuals (so also yielding the Web pages containing them):

(1) scientists working for u but not having a doctoral degree from u, or scientists having a doctoral degree from u but not working for u:

$$Q_1(x) = (\mathit{Scientist}(x) \wedge \mathit{worksFor}(x,u) \wedge \mathit{not}\ \mathit{doctoralDegree}(x,u)) \vee (\mathit{Scientist}(x) \wedge \mathit{doctoralDegree}(x,u) \wedge \mathit{not}\ \mathit{worksFor}(x,u))\,.$$

(2) professors who are also the head of a department:

$$Q_2(x) = \exists y\, (\mathit{Professor}(x) \wedge \mathit{headOf}(x,y) \wedge \mathit{Department}(y))\,.$$

(3) articles with an Italian author and published in 2007:

$$Q_3(x) = \exists y\, (\mathit{Article}(x) \wedge \mathit{yearOfPublication}(x, 2007) \wedge \mathit{hasWritten}(y,x) \wedge \mathit{Scientist}(y) \wedge \mathit{nationality}(y, \mathit{italian}))\,.$$

(4) scientists who are the authors of a journal and a conference paper published in 2007, or scientists who are the authors of a book published in 2007:

$$Q_4(x) = \exists y,z\, (\mathit{Scientist}(x) \wedge \mathit{hasWritten}(x,y) \wedge \mathit{JournalPaper}(y) \wedge \mathit{yearOfPublication}(y, 2007) \wedge \mathit{hasWritten}(x,z) \wedge \mathit{ConferencePaper}(z) \wedge \mathit{yearOfPublication}(z, 2007)) \vee \exists y\, (\mathit{Scientist}(x) \wedge \mathit{hasWritten}(x,y) \wedge \mathit{Book}(y) \wedge \mathit{yearOfPublication}(y, 2007))\,.$$

(5) Italian professors who are not heading any department:

$$Q_5(x) = \mathit{Professor}(x) \wedge \mathit{nationality}(x, \mathit{italian}) \wedge \mathit{not}\ \exists y\, (\mathit{headOf}(x,y) \wedge \mathit{Department}(y))\,.$$

(6) scientists who work for a university, but for no university from which they have the doctoral degree:

$$Q_6(x) = \exists z\,(Scientist(x) \wedge worksFor(x,z) \wedge University(z) \wedge \\ not\ \exists y\,(doctoralDegree(x,y) \wedge worksFor(x,y) \wedge University(y)))\,.$$

Precision and Recall of Semantic Web Search. Differently from conventional Boolean keyword-oriented Web search, the proposed Semantic Web search clearly empowers the user to precisely describe her information need for certain kinds of queries, resulting in a very precise result set and a very high precision and recall [2] for the query result. In particular, in many cases, Semantic Web search queries exactly describe the desired answer sets, resulting into a precision and a recall of 1. Some examples of such search queries (addressed to the CIA World Fact Book[5] relative to the WORLD-FACT-BOOK ontology[6]), which have a precision and a recall of 1 in our approach to Semantic Web search, are shown in Table 3, along with corresponding Google queries. For example, Query (1) asks for all countries having a common border with Austria, while Query (10) asks for all countries in which Arabic and not English is spoken. The corresponding Google queries, however, often cannot that precisely describe the desired answer sets, and are thus often resulting into a precision and a recall significantly below 1.

Table 3. Precision and recall of Google vs. Semantic Web search

	Semantic Web Search Query / Google Query	Results Google	Correct Results	Correct Results Google	Precision Google	Recall Google
1	$Country(x) \wedge borderCountries(x, Austria)$ / "border countries" Austria	17	8	8	0.47	1
2	$Country(x) \wedge exportsPartners(x, Bulgaria)$ / "exports - partners" Bulgaria	19	5	5	0.26	1
3	$Country(x) \wedge nationality(x, Italian)$ / nationality Italian	20	1	1	0.05	1
4	$Country(x) \wedge languages(x, Italian)$ / languages Italian	21	13	13	0.62	1
5	$Country(x) \wedge importsCommodities(x, tobacco)$ / "imports - commodities" tobacco	51	10	10	0.2	1
6	$Country(x) \wedge exportsCommodities(x, tobacco) \wedge languages(x, French)$ / "exports - commodities" tobacco languages French	24	4	4	0.17	1
7	$Country(x) \wedge importsCommodities(x, petroleum) \wedge government(x, monarchy)$ / "imports - commodities" petroleum government monarchy	30	6	6	0.2	1
8	$Country(x) \wedge not\ languages(x, Italian)$ / languages -Italian	229	253	229	1	0.91
9	$Country(x) \wedge languages(x, Arabic)$ / languages arabic	33	32	32	0.97	1
10	$Country(x) \wedge languages(x, Arabic) \wedge not\ languages(x, English)$ / languages arabic -English	11	13	11	1	0.85
11	$Country(x) \wedge importsCommodities(x, tobacco) \wedge importsCommodities(x, food)$ / "imports - commodities" food tobacco	45	7	7	0.16	1
12	$Country(x) \wedge importsCommodities(x, tobacco) \wedge not\ importsCommodities(x, food)$ / "imports - commodities" -food tobacco	6	3	1	0.17	0.33

8 Related Work

Related work on Semantic Web search (see especially [10] for a recent survey) can roughly be divided into (i) many approaches to search on the new representation formalisms for the Semantic Web, and (ii) some few approaches to search on the Web

[5] http://www.cia.gov/library/publications/the-world-factbook/

[6] http://www.ontoknowledge.org/oil/case-studies/

using Semantic Web data and knowledge. We briefly discuss the most closely related such approaches in this section. Clearly, our work has much different goals than all these approaches, as summarized below.

As for (i), the academic Semantic Web search engine Swoogle[7] [10] is among the earliest Semantic Web search engines. Swoogle is an indexing and retrieval system for the Semantic Web, which extracts metadata for each discovered resource in the Semantic Web and computes relations between the resources. The work also introduces an ontology rank as a measure of the importance of a resource. The following is a (non-exhaustive) list of some further existing or currently being developed Semantic Web search engines: Semantic Web Search Engine (SWSE)[8], Watson[9], Falcons[10], Semantic Web Search[11], Sindice[12], Yahoo! Microsearch [13], and Zitgist Search[14].

A representative of (ii) is the TAP system [12], which augments traditional keyword search by matching concepts. More concretely, in addition to traditional keyword search on a collection of resources, the keywords are matched against concepts in an RDF repository, and the matching concepts are then returned in addition to the located resources. Another representative of (ii) is the SemSearch system [15], which focuses especially on hiding the complexity of semantic search queries from end users.

9 Summary and Outlook

We have presented a novel approach to Semantic Web search, which allows for ontological conjunctive queries, combining standard Web search queries with ontological background knowledge. We have shown how the approach can be implemented on top of standard Web search engines and ontological inference technologies. We have developed the formal model behind this approach, and we have also generalized the PageRank technique to this approach. We have provided a technique for processing Semantic Web search queries, which consists of an offline ontological inference step and an online reduction to standard Web search queries (which can be implemented using efficient relational database technology), and we have proved it ontologically correct (and in many cases also ontologically complete). We have reported on a prototype implementation in desktop search, and provided very positive experimental results on the size of the completed semantic annotations, the running time of the online query processing step, and the precision and the recall of our approach to Semantic Web search.

In the future, we aim especially at extending the desktop implementation to a real Web implementation, using existing Web search engines. Another interesting topic is to explore how search expressions that are formulated as plain natural language sentences can be translated into the ontological conjunctive queries of our approach. It would also be interesting to investigate the use of probabilistic ontologies rather than classical ones.

[7] `http://swoogle.umbc.edu/`
[8] `http://swse.deri.org/`
[9] `http://watson.kmi.open.ac.uk/WatsonWUI/`
[10] `http://iws.seu.edu.cn/services/falcons/`
[11] `http://www.semanticwebsearch.com/query/`
[12] `http://www.sindice.com/`
[13] `http://www.yr-bcn.es/demos/microsearch/`
[14] `http://zitgist.com/`

Acknowledgments. G. Gottlob's work was supported by the EPSRC grant Number EP/E010865/1 "Schema Mappings and Automated Services for Data Integration." G. Gottlob, whose work was partially carried out at the Oxford-Man Institute of Quantitative Finance, gratefully acknowledges support from the Royal Society as the holder of a Royal Society-Wolfson Research Merit Award. T. Lukasiewicz's work was supported by the German Research Foundation (DFG) under the Heisenberg Programme.

References

[1] Baader, F., Calvanese, D., McGuinness, D.L., Nardi, D., Patel-Schneider, P.F. (eds.): The Description Logic Handbook. Cambridge University Press, Cambridge (2003)

[2] Baeza-Yates, R., Ribeiro-Neto, B.: Modern Information Retrieval. Addison-Wesley, Reading (1999)

[3] Bao, J., Kendall, E.F., McGuinness, D.L., Wallace, E.K.: OWL2 Web ontology language: Quick reference guide (2008), www.w3.org/TR/owl2-quick-reference/

[4] Berners-Lee, T.: Weaving the Web. Harper, San Francisco (1999)

[5] Berners-Lee, T., Hendler, J., Lassila, O.: The Semantic Web. Sci. Amer. 284, 34–43 (2001)

[6] Brin, S., Page, L.: The anatomy of a large-scale hypertextual web search engine. Comput. Netw. 30(1-7), 107–117 (1998)

[7] Calvanese, D., De Giacomo, G., Lembo, D., Lenzerini, M., Rosati, R.: Tractable reasoning and efficient query answering in description logics: The *DL-Lite* family. J. Autom. Reasoning 39(3), 385–429 (2007)

[8] Chirita, P.-A., Costache, S., Nejdl, W., Handschuh, S.: P-TAG: Large scale automatic generation of personalized annotation tags for the Web. In: Proceedings WWW 2007, pp. 845–854. ACM Press, New York (2007)

[9] Ding, L., Finin, T.: Characterizing the Semantic Web on the Web. In: Cruz, I., Decker, S., Allemang, D., Preist, C., Schwabe, D., Mika, P., Uschold, M., Aroyo, L.M. (eds.) ISWC 2006. LNCS, vol. 4273, pp. 242–257. Springer, Heidelberg (2006)

[10] Ding, L., Finin, T.W., Joshi, A., Peng, Y., Pan, R., Reddivari, P.: Search on the Semantic Web. IEEE Computer 38(10), 62–69 (2005)

[11] Fazzinga, B., Gianforme, G., Gottlob, G., Lukasiewicz, T.: From Web search to Semantic Web search. Technical Report INFSYS RR-1843-08-11, Institut für Informationssysteme, TU Wien (November 2008)

[12] Guha, R.V., McCool, R., Miller, E.: Semantic search. In: Proceedings WWW 2003, pp. 700–709. ACM Press, New York (2003)

[13] Horrocks, I., Patel-Schneider, P.F., van Harmelen, F.: From $\mathcal{SHIQ}$ and RDF to OWL: The making of a Web ontology language. J. Web Sem. 1(1), 7–26 (2003)

[14] Hustadt, U., Motik, B., Sattler, U.: Data complexity of reasoning in very expressive description logics. In: Proc. IJCAI 2005, pp. 466–471. Professional Book Center (2005)

[15] Lei, Y., Uren, V.S., Motta, E.: SemSearch: A search engine for the Semantic Web. In: Staab, S., Svátek, V. (eds.) EKAW 2006. LNCS (LNAI), vol. 4248, pp. 238–245. Springer, Heidelberg (2006)

[16] Leone, N., Pfeifer, G., Faber, W., Eiter, T., Gottlob, G., Perri, S., Scarcello, F.: The DLV system for knowledge representation and reasoning. ACM Trans. Comput. Log. 7(3), 499–562 (2006)

[17] Poggi, A., Lembo, D., Calvanese, D., De Giacomo, G., Lenzerini, M., Rosati, R.: Linking data to ontologies. J. Data Semantics 10, 133–173 (2008)

[18] W3C. OWL web ontology language overview, 2004. W3C Recommendation (February 10, 2004), www.w3.org/TR/2004/REC-owl-features-20040210/

Semantically Characterizing Collaborative Behavior in an Abstract Dialogue Framework*

M. Julieta Marcos, Marcelo A. Falappa, and Guillermo R. Simari

National Council of Scientific and Technical Research (CONICET)
Artificial Intelligence Research & Development Laboratory (LIDIA)
Universidad Nacional del Sur (UNS), Bahía Blanca, Argentina
{mjm,mfalappa,grs}@cs.uns.edu.ar

Abstract. A fundamental requirement of collaborative dialogue formal systems is ensuring both that all the relevant information will be exposed and also irrelevancies will be avoided. The challenge is to fulfill this requirement in the context of a distributed MAS where each agent is unaware of the private knowledge of the others. We argue that it is possible to give a general treatment to this problem in terms of *relevance notions*, and propose a partial solution which reduces the problem to that of finding adequate *potential relevance notions*. Specifically, we present in this work an *Abstract Dialogue Framework* which provides an environment for studying the behavior of collaborative dialogue systems in terms of abstract relevance notions, together with three *Collaborative Semantics* each of which defines a different collaborative behavior of the dialogues under the framework. One of these semantics describes an utopian, non practical, behavior which is approximated in different ways by the other two constructive semantics. Complete examples are provided in Propositional Logic Programming.

1 Introduction and Motivation

Multi-agent systems (MAS) provide solutions to problems in terms of autonomous interactive components (agents). A *dialogue* is a kind of interaction in which a sequence of messages, over the same topic, is exchanged among a group of agents, with the purpose of jointly drawing some sort of conclusion. This work is about modeling *collaborative* dialogues in MAS. By collaborative, we mean that the agents are willing to share any relevant knowledge, to the topic at issue. In order to design well-behaved models of dialogue, a formal specification of the expected behaviors would be useful. Besides, there is a need of practical behaviors, suitable to be implemented in a MAS (inherently distributed) where each agent has access only to the private knowledge of her own and to the public knowledge generated during the dialogue.

Most of the existent works in the area propose a formal system for some particular type of dialogue, based upon certain reasoning model (mostly argumentative systems) and identify properties of the generated dialogues, usually termination and properties of

* This research is partially supported by Sec. Gral. de Ciencia y Tecnología (Univ. Nac. del Sur), CONICET and Agencia Nac. de Prom. Científica y Técnica (ANPCyT).

S. Link and H. Prade (Eds.): FoIKS 2010, LNCS 5956, pp. 173–190, 2010.

the outcome, e. g. in [1], [2], [3], [4], [5]. We have observed that there are some desirable properties of these systems which are rarely satisfied. One such property is ensuring that, when the dialogue ends, there is no relevant information left unpublished, not even distributed among several participants. This property may not be easy to achieve if the underlying logic is complex. For example, argumentation-based dialogues usually consist of interchanging arguments *for* and *against* certain claim, but they do not consider other possible relevant contributions which are not necessarily arguments, such as parts of distributed arguments, or information which somehow invalidates a previously exposed argument (without being precisely a counter-argument), or information which changes the defeat relation between arguments, etc. In particular, in [5] this property is successfully achieved, but for a simplified version of a particular argumentative system. Another property which is in some cases overlooked, is ensuring that the final conclusion is coherent with all what has been said during the dialogue.

These observations motivated the present work, in which we intend to abstractly and formally specify the main requirements to be achieved by collaborative dialogue systems, as well as analyzing to what extent these can be fulfilled in a distributed environment where none of the participants has access to the entirety of the information. In this first approach, we will consider a restricted notion of collaborative dialogue which takes place among a fixed set of homogeneous agents equipped with finite and static knowledge bases expressed in a common knowledge representation language. The only possible move in the dialogue will be to make a contribution (to publish a subset of one's private knowledge base) and no other locution (such as questions, proposals, etc.) will be allowed. We will make no assumption regarding the nature of the underlying reasoning model, except for being a well defined function which computes a unique outcome, given a topic and a knowledge base. This will make our analysis suitable for a wide range of underlying logics, regardless whether they are monotonic or non-monotonic, and also including both argumentative and non-argumentative approaches. A preliminary reduced version of this work was presented in [6]. Here, important results and illustrative examples are added to the core framework.

2 Informal Requirements for Collaborative Dialogue Models

We take for granted that an ideal collaborative behavior of dialogues should satisfy the following, informally specified, requirements:

R_1: All the **relevant** information is exposed in the dialogue.
R_2: The exchange of **irrelevant** information is avoided.
R_3: The final conclusion **follows** from all what has been said.

On that basis, we will conduct our analysis of collaborative dialogue behavior in terms of two abstract elements: a *reasoning model* and a *relevance notion*[1], assuming that

[1] The term *relevance* appears in many research areas: *epistemology, belief revision, economics, information retrieval*, etc. In this work we intend to use it in its most general sense, which may be closer to the epistemic one: *pertinence in relation to a given question*, but it should not be tied to any particular interpretation, except for concrete examples given in this work.

the former gives a formal meaning to the word *follows*, and the latter to the word *relevant*. Both elements are domain-dependent and, as we shall see, they are not unattached concepts. It is important to mention that the relevance notion is assumed to work in a context of *complete information* (this will be clarified later). Also recall that our analysis will be intended to be suitable both for monotonic and non-monotonic logics.

We believe that the achievement of R_1-R_3 should lead to achieving other important requirements (listed below) and, hence, part of the contribution of this work will be to state the conditions under which this hypothesis actually holds.

R_4: The dialogue should always end.

R_5: Once the dialogue ends, if the agents added all their still private information, and reasoned from there, the previously drawn conclusions should not change.

In the task of simultaneously achieving requirements R_1 and R_2, in the context of a distributed MAS, a non-trivial problem arises: relevant information distributed in such a way that none of the parts is relevant by itself. A simple example illustrates this situation: suppose that A knows that a implies b, and also that c implies b, and B knows that a, as well as d, holds. If agents A and B engage in dialogue for determining whether b holds or not, then it is clear that the relevant information is: *a implies b*, and *a holds*. However, neither A knows that a holds, nor B knows that a implies b, making them unaware of the relevance of these pieces of information. It is true, though, that A could suspect the relevance of *a implies b* since the dialogue topic, b, is the consequent of the implication, but she has certainly no way of anticipating any difference between this and *c implies b*. This last means that, either she abstains from exposing any of the two implications (relegating R_1), or she tries with some or both of them (relegating R_2, in the case she chooses the wrong one first). In short, there is a tradeoff between requirements R_1 and R_2. Because of the nature of collaborative dialogues, we believe that R_1 may be mandatory in many application domains, and hence we will seek solutions which achieve it, even at the expense of relegating R_2 a bit. Although a concrete solution will depend on specific instances of the reasoning model and the relevance notion, we feel it is possible to analyze how could solutions be constructed for the abstract case. The basic idea will be to develop a new relevance notion (which will be called a *potential relevance notion*) able to detect parts of distributed relevant contributions (under the original notion). Furthermore, we will see how the concept of *abduction* in logic is related to the construction of this potential relevance notions.

The rest of this work is organized as follows. Sec. 3 introduces an *Abstract Dialogue Framework* useful for carrying out an abstract study of collaborative dialogues, and which includes the two elements mentioned above: the *reasoning model* and the *relevance notion*. In Sec. 4, we formalize requirements R_1-R_3 by defining an *Utopian Semantics* for the framework, and show why it is not in general implementable in a distributed MAS. In Sec. 5, we propose alternative, practical semantics which approximate the utopian behavior, by achieving one of the requirements, either R_1 or R_2, and relaxing the other. Examples throughout this work are given in Propositional Logic Programming, and its extension with Negation as Failure. In Sec. 6, we briefly discuss some complementary issues: an *interaction protocol* for allowing the practical semantics implementation, and some approaches for *handling inconsistencies* in dialogues. In Sec. 7, we comment on some existing works in the area. Finally in Sec. 8, we summarize the main contributions of this work, as well as pointing out some issues that have been left for further research in future work.

3 An Abstract Dialogue Framework

Three languages are assumed to be involved in a dialogue: the *Knowledge Representation Language* $\mathcal{L}$ for expressing the information exchanged by the agents, the *Topic Language* $\mathcal{L}_T$ for expressing the topic that gives rise to the dialogue, and the *Outcome Language* $\mathcal{L}_O$ for expressing the final conclusion (or *outcome*). These languages will be kept abstract in our formal definitions, but for the purpose of examples they will be instantiated in the context of Propositional Logic Programming (*PLP*) and its extension with Negation As Failure (PLP_{naf}). It is also assumed a language $\mathcal{L}_I$ for agent identifiers. As mentioned in Sec. 1, we consider a restricted notion of dialogue which is *based on contributions only*. The following is a *public view of dialogue*: agents' private knowledge is not taken into account.

Definition 1 (Move). *A* move *is a pair* $\langle id, \mathrm{X}\rangle$ *where* $id \in \mathcal{L}_I$ *is the identifier of the* speaker, *and* $\mathrm{X} \subseteq \mathcal{L}$ *is her* contribution.

Definition 2 (Dialogue). *A* dialogue *is a tuple* $\langle t, \langle m_j\rangle, o\rangle$ *where* $t \in \mathcal{L}_T$ *is the* dialogue topic, $\langle m_j\rangle$ *is a sequence of moves, and* $o \in \mathcal{L}_O$ *is the* dialogue outcome.

As anticipated in Sec. 2, we will study the behavior of dialogues in terms of two abstract concepts: *relevance* and *reasoning*. To that end, an *Abstract Dialogue Framework* is introduced, whose aim is to provide an environment under which dialogues take place. This framework includes: the languages involved in the dialogue, a set of participating *agents*, a *relevance notion* and a *reasoning model*. An *agent* is represented by a pair consisting of an agent identifier and a private knowledge base, providing in this way a *complete view* of dialogues. A *relevance notion*, in this article, is a criterion for determining, given certain already known information and a topic, whether it would be relevant to add certain other information (*i.e.*, to make a contribution). We emphasize that this criterion works under an assumption of *complete information*, to be contrasted with the situation of a dialogue where each agent is unaware of the private knowledge of the others. This issue will be revisited in Sec. 4. Finally, a *reasoning model* will be understood as a mechanism for drawing a conclusion about a topic, on the basis of an individual knowledge base.

Definition 3 (Agent). *An* agent *is a pair* $\langle id, \mathrm{K}\rangle$, *noted* K_{id}, *where* $\mathrm{K} \subseteq \mathcal{L}$ *is a private finite knowledge base, and* $id \in \mathcal{L}_I$ *is an agent identifier.*

Definition 4 (Abstract Dialogue Framework). *An* abstract dialogue framework $\mathfrak{F}$ *is a tuple* $\langle \mathcal{L}, \mathcal{L}_T, \mathcal{L}_O, \mathcal{L}_I, \mathcal{R}, \Phi, \mathrm{Ag}\rangle$ *where* $\mathcal{L}$, $\mathcal{L}_T$, $\mathcal{L}_O$ *and* $\mathcal{L}_I$ *are the languages involved in the dialogue,* Ag *is a finite set of agents,* $\mathcal{R} \subseteq 2^{\mathcal{L}} \times 2^{\mathcal{L}} \times \mathcal{L}_T$ *is a* relevance notion, *and* $\Phi : 2^{\mathcal{L}} \times \mathcal{L}_T \Rightarrow \mathcal{L}_O$ *is a* reasoning model. *The brief notation* $\mathfrak{F} = \langle \mathcal{R}, \Phi, \mathrm{Ag}\rangle$ *will be also used.*

Notation 1. *If* $(\mathrm{X}, \mathrm{S}, t) \in \mathcal{R}$, *we say that* X *is a* t-relevant contribution to S under $\mathcal{R}$, *and we note it* $\mathrm{X}\mathcal{R}_t\mathrm{S}$. *When it is clear what relevance notion is being used, we just say that* X *is a t-relevant contribution to* S. *For individual sentences* α *in* $\mathcal{L}$, *we also use the simpler notation* $\alpha\mathcal{R}_t\mathrm{S}$ *meaning that* $\{\alpha\}\mathcal{R}_t\mathrm{S}$.

Throughout this work we will make reference to the following partially instantiated dialogue frameworks. It is assumed that the reader is familiarized with the concept of derivation in *PLP* (noted $\vdash$) and PLP_{naf} (noted $\vdash_{naf}$)[2].

- $\mathfrak{F}^{lp} = \langle \mathcal{L}^{lp}, \mathcal{L}_{Facts}, \{\text{Yes}, \text{No}\}, \mathcal{L}_I, \mathcal{R}_t, \Phi^{lp}, \text{Ag}\rangle$ where $\mathcal{L}^{lp}$ is the set of rules and facts in *PLP*, $\mathcal{L}_{Facts} \subset \mathcal{L}^{lp}$ is the subset of facts (which in this case works as the Topic Language) and $\Phi^{lp}(\mathrm{S}, h) = \text{Yes}$ if $\mathrm{S} \vdash h$, and No otherwise.
- $\mathfrak{F}^{naf} = \langle \mathcal{L}^{naf}, \mathcal{L}_{Facts}, \{\text{Yes}, \text{No}\}, \mathcal{L}_I, \mathcal{R}_t, \Phi^{naf}, \text{Ag}\rangle$ where $\mathcal{L}^{naf}$ is the set of rules and facts in PLP_{naf} and $\Phi^{naf}(\mathrm{S}, h) = \text{Yes}$ if $\mathrm{S} \vdash_{naf} h$, and No otherwise.

Notice the existence of two different sets of knowledge involved in a dialogue: the *private knowledge* which is the union of the agents' knowledge bases, and the *public knowledge* which is the union of all the contributions already made, up to certain step. The former is a static set, whereas the latter grows as the dialogue progresses.

Definition 5 (Public Knowledge). *Let d be a dialogue consisting of a sequence* $\langle\langle id_1, \mathrm{X}_1\rangle \dots \langle id_m, \mathrm{X}_m\rangle\rangle$ *of moves. The* public knowledge *associated to d at step* j *(*$j \leq m$*) is the union of the first* j *contributions of the sequence and is noted* $\mathbf{PU}_d^j$ *(*$\mathbf{PU}_d^j = \mathrm{X}_1 \cup \dots \cup \mathrm{X}_j$*).*

Definition 6 (Private Knowledge). *Let* $\mathfrak{F}$ *be an abstract dialogue framework including a set* Ag *of agents. The* private knowledge *associated to* $\mathfrak{F}$ *(and to any admissible dialogue under* $\mathfrak{F}$*) is the union of the knowledge bases of the agents in* Ag, *and is noted* $\mathbf{PR}_{\mathfrak{F}}$ *(*$\mathbf{PR}_{\mathfrak{F}} = \bigcup_{\mathrm{K}_{id} \in \text{Ag}} \mathrm{K}$*).*

In our restricted notion of dialogue, agents' contributions are subsets of their private knowledge. We define next a set of *admissible dialogues* under a given framework.

Definition 7 (Admissible Dialogues). *Let* $\mathfrak{F} = \langle \mathcal{L}, \mathcal{L}_T, \mathcal{L}_O, \mathcal{L}_I, \mathcal{R}_t, \Phi, \text{Ag}\rangle$ *be an abstract dialogue framework,* $t \in \mathcal{L}_T$ *and* $o \in \mathcal{L}_O$. *A dialogue* $\langle t, \langle m_j\rangle, o\rangle$ *is* admissible *under* $\mathfrak{F}$ *if, and only if, for each move* $m = \langle id, \mathrm{X}\rangle$ *in the sequence, there is an agent* $\mathrm{K}_{id} \in \text{Ag}$ *such that* $\mathrm{X} \subseteq \mathrm{K}$. *The set of admissible dialogues under* $\mathfrak{F}$ *is noted* $d(\mathfrak{F})$.

Remark 1. For any step j of any dialogue $\mathrm{d} \in \mathrm{d}(\mathfrak{F})$, it holds that $\mathbf{PU}_{\mathrm{d}}^j \subseteq \mathbf{PR}_{\mathfrak{F}}$.

Returning to the notions of relevance and reasoning, it was mentioned in Sec. 2 that these were not unattached concepts. A coherent dialogue must exhibit some connection between them. A natural connection is to consider that a contribution is relevant if its addition alters the conclusion achieved by the reasoning model, as defined below.

---

[2] We will assume that *PLP* contains the following binary connectives: $\wedge$ and $\leftarrow$, and PLP_{naf} also includes the unary prefix connective **not** . Examples in PLP_{naf} will have exactly one stable expansion, so there will be no confusion regarding their semantics.

Definition 8 (Natural Relevance Notion). *Let Φ be a reasoning model. The* natural relevance notion *associated to Φ is a relation $\mathcal{N}_t^{\Phi}$ such that:* $\mathrm{X}\mathcal{N}_t^{\Phi}\mathrm{S}$ *iff* $\Phi(\mathrm{S}, t) \neq \Phi(\mathrm{S} \cup \mathrm{X}, t)$. *If* $\mathrm{X}\mathcal{N}_t^{\Phi}\mathrm{S}$, *we say that* X *is a* natural t-relevant contribution to S under Φ.

It will be seen later that this connection can be relaxed, *i.e.*, other relevance notions which are not *exactly* the natural one, will also be accepted. We distinguish the subclass of abstract dialogue frameworks in which the relevance notion is the natural one associated to the reasoning model. We refer to them as *Inquiry Dialogue Frameworks*[3], and the relevance notion is omitted in their formal specification.

Definition 9 (Inquiry Dialogue Framework). *An abstract dialogue framework* $\mathfrak{I} = \langle \mathcal{R}_t, \Phi, \mathrm{Ag} \rangle$ *is an* Inquiry Dialogue Framework *if, and only if, it holds that* $\mathcal{R}_t = \mathcal{N}_t^{\Phi}$. *The brief notation* $\mathfrak{I} = \langle \Phi, \mathrm{Ag} \rangle$ *will be used.*

Throughout this work we will make reference to the natural relevance notions $\mathcal{N}_h^{lp}$ and $\mathcal{N}_h^{naf}$, associated to the reasoning models Φ^{lp} and Φ^{naf}, and also to the inquiry frameworks $\mathfrak{I}^{lp}$ and $\mathfrak{I}^{naf}$, which result from $\mathfrak{F}^{lp}$ and $\mathfrak{F}^{naf}$, by instantiating the abstract relevance notions with the natural ones.

4 Utopian Collaborative Semantics

A *semantics* for an abstract dialogue framework, in this work, is a subset of the admissible dialogues, whose elements satisfy certain properties, representing a particular *dialogue behavior*. We are interested in specifying which, from all the admissible dialogues under a given framework, have an acceptable *collaborative behavior*. Recall that by *collaborative* we mean that the participants are willing to share any relevant knowledge to the topic under discussion, having no other ambition than achieving the right conclusion on the basis of all the information they have. In Sec. 2, we identified three requirements, R_1-R_3, to be ideally achieved by collaborative dialogue systems. In this section, we will define an *Utopian Collaborative Semantics* which gives a formal characterization of such ideal behavior, in terms of the elements of the framework.

In order to translate requirements R_1-R_3 into a formal specification, some issues need to be considered first. In particular, the notion of *relevant contribution* needs to be adjusted. On the one hand, there may be contributions which does not qualify as relevant but it would be adequate to allow. To understand this, it should be noticed that, since relevance notions are related to reasoning models, and reasoning models may be nonmonotonic, then it is possible for a contribution to contain a relevant subset, without being relevant itself. Consider, for instance, the following set in the context of the $\mathfrak{I}^{naf}$ framework: $\{\, a \leftarrow \mathbf{not}\ c \,\}$, which is a natural a-relevant contribution to the empty set, but if we added the fact c, or the rule $c \leftarrow \mathbf{not}\ d$, then it would not. The possibility of

[3] The term *Inquiry* is inspired on the popularized typology of dialogues proposed in [7], since we believe that the natural relevance notion captures the essence of this type of interaction: *collaboration to answer some question*. However, the term will be used in a broader sense here, since nothing is assumed regarding the degree of knowledge of the participants.

some other agent knowing, for instance, that d holds, explains why it would be useful to allow the whole contribution $\{ a \leftarrow \textbf{not}\ c, c \leftarrow \textbf{not}\ d \}$. In these cases, we say that the relevance notion fails to satisfy *left-monotonicity*[4] and that the whole contribution is *weakly relevant*[5]. The formal definitions are given below.

Definition 10 (Left Monotonicity). *Let $\mathcal{R}_t$ be a relevance notion. We say that $\mathcal{R}_t$ satisfies* left monotonicity *iff the following condition holds: if* $\mathrm{X}\mathcal{R}_t\mathrm{S}$ *and* $\mathrm{X} \subseteq \mathrm{Y}$ *then* $\mathrm{Y}\mathcal{R}_t\mathrm{S}$.

Definition 11 (Weak Contribution). *Let $\mathcal{R}_t$ be a relevance notion. We say that* X *is a* weak t-relevant contribution *to* S *iff there exists* $\mathrm{Y} \subseteq \mathrm{X}$ *such that* $\mathrm{Y}\mathcal{R}_t\mathrm{S}$.

Proposition 1. *Let $\mathcal{R}_t$ be a relevance notion that satisfies left monotonicity. Then,* X *is a t-relevant contribution to* S *iff* X *is a weak t-relevant contribution to* S.

On the other hand, there may be contributions which qualify as relevant but they are not *purely* relevant. Consider, for example, the following set in the context of any of the two instantiated inquiry frameworks: $\{ a \leftarrow b, b, e \}$, which is a natural a-relevant contribution to the empty set, although the fact e is clearly irrelevant. These impure relevant contributions must be avoided in order to obey requirement R_2. For that purpose, *pure relevant contributions* impose a restriction over weak relevant ones, disallowing absolutely irrelevant sentences within them, as defined below.

Definition 12 (Pure Contribution). *Let $\mathcal{R}_t$ be a relevance notion, and* X *a weak t-relevant contribution to* S. *We say that* X *is a* pure t-relevant contribution *to* S *iff the following condition holds for all $\alpha \in \mathrm{X}$: there exists $\mathrm{Y} \subset \mathrm{X}$ such that $\alpha\mathcal{R}_t(\mathrm{S} \cup \mathrm{Y})$.*

Finally, it has been mentioned that the relevance notion works under an assumption of *complete information*, and thus it will be necessary to inspect the private knowledge of the others for determining the actual relevance of a given move. Now we are able to give a formal interpretation of requirements R_1-R_3 in terms of the framework elements:

Definition 13 (Utopian Collaborative Semantics). *Let $\mathfrak{F} = \langle \mathcal{R}_t, \Phi, \mathrm{Ag} \rangle$ be an abstract dialogue framework. A dialogue $d = \langle t, \langle m_j \rangle, o \rangle \in d(\mathfrak{F})$ belongs to the* Utopian Collaborative Semantics *for $\mathfrak{F}$ (noted Utopian($\mathfrak{F}$)) if, and only if:*

Correctness: *if m_j is the last move in the sequence, then $\Phi(\mathbf{PU}_d^j, t) = o$.*

Global Progress: *for each move $m_j = \langle id_j, \mathrm{X}_j \rangle$ in the sequence, there exists $\mathrm{Y} \subseteq \mathbf{PR}_{\mathfrak{F}}$ such that $\mathrm{X}_j \subseteq \mathrm{Y}$ and Y is a pure t-relevant contribution to $\mathbf{PU}_d^{j-1}$.*

Global Completeness: *if m_j is the last move in the sequence, then $\mathbf{PR}_{\mathfrak{F}}$ is not a weak t-relevant contribution to $\mathbf{PU}_d^j$.*

Requirement R_3 is achieved by the *Correctness* condition, which states that the dialogue outcome coincides with the application of the reasoning model to the public knowledge at the final step of the dialogue (*i.e.*, the outcome of the dialogue can be obtained by reasoning from all that has been said). Requirement R_2 is achieved by the *Global Progress*

[4] The name of this property is inspired in [8].

[5] The term *weak relevance* is used in [9] in a different sense, which should not be related to the one introduced here.

condition, which states that each move in the sequence is part of a distributed pure relevant contribution to the public knowledge generated so far. Finally, requirement R_1 is achieved by the *Global Completeness* condition, which states that there are no more relevant contributions, not even distributed among different knowledge bases, after the dialogue ends. Notice that the three conditions are simultaneously satisfiable by any dialogue framework and topic, *i.e.,* there always exists at least one dialogue which belongs to this semantics, as stated in the following proposition.

Proposition 2 (Satisfiability). *For any dialogue framework* $\mathfrak{F} = \langle \mathcal{R}_t, \Phi, \mathrm{Ag} \rangle$, *the set Utopian*($\mathfrak{F}$) *contains at least one element.*

Furthermore, any sequence of moves satisfying *global progress* can be completed to a dialogue belonging to the semantics. This means that a system implementation under this semantics would not need to do *backtracking*. Although this property is useless for the case of the utopian semantics which, as will be seen in short, is not implementable in a distributed system, it will be useful in the case of the two practical semantics that will be presented in Sec. 5.

Definition 14. *A dialogue* d_2 *over a topic* t *is a* continuation *of a dialogue* d_1 *over the same topic* t *if, and only if, the sequence of moves of* d_2 *can be obtained by adding zero or more elements to the sequence of moves of* d_1.

Proposition 3 (No Backtracking). *Let* $\mathfrak{F} = \langle \mathcal{R}_t, \Phi, \mathrm{Ag} \rangle$ *be an abstract dialogue framework, and* $d_1 \in d(\mathfrak{F})$. *If* d_1 *satisfies* global progress *under* $\mathfrak{F}$, *then there exists a dialogue* $d_2 \in$ *Utopian*($\mathfrak{F}$) *which is a* continuation *of* d_1.

Note that the truth of the previous statements (regarding *satisfiability* and *no backtracking*) comes from the following facts, which can be easily proven: (1) if *global completeness* is not achieved, then there exists at least one possible move that can be added to the sequence according to *global progress*, and (2) the *correctness* condition is orthogonal to the other two. Next, an illustrative example of the dialogues generated under the Utopian Semantics is given.

Example 1. *Consider an instance of the* $\mathfrak{I}^{lp}$ *framework, where the set* Ag *is composed by* $\mathrm{K_A} = \{a \leftarrow b, e\}$, $\mathrm{K_B} = \{b \leftarrow c, b \leftarrow d, f\}$ *and* $\mathrm{K_C} = \{c, g\}$. *The dialogue* d_1 *shown on the right, over topic* a, *and also all the permutations of its moves with the same topic and outcome, belong to the Utopian Collaborative Semantics for the framework. The chart traces the dialogue, showing the partial results of reasoning from the public knowledge so far generated. The last of these results (underlined) is the final dialogue outcome.*

step	A	B	C	$\Phi(\mathbf{PU}_{d_1}^{\mathrm{step}}, a)$
1	$a \leftarrow b$			No
2		$b \leftarrow c$		No
3			c	<u>Yes</u>

An essential requirement of dialogue systems is ensuring the termination of the generated dialogues. This is intuitively related to requirement R_2 (achieved by *global progress*) since it is expected that agents will eventually run out of relevant contributions, given that their private knowledge bases are finite. This is actually true as long as the relevance notion satisfies an intuitive property, defined below, which states that a relevant contribution must add some new information to the public knowledge.

Definition 15 (Novelty). *A relevance notion $\mathcal{R}_t$ satisfies* novelty *iff the following condition holds: if* $x\mathcal{R}_t s$ *then* $x \not\subseteq s$.

Proposition 4 (Termination). *Let $\mathfrak{F} = \langle \mathcal{R}_t, \Phi, \mathrm{Ag} \rangle$ be an abstract dialogue framework, and $d = \langle t, \langle m_j \rangle, o \rangle \in d(\mathfrak{F})$. If the notion $\mathcal{R}_t$ satisfies* novelty *and dialogue d satisfies* global progress *under $\mathfrak{F}$, then $\langle m_j \rangle$ is a finite sequence of moves.*

It is easy to see that any natural relevance notion satisfies novelty, since it is not possible for the conclusion achieved by the reasoning model to change without changing the topic nor the knowledge base.

Proposition 5. *For any reasoning model Φ, it holds that its associated natural relevance notion, $\mathcal{N}_t^{\Phi}$, satisfies novelty.*

Another desirable property of collaborative dialogue models is ensuring it is not possible to draw different conclusions, for the same set of agents and topic. In other words, from the entirety of the information, it should be possible to determine the outcome of the dialogue, no matter what sequence of steps are actually performed[6]. Furthermore, this outcome should coincide with the result of applying the reasoning model to the private knowledge involved in the dialogue. We emphasize that this is required for *collaborative* dialogues (and probably not for non-collaborative ones). For instance, in Ex. 1, the conclusion achieved by all the possible dialogues under the semantics is Yes, which is also the result of reasoning from $K_A \cup K_B \cup K_C$. This is intuitively related to requirements R_1 (achieved by *global completeness*) and R_3 (achieved by *correctness*) since it is expected that the absence of relevant contributions implies that the current conclusion cannot be changed by adding more information. This is actually true as long as the relevance notion is the natural one associated to the reasoning model, or a *weaker* one, as stated below.

Definition 16 (Stronger Relevance Notion). *Let $\mathcal{R}_t$ and $\mathcal{R}'_t$ be two relevance notions. We say that the notion $\mathcal{R}_t$ is* stronger or equal *than the notion $\mathcal{R}'_t$ iff the following holds: if $x\mathcal{R}_t s$ then $x\mathcal{R}'_t s$ (i.e., $\mathcal{R}_t \subseteq \mathcal{R}'_t$). We will also say that $\mathcal{R}'_t$ is* weaker or equal *than $\mathcal{R}_t$.*

Observe that here we use the term *weaker*, as the opposite of *stronger*, denoting a binary relation between relevance notions, and this should not be confused with its previous use in Def. 11 of *weak relevant contribution*.

Proposition 6 (Outcome Determinism). *Let $\mathfrak{F} = \langle \mathcal{R}_t, \Phi, \mathrm{Ag} \rangle$ be an abstract dialogue framework and $d = \langle t, \langle m_j \rangle, o \rangle \in d(\mathfrak{F})$. If d satisfies* correctness *and* global completeness *under $\mathfrak{F}$, and $\mathcal{R}_t$* is weaker or equal than *$\mathcal{N}_t^{\Phi}$, then $o = \Phi(\mathbf{PR}_{\mathfrak{F}}, t)$.*

For example, in *PLP*, a relevance notion which detects the generation of new derivations for a given literal, would be *weaker* than the natural one. It is easy to see that this weaker relevance notion would also achieve *outcome determinism*.

The following corollaries summarize the results regarding the Utopian Collaborative Semantics for an abstract dialogue framework, and also for the particular case of inquiry dialogue frameworks.

[6] This property, which we will call *outcome determinism*, has been studied in various works under different names. For instance in [10] it was called *completeness*. Notice that we use that term for another property, which is not the same but is related to the one under discussion.

Corollary 1. *Let $\mathfrak{F} = \langle \mathcal{R}_t, \Phi, \text{Ag}\rangle$ be an abstract dialogue framework. The dialogues in Utopian($\mathfrak{F}$) satisfy* termination *and* outcome determinism, *provided that the relevance notion $\mathcal{R}_t$ satisfies* novelty *and is* weaker or equal *than $\mathcal{N}_t^{\Phi}$.*

Corollary 2. *Let $\mathfrak{I}$ be an inquiry framework. The dialogues in Utopian($\mathfrak{I}$) satisfy* termination *and* outcome determinism.

It is clear that Def. 13 of the Utopian Collaborative Semantics is not constructive, since both *global progress* and *global completeness* are expressed in terms of the private knowledge $\mathbf{PR}_{\mathfrak{F}}$, which is not entirely available to any of the participants. The following example shows that, it is not only not constructive, but also in many cases not even implementable in a distributed MAS.

Example 2. *Consider the $\mathfrak{I}^{lp}$ framework instantiated in Ex. 1. The dialogue d_2 shown on the right, does not belong to the Utopian Semantics because step 2 violates the* global progress *condition. However, it would not be possible to design a dialogue system which allows dialogue d_1 (presented in Ex. 1) but disallows d_2, since agent* B *can not know in advance that* c*, rather than* d*, holds.*

step	A	B	C	$\Phi(\mathbf{PU}_{d_2}^{step}, a)$
1	$a \leftarrow b$			No
2		$b \leftarrow d$		No
3		$b \leftarrow c$		No
4			c	Yes

The undesired situation is caused by a relevant contribution distributed among several agents, in such a way that none of the parts is relevant by itself, leading to a tradeoff between requirements R_1 and R_2 (*i.e.*, between *global progress* and *global completeness*). In the worst case, each sentence of the contribution resides in a different agent. Thus, to avoid such situations, it would be necessary for the relevance notion to warrant that every relevant contribution contains at least one individually relevant sentence. When this happens, we say that the relevance notion satisfies *granularity*, defined next.

Definition 17 (Granularity). *Let $\mathcal{R}_t$ be a relevance notion. We say that $\mathcal{R}_t$ satisfies* granularity *iff the following holds: if $\text{X}\mathcal{R}_t\text{S}$ then there exists $\alpha \in \text{X}$ such that $\alpha\mathcal{R}_t\text{S}$.*

Unfortunately, the relevance notions we are interested in, fail to satisfy granularity. It does not hold in general for the natural notions associated to deductive inference mechanisms. It has been shown, in Ex. 2, that it does not hold for the simple case of *PLP*, and clearly neither for PLP_{naf}. Just as an example, we will show a relevance notion (not logic-based) which satisfies granularity.

Example 3. *Suppose a set of items, each of which an agent may have associated incomes and/or expenses to, and suppose a dialogue among a set of agents with the purpose of determining the final balance on a certain item. The $\mathcal{L}$ language is the set of pairs* (item, amount) *where* amount *is a non-zero integer, the $\mathcal{L}_T$ language is the set of items, and the $\mathcal{L}_O$ language is the set of integers (including zero). The reasoning model obtains the final balance on a certain item, i.e., does the sum of all the amounts associated to this item. It is easy to see that the natural relevance notion corresponding to this reasoning model satisfies* granularity. *In fact, any particular income or expense associated to that item (these were assumed non-zero values) would be a natural relevant contribution to any given set.*

5 Practical Collaborative Semantics

The lack of granularity of relevance notions motivates the definition of alternative semantics which approach the utopian one, and whose distributed implementation is viable. The simplest approach is to relax requirement R_1 by allowing distributed relevant contributions to be missed, as follows.

Definition 18 (Basic Collaborative Semantics). *Let $\mathfrak{F} = \langle \mathcal{R}_t, \Phi, \mathrm{Ag}\rangle$ be an abstract dialogue framework. A dialogue $d = \langle t, \langle m_j\rangle, o\rangle \in d(\mathfrak{F})$ belongs to the* Basic Collaborative Semantics *for $\mathfrak{F}$ (noted Basic($\mathfrak{F}$)) if, and only if, the following conditions, as well as* **Correctness** *(Def. 13), hold:*

Local Progress: *for each move $m_j = \langle id_j, \mathrm{X}_j\rangle$ in the sequence, X_j is a pure t-relevant contribution to $\mathbf{PU}_d^{j-1}$.*

Local Completeness: *if m_j is the last move in the sequence, then it does not exist an agent $\mathrm{K}_{id} \in \mathrm{Ag}$ such that K is a weak t-relevant contribution to $\mathbf{PU}_d^j$.*

In the above definition, requirement R_2 is achieved by the *local progress* condition which states that each move in the sequence constitutes a pure relevant contribution to the public knowledge generated so far. Notice that this condition implies global progress (enunciated in Sec. 4), as stated below.

Proposition 7. *Let $\mathfrak{F} = \langle \mathcal{R}_t, \Phi, \mathrm{Ag}\rangle$ be an abstract dialogue framework, and $d \in d(\mathfrak{F})$. If the dialogue d satisfies* local progress, *then it satisfies* global progress *under $\mathfrak{F}$.*

Requirement R_1 is now compromised. The *local completeness* condition states that each agent has no more relevant contributions to make after the dialogue ends. Unless the relevance notion satisfies granularity, this is not enough for ensuring global completeness (enunciated in Sec. 4), since there could be a relevant contribution distributed among several agents, in such a way that none of the parts is relevant by itself.

Proposition 8. *Let $\mathfrak{F} = \langle \mathcal{R}_t, \Phi, \mathrm{Ag}\rangle$ be an abstract dialogue framework, and $d \in d(\mathfrak{F})$. If the dialogue d satisfies* global completeness, *then it satisfies* local completeness *under $\mathfrak{F}$. The reciprocal holds if, and only if, the relevance notion $\mathcal{R}_t$ satisfies* granularity.

As a result, requirement R_4 (termination) is achieved, given the same condition as in Sec. 4, whereas requirement R_5 (outcome determinism) cannot be warranted. These results are summarized in the corollary below.

Corollary 3. *Let $\mathfrak{F} = \langle \mathcal{R}_t, \Phi, \mathrm{Ag}\rangle$ be an abstract dialogue framework. The dialogues in Basic($\mathfrak{F}$) satisfy* termination, *provided that the relevance notion $\mathcal{R}_t$ satisfies* novelty.

Considering the same scenario as in Ex. 1 for the $\mathfrak{I}^{lp}$ framework, it is easy to see that the only possible dialogue under the Basic Collaborative Semantics is the empty one (*i.e.,* no moves are performed), with outcome = No. Furthermore, it can be shown that any dialogue in Basic($\mathfrak{I}^{lp}$) consists in zero moves, or one move (when at least one of the participants already has a derivation for the literal at issue). More interesting examples, of dialogues with more than one step under the Basic Semantics, can be developed in the context of non-monotonic reasoning models, as shown in the following example in the context of the $\mathfrak{I}^{naf}$ framework.

Example 4. *Consider an instance of the $\mathfrak{I}^{naf}$ inquiry framework, where the set* Ag *is composed by* $K_A = \{a \leftarrow b \wedge \textbf{not } c,\ b\}$, $K_B = \{d \leftarrow \textbf{not } e,\ f\}$, *and* $K_C = \{c \leftarrow \textbf{not } d, e \leftarrow f\}$. *The dialogue traced on the right, over topic a, belongs to the Basic Semantics for the framework instantiated above. Note that* global completeness *is not achieved, since there still exists a distributed relevant contribution when the dialogue ends:* $\{e \leftarrow f, f\}$. *Consequently, outcome determinism is not achieved: the dialogue outcome is* Yes *whereas the result of reasoning from* $K_A \cup K_B \cup K_C$ *is* No.

step	A	B	C	$\Phi(\mathbf{PU}_d^{step}, a)$
1	$a \leftarrow b \wedge \textbf{not } c$ b			Yes
2			$c \leftarrow \textbf{not } d$	No
3		$d \leftarrow \textbf{not } e$		Yes

In Sec. 2 we argued that requirement R_1 may be mandatory in many domains, but the Basic Semantics does not achieve it unless the relevance notion satisfies granularity, which does not usually happen. In order to make up for this lack of granularity, we propose to build a new notion (say $\mathcal{P}$) based on the original one (say $\mathcal{R}$) which ensures that, in the presence of a distributed relevant contribution under $\mathcal{R}$, *at least one* of the parts will be relevant under $\mathcal{P}$. We will say that $\mathcal{P}$ is a *potential relevance notion for* $\mathcal{R}$, since its aim is to detect contributions that could be relevant within certain *context*, but it is uncertain whether that context actually exists or not. Observe that the *context* is given by other agents' private knowledge, which has not been exposed yet. Just for clarifying the idea, a simple example (not logic-based) is showed next.

Example 5. *Consider the scenario of Ex. 3, but suppose now that the agents only need to know whether the balance on certain item is non-negative, so the reasoning model answers* Yes *if the sum is positive or zero, and* No *otherwise. In this case, the associated natural relevance notion does not satisfy granularity. A simple potential relevance notion would consider the incomes to be relevant when the current balance is negative, and the expenses to be relevant when the current balance is positive or zero.*

Below we define the binary relation (*"is a potential for"*) between relevance notions, and also its propagation to dialogue frameworks.

Definition 19 (Potential Relevance Notion). *Let $\mathcal{R}_t$ and $\mathcal{P}_t$ be relevance notions. We say that $\mathcal{P}_t$ is a* potential (relevance notion) for $\mathcal{R}_t$ *iff the following conditions hold: (1)* $\mathcal{R}_t$ is stronger or equal *than $\mathcal{P}_t$, and (2) if* $X\mathcal{R}_t S$ *then there exists* $\alpha \in X$ *such that* $\alpha\mathcal{P}_t S$. *If* $X\mathcal{P}_t S$ *and $\mathcal{P}_t$ is a potential for $\mathcal{R}_t$, we say that* X *is a* potential t-relevant contribution to S under $\mathcal{R}_t$.

Definition 20 (Potential Dialogue Framework). *Let* $\mathfrak{F} = \langle \mathcal{R}_t, \Phi, \mathrm{Ag}\rangle$ *and* $\mathfrak{F}^* = \langle \mathcal{P}_t, \Phi, \mathrm{Ag}\rangle$ *be abstract dialogue frameworks. We say that $\mathfrak{F}^*$ is a* potential (framework) *for $\mathfrak{F}$ if, and only if, $\mathcal{P}_t$ is a potential relevance notion for $\mathcal{R}_t$.*

Clearly, if a relevance notion already satisfies granularity then nothing needs to be done. Indeed, it would work as a potential relevance notion for itself, as stated in Prop. 9. Another useful property, stated in Prop. 10, is that if a relevance notion satisfies granularity and is weaker or equal than the original one, then it works as a potential for the latter.

Proposition 9. *If the relevance notion $\mathcal{R}_t$ satisfies* granularity, *then $\mathcal{R}_t$ is a potential relevance notion for itself.*

Proposition 10. *If the relevance notion $\mathcal{R}_t$ is* stronger or equal *than the relevance notion $\mathcal{P}_t$, and $\mathcal{P}_t$ satisfies* granularity, *then $\mathcal{P}_t$ is a potential relevance notion for $\mathcal{R}_t$.*

Now we will show a more interesting potential relevance notion, in the context of the $\mathfrak{I}^{lp}$ framework. The basic idea is to detect contributions that would be relevant given a certain *context of facts* (which are currently uncertain), as exemplified below.

Example 6. *Consider the $\mathfrak{I}^{lp}$ framework. In the chart on the right, the set in the first column would be a natural a-relevant contribution to the set in the second column, given the context of the third column. Note that these contexts (except for the empty one) would be natural relevant contributions to* S *after adding* X*, but not before.*

X	S	Context
$\{a \leftarrow b\}$	$\{\}$	$\{b\}$
$\{b \leftarrow c \wedge d\}$	$\{a \leftarrow b\}$	$\{c, d\}$
$\{a\}$	$\{\}$	$\{\}$

In order to define this potential relevance notion, we first define the *abduction set*[7] associated to a given fact h and a given set S. In short, the abduction set of h from S is the set of all the minimal sets of facts that could be added to S in order to derive h.

Definition 21 (Abduction Set). *Let* $\mathrm{S} \subseteq \mathcal{L}^{lp}$ *and* $h \in \mathcal{L}_{Facts}$. *The* abduction set *of h from* S *is defined as follows:*

$$AB(\mathrm{S}, h) = \{\mathrm{H} \subseteq \mathcal{L}_{Facts} : (\mathrm{S} \cup \mathrm{H}) \vdash h \text{ and } \nexists \mathrm{H}' \subset \mathrm{H} \text{ s. t. } (\mathrm{S} \cup \mathrm{H}') \vdash h\}$$

Example 7. *Consider the $\mathfrak{I}^{lp}$ framework. In the chart on the right, the second column shows the abduction set of the fact a, from the set* S *on the first column.*

S	AB(S, a)
$\{a \leftarrow b, b \leftarrow c \wedge d\}$	$\{\{a\}\{b\}\{c, d\}\}$
$\{a\}$	$\{\{\}\}$
$\{\}$	$\{\{a\}\}$

Now we are able to introduce an *abductive relevance notion* $\mathcal{A}_h^{lp}$. Basically, X is an h-relevant contribution to S under $\mathcal{A}_h^{lp}$ if, and only if, its addition generates a new element in the abduction set of h. This means that either a new fact-composed h-relevant contribution to S under $\mathcal{N}_h^{lp}$ arises, or h is actually derived. It can be shown (proof is omitted due to space reasons) that $\mathcal{A}_h^{lp}$ is a potential relevance notion for $\mathcal{N}_h^{lp}$.

Definition 22 (Abductive Relevance). *Let* $\mathrm{S} \subseteq \mathcal{L}^{lp}$ *and* $h \in \mathcal{L}_{Facts}$. *A set* $\mathrm{X} \subseteq \mathcal{L}^{lp}$ *is an h-relevant contribution to* S *under $\mathcal{A}_h^{lp}$ iff there exists* $\mathrm{H} \subseteq \mathcal{L}_{Facts}$ *such that: (1)* $\mathrm{H} \in AB(\mathrm{S} \cup \mathrm{X}, h)$ *and (2)* $\mathrm{H} \notin AB(\mathrm{S}, h)$.

Example 8. *Consider the $\mathcal{A}_h^{lp}$ relevance notion, introduced in Def. 22. In the chart on the right, the set* X *in the first column is an a-relevant contribution to the set* S *in the second column.*

X	S
$\{b \leftarrow c \wedge d\}$	$\{a \leftarrow b\}$
$\{b\}$	$\{a \leftarrow b \wedge c\}$
$\{a \leftarrow b\}$	$\{\}$
$\{a\}$	$\{\}$

Returning to the semantics definition, the idea is to use the potential framework under the Basic Semantics, resulting in a new semantics for the original framework. Next we introduce the *Full Collaborative Semantics*, which is actually a family of semantics: each possible potential framework defines a different semantics of the family.

[7] *Abduction* has been widely used for finding *minimal explanations* for a certain result. A survey of works on the extension of Logic Prog. to perform abductive reasoning is provided in [11].

Definition 23 (Full Collaborative Semantics). *Let $\mathfrak{F} = \langle \mathcal{R}_t, \Phi, \mathrm{Ag} \rangle$ be an abstract dialogue framework. A dialogue $d = \langle t, \langle m_j \rangle, o \rangle \in d(\mathfrak{F})$ belongs to the* Full Collaborative Semantics *for $\mathfrak{F}$ (noted Full($\mathfrak{F}$)) iff $d \in$ Basic($\mathfrak{F}^*$) for some framework $\mathfrak{F}^* = \langle \mathcal{P}_t, \Phi, \mathrm{Ag} \rangle$ which is a potential for $\mathfrak{F}$. We will also use the more specific notation $d \subset$ Full($\mathfrak{F}, \mathcal{P}_t$).*

In this way, each agent would be able to autonomously determine that she has no more potential relevant contributions to make, ensuring there cannot be any distributed relevant contribution when the dialogue ends, and hence achieving R_1. In other words, achieving local completeness under the potential relevance notion implies achieving global completeness under the original one, as stated below.

Proposition 11. *Let $\mathfrak{F} = \langle \mathcal{R}_t, \Phi, \mathrm{Ag} \rangle$ and $\mathfrak{F}^* = \langle \mathcal{P}_t, \Phi, \mathrm{Ag} \rangle$ be abstract dialogue frameworks such that $\mathfrak{F}^*$ is a potential for $\mathfrak{F}$, and $d \in d(\mathfrak{F})$. If dialogue d satisfies* local completeness *under $\mathfrak{F}^*$, then it satisfies* global completeness *under $\mathfrak{F}$.*

Requirement R_2 is in now compromised, since the context we have mentioned may not exist. In other words, achieving local progress under the potential relevance notion does not ensure achieving global progress under the original one. The challenge is to design *good* potential relevance notions which considerably reduce the amount of cases in which a contribution is considered potentially relevant but, eventually, it is not. Observe that a relevance notion which considers any sentence of the language as relevant, works as a potential for any given relevance notion, but it is clearly not a good one.

Next we summarize the results for the dialogues generated under the Full Collaborative Semantics. By achieving global completeness these dialogues achieve outcome determinism under the same condition as before. Although global progress is not achieved under the original relevance notion, it is achieved under the potential one, and thus termination can be ensured as long as the latter satisfies novelty.

Corollary 4. *Let $\mathfrak{F} = \langle \mathcal{R}_t, \Phi, \mathrm{Ag} \rangle$ be an abstract dialogue framework, and $\mathcal{P}_t$ a potential for $\mathcal{R}_t$. The dialogues in Full($\mathfrak{F}, \mathcal{P}_t$) satisfy* termination *and* outcome determinism, *provided that $\mathcal{P}_t$ satisfies* novelty *and $\mathcal{R}_t$ is* weaker or equal *than $\mathcal{N}_t^{\Phi}$.*

Example 9. *Consider the same scenario as in Ex. 1 for the $\mathfrak{I}^{lp}$ framework. Both dialogues d_1 and d_2, presented in Ex. 1 and Ex. 2 respectively, belong to Full($\mathfrak{I}^{lp}, \mathcal{A}_h^{lp}$). Also belongs to this semantics the dialogue d_3 traced on the right, which results from dialogue d_2 by interchanging steps 2 and 3. The fifth column of the chart shows the evolution of the abduction set of the fact a from the generated public knowledge. An additional step **0** is added, in order to show the initial state of this abduction set (i.e., when the public knowledge is still empty). Also belongs to Full($\mathfrak{I}^{lp}, \mathcal{A}_h^{lp}$) the dialogue which results from d_2 by merging steps **2** and **3** together in a single one. Note that all these dialogues achieve* global completeness, *although* global progress *is achieved only by dialogue d_1.*

step	A	B	C	$\mathrm{AB}(\mathbf{PU}_{d_3}^{step}, a)$	$\Phi(\mathbf{PU}_{d_3}^{step}, a)$
0				$\{\{a\}\}$	No
1	$a \leftarrow b$			$\{\{a\}\{b\}\}$	No
2		$b \leftarrow c$		$\{\{a\}\{b\}\{c\}\}$	No
3		$b \leftarrow d$		$\{\{a\}\{b\}\{c\}\{d\}\}$	No
4			c	$\{\{\}\}$	Yes

Results regarding *satisfiability* and *no-backtracking* also hold under the two practical semantics we have presented in this section, as stated below.

Proposition 12. *For any dialogue framework* $\mathfrak{F} = \langle \mathcal{R}_t, \Phi, \mathrm{Ag}\rangle$*, each one of the sets* $Basic(\mathfrak{F})$ *and* $Full(\mathfrak{F}, \mathcal{P}_t)$*, contains at least one element.*

Proposition 13. *Let* $\mathfrak{F} = \langle \mathcal{R}_t, \Phi, \mathrm{Ag}\rangle$ *and* $\mathfrak{F}^* = \langle \mathcal{P}_t, \Phi, \mathrm{Ag}\rangle$ *be abstract dialogue frameworks such that* $\mathfrak{F}^*$ *is a potential for* $\mathfrak{F}$*, and let* $d_1 \in d(\mathfrak{F})$*. If* d_1 *satisfies* local progress *under* $\mathfrak{F}$ *(*$\mathfrak{F}^*$*), then there exists a dialogue* $d_2 \in Basic(\mathfrak{F})$ *(*$d_2 \in Full(\mathfrak{F}, \mathcal{P}_t)$*) which is a* continuation *of* d_1.

Finally, a result showing the relation among the three collaborative semantics, for the case in which the relevance notion satisfies *granularity*, is stated.

Proposition 14. *Let* $\mathfrak{F} = \langle \mathcal{R}_t, \Phi, \mathrm{Ag}\rangle$ *be an abstract dialogue framework. If the relevance notion* $\mathcal{R}_t$ *satisfies* granularity, *then it holds that:*

$$Basic(\mathfrak{F}) = Full(\mathfrak{F}, \mathcal{R}_t) \subseteq Utopian(\mathfrak{F})$$

To sum up, we have defined three collaborative semantics for an abstract dialogue framework. The Utopian Semantics describes an idealistic, in most cases impractical behavior of a collaborative dialogue. Its usefulness is theoretical. It is approximated, in different ways, by the other two practical semantics. The Basic Semantics, on the other side, describes a straightforward implementable behavior of a collaborative dialogue. The weak point of this semantics is that it does not ensure *global completeness* (neither *outcome determinism*, consequently). The Full Collaborative Semantics is actually a family of semantics: each potential relevance notion $\mathcal{P}_t$ associated to $\mathcal{R}_t$ defines a semantics of the family. Thus, the constructiveness of these semantics is reduced to the problem of finding a potential relevance notion for $\mathcal{R}_t$. These semantics succeed in achieving *global completeness*, at the price of allowing moves which may not be allowed by the Utopian Semantics. The goodness of a given potential relevance notion increases as it minimizes the amount of such moves.

6 Discussion

In this section we briefly discuss some issues which complement the formalism presented in the previous sections. A deep analysis of these items is left for future works. Note that the core framework is suitable and helpful for handling all these aspects.

The Interaction Protocol. As defined earlier, a semantics for our dialogue framework describes a set of acceptable dialogues. We say that a dialogue system implementation *respects* a certain semantics, if any dialogue generated under this implementation belongs to the semantics. In order to design a dialogue system which respects a Basic Semantics, as defined in Sec. 5, an *interaction protocol* is needed, which allows the agents to coordinate and synchronize for making relevant contributions, until no more of those exist. Recall that, as mentioned in Sec. 1, agents' private knowledge bases are static during the dialogue. Some issues to consider are the following: (1) When/How often do the agents check for relevant contributions (*relevance-checking*)? (2) How to avoid interruptions? (3) Who gets the right to speak when several have relevant contributions to make? (4) How to ensure that the public knowledge is not modified in the lag between relevance-checking and speaking? In other words, how to ensure that contributions do not become obsolete by the time of speaking? (5) Last but not least, how to signal termination? That is, how do agents realize that no one has more relevant contributions to make, and thus the dialogue should end?

One possible solution could be as follows. Assume a *right-to-speak token*, and also a shared *counter* for termination detection. This counter should be initialized with the number of participants, at the beginning of each dialogue cycle (or step), and it should be decremented by each agent who has no relevant contributions in that cycle. Full details are not discussed here, but it is easy to see that a protocol could be implemented such that: (i) The agents speak one at a time (only one agent per cycle gets the token). (ii) The agent who first finish the relevance-checking, having something to say, gets the right to speak (first request acquires the token). (iii) Relevance-checking occurs at the beginning of the dialogue and after each contribution (when the token is released). (iv) All the agents are signaled when none of them has relevant contributions to make, *i.e.*, the dialogue has ended (counter value reaches zero).

Another alternative could be to implement the dialogue in a *round-robin* fashion (*i.e.*, agents have fixed turns for making moves), but this would be a much more restrictive protocol. Finally, it is important to mention that the above discussion also applies to a Full Collaborative Semantics, since this last corresponds to the Basic Semantics under a potential relevance notion.

Inconsistency Handling. An important feature of dialogue formal models is being capable of handling inconsistencies, since these last are very likely to appear when merging knowledge from different sources (agents). The framework presented earlier abstracts from this issue, relying on the reasoning model (one of the framework parameters) to handle it. This allows for a flexible and transparent choice of the inconsistency handling policy, which should be encapsulated within the reasoning mechanism. For the aim of simplicity, all the examples in the previous sections were developed in the context of Propositional Logic Programming, which does not allow for the representation of inconsistent knowledge, so no handling was required. Systems which inherently deal with inconsistencies (either by resolving them, such as a logic for *defeasible argumentation*, or by eluding them, such as a *paraconsistent logic*) could be used for doing the reasoning. Otherwise (if the chosen underlying logic does not handle inconsistencies) a prior *consolidation step* could be put before the proper *inference step*, in order to erase possible inconsistencies, and to build a more robust reasoning mechanism. Note that the framework presented in this work assumes a unified reasoning model for all the participating agents, and therefore a unified policy for handling inconsistencies.

7 Related Work

There are some works particulary related to our proposed approach, due to any of the following: (a) an explicit treatment of the notion of relevance in dialogue, (b) the search of the *global completeness* property, as we called it in this work, or (c) a tendency to examine general properties of dialogues rather than designing particular systems.

Regarding category (a), in [12], [13] and [9], the importance of a precise relevance notion definition is emphasized. However, these works focus on argumentation-based persuasion dialogues (actually a subset of those, which the author called *disputes*), which belong to the non-collaborative class, and thus *global completeness* is not pursued. Instead, the emphasis is put on properties with similar spirit to our properties of *correctness* and *local progress* (*i.e.*, only the *public knowledge* involved in the dialogue is given importance). In [13] the author considers dynamic disputes in which two participants (proponent and opponent) interchange arguments and counter-arguments, and studies two properties of protocols (namely *soundness* and *fairness*) regarding the relation between the generated public knowledge and the conclusion achieved (in this case, the *winner* of the dispute). The author also gives a natural definition of when a move is relevant: *"iff it changes the status of the initial move of the dispute"* whose spirit is similar to our definition of *natural relevance notion* but taken to the particular case in which the reasoning model is a logic for defeasible argumentation. In [9] the author considers more flexible protocols for disputes, allowing alternative sets of locutions, such as *challenge* and *concede*, and also a more flexible notion of relevance.

Another work in which *relevance* receives an explicit treatment is [10], where the authors investigate the relevance of utterances in an argumentation-based dialogue. However, our *global completeness* property is not pursued, so they do not consider the problematic of distributed contributions (distributed arguments in this case). They study three notions of relevance showing how they can affect the dialogue outcome.

Regarding category (b), in [5] an inquiry dialogue protocol which successfully pursues our idea of *global completeness* is defined. However, the protocol is set upon a particular argumentative system, with the design methodology implicit. They take a simplified version of the *DeLP*[8] system, and define an *argument inquiry dialogue* which allows exactly two agents to jointly construct arguments for a given claim. In the present work, we not only explicitly and abstractly analyze the distributed relevance issue, but also consider the complete panorama of collaborative dialogue system behavior, including *correctness* and *progress* properties.

Regarding category (c), different measures for analyzing argumentation-based persuasion are proposed in [15]: measures of the quality of the exchanged arguments, of the behavior of each agent, and of the quality of the dialog itself in terms of the relevance and usefulness of its moves. The analysis is done from the point of view of an external agent (*i.e.*, *private knowledge* is not considered), and it is focused in a non-collaborative dialogue type, so they are not concerned with the main problematic of our work.

8 Conclusions and Future Work

From a theoretical view point, we have made progress towards a formal understanding of collaborative dialogues: ideal behavior, main disallowance for its consecution in a distributed environment, and viable approximations. In a practical sense, we have provided a methodology for developing formal models of collaborative dialogues in MAS, as follows: (1) establish suitable instances of the *knowledge representation language* and the *reasoning model* (depending on the particular dialogue type); (2) choose an adequate *relevance notion* (a *natural* relevance notion is generally suitable, although a *weaker* one could also be chosen, provided it still satisfies *novelty*); (3) if the selected

[8] See [14] for details.

relevance notion does not satisfy *granularity*, and a *full collaborative semantics* is pursued, then build a *potential relevance notion* for the original one.

By instantiating the abstract dialogue framework with the elements listed above, a formal model for a specific type of dialogue is obtained, and for which all the properties stated in the previous sections hold. The generated dialogue models are still restricted in some aspects. We aim at relaxing these restrictions by future research. Specifically, we plan to: (1) extend the present analysis to not purely collaborative dialogue types (such as persuasion and negotiation), analyzing which elements need to be added to the framework, how the utopian behavior changes, and which properties hold for the generated dialogues; (2) integrate the abstract dialogue framework with mechanisms for handling inconsistencies, as discussed in Sec. 6; and (3) explore the possibility of extending the framework to allow different types of locutions in dialogues.

References

1. Kraus, S., Sycara, K., Evenchik, A.: Reaching agreements through argumentation: A logical model and implementation. Artificial Intelligence (1998)
2. Parsons, S., Amgoud, L., Wooldridge, M.: An analysis of formal inter-agent dialogues. In: AAMAS 2002, Bologna, Italy (2002)
3. Amgoud, L., Prade, H., Belabbes, S.: Towards a formal framework for the search of a consensus between autonomous agents. In: AAMAS 2005, Utrecht (2005)
4. Amgoud, L., Dimopoulos, Y., Moraitis, P.: A unified and general framework for argumentation based negotiation. In: AAMAS 2007, Honolulu, Hawai'i (2007)
5. Black, E., Hunter, A.: A generative inquiry dialogue system. In: AAMAS 2007, Honolulu, Hawai'i (2007)
6. Marcos, M.J., Falappa, M.A., Simari, G.R.: A set of collaborative semantics for an abstract dialogue framework. In: The 6th IJCAI Workshop on Knowledge and Reasoning in Practical Dialogue Systems (KRPD 2009), Pasadena, California, USA (2009)
7. Walton, D., Krabbe, E.: Commitment in Dialogue: Basic Concepts of Interpersonal Reasoning. State University of New York Press, Albany (1995)
8. Makinson, D.: General patterns in nonmonotonic reasoning. In: Handbook of Logic in Artificial Intelligence and Logic Programming (1994)
9. Prakken, H.: Coherence and flexibility in dialogue games for argumentation. J. Log. Comput. (2005)
10. Parsons, S., McBurney, P., Sklar, E., Wooldridge, M.: On the relevance of utterances in formal inter-agent dialogues. In: AAMAS 2007, Honolulu, Hawai'i (2007)
11. Kakas, A.C., Kowalski, R.A., Toni, F.: Abductive logic programming. J. Log. Comput. (1992)
12. Prakken, H.: On dialogue systems with speech acts, arguments, and counterarguments. In: Brewka, G., Moniz Pereira, L., Ojeda-Aciego, M., de Guzmán, I.P. (eds.) JELIA 2000. LNCS (LNAI), vol. 1919, p. 224. Springer, Heidelberg (2000)
13. Prakken, H.: Relating protocols for dynamic dispute with logics for defeasible argumentation. Synthese (2001)
14. García, A.J., Simari, G.R.: Defeasible logic programming: An argumentative aproach. Theory and Practice of Logic Programming 4(1), 95–138 (2004)
15. Amgoud, L., de Saint-Cyr, F.D.: Measures for persuasion dialogs: A preliminary investigation. In: COMMA 2008, Toulouse, France (2008)

The Relationship of the Logic of Big-Stepped Probabilities to Standard Probabilistic Logics*

Christoph Beierle[1] and Gabriele Kern-Isberner[2]

[1] Dept. of Computer Science, FernUniversität in Hagen, 58084 Hagen, Germany
christoph.beierle@fernuni-hagen.de
[2] Dept. of Computer Science, TU Dortmund, 44221 Dortmund, Germany
gabriele.kern-isberner@cs.uni-dortmund.de

Abstract. Different forms of semantics have been proposed for conditionals of the form "Usually, if A then B", ranging from quantitative probability distributions to qualitative approaches using plausibility orderings or possibility distributions. Atomic-bound systems, also called big-stepped probabilities, allow qualitative reasoning with probabilities, aiming at bridging the gap between qualitative and quantitative argumentation and providing a model for the nonmonotonic reasoning system P. By using Goguen and Burstall's notion of institutions for the formalization of logical systems, we elaborate precisely which formal connections exist between big-stepped probabilities and standard probabilities, thereby establishing the exact relationships among these logics.

Keywords: conditional logic, probabilistic logic, big-stepped probability, institution, institution morphism.

1 Introduction

Conditionals of the form "usually, if A then B" establishing a plausible, probable, possible etc. connection between the antecedent A and the consequent B, but still allowing exceptions, can be viewed as default rules and are a powerful tool for logic-based knowledge representation. Different forms of semantics have been proposed for such conditionals, ranging from quantitative probability distributions to pure qualitative approaches (see e.g., [1, 6, 16]).

Ernest Adams was the first to present a probabilistic framework for qualitative default reasoning. In his work [1], he used an infinitesimal approach to define "reasonable (probabilistic) consequences". On these ideas, Pearl later based his ϵ-*semantics* [17] which turned out to be the same as *preferential semantics* and can be characterized by the axioms for nonmonotonic inference relations $\mid\sim$ which have become known as *system P* [12]. Therefore, the infinitesimal ϵ-semantics provides a probabilistic semantics for system P.

* The research reported here was supported by the Deutsche Forschungsgemeinschaft (grants BE 1700/7-1 and KE 1413/2-1).

S. Link and H. Prade (Eds.): FoIKS 2010, LNCS 5956, pp. 191–210, 2010.

This seemed hardly possible to realize within a standard probabilistic framework. An obvious way to interpret a default rule "usually, if A then B", or "from A, defeasibly infer B" (written as $A \mathrel{|\!\sim} B$) by a probability distribution P would be to postulate $P(AB) > P(A\overline{B})$ (which is equivalent to $P(B|A) > 0.5$). I.e., given A, the presence of B should be more probable than its absence. This interpretation, however, is not generally compatible with system P, it may conflict, for instance, with the *OR*-postulate of system P. Indeed, it is easy to find counterexamples where $P(AC) > P(A\overline{C})$ and $P(BC) > P(B\overline{C})$, but $P((A \vee B)C) < P((A \vee B)\overline{C})$. So, in order to give reasonable probabilistic meanings to defaults, one has to focus on special subclasses of probability distributions.

Atomic bound systems, introduced by Snow in [18], turned out to be such proper subclasses. Their distributions are also known as *big-stepped probabilities* (this more intuitive name was coined by Benferhat, Dubois & Prade, see [6]). Thus, big-stepped probabilities allow qualitative reasoning with probabilities, aiming to bridge the gap between qualitative and quantitative argumentation.

In this paper, we provide a systematic, formal comparison of big-stepped probabilities to purely qualitative logics (where plausibility preorders on possible worlds realize the *system-of-spheres* model of Lewis [15]) and to standard probabilistic logics. The different logics are formalized as institutions [7], and the relationships among the logical systems are expressed by institution morphisms [8]. In particular, we adress the following questions:

- What are the precise relationships among big-stepped probabilities and standard probabilistic and probabilistic conditional logic? Which institution morphisms exist among the corresponding institutions?
- Are there different ways of interpreting big-stepped probabilities in a purely qualitative setting?
- In a full picture of qualitative and probabilistic semantics for conditionals, what is the precise position of big-stepped probabilities? Can we underpin its qualitative approach to the quantitative flair of probabilities as an intermediary between these two opposite realms not only on the grounds of ad-hoc comparisons, but also on the formal grounds of comparing logical systems by institution morphisms?

In our previous work [3, 4, 5], we already advocated the use of institutions for studying conditionals and formalized different conditional logics as institutions, but did not address the specific questions outlined above. In this paper, we give detailed answers to these questions. In particular, we will show that a connection between the logic of big-stepped probabilities and standard probabilistic conditional logic via morphisms requires syntactical and semantical deformations. So, in spite of superficial similarities between both logics, our analysis reveals essential differences with respect to the representation and interpretation of conditional knowledge. Hence, the logic of big-stepped probabilities can by no means be considered as some kind of "sublogic" of the full probabilistic conditional logic. Furthermore, also links between big-steppend probabilities and the qualitative conditional approach where worlds are ordered according to their plausibility can not be established easily. There is no way to recover the fine-grained semantics of big-stepped conditionals via total preorders in a rigidly formal way, while purely qualitative conditionals might be interpreted in different ways by big-stepped probabilities, even allowing the propositionalisation of conditionals. Therefore, as a consequence of our investigations, the logic of big-stepped probabilties can not be looked upon as some intermediary between qualitative and probabilistic logics, but rather occupies a singular position in this picture.

The rest of this paper is organized as follows: In Sec. 2 we discuss related work investigating the borderline between qualitative and probabilistic reasoning with conditionals. In Sec. 3 we present the definitions of institutions and their morphisms introduced in [7], and fix our notation by defining propositional logic as an institution as done in [4]. Section 4 recalls the institutions of big-stepped probabilities [5] and other semantics for conditionals (from [3, 4]) as far as they are needed here. In Sec. 5, we elaborate in detail the precise relationships among the four different probabilistic and conditional logics studied in this paper, while Sec. 6 concludes and points out further work.

2 Related Work

The formalization of uncertain reasoning can be approached both from a qualitative and from a quantitative point of view. For qualitative reasoning, using unquantified default rules on the syntactic side, the postulates of system P given in [12] provide what most researchers regard as a core any nonmonotonic system should satisfy; in [10], Hawthorne and Makinson call these axioms the "industry standard" for qualitative nonmonotonic inference:

Reflexivity or *Inclusion*	:	$A \mid\!\sim A$
Cut	:	$\dfrac{A \wedge B \mid\!\sim C,\ A \mid\!\sim B}{A \mid\!\sim C}$
Cautious Monotony	:	$\dfrac{A \mid\!\sim B,\ A \mid\!\sim C}{A \wedge B \mid\!\sim C}$
Right Weakening	:	$\dfrac{A \mid\!\sim B,\ B \models C}{A \mid\!\sim C}$
Left Logical Equivalence	:	$\dfrac{\models A \equiv B,\ A \mid\!\sim C}{B \mid\!\sim C}$
Or	:	$\dfrac{A \mid\!\sim C,\ B \mid\!\sim C}{A \vee B \mid\!\sim C}$

In [6] it is shown that big-stepped probabilities provide a model class for system P. Using system P and a corresponding weaker set of axioms O tailored for probabilistic reasoning [9], in [10] Hawthorne and Makinson study representation and completeness problems for qualitative and quantitative reasoning.

Using sentences employing intervals of probabilistic values and modeling epistemic states based upon acceptance of sentences, Kyburg et al. [13] argue that system P is not a conservative core of nonmonotonic logic, and they weaken some of the axioms of system P, e.g., by replacing $A \mid\!\sim B$ by $A \vdash B$ in the promise of *Cautious Monotony*.

Another modification of system P given by adding the rule

$$\textit{Rational Monotony} \quad : \quad \frac{A \mid\!\sim C,\ \neg(A \mid\!\sim \neg B)}{A \wedge B \mid\!\sim C}$$

yields the preferential system $\mathcal{R}$ [14]. Arló Costa and Parikh [2] provide a probabilistic model of $\mathcal{R}$ and use this model for relating it to qualitative models of nonmonotonic relations, based on a notion saying that an event B is expected given A.

While these and most other papers on comparing different semantics for conditionals deal with the semantics in general, our approach is more finely grained. We will investigate with respect to each model and each sentence, whether transformations between different logical frameworks exist that respect logical entailment. By making use of the formal means of institution morphisms, in cases where they exist, we will be able to make precise which syntactical adjustment has to be made under obvious or plausible mappings between models, and conversely, which semantical change is required to correspond to mappings between sentences.

3 Institutions and Institution Morphisms

As institutions are formalized by using category theory, we will also very briefly recall some basic notions of category theory; for more information about categories, see e.g. [11].

If C is a category, $|C|$ denotes the objects of C and $/C/$ its morphisms; for both objects $c \in |C|$ and morphisms $\varphi \in /C/$, we also write just $c \in C$ and $\varphi \in C$, respectively. C^{op} is the opposite category of C, with the direction of all morphisms reversed. The composition of two functors $F : C \to C'$ and $G : C' \to C''$ is denoted by $G \circ F$ (first apply F, then G). For functors $F, G : C \to C'$, a natural transformation η from F to G, denoted by $\eta : F \Longrightarrow G$, assigns to each object $c \in |C|$ a morphism $\eta_c : F(C) \to G(C) \in /C'/$ such that for every morphism $\varphi : c \to d \in /C/$ we have $\eta_d \circ F(\varphi) = G(\varphi) \circ \eta_c$. $\mathcal{SET}$ and $\mathcal{CAT}$ denote the categories of sets (with functions as morphisms) and of categories (with functors as morphisms), respectively.

Definition 1 ([7]). *An institution is a quadruple $Inst = \langle Sig, Mod, Sen, \models \rangle$ with a category Sig of signatures as objects, a functor $Mod : Sig \to \mathcal{CAT}^{op}$ yielding the category of Σ-models for each signature Σ, a functor $Sen : Sig \to \mathcal{SET}$ yielding the sentences over a signature, and a $|Sig|$-indexed relation $\models_\Sigma \subseteq |Mod(\Sigma)| \times Sen(\Sigma)$ such that for each signature morphism $\varphi : \Sigma \to \Sigma' \in /Sig/$,*

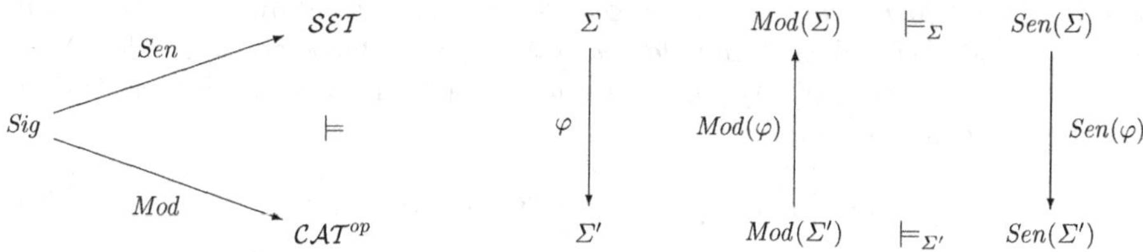

Fig. 1. Relationships within an institution $Inst = \langle\, Sig,\ Mod,\ Sen,\ \models\, \rangle$ [7]

for each $m' \in |\,Mod(\Sigma')\,|$, *and for each* $f \in Sen(\Sigma)$ *the following* satisfaction condition *holds (cf. Fig. 1):*

$$m' \models_{\Sigma'} Sen(\varphi)(f) \quad \textit{iff} \quad Mod(\varphi)(m') \models_\Sigma f \tag{1}$$

For sets F, G of Σ-sentences and a Σ-model m we write $m \models_\Sigma F$ iff $m \models_\Sigma f$ for all $f \in F$. The satisfaction relation is lifted to semantical entailment $\models_\Sigma$ between sentences by defining $F \models_\Sigma G$ iff for all Σ-models m with $m \models_\Sigma F$ we have $m \models_\Sigma G$. Entailment is preserved under change of notation carried out by a signature morphism, i.e. $F \models_\Sigma G$ implies $\varphi(F) \models_{\varphi(\Sigma)} \varphi(G)$ (but not vice versa).

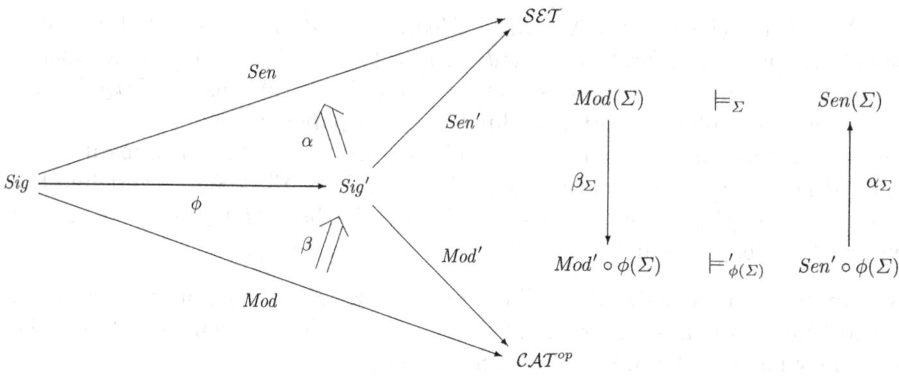

Fig. 2. Relationships within an institution morphism $\langle\, \phi,\ \alpha,\ \beta\, \rangle$: $\langle\, Sig,\ Mod,\ Sen,\ \models\, \rangle \longrightarrow \langle\, Sig',\ Mod',\ Sen',\ \models'\, \rangle$

An institution morphism Φ expresses a relation between two institutions *Inst* und *Inst'* such that the satisfaction condition of *Inst* may be computed by the satisfaction condition of *Inst'* if we translate it according to Φ. The translation is done by relating every *Inst*-signature Σ to an *Inst'*-signature Σ', each Σ'-sentence to a Σ-sentence, and each Σ-model to a Σ'-model.

Definition 2 ([7, 8]). *Let* $Inst, Inst'$ *be two institutions,* $Inst = \langle\, Sig,\ Mod,\ Sen,\ \models\, \rangle$ *and* $Inst' = \langle\, Sig',\ Mod',\ Sen',\ \models'\, \rangle$. *An* institution morphism Φ *from Inst to Inst' is a triple* $\langle\, \phi,\ \alpha,\ \beta\, \rangle$ *with a functor* $\phi : Sig \rightarrow Sig'$, *a natural transformation* $\alpha : Sen' \circ \phi \Longrightarrow Sen$*, and a natural transformation* $\beta : Mod \Longrightarrow Mod' \circ \phi$ *such that for each* $\Sigma \in |Sig|$*, for each* $m \in |Mod(\Sigma)|$*, and for each* $f' \in Sen'(\phi(\Sigma))$ *the following* satisfaction condition (for institution morphisms) *holds (cf. Fig. 2):*

$$m \models_{\Sigma} \alpha_{\Sigma}(f') \quad \textit{iff} \quad \beta_{\Sigma}(m) \models'_{\phi(\Sigma)} f' \tag{2}$$

The Institution of Propositional Logic: As a basic example that will be useful throughout the paper, the components of the institution $Inst_{\mathcal{B}} = \langle Sig_{\mathcal{B}}, Mod_{\mathcal{B}}, Sen_{\mathcal{B}}, \models_{\mathcal{B}} \rangle$ of classical propositional logic will be defined in the following.

Signatures: $Sig_{\mathcal{B}}$ is the category of propositional signatures. A propositional signature $\Sigma \in |Sig_{\mathcal{B}}|$ is a (finite) set of propositional variables, $\Sigma = \{a_1, \ldots, a_n\}$. A propositional signature morphism $\varphi : \Sigma \to \Sigma' \in /Sig_{\mathcal{B}}/$ is an injective function mapping propositional variables to propositional variables. Note that for $Inst_{\mathcal{B}}$, injectivity of signature morphisms is not needed, but for the conditional semantics of big-stepped probabilities, injectivity of φ will ensure that the respective model functor $Mod(\varphi)$ is well-defined.

All other institutions $Inst_X$ defined later in this paper will use the same signatures, i.e, we will tacitly assume $Sig_X = Sig_{\mathcal{B}}$.

Models: For each signature $\Sigma \in Sig_{\mathcal{B}}$, $Mod_{\mathcal{B}}(\Sigma)$ contains the set of all propositional interpretations for Σ, i.e. $|Mod_{\mathcal{B}}(\Sigma)| = \{I \mid I : \Sigma \to Bool\}$ where $Bool = \{true, false\}$. Due to its simple structure, the only morphisms in $Mod_{\mathcal{B}}(\Sigma)$ are the identity morphisms; correspondingly, in this paper, we will also not investigate other morphisms within the model categories of the other institutions studied here. For each signature morphism $\varphi : \Sigma \to \Sigma' \in Sig_{\mathcal{B}}$, we define the functor $Mod_{\mathcal{B}}(\varphi) : Mod_{\mathcal{B}}(\Sigma') \to Mod_{\mathcal{B}}(\Sigma)$ by $(Mod_{\mathcal{B}}(\varphi)(I'))(a_i) := I'(\varphi(a_i))$ where $I' \in Mod_{\mathcal{B}}(\Sigma')$ and $a_i \in \Sigma$.

Sentences: For each signature $\Sigma \in Sig_{\mathcal{B}}$, the set $Sen_{\mathcal{B}}(\Sigma)$ contains the usual propositional formulas constructed from the propositional variables in Σ and the logical connectives $\wedge$ (and), $\vee$ (or), and $\neg$ (not).

For each signature morphism $\varphi : \Sigma \to \Sigma' \in Sig_{\mathcal{B}}$, the function $Sen_{\mathcal{B}}(\varphi) : Sen_{\mathcal{B}}(\Sigma) \to Sen_{\mathcal{B}}(\Sigma')$ is defined by straightforward inductive extension on the structure of the formulas; e.g., $Sen_{\mathcal{B}}(\varphi)(A \wedge B) = Sen_{\mathcal{B}}(\varphi)(A) \wedge Sen_{\mathcal{B}}(\varphi)(B)$. We may abbreviate $Sen_{\mathcal{B}}(\varphi)(A)$ by just writing $\varphi(A)$.

In order to simplify notations, we will often replace conjunction by juxtaposition and indicate negation of a formula by overlining it, i.e. $AB = A \wedge B$ and $\overline{A} = \neg A$. An *atomic formula* is a formula consisting of just a propositional variable, a *literal* is a positive or a negated atomic formula, an *elementary conjunction* is a conjunction of literals, and a *complete conjunction* is an elementary conjunction containing each atomic formula either in positive or in negated form. Ω_{Σ} denotes the set of all complete conjunctions over a signature Σ; if Σ is clear from the context, we may drop the index Σ. Note that there is an obvious bijection between $|Mod_{\mathcal{B}}(\Sigma)|$ and Ω_{Σ}, associating with $I \in |Mod_{\mathcal{B}}(\Sigma)|$ the complete conjunction $\omega_I \in \Omega_\Sigma$ in which an atomic formula $a_i \in \Sigma$ occurs in positive form iff $I(a_i) = true$.

Satisfaction relation: For any $\Sigma \in |\mathit{Sig}_\mathcal{B}|$, the satisfaction relation is defined as expected for propositional logic, e.g. $I \models_{\mathcal{B},\Sigma} a_i$ iff $I(a_i) = true$ and $I \models_{\mathcal{B},\Sigma} A \wedge B$ iff $I \models_{\mathcal{B},\Sigma} A$ and $I \models_{\mathcal{B},\Sigma} B$ for $a_i \in \Sigma$ and $A, B \in \mathit{Sen}_\mathcal{B}(\Sigma)$.

Example 1. Let $\Sigma = \{s, u\}$ and $\Sigma' = \{a, b, c\}$ be two propositional signatures with the atomic propositions s – *being a scholar*, u – *being single* and a – *being a student*, b – *being young*, c – *being unmarried*. Let I' be the Σ'-model with $I'(a) = true$, $I'(b) = true$, $I'(c) = false$. Let $\varphi : \Sigma \to \Sigma' \in \mathit{Sig}_\mathcal{B}$ be the signature morphism with $\varphi(s) = a$, $\varphi(u) = c$. The functor $\mathit{Mod}_\mathcal{B}(\varphi)$ takes I' to the Σ-model $I := \mathit{Mod}_\mathcal{B}(\varphi)(I')$, yielding $I(s) = I'(a) = true$ and $I(u) = I'(c) = false$.

Note that a signature morphism φ not being surjective makes the model functor $\mathit{Mod}_\mathcal{B}(\varphi)$ a forgetful functor: In Example 1, any information about b (being young) in I' is forgotten in I since b is not in the codomain of φ. On the other hand, in this paper, we require signature morphims in $\mathit{Inst}_\mathcal{B}$ to be injective. This avoids redundancies among the propositional variables that might cause ambiguities when two different variables finally turn out to have the same meaning. (Any two variables mapped to the same variable under φ would always be interpreted identically in any model reached by $\mathit{Mod}_\mathcal{B}(\varphi)$.) A consequence of φ being injective is that a complete conjunction will not be mapped to $\bot$ under φ, i.e. $\varphi(\omega) \not\equiv \bot$ for all $\omega \in \Omega_\Sigma$. Furthermore, for two distinct complete conjunctions $\omega_1, \omega_2 \in \Omega_\Sigma$, the formulas $\varphi(\omega_1)$ and $\varphi(\omega_2)$ are exclusive, i.e. $\varphi(\omega_1)\varphi(\omega_2) \equiv \bot$.

Probability distributions: Let $\Sigma \in |\mathit{Sig}_\mathcal{B}|$ be a propositional signature. A *probability distribution* (or *probability function*) *over* Σ is a function $P : \mathit{Sen}_\mathcal{B}(\Sigma) \to [0,1]$ such that $P(\top) = 1$, $P(\bot) = 0$, and $P(A \vee B) = P(A) + P(B)$ for any formulas $A, B \in \mathit{Sen}_\mathcal{B}(\Sigma)$ with $AB = \bot$. Each probability distribution P is determined uniquely by its values on the complete conjunctions $\omega \in \Omega_\Sigma$, since $P(A) = \sum_{\omega \in \Omega_\Sigma, \omega \models_{\mathcal{B},\Sigma} A} P(\omega)$. For two propositional formulas $A, B \in \mathit{Sen}_\mathcal{B}(\Sigma)$ with $P(A) > 0$, the *conditional probability of* B *given* A is $P(B|A) := \frac{P(AB)}{P(A)}$.

4 Conditional Logics

We recall the formalizations of big-stepped probabilities ($\mathit{Inst}_\mathcal{S}$) [5], propositional probabilistic ($\mathit{Inst}_\mathcal{P}$) [4], conditional probabilistic ($\mathit{Inst}_\mathcal{C}$) [4] and a purely qualitative conditional logic ($\mathit{Inst}_\mathcal{K}$) [3] as institutions.

4.1 The Institution of Big-Stepped Probabilities

Definition 3. *A* big-stepped probability distribution P over a signature Σ *is a probability distribution on* Ω_Σ *such that the following conditions are satisfied for all* $\omega, \omega_0, \omega_1, \omega_2 \in \Omega_\Sigma$*:*

$$P(\omega) > 0 \tag{3}$$
$$P(\omega_1) = P(\omega_2) \quad \textit{iff} \quad \omega_1 = \omega_2 \tag{4}$$
$$P(\omega_0) \; > \sum_{\omega : P(\omega_0) > P(\omega)} P(\omega) \tag{5}$$

The set of all big-stepped probability distributions over Σ is denoted by $\mathcal{P}_{BS}(\Sigma)$.

Each big-stepped probability distribution $P : \Omega_\Sigma \to [0,1]$ uniquely extends to general propositional formulas $A \in \mathit{Sen}_\mathcal{B}(\Sigma)$ by setting $P(A) = \sum_{\omega \in \Omega_\Sigma, \omega \models_{\mathcal{B},\Sigma} A} P(\omega)$ as usual for standard probability distributions (cf. Sec. 3).

Big-stepped probabilities are quite artificial constructions which are rarely to never found in applications. They are built in such a way as to allow qualitative reasoning with probabilities. The requirements (4) and (5) are particularly crucial for this aim. (4) undercuts the number argument of probabilistic reasoning that takes into account how many models of the same probability are considered. Condition (5) inhibits that many small probabilities might have more impact than a big probability; it explains the attribute "big-stepped": In a big-stepped probability distribution, the probability of each possible world is bigger than the sum of all probabilities of less probable worlds. Big-stepped probabilities actually provide a standard probabilistic semantics for system P [12], as was shown in [6].

The following two lemmata give a more detailed impression of properties of big-stepped probabilities, and of their presentations of belief.

Lemma 1. *Let $P \in \mathcal{P}_{BS}(\Sigma)$ be a big-stepped probability distribution over a signature Σ, and let A, B be two propositional formulas. Then $P(A) = P(B)$ iff $A \equiv B$.*

Lemma 2. *Let $P \in \mathcal{P}_{BS}(\Sigma)$ be a big-stepped probability distribution over a signature Σ, and let A, B be two non-contradictory, exclusive propositional formulas. Then $P(A) > P(B)$ iff $\exists \omega_0 \models A, \forall \omega \models B : P(\omega_0) > P(\omega)$.*

The institution $\mathit{Inst}_\mathcal{S}$ of big-stepped probabilities is given by the following components:

Models: $\mathit{Mod}_\mathcal{S}(\Sigma)$ is obtained from the set of probability distributions by restriction to the big-stepped probability distributions, i.e., $|\mathit{Mod}_\mathcal{S}(\Sigma)| = \mathcal{P}_{BS}(\Sigma)$.

For each signature morphism $\varphi : \Sigma \to \Sigma'$, we define a functor $\mathit{Mod}_\mathcal{S}(\varphi) : \mathit{Mod}_\mathcal{S}(\Sigma') \to \mathit{Mod}_\mathcal{S}(\Sigma)$ by mapping each big-stepped probability distribution P' over Σ' to a distribution $\mathit{Mod}_\mathcal{S}(\varphi)(P')$ over Σ. $\mathit{Mod}_\mathcal{S}(\varphi)(P')$ is defined by

$$(\mathit{Mod}_\mathcal{S}(\varphi)(P'))(\omega) \;:=\; P'(\varphi(\omega)) = \sum_{\omega' : \omega' \models_{\mathcal{B},\Sigma'} \varphi(\omega)} P'(\omega')$$

where ω and ω' are complete conjunctions over Σ and Σ', respectively. This generalizes to propositional formulas $A \in \mathit{Sen}_\mathcal{B}(\Sigma)$, i.e., $(\mathit{Mod}_\mathcal{P}(\varphi)(P'))(A) = P'(\varphi(A))$.

Sentences: For each signature Σ, the set $Sen_{\mathcal{S}}(\Sigma)$ contains (propositional) *conditionals* of the form $(B|A)$ where $A, B \in Sen_{\mathcal{B}}(\Sigma)$ are propositional formulas from $Inst_{\mathcal{B}}$. For $\varphi : \Sigma \to \Sigma'$, the extension $Sen_{\mathcal{S}}(\varphi)$ is defined as usual by $Sen_{\mathcal{S}}(\varphi)((B|A)) = (\varphi(B)|\varphi(A))$.

Satisfaction relation: The satisfaction relation $\models_{\mathcal{S},\Sigma}$ between $|\,Mod_{\mathcal{S}}(\Sigma)\,|$ and $Sen_{\mathcal{S}}(\Sigma)$ is defined, for any $\Sigma \in |\,Sig_{\Pi}\,|$, by: $P \models_{\mathcal{S},\Sigma} (B|A)$ iff $P(AB) > P(A\overline{B})$. Hence a conditional $(B|A)$ is accepted by P iff its confirmation AB is more probable than its refutation $A\overline{B}$.

Example 2. When specifying our Example 1 by big-stepped probabilities, we have to observe carefully the crucial property (5). In principle, it is straightforward to obtain a big-stepped probability from a given ordering of the possible worlds; more details can be found e.g. in [6]. Here, we model our example in the following way:

ω'	$P'(\omega')$	ω'	$P'(\omega')$
$\overline{a}\overline{b}\overline{c}$	0.501	$a\overline{b}\overline{c}$	0.004
$\overline{a}\overline{b}c$	0.063	$a\overline{b}c$	0.008
$\overline{a}b\overline{c}$	0.032	$ab\overline{c}$	0.016
$\overline{a}bc$	0.126	abc	0.250

From this, one computes easily $P'(ac) = 0.250 + 0.008 = 0.258$ and $P'(a\overline{c}) = 0.016 + 0.004 = 0.020$; hence the conditional $(c|a)$ is accepted by P'.

The induced distribution $P := Mod_{\mathcal{S}}(\varphi)(P')$ is:

ω	$P(\omega)$
$\overline{s}\,\overline{u}$	$P(\overline{s}\,\overline{u}) = P'(\overline{a}\,\overline{c}) = 0.533$
$\overline{s}u$	$P(\overline{s}u) = P'(\overline{a}c) = 0.189$
$s\overline{u}$	$P(s\overline{u}) = P'(a\overline{c}) = 0.020$
su	$P(su) = P'(ac) = 0.258$

Since $P(su) > P(s\overline{u})$, P accepts the conditional $(u|s)$ corresponding to $(c|a)$ in $Sen_{\mathcal{S}}(\Sigma')$ under φ.

4.2 Standard Probabilistic Logics

For the **Probabilistic Propositional Logic** $Inst_{\mathcal{P}}$, the models are given by $|\,Mod_{\mathcal{P}}(\Sigma)\,| = \{P \mid P \text{ is a probability distribution over } \Sigma\}$.

For each signature morphism $\varphi : \Sigma \to \Sigma'$, we define a functor $Mod_{\mathcal{P}}(\varphi) : Mod_{\mathcal{P}}(\Sigma') \to Mod_{\mathcal{P}}(\Sigma)$ analogously to $Mod_{\mathcal{S}}(\varphi)$ (see Sec. 4.1).

Sentences: For each signature Σ, the set $Sen_{\mathcal{P}}(\Sigma)$ contains *probabilistic facts* of the form $A[x]$ where $A \in Sen_{\mathcal{B}}(\Sigma)$ is a propositional formula and $x \in [0, 1]$ is a probability value indicating the degree of certainty for the occurrence of A.

For each signature morphism $\varphi : \Sigma \to \Sigma'$, the extension $Sen_{\mathcal{P}}(\varphi) : Sen_{\mathcal{P}}(\Sigma) \to Sen_{\mathcal{P}}(\Sigma')$ is defined by $Sen_{\mathcal{P}}(\varphi)(A[x]) = \varphi(A)[x]$.

Satisfaction relation: The satisfaction relation $\models_{\mathcal{P},\Sigma} \subseteq |Mod_{\mathcal{P}}(\Sigma)| \times Sen_{\mathcal{P}}(\Sigma)$ is defined, for any $\Sigma \in |Sig_{\mathcal{P}}|$, by

$$P \models_{\mathcal{P},\Sigma} A[x] \quad \text{iff} \quad P(A) = x.$$

Note that, since $P(\overline{A}) = 1 - P(A)$ for each formula $A \in Sen_{\mathcal{B}}(\Sigma)$, it holds that $P \models_{\mathcal{P},\Sigma} A[x]$ iff $P \models_{\mathcal{P},\Sigma} \overline{A}[1-x]$.

For the **Probabilistic Conditional Logic** $Inst_{\mathcal{C}}$, the models are again probability distributions over the propositional variables; therefore, $Mod_{\mathcal{C}} = Mod_{\mathcal{P}}$.

Sentences: For each signature Σ, the set $Sen_{\mathcal{C}}(\Sigma)$ contains *probabilistic conditionals* (sometimes also called *probabilistic rules*) of the form $(B|A)[x]$ where $A, B \in Sen_{\mathcal{B}}(\Sigma)$ are propositional formulas and $x \in [0,1]$ is a probability value indicating the degree of certainty for the occurrence of B under the condition A.

Note that the sentences from $Inst_{\mathcal{P}}$ are included implicitly since a probabilistic fact of the form $B[x]$ can easily be expressed as a conditional $(B|\top)[x]$ with a tautology as trivial antecedent.

For each signature morphism $\varphi : \Sigma \to \Sigma'$, the extension $Sen_{\mathcal{C}}(\varphi) : Sen_{\mathcal{C}}(\Sigma) \to Sen_{\mathcal{C}}(\Sigma')$ is defined by straightforward inductive extension on the structure of the formulas: $Sen_{\mathcal{C}}(\varphi)((B|A)[x]) = (\varphi(B)|\varphi(A))[x]$.

Satisfaction relation: The satisfaction relation $\models_{\mathcal{C},\Sigma} \subseteq |Mod_{\mathcal{C}}(\Sigma)| \times Sen_{\mathcal{C}}(\Sigma)$ is defined, for any $\Sigma \in |Sig_{\mathcal{C}}|$, by

$$P \models_{\mathcal{C},\Sigma} (B|A)[x] \quad \text{iff} \quad P(A) > 0 \text{ and } P(B \mid A) = \frac{P(AB)}{P(A)} = x$$

Note that for probabilistic facts we have $P \models_{\mathcal{C},\Sigma} (B|\top)[x]$ iff $P(B) = x$ from the definition of the satisfaction relation since $P(\top) = 1$. Thus, $P \models_{\mathcal{P},\Sigma} B[x]$ iff $P \models_{\mathcal{C},\Sigma} (B|\top)[x]$.

4.3 Qualitative Logics

Besides the three probabilistic semantics introduced above, there are various types of models that have been proposed to interpret conditionals $(B|A)$ adequately within a logical system (cf. e.g. [16]) in a purely qualitative way. One of the most prominent approaches is the *system-of-spheres* model of Lewis [15] which makes use of a notion of similarity between possible worlds. This idea of comparing worlds and evaluating conditionals with respect to the "nearest" or "best" worlds (which are somehow selected) is common to very many approaches in conditional logics. A basic implementation of this can be achieved by plausibility preorders.

From this purely qualitative point of view, proper models of conditionals are provided by total preorders R (i.e. R is a total, reflexive and transitive relation) over classical propositional interpretations, or possible worlds, respectively. Possible worlds are ordered according to their *plausibility*; by convention, the least worlds are the most plausible worlds.

For a preorder R, we use the infix notation $\omega_1 \preceq_R \omega_2$ instead of $(\omega_1, \omega_2) \in R$. As usual, we introduce the $\prec_R$-relation by saying that $\omega_1 \prec_R \omega_2$ iff $\omega_1 \preceq_R \omega_2$ and not $(\omega_2 \preceq_R \omega_1)$. Furthermore, $\omega_1 \approx_R \omega_2$ means that both $\omega_1 \preceq_R \omega_2$ and $\omega_2 \preceq_R \omega_1$ hold.

Each total preorder R induces a partitioning $\Omega_0, \Omega_1, \ldots$ of Ω, such that all worlds in the same partitioning subset are considered equally plausible ($\omega_1 \approx_R \omega_2$ for $\omega_1, \omega_2 \in \Omega_j$), and whenever $\omega_1 \in \Omega_i$ and $\omega_2 \in \Omega_k$ with $i < k$, then $\omega_1 \prec_R \omega_2$. Moreover, a total preorder on Ω extends to a total preorder on propositional formulas A, B via: $A \preceq_R B$ iff there exists $\omega_1 \in \Omega$ with $\omega_1 \models_{\mathcal{B},\Sigma} A$ such that for all $\omega_2 \in \Omega$ with $\omega_2 \models_{\mathcal{B},\Sigma} B$, we have $\omega_1 \preceq_R \omega_2$.

Again, $A \prec_R B$ means both $A \preceq_R B$ and not $B \preceq_R A$.

Models: In correspondence with Lewis' system-of-spheres semantics, we have $|\mathit{Mod}_{\mathcal{K}}(\Sigma)| = \{R \mid R \text{ is a total preorder on } |\mathit{Mod}_{\mathcal{B}}(\Sigma)|\}$.

For each signature morphism $\varphi : \Sigma \to \Sigma'$, we define a functor $\mathit{Mod}_{\mathcal{K}}(\varphi) : \mathit{Mod}_{\mathcal{K}}(\Sigma') \to \mathit{Mod}_{\mathcal{K}}(\Sigma)$ by mapping a (total) preorder R' over $\mathit{Mod}_{\mathcal{B}}(\Sigma')$ to a (total) preorder $\mathit{Mod}_{\mathcal{K}}(\varphi)(R')$ over $\mathit{Mod}_{\mathcal{B}}(\Sigma)$ in the following way:

$$\omega_1 \preceq_{\mathit{Mod}_{\mathcal{K}}(\varphi)(R')} \omega_2 \quad \text{iff} \quad \varphi(\omega_1) \preceq_{R'} \varphi(\omega_2) \tag{6}$$

Sentences: The sentences are (unquantified) conditionals $(B|A)$, i.e., $\mathit{Sen}_{\mathcal{K}} = \mathit{Sen}_{\mathcal{S}}$.

Satisfaction relation: The satisfaction relation $\models_{\mathcal{K},\Sigma} \subseteq |\mathit{Mod}_{\mathcal{K}}(\Sigma)| \times \mathit{Sen}_{\mathcal{K}}(\Sigma)$ is defined, for any $\Sigma \in |\mathit{Sig}_{\mathcal{K}}|$, by

$$R \models_{\mathcal{K},\Sigma} (B|A) \quad \text{iff} \quad AB \prec_R A\overline{B}$$

Therefore, a conditional $(B|A)$ is satisfied (or accepted) by the plausibility preorder R iff its confirmation AB is more plausible than its refutation $A\overline{B}$.

Example 3. We continue our student example in this qualitative conditional environment, so let Σ, Σ', φ be as defined in Example 1. Let R' be the following total preorder on Ω': $\overline{a}\overline{b}\overline{c} \prec_{R'} abc \approx_{R'} \overline{a}bc \prec_{R'} ab\overline{c} \approx_{R'} a\overline{b}c \approx_{R'} a\overline{b}\overline{c} \approx_{R'} \overline{a}b\overline{c} \approx_{R'} \overline{a}\overline{b}c$.

For instance $R' \models_{\mathcal{K},\Sigma'} (c|a)$ – *students* are supposed to be *unmarried* since under R', ac is more plausible than $a\overline{c}$.

5 Singularity of Big-Stepped Probabilities

We are now ready to investigate the precise relationships between the institution of big-stepped probabilities to any of the other three conditional logics defined in Sec. 4. After recalling the relationships from $\mathit{Inst}_{\mathcal{C}}$ to $\mathit{Inst}_{\mathcal{P}}$ and $\mathit{Inst}_{\mathcal{K}}$ (Sec. 5.1), we prove detailed existence and nonexistence results of institution morphisms between $\mathit{Inst}_{\mathcal{S}}$ and $\mathit{Inst}_{\mathcal{C}}$ (Sec. 5.2) and between $\mathit{Inst}_{\mathcal{S}}$ and $\mathit{Inst}_{\mathcal{P}}$ (Sec. 5.3), establishing essentially unique morphisms from $\mathit{Inst}_{\mathcal{S}}$ to both $\mathit{Inst}_{\mathcal{C}}$ and $\mathit{Inst}_{\mathcal{P}}$. In Sec. 5.4, we show that there are two rather different institution morphisms from $\mathit{Inst}_{\mathcal{S}}$ to $\mathit{Inst}_{\mathcal{K}}$, thus completing the picture with institution morphisms starting from $\mathit{Inst}_{\mathcal{S}}$, but none going back to $\mathit{Inst}_{\mathcal{S}}$.

5.1 Relationships among Standard Probabilistic and Qualitative Logics

When mapping a probabilistic proposition $A[x]$ to a probabilistic conditional, the obvious correspondence $\alpha_{\mathcal{P}/\mathcal{C},\Sigma}(A[x]) = (A|\top)[x]$ together with the model identity between $Inst_{\mathcal{P}}$ and $Inst_{\mathcal{C}}$ yields an institution morphism.

Proposition 1 ([4]). *$\langle\, id_{Sig_{\mathcal{B}}}, \alpha_{\mathcal{P}/\mathcal{C}}, \beta \,\rangle : Inst_{\mathcal{C}} \longrightarrow Inst_{\mathcal{P}}$ is an institution morphism iff $\beta = id_{Mod_{\mathcal{C}}}$.*

In the other direction, there is no institution morphism from $Inst_{\mathcal{P}}$ to $Inst_{\mathcal{C}}$ which leaves the models unchanged.

Proposition 2. *There is no α such that $\langle\, id_{Sig_{\mathcal{B}}}, \alpha, id_{Mod_{\mathcal{P}}} \,\rangle : Inst_{\mathcal{P}} \longrightarrow Inst_{\mathcal{C}}$ is an institution morphism.*

Proof. If there were such a morphism, the satisfaction condition would require every $\alpha_{\Sigma}((B|A)[x])$ to be probabilistic equivalent to $(B|A)[x]$. However, this is not possible since in general, a probabilistic fact can not be probabilistically equivalent to a probabilistic conditional. □

For relating the sentences of $Inst_{\mathcal{K}}$ and $Inst_{\mathcal{C}}$, a qualitative conditional $(B|A)$ is mapped to a quantitative conditional by $\alpha_{\mathcal{K}/\mathcal{C},\Sigma}((B|A)) = (B|A)[1]$. For the other direction, we can safely assume at least $\alpha((B|A)[1]) = (B|A)$ for any sentence $(B|A)[1]$ having trivial probability 1.

Relating the models of $Inst_{\mathcal{K}}$ and $Inst_{\mathcal{C}}$, however, is far less obvious. Many different ways can be devised to map preorders and probability distributions to one another. As a first approach, we define a mapping sending a probability distribution P to a conditional logic model $\beta_{\mathcal{C}/\mathcal{K},\Sigma}(P) = R_P$. Under R_P, all complete conjunctions with a positive probability are considered most plausible, and all complete conjunctions with zero probability are taken as less (yet equally) plausible. Thus, R_P partitions Ω into two sets, namely $\{\omega \in \Omega \mid P(\omega) > 0\}$ and $\{\omega \in \Omega \mid P(\omega) = 0\}$:

$$\omega_1 \preceq_{R_P} \omega_2 \quad \text{iff} \quad P(\omega_2) = 0 \text{ or } (P(\omega_1) > 0 \text{ and } P(\omega_2) > 0) \tag{7}$$

Although the preordering concept would allow a rather fine-grained hierarchy of plausibilities, it is surprising to see that the somewhat simplistic two-level approach of R_P is the *only* possibility to augment $\alpha_{\mathcal{K}/\mathcal{C}}$ towards an institution morphism.

Proposition 3 ([3]). *$\langle\, id_{Sig_{\mathcal{B}}}, \alpha_{\mathcal{K}/\mathcal{C}}, \beta \,\rangle : Inst_{\mathcal{C}} \longrightarrow Inst_{\mathcal{K}}$ is an institution morphism iff $\beta = \beta_{\mathcal{C}/\mathcal{K}}$, and there is no institution morphism $\langle\, id_{Sig_{\mathcal{B}}}, \alpha, \beta \,\rangle : Inst_{\mathcal{K}} \longrightarrow Inst_{\mathcal{C}}$ such that $\alpha_{\Sigma}((B|A)[1]) = (B|A)$ for all signatures Σ.*

5.2 Relating $Inst_{\mathcal{S}}$ to Probabilistic Conditional Logic

When looking for an institution morphism from $Inst_{\mathcal{C}}$ to $Inst_{\mathcal{S}}$, a natural requirement is to attach probability 1 to a qualitative conditional. However, this is not possible within an institution morphism.

Proposition 4. *There is no* β *such that* $\langle\, id_{Sig_{\mathcal{B}}},\, \alpha,\, \beta\,\rangle : Inst_{\mathcal{C}} \longrightarrow Inst_{\mathcal{S}}$ *is an institution morphism with* $\alpha_\Sigma((B|A)) = (B|A)[1]$ *for every* $(B|A) \in Sen_{\mathcal{S}}(\Sigma)$.

Proof. Assume that there were such a β. Then the satisfaction condition for institution morphisms would require $P \models_{\mathcal{C},\Sigma} (B|A)[1]$ iff $\beta_\Sigma(P) \models_{\mathcal{S},\Sigma} (B|A)$ for every probability distribution $P \in Mod_{\mathcal{C}}(\Sigma)$.

Let $\Sigma = \{a, b\}$ and let P be the distribution with $P(\omega) = 0.25$ for every $\omega \in \{ab, a\overline{b}, \overline{a}b, \overline{a}\overline{b}\}$. Since every big-stepped probability distribution either satisfies a conditional $(B|A)$ or its counterpart $(\overline{B}|A)$, we have either $\beta_\Sigma(P) \models_{\mathcal{S},\Sigma} (b|a)$ or $\beta_\Sigma(P) \models_{\mathcal{S},\Sigma} (\overline{b}|a)$. Thus, since $P \not\models_{\mathcal{C},\Sigma} (b|a)[1]$ and $P \not\models_{\mathcal{C},\Sigma} (\overline{b}|a)[1]$ the needed satisfaction condition is not satisfied. □

When looking in the other direction from $Inst_{\mathcal{S}}$ to $Inst_{\mathcal{C}}$, the special situation that all models of $Inst_{\mathcal{S}}$ are also models of $Inst_{\mathcal{C}}$ yields the identity $id_{Mod_{\mathcal{S}}}$ on models as an obvious choice for the model translation. Moreover, requiring at least the trivial correspondence $\alpha_\Sigma((B|A)[1]) = (B|A)$ for the sentence translation seems obvious. However, the following two propositions show that both of theses choices do not enable us to find an institution morphism going from big-stepped probabilities to probabilistic conditional logic.

Proposition 5. *There is no* β *such that* $\langle\, id_{Sig_{\mathcal{B}}},\, \alpha,\, \beta\,\rangle : Inst_{\mathcal{S}} \longrightarrow Inst_{\mathcal{C}}$ *is an institution morphism with* $\alpha_\Sigma((B|A)[1]) = (B|A)$ *for every* $(B|A)[1] \in Sen_{\mathcal{C}}(\Sigma)$.

Proof. Assume that there were such a β. Then the satisfaction condition for institution morphisms would require $\beta_\Sigma(P) \models_{\mathcal{C},\Sigma} (B|A)[1]$ iff $P \models_{\mathcal{S},\Sigma} (B|A)$ for any big-stepped probability distribution $P \in Mod_{\mathcal{S}}(\Sigma)$.

Let $\Sigma = \{a, b\}$ and let P be given by $P(ab) = 0.51$, $P(a\overline{b}) = 0.26$, $P(\overline{a}b) = 0.14$, and $P(\overline{a}\overline{b}) = 0.09$. Since $P \models_{\mathcal{S},\Sigma} (a|\top)$ and $P \models_{\mathcal{S},\Sigma} (b|\top)$, the satisfaction condition gives us $\beta_\Sigma(P) \models_{\mathcal{C},\Sigma} (a|\top)[1]$ and $\beta_\Sigma(P) \models_{\mathcal{C},\Sigma} (b|\top)[1]$ and hence necessarily $\beta_\Sigma(P)(ab) = 1$ and $\beta_\Sigma(P)(a\overline{b}) = \beta_\Sigma(P)(\overline{a}b) = \beta_\Sigma(P)(\overline{a}\overline{b}) = 0$. Now the satisfaction condition is violated since $P \models_{\mathcal{S},\Sigma} (\overline{a}|b)$, but $\beta_\Sigma(P) \not\models_{\mathcal{C},\Sigma} (\overline{a}|b)[1]$. □

Proposition 6. *There is no* α *such that* $\langle\, id_{Sig_{\mathcal{B}}},\, \alpha,\, id_{Mod_{\mathcal{S}}}\,\rangle : Inst_{\mathcal{S}} \longrightarrow Inst_{\mathcal{C}}$ *is an institution morphism.*

Proof. Assume that there were such an α. Let $\Sigma = \{a\}$ and $P_1, P_2 \in Mod_{\mathcal{S}}(\Sigma)$ be the big-stepped probability distributions uniquely determined by $P_1(a) = 0.9$ and $P_2(a) = 0.7$. Since $P_1 \models_{\mathcal{C},\Sigma} (a|\top)[0.9]$ and $P_2 \not\models_{\mathcal{C},\Sigma} (a|\top)[0.9]$, the satisfaction condition requires $P_1 \models_{\mathcal{S},\Sigma} \alpha_\Sigma((a|\top)[0.9])$ and $P_2 \not\models_{\mathcal{S},\Sigma} \alpha_\Sigma((a|\top)[0.9])$. However, since as big-stepped probabilities P_1 and P_2 are equivalent in the sense that they satisfy exactly the same conditionals in $Sen_{\mathcal{S}}(\Sigma)$, this is a contradiction. □

Thus, it is not obvious how the logic of big-stepped probabilities can formally be related to standard probabilistic conditional logic by means of an institution morphism, despite the fact that there are obvious and straight-forward relationships both on the semantical level of models as well as on the syntactical level of sentences: There is a simple subset relationship between their sets of models, and their sentences can be plausibly related using the correspondence between $(B|A)$ and $(B|A)[1]$. Summarizing Propositions 4 – 6, we have shown the nonexistence of institution morphisms for the following given sentence and model translations:

Institutions	Sentence translation	Model translation
$Inst_{\mathcal{C}} \not\rightarrow Inst_{\mathcal{S}}$	$(B\|A) \mapsto (B\|A)[1]$	any β
$Inst_{\mathcal{S}} \not\rightarrow Inst_{\mathcal{C}}$	any α with $(B\|A)[1] \mapsto (B\|A)$	any β
$Inst_{\mathcal{S}} \not\rightarrow Inst_{\mathcal{C}}$	any α	$P \mapsto P$

Despite the fact that neither mapping $(B|A)[1]$ to $(B|A)$ nor sticking to the identity for the model translation yields a morphism from $Inst_{\mathcal{S}}$ to $Inst_{\mathcal{C}}$, we will show that there is still another possibility. This requires morphing a probability distribution into another one: Instead of the obvious choice for using the model identity, we can transform any big-stepped probability distribution $P \in Mod_{\mathcal{S}}(\Sigma)$ to the simplistic probability distribution $\beta_{\mathcal{S}/\mathcal{C}}(P) = P_S$ with

$$P_S(\omega) = \begin{cases} 1 & \text{if } \omega = \arg\max_{\omega \in \Omega} P(\omega) \\ 0 & \text{otherwise} \end{cases} \tag{8}$$

As a consequence, we must also drop the straightforward requirement on the sentence translation. Since in the simplistic model P_S the probabilty of the maximal world under P is increased to 1 and the probabilty of all other worlds is decreased to 0, also the interpretation of a conditional changes. Instead of mapping the probabilistic conditional $(B|A)[1]$ to the qualitative conditional $(B|A)$, we use $(AB|\top)$ instead, with AB confirming the conditional. Completing this dually by interpreting probability 0 as negation and all other probabilities between 0 and 1 as inconsistencies, we get:

$$\alpha_{\mathcal{C}/\mathcal{S},\Sigma}((B|A)[x]) = \begin{cases} (AB|\top) & \text{if } x = 1 \\ (A\overline{B}|\top) & \text{if } x = 0 \\ (\bot|\top) & \text{otherwise} \end{cases} \tag{9}$$

Proposition 7. $\langle id_{Sig_{\mathcal{B}}}, \alpha_{\mathcal{C}/\mathcal{S}}, \beta_{\mathcal{S}/\mathcal{C}} \rangle : Inst_{\mathcal{S}} \longrightarrow Inst_{\mathcal{C}}$ *is an institution morphism.*

Proof. We have to show

$$P_S \models_{\mathcal{C},\Sigma} (B|A)[x] \quad \text{iff} \quad P \models_{\mathcal{S},\Sigma} \alpha_{\mathcal{C}/\mathcal{S},\Sigma}((B|A)[x]) \tag{10}$$

for any big-stepped probability distribution $P \in Mod_{\mathcal{S}}(\Sigma)$ and any probabilistic conditional $(B|A)[x]$. Let $\omega_P = \max_{\omega \in \Omega} P(\omega)$.

If $x = 1$, we have $P \models_{\mathcal{S},\Sigma} \alpha_{\mathcal{C}/\mathcal{S},\Sigma}((B|A)[1])$ iff $P \models_{\mathcal{S},\Sigma} (AB|\top)$ iff $P(AB) > P(\neg(AB))$ iff $\omega_P \models_{\mathcal{B},\Sigma} AB$ iff – since P is big-stepped – $P_S(AB) = 1$ iff $P_S(AB) = 1$ and $P_S(A) = 1$ iff $P_S \models_{\mathcal{C},\Sigma} (B|A)[1]$.

If $x = 0$, we have $P \models_{\mathcal{S},\Sigma} \alpha_{\mathcal{C}/\mathcal{S},\Sigma}((B|A)[0])$ iff $P \models_{\mathcal{S},\Sigma} (A\overline{B}|\top)$ iff $P(A\overline{B}) > P(\neg(A\overline{B}))$ iff $\omega_P \models_{\mathcal{B},\Sigma} A\overline{B}$ iff $P_S(A\overline{B}) = 1$ iff $P_S(A\overline{B}) = 1, P_S(A) = 1$ and $P_S(\overline{B}) = 1$ iff $P_S \models_{\mathcal{C},\Sigma} (B|A)[0]$.

If $x \neq 0$ and $x \neq 1$ we have $P \not\models_{\mathcal{S},\Sigma} \alpha_{\mathcal{C}/\mathcal{S},\Sigma}((B|A)[x])$ since $P \not\models_{\mathcal{S},\Sigma} \alpha_{\mathcal{C}/\mathcal{S},\Sigma}((\bot|\top))$ because $P(\bot) = 0$. At the same time, $P_S \not\models_{\mathcal{C},\Sigma} (B|A)[x]$ since $P_S((B|A)) \neq x$ because $P_S(t)$ can only be 0 or 1 for any sentence t. □

Thus, there is indeed an institution morphism from big-stepped probabilities to standard probabilistic conditional logic: Any big-stepped probability P must be mapped to its two-valued abstraction P_S given by (8). The corresponding syntactical adjustment transforms a quantitative conditional $(B|A)[1]$ to the qualitative conditional $(AB|\top)$ as given by (9). That means that $(B|A)[1]$ holds on the abstract level of P_S iff AB is plausible with respect to P. Moreover, from general properties of institition morphisms [8] we conclude that under the given model translation, the sentence translation given by (9) is determined uniquely up to semantic equivalence.

5.3 Relating *Inst*$_{\mathcal{S}}$ to Probabilistic Propositional Logic

For an institution morphism from $Inst_{\mathcal{P}}$ to $Inst_{\mathcal{S}}$ we could map a qualitative conditional $(B|A)$ to a probabilistic material implication $(A \Rightarrow B)[1]$ or to the verifying world $(AB)[1]$. Both choices are not successful.

Proposition 8. *There is no β such that $\langle id_{Sig_{\mathcal{B}}}, \alpha, \beta \rangle : Inst_{\mathcal{P}} \longrightarrow Inst_{\mathcal{S}}$ is an institution morphism with $\alpha_\Sigma((B|A)) = (A \Rightarrow B)[1]$ for every $(B|A) \in Sen_{\mathcal{S}}(\Sigma)$.*

Proposition 9. *There is no β such that $\langle id_{Sig_{\mathcal{B}}}, \alpha, \beta \rangle : Inst_{\mathcal{P}} \longrightarrow Inst_{\mathcal{S}}$ is an institution morphism with $\alpha_\Sigma((B|A)) = (AB)[1]$ for every $(B|A) \in Sen_{\mathcal{S}}(\Sigma)$.*

In the other direction, similar to the situation with $Inst_{\mathcal{C}}$, exploiting the model inclusion property between $Inst_{\mathcal{S}}$ and $Inst_{\mathcal{P}}$ yields a morphism.

Proposition 10. *There is no α such that $\langle id_{Sig_{\mathcal{B}}}, \alpha, id_{Mod_{\mathcal{S}}} \rangle : Inst_{\mathcal{S}} \longrightarrow Inst_{\mathcal{P}}$ is an institution morphism.*

Proof. By using $a[0.9]$ and $\models_{\mathcal{P},\Sigma}$ instead of $(a|\top)[0.9]$ and $\models_{\mathcal{C},\Sigma}$, the proof follows from the argumentation given in the proof of Prop. 6. □

Since sticking to the model identity is not successful, we try the same model transformation as in the previous section:

$$\beta_{\mathcal{S}/\mathcal{P},\Sigma}(P) := \beta_{\mathcal{S}/\mathcal{C},\Sigma}(P) = P_S$$

Correspondingly, we have to adapt the sentence translation. In the simplistic model P_S, all probabilities except 0 and 1 are viewed as indicating an inconsistency:

$$\alpha_{\mathcal{P}/\mathcal{S},\Sigma}(A[x]) = \begin{cases} (A|\top) & \text{if } x = 1 \\ (\overline{A}|\top) & \text{if } x = 0 \\ (\bot|\top) & \text{otherwise} \end{cases}$$

Proposition 11. *$\langle id_{Sig_{\mathcal{B}}}, \alpha_{\mathcal{P}/\mathcal{S}}, \beta_{\mathcal{S}/\mathcal{P}} \rangle : Inst_{\mathcal{S}} \longrightarrow Inst_{\mathcal{P}}$ is an institution morphism.*

Proof. The proof follows immediately from the observation that $\langle id_{Sig_{\mathcal{B}}}, \alpha_{\mathcal{P}/\mathcal{S}}, \beta_{\mathcal{S}/\mathcal{P}} \rangle$ is the composition of the institution morphisms from $Inst_{\mathcal{S}}$ to $Inst_{\mathcal{C}}$ and from $Inst_{\mathcal{C}}$ to $Inst_{\mathcal{P}}$ established in Propositions 7 and 1. □

5.4 Relating $Inst_{\mathcal{S}}$ to Purely Qualitative Logic

Since $Inst_{\mathcal{K}}$ and $Inst_{\mathcal{S}}$ share the same sentences, the natural choice for mapping sentences between these logics is of course the identity. With regard to models, we can associate to each big-stepped probability $P_{BS} \in Mod_{\mathcal{S}}(\Sigma)$ a total preorder $\beta_{\mathcal{S}/\mathcal{K},\Sigma}(P_{BS}) = R_{P_{BS}} \in Mod_{\mathcal{K}}(\Sigma)$ via

$$\omega_1 \preceq_{R_{P_{BS}}} \omega_2 \quad \text{iff} \quad P_{BS}(\omega_1) \geqslant P_{BS}(\omega_2) \tag{11}$$

(cf. [6]), and indeed, this yields an institution morphism. We can also show that there is no model translation other than $\beta_{\mathcal{S}/\mathcal{K}}$ that can be used to augment the sentence identity $id_{Sen_{\mathcal{K}}}$ towards an institution morphism. However, in the other direction, there is no such institution morphism:

Proposition 12 ([5]). *$\langle id_{Sig_{\mathcal{B}}}, id_{Sen_{\mathcal{K}}}, \beta \rangle : Inst_{\mathcal{S}} \rightarrow Inst_{\mathcal{K}}$ is an institution morphism iff $\beta = \beta_{\mathcal{S}/\mathcal{K}}$, and there is no institution morphism $\langle id_{Sig_{\mathcal{B}}}, id_{Sen_{\mathcal{K}}}, \beta \rangle : Inst_{\mathcal{K}} \rightarrow Inst_{\mathcal{S}}$.*

Since using the identity for sentence translation as done in [5] is certainly justified by the fact that $Inst_{\mathcal{K}}$ and $Inst_{\mathcal{S}}$ have the same sentences, so far no other sentence translation has been considered for defining an institution morphism from the logic of big-stepped probabilites to qualitative logics with preorders. However, surprisingly, there is still another possibility. Our findings concerning the interpretation of conditionals with big-stepped probabilities as semantics in a standard pobabilistic setting given in Sec. 5.2, yield an alternative sentence translation obtained by composing the sentence translations used in the institution morphisms from $Inst_{\mathcal{S}}$ to $Inst_{\mathcal{C}}$ and from $Inst_{\mathcal{C}}$ to $Inst_{\mathcal{K}}$. Under this alternative translation, a conditional $(B|A)$ is mapped to a conditional with trivial antecedance and the proposition that confirms the conditional as consequence:

$$\alpha'_{\mathcal{K}/\mathcal{S},\Sigma}((B|A)) = (AB|\top).$$

Together with the corresponding model translation

$$\beta^{bin}_{\mathcal{S}/\mathcal{K},\Sigma}(P_{BS}) = R^{bin}_{P_{BS}} \tag{12}$$

where $R^{bin}_{P_{BS}}$ is the preorder given by the binary partitioning of Ω_Σ into

$$\{\omega_{max}\} \preceq_{R^{bin}_{P_{BS}}} \Omega_\Sigma \backslash \{\omega_{max}\}$$

with $\omega_{max} = \arg\max_\omega P_{BS}(\omega)$, we obtain a second, rather different institution morphism from $Inst_{\mathcal{S}}$ to $Inst_{\mathcal{K}}$ by composition of the institution morphisms from $Inst_{\mathcal{S}}$ to $Inst_{\mathcal{C}}$ (Prop. 7) and from $Inst_{\mathcal{C}}$ to $Inst_{\mathcal{K}}$ (Prop. 3):

Proposition 13. *$\langle\, id_{Sig_{\mathcal{B}}}, \alpha'_{\mathcal{K}/\mathcal{S}}, \beta^{bin}_{\mathcal{S}/\mathcal{K}} \rangle : Inst_{\mathcal{S}} \longrightarrow Inst_{\mathcal{K}}$ is an institution morphism.*

Furthermore, it is straightforward to prove that the model translation $\beta^{bin}_{\mathcal{S}/\mathcal{K}}$ is uniquely determined by the sentence translation $\alpha'_{\mathcal{K}/\mathcal{S}}$.

Figures 3 and 4 summarize our findings with respect to the existence resp. nonexistence of institutions morphisms between the four logics. The special characteristics of big-stepped probabilities are underlined by the position of $Inst_{\mathcal{S}}$ in the graph of institutions: There are morphisms from $Inst_{\mathcal{S}}$ to each of the other logics, both standard probabilistic and qualitative, but none in the other direction. This is due to the overspecialized semantics of $Inst_{\mathcal{S}}$. Furthermore, moving from the logic $Inst_{\mathcal{S}}$ of big-stepped probabilities to the standard probabilistic logic $Inst_{\mathcal{C}}$ necessarily requires seemingly unexpected transformations of models and sentences, and also surprisingly, there are two rather different alternatives of going from $Inst_{\mathcal{S}}$ to the system-of-spheres logic $Inst_{\mathcal{K}}$.

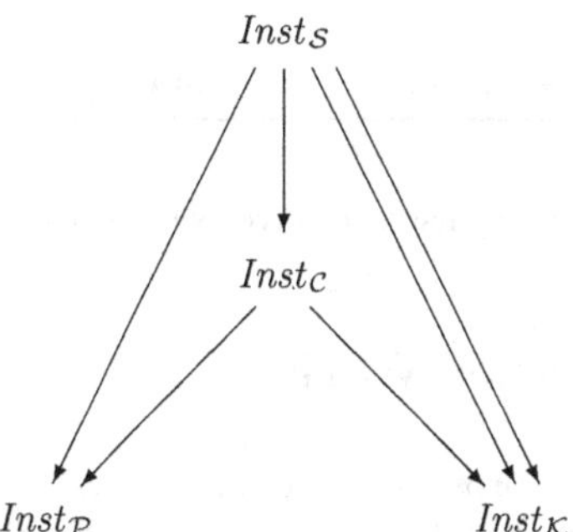

Morphism	Sentence translation		Model translation		Prop.
$Inst_{\mathcal{C}} \longrightarrow Inst_{\mathcal{P}}$	$A[x]$	$\mapsto (A\|\top)[x]$	$P \mapsto P$		1
$Inst_{\mathcal{C}} \longrightarrow Inst_{\mathcal{K}}$	$(B\|A)$	$\mapsto (B\|A)[1]$	$P \mapsto R_P$	(7)	3
$Inst_{\mathcal{S}} \longrightarrow Inst_{\mathcal{C}}$	$(B\|A)[x]$	$\mapsto \begin{cases} (AB\|\top) & \text{if } x=1 \\ (A\overline{B}\|\top) & \text{if } x=0 \\ (\bot\|\top) & \text{otherwise} \end{cases}$	$P \mapsto P_S$	(8)	7
$Inst_{\mathcal{S}} \longrightarrow Inst_{\mathcal{P}}$	$A[x]$	$\mapsto \begin{cases} (A\|\top) & \text{if } x=1 \\ (\overline{A}\|\top) & \text{if } x=0 \\ (\bot\|\top) & \text{otherwise} \end{cases}$	$P \mapsto P_S$	(8)	11
$Inst_{\mathcal{S}} \longrightarrow Inst_{\mathcal{K}}$	$(B\|A)$	$\mapsto (B\|A)$	$P_{BS} \mapsto R_{P_{BS}}$	(11)	12
$Inst_{\mathcal{S}} \longrightarrow Inst_{\mathcal{K}}$	$(B\|A)$	$\mapsto (AB\|\top)$	$P_{BS} \mapsto R^{bin}_{P_{BS}}$	(12)	13

Fig. 3. Institution morphisms among $Inst_{\mathcal{S}}$, $Inst_{\mathcal{P}}$, $Inst_{\mathcal{C}}$, and $Inst_{\mathcal{K}}$

Institutions	Sentence translation	Model translation	
$Inst_{\mathcal{P}} \not\mapsto Inst_{\mathcal{C}}$	any α	$P \mapsto P$	Prop. 2
$Inst_{\mathcal{K}} \not\mapsto Inst_{\mathcal{C}}$	any α with $(B\|A)[1] \mapsto (B\|A)$	any β	Prop. 3
$Inst_{\mathcal{C}} \not\mapsto Inst_{\mathcal{S}}$	$(B\|A) \mapsto (B\|A)[1]$	any β	Prop. 4
$Inst_{\mathcal{S}} \not\mapsto Inst_{\mathcal{C}}$	any α with $(B\|A)[1] \mapsto (B\|A)$	any β	Prop. 5
$Inst_{\mathcal{S}} \not\mapsto Inst_{\mathcal{C}}$	any α	$P \mapsto P$	Prop. 6
$Inst_{\mathcal{P}} \not\mapsto Inst_{\mathcal{S}}$	any α with $(B\|A) \mapsto (A \Rightarrow B)[1]$	any β	Prop. 8
$Inst_{\mathcal{P}} \not\mapsto Inst_{\mathcal{S}}$	any α with $(B\|A) \mapsto AB[1]$	any β	Prop. 9
$Inst_{\mathcal{S}} \not\mapsto Inst_{\mathcal{P}}$	any α	$P \mapsto P$	Prop. 10
$Inst_{\mathcal{K}} \not\mapsto Inst_{\mathcal{S}}$	$(B\|A) \mapsto (B\|A)$	any β	Prop. 12

Fig. 4. Nonexistence of institution morphisms for given sentence and model translations

6 Conclusions and Further Work

Big-stepped probabilities [6, 18] provide a model for system P [12] and allow for qualitative reasoning with probabilities. Using formalizations of big-stepped probabilities, standard probabilities and a purely qualitative logic as institutions, we investigated the exact formal relationships between these logics in terms of institutions morphisms [8], comprising a fine-grained comparison taking into account mappings between sentences and corresponding transformations between models while respecting logical entailment. Instead of using possibly different ad-hoc comparisons, the institution framework enabled us not only to base our investigations on common, well-established grounds, but also allowed us to exploit general properties like composition arguments for institution morphisms.

In this paper, we focus on elaborating the probabilistic logical qualities of big-stepped probabilities. In [5], we investigated their qualitative aspects in more detail by studying relationships among ordinal conditional logic, possibilistic conditional logic, and big-stepped probabilities. While ordinal and possibilistic logic provide, as expected, more fine-grained frameworks to give semantics to conditionals than the purely qualitative logic $Inst_{\mathcal{K}}$, the big-stepped probability approach keeps its singular position in this broadened qualitative picture. There exist institution morphisms in both directions between each pair of purely qualitative conditional logic, ordinal conditional logic, and possibilistic conditional logic, while none of them can be embedded via a morphism in the institution of big-stepped probabilities. So basically, we have the same situation for all these qualitative conditional logics as for $Inst_{\mathcal{K}}$, when comparing them to $Inst_{\mathcal{S}}$ (see Figures 3 and 4), while no essential logical differences prevail among them.

Formalizing big-stepped probabilities as an institution requires the signature morphisms taken from the underlying propositional logic $Inst_{\mathcal{B}}$ to be injective (otherwise, the corresponding forgetful functor might not be well-defined). In our further work, we will elaborate the exact details which effects occur among qualitative and quantiative logics when allowing non-injective signature morphisms.

Besides providing a formal framework for comparing logics, institution morphisms are also a powerful concept supporting e.g. the moving of deduction tasks from one institution to another, or the building up of knowledge bases involving different logics [8]. We are currently working on methods and tools exploiting these options for conditional knowledge bases. Another line of further work is to investigate the benefits of adding more structure to the model categories $Mod(\Sigma)$ for which in this paper, for simplicity reasons we considered only the identity morphisms.

References

1. Adams, E.W.: The Logic of Conditionals. D. Reidel, Dordrecht (1975)
2. Arló Costa, H., Parikh, R.: Conditional probability and defeasible inference. Journal of Philosophical Logic 34, 97–119 (2005)
3. Beierle, C., Kern-Isberner, G.: Using institutions for the study of qualitative and quantitative conditional logics. In: Flesca, S., Greco, S., Leone, N., Ianni, G. (eds.) JELIA 2002. LNCS (LNAI), vol. 2424, pp. 161–172. Springer, Heidelberg (2002)
4. Beierle, C., Kern-Isberner, G.: Looking at probabilistic conditionals from an institutional point of view. In: Kern-Isberner, G., Rödder, W., Kulmann, F. (eds.) WCII 2002. LNCS (LNAI), vol. 3301, pp. 162–179. Springer, Heidelberg (2005)
5. Beierle, C., Kern-Isberner, G.: Formal similarities and differences among qualitative conditional semantics. International Journal of Approximate Reasoning (2009) (to appear)
6. Benferhat, S., Dubois, D., Prade, H.: Possibilistic and standard probabilistic semantics of conditional knowledge bases. Journal of Logic and Computation 9(6), 873–895 (1999)
7. Goguen, J., Burstall, R.: Institutions: Abstract model theory for specification and programming. Journal of the ACM 39(1), 95–146 (1992)
8. Goguen, J.A., Rosu, G.: Institution morphisms. Formal Aspects of Computing 13(3-5), 274–307 (2002)
9. Hawthorne, J.: On the logic of nonmonotonic conditionals and conditional probabilities. Journal of Philosophical Logic 25(2), 185–218 (1996)
10. Hawthorne, J., Makinson, D.: The quantitative/qualitative watershed for rules of uncertain inference. Studia Logica 86(2), 247–297 (2007)
11. Herrlich, H., Strecker, G.E.: Category theory. Allyn and Bacon, Boston (1973)
12. Kraus, S., Lehmann, D., Magidor, M.: Nonmonotonic reasoning, preferential models and cumulative logics. Artificial Intelligence 44, 167–207 (1990)
13. Kyburg Jr., H.E., Teng, C.-M., Wheeler, G.R.: Conditionals and consequences. J. Applied Logic 5(4), 638–650 (2007)

14. Lehmann, D., Magidor, M.: What does a conditional knowledge base entail? Artificial Intelligence 55, 1–60 (1992)
15. Lewis, D.: Counterfactuals. Harvard University Press, Cambridge (1973)
16. Nute, D.: Topics in Conditional Logic. D. Reidel Publishing Company, Dordrecht (1980)
17. Pearl, J.: Probabilistic semantics for nonmonotonic reasoning: A survey. In: Shafer, G., Pearl, J. (eds.) Readings in uncertain reasoning, pp. 699–710. Morgan Kaufmann, San Mateo (1989)
18. Snow, P.: The emergence of ordered belief from initial ignorance. In: Proceedings AAAI 1994, Seattle, WA, pp. 281–286 (1994)

Theoretical Foundations for Enabling a Web of Knowledge

David W. Embley* and Andrew Zitzelberger*

Brigham Young University, Provo, Utah 84602, U.S.A.

Abstract. The current web is a web of linked pages. Frustrated users search for facts by guessing which keywords or keyword phrases might lead them to pages where they can find facts. Can we make it possible for users to search directly for facts embedded in web pages? Instead of a web of human-readable pages containing machine-inaccessible facts, can the web be a web of machine-accessible facts superimposed over a web of human-readable pages? Ultimately, can the web be a web of knowledge that can provide direct answers to factual questions and support these answers by referencing and highlighting relevant base facts embedded in source pages? Answers to these questions call for distilling knowledge from the web's wealth of heterogeneous digital data into a web of knowledge. But how? Or, even more fundamentally, what, precisely, is this web of knowledge, and what is required to enable it? To answer these questions, we proffer a theoretical foundation for a web of knowledge: We formally define a computational view of knowledge in a way that enables practical construction and use of a web of knowledge.

1 Introduction

The web contains a wealth of knowledge. Unfortunately, most of the knowledge is not encoded in a way that enables direct user query. We cannot, for example, directly google for a car that is a 2003 or newer selling for under 15 grand; or for the names of the parents of great-grandpa Schnitker; or for countries whose population will likely decrease by more than 10% in 50 years.

A way to enable direct query for facts embedded in web pages and facts implied by these stated facts is to annotate stated facts with respect to ontologies. Annotating facts with respect to ontologies, implicitly populates these ontologies, turning them into a database over which structured queries can be executed. Annotation links also provide a form of provenance and authentication, allowing users to verify query results by checking original sources. Furthermore, facts and ontological concepts may appear in more than one populated ontology. Linking facts and ontological concepts across ontologies can provide navigation paths to explore additional, related knowledge. The web with a superimposed layer of

* Supported in part by the National Science Foundation under Grant #0414644.

S. Link and H. Prade (Eds.): FoIKS 2010, LNCS 5956, pp. 211–229, 2010.

interlinked ontologies each annotating a myriad of facts on the underlying web becomes a *Web of Knowledge*, a *WoK*.[1]

Although this vision of a WoK is appealing, there are significant barriers preventing both its creation and its use. Ontology languages exist, with OWL being the de facto standard. RDF files can provide data for these ontologies and can also store annotation information linking data to facts in web pages and linking equivalent information in RDF files to one another. The SPARQL query language is a standard for querying RDF data. Thus, all constituent components for a WoK are industry standards in common use, and they even all work together allowing for immediate WoK development and usage. Nevertheless, the barriers of creation and usage remain high and effectively prevent WoK deployment. The creation barrier is high because of the cost involved in developing OWL ontologies and annotating web pages by linking RDF-encoded facts in web pages to these OWL ontologies. The usage barrier is high because untrained users cannot write SPARQL queries.

Extraction ontologies provide a way to solve both creation and usage problems [13]. But, what exactly are extraction ontologies, and how do they resolve creation and usage problems? We answer these questions in this paper by formalizing extraction ontologies (Section 2), formalizing the notion of a WoK (Section 3), formalizing WoK construction procedures (Section 4), and formalizing user-friendly, WoK query-processing procedures (Section 5).

The success of the WoK vision depends on a solid theoretical foundation. Thus, the contributions of this paper are: (1) the formalization of extraction ontologies, knowledge bundles, and interconnected knowledge bundles as a WoK; (2) the formalization of WoK construction tools; and (3) the formalization of WoK usage tools. Indirectly, further contributions include: (1) the basis for an open architecture for pluggable tools for realizing a WoK along with implemented examples of construction and usage tools; and (2) the identification of strengths and weaknesses of WoK construction and usage tools and thus the identification of where opportunities lie for future research and development.

2 Extraction Ontologies

We base our foundational conceptualization for a web of knowledge on the conceptual modeling language OSM (Object-oriented Systems Model) [17]. OSM, however, simply provides a graphical representation of a first-order-logic language. Here we restrict OSM to be decidable, yet powerful enough to represent desired ontological concepts and constraints. We call our restriction *OSM-O*, short for *OSM-Ontology*. We thus base our foundational conceptualization directly on an appropriate restriction of first-order logic. This WoK foundation should be no surprise since it is the basis for modern information systems and has been the basis for formalizing information since the days of Aristotle [4].

[1] To many, this vision of a WoK constitutes the semantic web [37].

Definition 1. OSM-O *is a triple (O, R, C):*

- *O is a set of object sets; each is a one-place predicate; and each predicate has a* lexical *or a* non-lexical *designation.*[2]
- *R is a set of n-ary relationship sets ($n \geq 2$); each is an n-place predicate.*
- *C is a set of constraints:*
  - *Referential integrity:* $\forall x_1 ... \forall x_n(R(x_1, ..., x_n) \Rightarrow S_1(x_1) \wedge ... \wedge S_n(x_n)$ *for each n-ary relationship set R connecting object sets S_1, ..., S_n.*
  - *Participation constraint min:max cardinality: for every connection of an object set S to an n-ary relationship set R,* $\forall x_i(S(x_i) \Rightarrow \exists^{\geq min} <x_1, ..., x_{i-1}, x_{i+1}, ..., x_n>(R(x_1, ..., x_n)))$ *if min>0, and* $\forall x_i(S(x_i) \Rightarrow \exists^{\leq max} <x_1, ..., x_{i-1}, x_{i+1}, ..., x_n>(R(x_1, ..., x_n)))$ *if max is not * (the symbol denoting an unbounded maximum).*
  - *Generalization/specialization:* $\forall x(S_1(x) \vee ... \vee S_n(x) \Rightarrow G(x))$ *for each generalization object set G of specialization object sets S_1, ..., S_n in an is-a hierarchy. In addition,* $\forall x(S_i(x) \Rightarrow \neg S_j(x))$ *for $1 \leq i,j \leq n$ and $i \neq j$ if the specialization object sets are disjoint and* $\forall x(G(x) \Rightarrow S_1(x) \vee ... \vee S_n(x))$ *if the generalization object set is complete—is a union of the specialization object sets.*
  - *Aggregation: meronym-holonym relationship sets grouped as an aggregation in an is-part-of hierarchy.* □

Example 1. Figure 1 shows an OSM-O model instance. Rectangular boxes are object sets—dashed if lexical and solid if non-lexical. Lines between object sets denote relationship sets. Participation constraints are next to relationship-set/object-set connections in a *min:max* format. (We use *1* as shorthand for *1:1*.) A white triangle denotes generalization/specialization with the generalization connected to the apex of the triangle and the specializations connected to the base. A black triangle denotes aggregation with the super-part connected to the apex of the triangle and the sub-parts connected to the base. Although graphical in appearance, an OSM-O diagram is merely a two-dimensional rendition of predicates and closed formulas as defined in Definition 1. □

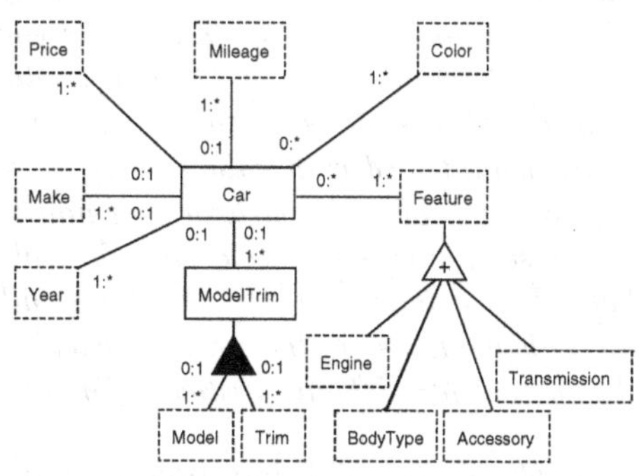

Fig. 1. OSM-O Model Instance

[2] In forthcoming definitions, lexical predicates will be restricted to literal domain-value substitutions like strings, integers, phone numbers, email addresses, and account numbers. Non-lexical predicates will be restricted to substitutions of object identifiers that represent real-world objects like people, geopolitical entities, and buildings, or abstract concepts like a marriage or ownership.

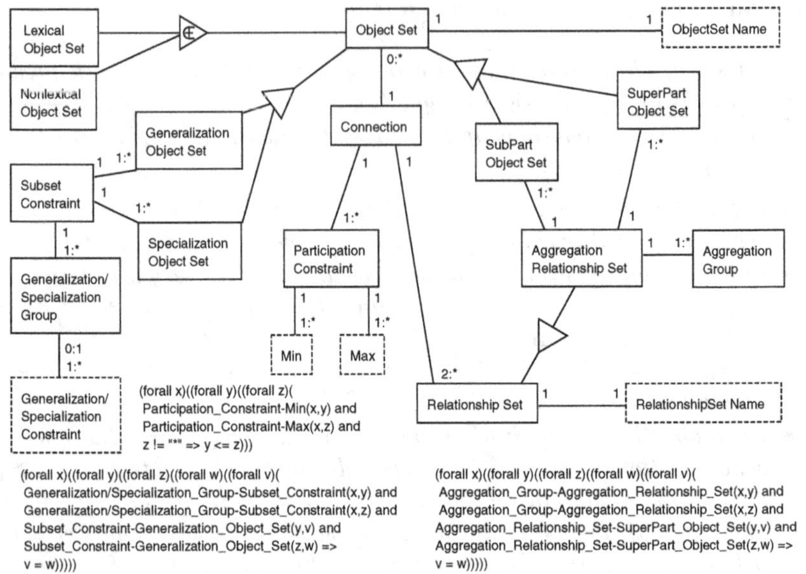

Fig. 2. OSM-O Meta-model

Definition 2. *Let $M = (O, R, C)$ be an OSM-O model instance. Let I be an interpretation for M that has a domain $D = L_{ID} \cup O_{ID}$ ($L_{ID} \cap O_{ID} = \emptyset$) and a declaration of* **True** *or* **False** *for each valid instantiation of each predicate in $O \cup R$ with values in D. For predicates in O, valid instantiations require lexical predicates to be instantiated with values in L_{ID} and non-lexical predicates to be instantiated with values in O_{ID}. For predicates in R, valid instantiations require each value v to be lexical or non-lexical according to whether the connected object set for v is lexical or non-lexical respectively. If all the constraints of C hold, I is a* model *of M, which we call a* valid interpretation *of M (to avoid an ambiguous use of the word "model" when also discussing conceptual models). An instantiated,* **True** *predicate for a valid interpretation is a* fact. □

Example 2. A valid interpretation of the OSM-O model instance in Figure 1 contains facts about cars. A possible valid interpretation might include the facts $Car(Car_3)$, $Year(2003)$, $Car\text{-}Year(Car_3,\ 2003)$, $Model(\text{"Accord"})$, $Trim(\text{"LX"})$, $ModelTrim(ModelTrim_{17})$, $Trim\text{-}isPartOf\text{-}ModelTrim(\text{"LX"},\ ModelTrim_{17})$, and $Car\text{-}ModelTrim(Car_3,\ ModelTrim_{17})$. Note that the object sets *Car* and *ModelTrim*, being non-lexical, have identifiers for their domain-value substitutions. Constraints, such as $\forall x(Car(x) \Rightarrow \exists^{\leq 1} y(Car\text{-}Year(x,y)))$, all hold. □

The *OSM-O model instance* in Figure 2 is a meta-model that defines valid OSM-O model instances. Note that since OSM-O is predicate calculus ("disguised" in graphical notation), the three additional closed formulas in Figure 2 are appropriate well-formed-formula additions to the statements in the meta-model. Two of them ensure that the group members of any generalization/specialization and any aggregation have the same parent. The third ensures that min-cardinality values are no greater than max-cardinality values.

Definition 3. *If M is a valid interpretation for the OSM-O meta-model instance, then M is a* valid *model instance.* □

Example 3. The car ontology in Figure 1 is a valid interpretation of the OSM-O meta-model in Figure 2. Part of the interpretation would include *ObjectSet_Name*("Feature"), and if the *Feature* object set has the identifier O_{13}, then also *Object_Set*(O_{13}), *Generalization_Set*(O_{13}), *Lexical_Object_Set*(O_{13}), and *Object_Set-ObjectSet_Name*(O_{13}, "Feature"). Note that the meta-model does not dictate names for object and relationship sets. We want them to be mnemonic, of course, as they are here and in Example 2, but we can make them mnemonic in any way we wish. This also solves the problem of naming a second binary relationship set between the same two object sets. In Figure 1, for example we could have two relationship sets between *Car* and *Price*: one of the predicate names could be *Car_sellsFor_Price* while the other could be *Car_hasManufacturerSuggested_Price*. □

Theorem 1. *OSM-O is decidable.* □

Proof. Let M = (O, R, C) be an OSM-O model instance. We show how to translate each component of M to an OWL-DL model instance Z. Then, since OWL-DL is decidable, OSM-O is decidable.

- For each object set $S \in O$ create a class (owl:Class) in Z.
- For each lexical object set $S \in O$ create a data-type property (owl:Datatype-Property) in Z with domain S and with the range being any appropriate data type.
- To make referential-integrity constraints enforceable, create for each connection c of an object set S with a relationship set a class S_c in Z and make S_c a subclass of S.
- For each generalization/specialization group H with G as the generalization object set and S_1, ..., S_n as the specialization object sets, process constraints in C as follows:
  - Make each class created from S_1, ..., S_n a subclass of (rdfs:subClassOf) the class created for G.
  - If H has a union constraint, make the class created for G the union of (owl:unionOf) the classes S_1, ..., S_n.
  - If H has a disjoint constraint, make S_i disjoint with (owl:disjointWith) S_j, $(1 \leq i, j \leq n; i \neq j)$.
- Process relationship sets in R as follows:
  - For each n-ary relationship set $Q \in R$ with $n > 2$, for purposes of the translation, replace Q with a non-lexical object set S and binary relationships to each participating object set of Q. Add participation

constraints *1* (= *1:1*) for each S-connection and let the participation constraints for connections to other object sets be as originally specified in Q. Create a class for S in Z.

- (After translating n-ary relationship sets, $n > 2$, to binary relationship sets, all relationship sets in R are binary. These relationship sets include the binary relationship sets originally in R, the binary relationship sets created from n-ary relationship sets for purposes of translation, and all aggregation relationship sets.) For each relationship set in R, create an object property (owl:ObjectProperty) and a second object property and make each an inverse of (owl:inverseOf) the other.
- For each participation constraint in C, create a restriction with min and max constraints (owl:Restriction, owl:minCardinality, and owl:maxCardinality) for the appropriate object property created from R. □

Similar to the work by Buitelaar, et al. [8], we now show how to linguistically ground OSM-O. Linguistically grounding OSM-O turns OSM-O model instances into *OSM-Extraction-Ontology* model instances (*OSM-EO* model instances). We begin by defining an ordinary abstract data type for each object set. We then add linguistic recognizers for instance values, operators, operator parameters, and relationships.

Definition 4. *An* abstract data type *is a pair* (V, O) *where* V *is a set of values and* O *is a set of operations.* □

Definition 5. *A* data frame *is an abstract data type augmented as follows:*

1. *The data frame has a name* N *designating the set of values* V*, and it may have a list of synonyms for* N*.*
2. *The value set* V *has an instance recognizer that identifies lexical patterns denoting values in* V*.*
3. *The operations set* O *includes two (additional) operators for each lexical object set: (1) an input operator to convert identified instances to the internal representation for* V *and (2) an output operator to convert instances in* V *to strings.*
4. *Each operation* o *in* O *has an operator recognizer that identifies lexical patterns as indicators that* o *applies. Further, the recognizer identifies lexical patterns that, along with instance recognizers, identify parameters for* o*.* □

As is standard, implementations of abstract data types are hidden, and we hide implementations for data frames as well. Similar to data independence in database systems, this approach accommodates any implementation; in particular it allows for new and better recognizers.[3] To be complete and precise, however, we give examples to illustrate how we have implemented them.

[3] Many kinds of recognizers are possible. Much work has been done on named entity recognition, and entire communities and conferences/workshops are devoted to these issues. We want the WoK vision to be able to include and make use of this work.

Price
  internal representation: Integer
  external representation: \$[1-9]\d{0,2},?\d{3} | \d?\d [Gg]rand | ...
  context keywords: price|asking|obo|neg(\.|otiable)| ...
  ...
  LessThan(p1: Price, p2: Price) **returns** (Boolean)
  context keywords: (less than | < | under | ...)\s*{p2} | ...
  ...
Make
  ...
  external representation: CarMake.lexicon
  ...

Fig. 3. Data Frames

Example 4. Figure 3 shows two partial data frames for object sets in the OSM-O model instance in Figure 1, one for *Price* and one for *Make*. The *Price* data frame uses regular expressions for its recognizers, whereas the *Make* data frame uses a lexicon. In our implementation, we can use either or both together. The *Price* data frame also shows a recognizer for an operator. The *p2* within curly braces indicates the expected appearance of a *Price* parameter *p2*. Thus a phrase like "under 15 grand" is recognized as indicating that the price of the car—parameter *p1*—should be less than $15,000. □

A data frame for a non-lexical object set is typically degenerate: Its value set is a set of object identifiers. Its operation set consists only of operators that add and remove object identifiers. Its name and synonyms and its recognizers that identify object instances, however, can be quite rich.

For relationship sets, the definition of a data frame does not change, but a typical view of the definition shifts as we allow value sets to be n-tuples of values rather than scalar values. Further, like recognizers for operators, they rely on instance recognizers from the data frames of their connected object sets.

Example 5. Suppose the *Car* object set in Figure 1 has a relationship set to a *Person* object set. The relationship-set data frame may have recognizers for any one of several possible relationships such as *{Person} is selling {Car}*, *{Person} posted {Car} ad*, or *{Person} is inquiring about {Car}*. □

Definition 6. *If M is an OSM-O model instance with a data frame for each object set and relationship set, M is an* OSM-EO *model instance.* □

An OSM-EO model instance is linguistically grounded in the sense that it can both "read" and "write" in some natural language. To "read" means to be able to recognize facts in natural language text and to extract fact instances with respect to the ontology in the OSM-EO model instance. To "write" means to display fact instances so that they are human-readable.

How well a particular OSM-EO model instance can "read" and "write" makes a difference in how well it performs. Our experience is that OSM-EO model instances can "read" some documents well (over 95% precision and recall [16]), but it is clear that opportunities abound for further research and development. Writing human-understandable descriptions is less difficult to achieve—just select any one of the phrases for each object set and relationship set (e.g., $Person(Person_{17})$ *is selling* $Car(Car_{734})$, $Car(Car_{734})$ *has* $Make(Honda)$). Making the written description pleasing is, of course, more difficult.

3 Web of Knowledge

Ontology is the study of "the nature of existence." Epistemology is the study of "the origin, nature, methods, and limits of human knowledge." In the previous section we have given a computational view of ontology—a view that lets us work with ontologies in information systems. We similarly give a computational view of epistemology. The computational view of epistemology we give here constitutes a formal foundation for a web of knowledge.

Definition 7. *The collection of facts in an OSM-O model instance constitutes the* extensional knowledge *of the OSM-O model instance. The collection of implied facts derived from the extensional knowledge by inference rules*[4] *constitutes the* intentional knowledge. *The extensional and intentional knowledge together constitute the* knowledge *of the OSM-O model instance.* □

Although this view of knowledge is common in computing, Plato, and those who follow his line of thought, also demand of knowledge that it be a "justified true belief" [26]. "Knowledge" without some sort of truth authentication can be confusing and misleading. But how can we attain truth authentication? We see three possibilities: (1) truth as community agreement—e.g., Wikipedia style; (2) probabilistic truth; and (3) truth derived from proper reasoning chains grounded in original sources. All three, unfortunately, are problematic: community agreement depends on the willingness of individuals to participate and to agree; probabilistic truth depends on establishing probabilities and on being able to derive probabilities for answers to queries—hard problems that do not scale well [12]; and reasoning with rules and fact sources depends on acceptance of the rules and fact sources as genuine.

For our vision of a WoK, we attempt to establish truth via provenance and authentication. We provide for reasoning with rules and for ground facts in sources. We cannot, however, guarantee that rules and facts in sources are genuine. We thus compensate by simply exposing them. When an extraction ontology extracts a fact from a source document, it retains a link to the fact; and when a query answer requires reasoning over rules, the system records the reasoning chain. Users can ask to see fact sources and rule chains, and in this way they can authenticate facts and reasoning the way we usually do—by checking sources and fact-derivation rules.

Definition 8. *A* knowledge bundle *is a 5-tuple (O, E, S, I, R) where O is an OSM-O model instance; E is an OSM-EO instance whose OSM-O instance is O;*

[4] We limit our inference rules to safe, positive datalog rules, which are decidable [27].

S is a set of source documents from which facts for E are extracted; I is a valid interpretation for O whose facts are extracted from the documents in S; and R is a rule set where each rule is a safe, positive, horn clause whose body-predicates are either predicates in O or are head-predicates of other rules in R. □

Definition 9. *A* Web of Knowledge *(*WoK*) is a collection of knowledge bundles interconnected with binary links, <x, y>, of two types: (1)* object identity: *non-lexical object identifier x in knowledge bundle B_1 refers to the same real-world object as non-lexical object identifier y in knowledge bundle B_2. (2)* Object-set identity: *object set x in knowledge bundle B_1 designates the same set of real-world objects as object set y in knowledge bundle B_2.* □

4 WoK Construction

To construct a WoK, we must be able to construct a knowledge bundle, and we must be able to establish links among knowledge bundles. We can construct knowledge bundles and establish links among them by hand (and this should always be an option). However, scaling WoK construction demands semi-automatic procedures, with as much of the construction burden placed on the system as possible—all of it when possible. For knowledge bundles, our automated construction tools identify applicable source information and transform it into knowledge-bundle components. For links among knowledge bundles, we apply record-linkage and schema-mapping tools.

Definition 10. *A* transformation *is a 5-tuple (R, S, T, Σ, Π), where R is a set of resources, S is the source conceptualization, T is the target conceptualization for an S-to-T transformation, Σ is a set of declarative source-to-target transformation statements, and Π is a set of procedural source-to-target transformation statements.* □

Definition 10 leaves several of its components open—to take on specific meanings in a variety of knowledge-bundle building tools. The "set of resources" is undefined, but we intend this to mean resources such as WordNet and a data-frame library. "Target conceptualizations" are knowledge bundles. "Source conceptualizations" depend on sources whose fact conceptualizations can be formal, semi-formal, or informal. "Declarative" and "procedural" "source-to-target transformation statements" can be written in a variety of formal languages. Our goal here has been and is to successfully develop automatic and good semi-automatic transformations over a broad spectrum of documents for a variety of ontological contexts.

To be specific about some of the possibilities, we provide some examples. We first give some restrictive transformations guaranteed to capture all facts and constraints in the source and then mention some less restrictive transformations.

Definition 11. *Let S be a predicate calculus theory with a valid interpretation, and let T be a populated OSM-O model instance constructed from S by a transformation t. Transformation t* preserves information *if there exists a procedure to* *compute S from T. Let C_S be the closed, well formed formulas of S, and let C_T be the closed, well formed formulas of T. Transformation t* preserves constraints *if $C_T \Rightarrow C_S$.* □

Theorem 2. *Let S be a nested table with a single label path to each data item, and let T be an OSM-O model instance. A transformation from S to T exists that preserves information and constraints.* □

Proof. (*sketch*) We show how to construct and populate T from S such that an inverse procedure exists and such that the constraints of T imply the constraints of S. (The transformation of the nested table in Figure 4 to the OSM-O model instance in Figure 5 illustrates the transformation.) To initialize T, we create a non-lexical object set to represent the table as a concept in the OSM-O ontology and label it "Table". Each label in S becomes an object set in T. Because of the restriction of S having a single label path to each data item, only the leaf nodes of T are lexical object sets (they contain the data). Relationship sets mark the path between labels down to the data items. We infer participation constraints on the parent sides in the tree of T from the number of items in the object sets of S and assume participation constraints to be *1:** on the child sides. The reverse transformation is straightforward as the ontology is a hierarchical view of the table starting at the "Table" object set. The parent participation constraints of T imply the inferred participation constraints of S. □

We can exploit the direct correspondence between the nested tables of Theorem 2 and OSM-O in several ways. First, as intended, we can convert facts in nested tables into knowledge bundles for a WoK. Second, nested tables from the WormBase site are all sibling tables—they have identical, or nearly identical structure. We have shown elsewhere that we can automatically construct an ontology for the site (and any other site with sibling tables) and extract the information in all the tables to populate the ontology [32]. Third we can create extraction ontologies automatically (although they likely need some enhancement) [32]. Fourth we can turn the process around and let users specify ontologies via nested forms [33].

Theorem 3. *Let S be a relational database with its schema restricted as follows: (1) the only declared constraints are single-attribute primary key constraints and single-attribute foreign-key constraints, (2) every relation schema has a primary key, (3) all foreign keys reference only primary keys and have the same name as the primary key they reference, (4) except for attributes referencing foreign keys, all attribute names are unique throughout the entire database schema, (5) all relation schemas are in 3NF. Let T be an OSM-O model instance. A transformation from S to T exists that preserves information and constraints.* □

Proof (sketch). For every relation, create a non-lexical object set and make its name be the name of the relation. Except for attributes that reference foreign keys, create a lexical object set for each attribute and make its name be the attribute name. For each schema S, create a binary relationship set between the

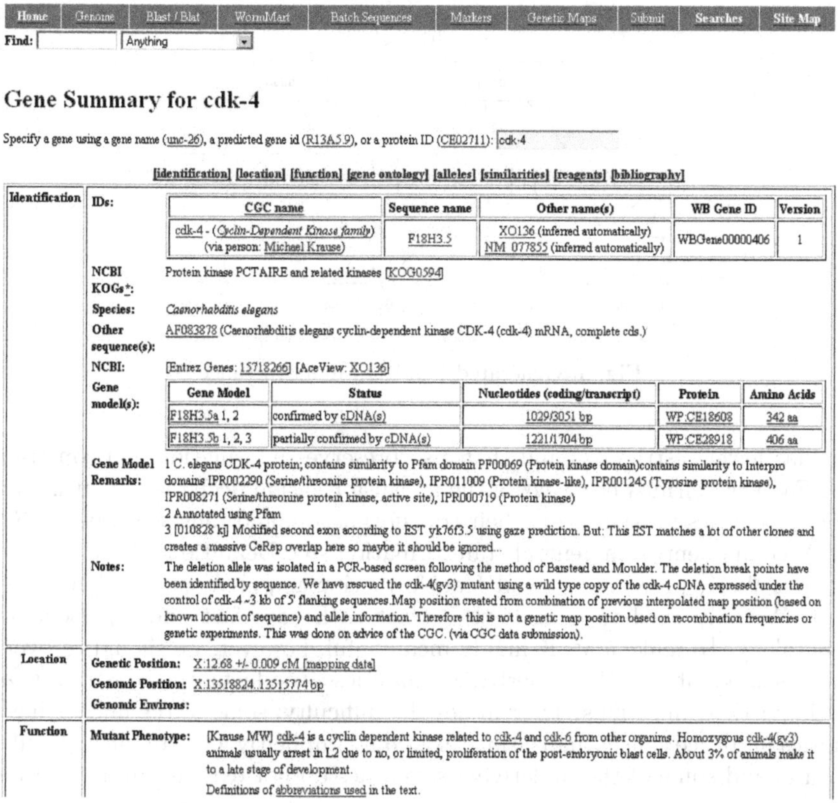

Home | Genome | Blast / Blat | WormMart | Batch Sequences | Markers | Genetic Maps | Submit | Searches | Site Map

Find: Anything

Gene Summary for cdk-4

Specify a gene using a gene name (unc-26), a predicted gene id (R13A5.9), or a protein ID (CE02711): cdk-4

[identification] [location] [function] [gene ontology] [alleles] [similarities] [reagents] [bibliography]

Identification

IDs:

CGC name	Sequence name	Other name(s)	WB Gene ID	Version
cdk-4 - (*Cyclin-Dependent Kinase family*) (via person: Michael Krause)	F18H3.5	XO136 (inferred automatically) NM_077855 (inferred automatically)	WBGene00000406	1

NCBI KOGs*: Protein kinase PCTAIRE and related kinases [KOG0594]

Species: *Caenorhabditis elegans*

Other sequence(s): AF083878 (Caenorhabditis elegans cyclin-dependent kinase CDK-4 (cdk-4) mRNA, complete cds.)

NCBI: [Entrez Genes: 15718266] [AceView: XO136]

Gene model(s):

Gene Model	Status	Nucleotides (coding/transcript)	Protein	Amino Acids
F18H3.5a 1, 2	confirmed by cDNA(s)	1029/3051 bp	WP:CE18608	342 aa
F18H3.5b 1, 2, 3	partially confirmed by cDNA(s)	1221/1704 bp	WP:CE28918	406 aa

Gene Model Remarks: 1 C. elegans CDK-4 protein; contains similarity to Pfam domain PF00069 (Protein kinase domain)contains similarity to Interpro domains IPR002290 (Serine/threonine protein kinase), IPR011009 (Protein kinase-like), IPR001245 (Tyrosine protein kinase), IPR008271 (Serine/threonine protein kinase, active site), IPR000719 (Protein kinase)
2 Annotated using Pfam
3 [010828 kj] Modified second exon according to EST yk76f3.5 using gaze prediction. But: This EST matches a lot of other clones and creates a massive CeRep overlap here so maybe it should be ignored...

Notes: The deletion allele was isolated in a PCR-based screen following the method of Barstead and Moulder. The deletion break points have been identified by sequence. We have rescued the cdk-4(gv3) mutant using a wild type copy of the cdk-4 cDNA expressed under the control of cdk-4 ~3 kb of 5' flanking sequences.Map position created from combination of previous interpolated map position (based on known location of sequence) and allele information. Therefore this is not a genetic map position based on recombination frequencies or genetic experiments. This was done on advice of the CGC. (via CGC data submission).

Location

Genetic Position: X:12.68 +/- 0.009 cM [mapping data]

Genomic Position: X:13518824..13515774 bp

Genomic Environs:

Function

Mutant Phenotype: [Krause MW] cdk-4 is a cyclin dependent kinase related to cdk-4 and cdk-6 from other organims. Homozygous cdk-4(gv3) animals usually arrest in L2 due to no, or limited, proliferation of the post-embryonic blast cells. About 3% of animals make it to a late stage of development.

Definitions of abbreviations used in the text.

Fig. 4. Nested Table in a Molecular-Biology Web Page

non-lexical object set created for S and each lexical object set created from attributes in S. The participation constraint on the non-lexical side of all these relationship sets is *1*. For lexical object sets corresponding to primary-key attributes the participation constraints on the lexical sides are also *1*; otherwise the participation constraints on the lexical sides are *1:**. For each primary-key foreign-key reference, create a generalization/specialization between the schema's non-lexical object sets with the object set of the referenced key as the generalization and the object set of the referencing key as the specialization. For all other foreign-key references, create a binary relationship set between the non-lexical object sets of the two schema's with *0:1* as the participation constraint on the referencing side and *1:** as the participation constraint on the referenced side. The reverse transformation is a standard transformation of an OSM model instance to a relational database [15]. Since the standard transformation generates all constraints of S, the constraints of T imply the constraints of S. □

Theorem 3 is overly restrictive for most practical applications, but its restrictions allow us to illustrate the process well in a short description. Recently, Tirmizi et al. defined a transformation from a relational database directly to OWL

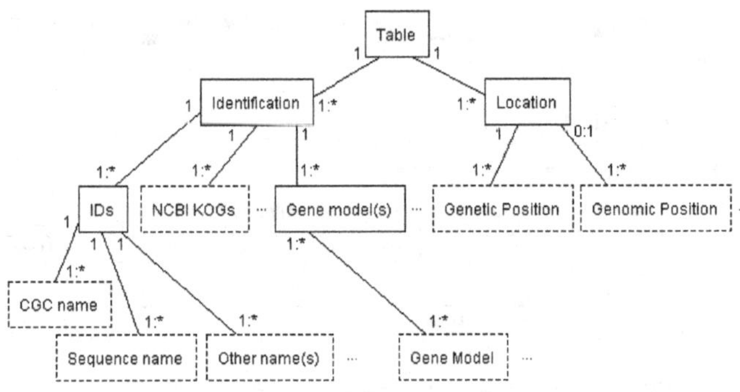

Fig. 5. Generated OSM-O Model Instance

that has fewer restrictions and that also preserves information and constraints [35]. Further, Tirmizi et al. write their transformation declaratively, using only Prolog-like rules for specifying their transformation, and thus use only the S, T, and Σ components of a general transformation (see Definition 10).

For a WoK, preservation and transformation requirements need not be so strict. We only need to be able to extract each base fact and represent it in an ontology. To make a WoK highly meaningful, however, we should recover as much as is possible of the underlying semantics—the facts, the constraints, and the linguistic connections. Therein lies the difficulty: some of the underlying semantics in source conceptualizations exist only implicitly and are thus difficult to capture, and some of the underlying semantics do not exist at all, having been discarded in the abstraction process of producing the conceptualization. But therein also lies the research opportunities. Many researchers are endeavoring to create automatic and semi-automatic procedures to capture richer semantics—making extensive use of the Π component of Definition 10. And some are making use of the R component of Definition 10 to recover semantics lost in the abstraction process. In general, there is an effort to recover as much of the semantics as possible from many different source genres. For example, researchers have investigated semantic recovery from relational databases [5, 6], XML [1, 38], human-readable tables [24, 25, 34], forms [22, 31], and free-running text [10].

For the last part of WoK creation—creating links among knowledge bundles—we rely on *record linkage* and *schema mapping*, also called *ontology alignment* or *ontology matching*. *Record linkage* is the task of finding entries that refer to the same entity in two or more data sources, and *ontology matching* is the task of finding the semantic correspondences between elements of two ontology schemas. An extensive body of work exists for both record linkage [14] and ontology matching [21]. Unfortunately, both problems are extremely hard. General solutions do not exist and may never exist. However, "best-effort" methods do exist and perform reasonably well. For the WoK vision, we can use these best-effort methods to initialize and update *same-as* links for object identity and *equivalence-class* links for concept identity. Using "best-effort" methods in a "pay-as-you-go" fashion appears to be a reasonable way to enable a WoK.

5 WoK Usage

The construction of extraction ontologies leads to "understanding" within a WoK. This "understanding" leads to the ability to answer a free-form query because, as we explain in this section, a WoK system can identify an extraction ontology that applies to a query and match the query to the ontology. Hence, a WoK system can reformulate the free-form query as a formal query, so that it can be executed over a knowledge bundle.

Definition 12. *Let S be a source conceptualization and let T be a target conceptualization formalized as an OSM-EO. We say that T* understands *S if there exists an S-to-T transformation that maps each one-place predicate of S to an object set of T, each n-place predicate of S to an n-place relationship set of T ($n \geq 2$), each fact of S to a fact of T with respect to the predicate mappings, and each operator of S to an operator in a data frame of T, such that the constraints of T all hold over the transformed predicates and facts.* □

Observe that although Definition 12 states how T is formalized, it does not state how S is formalized. Thus, the predicates and operators of S may or may not be directly specified. This is the hard part of "understanding"—to recognize the applicable predicates and operators. But this is exactly what extraction ontologies are meant to do. If an OSM-EO is linguistically well grounded, then it can "understand" so long as what is stated in S is within the context of T—that is if there is an object set or relationship set in T for every predicate in S and if there is an operator in a data frame of T for every operator in S.

Applications of understanding include free-form query processing, advanced form-query processing, knowledge augmentation, and knowledge-bundle building for research studies.

Example 6. Free-Form Query Processing: Figure 6 illustrates free-form query processing within our WoK prototype. To "understand" a user query, our WoK prototype first determines which OSM-EO applies to the query by seeing which one recognizes the most instances, predicates, and operators in the query request. Having chosen the *Car* extraction ontology illustrated in Figures 1 and 3, the WoK applies the S-to-T transformation highlighting what it "understands" ("Find me a honda, 2003 or newer for under 15 grand"). Figure 7 shows the result of this transformation—each predicate and each operation is mapped correctly and the constraints of the OSM-EO model instance all hold. Given this "understanding," it is straightforward to generate a SPARQL query. Before executing the query, our WoK prototype augments it so that it also obtains the stored annotation links. Then, when our WoK prototype displays the results of the query in the lower-left box in Figure 6, it makes returned values clickable. Clicking on a value, causes our WoK prototype to find the page from which the value was extracted, highlight it, and display the page appropriately scrolled to the location that includes the value. The right panel of Figure 6 shows several highlighted values, which happens when the user checks one or more check-boxes before clicking. □

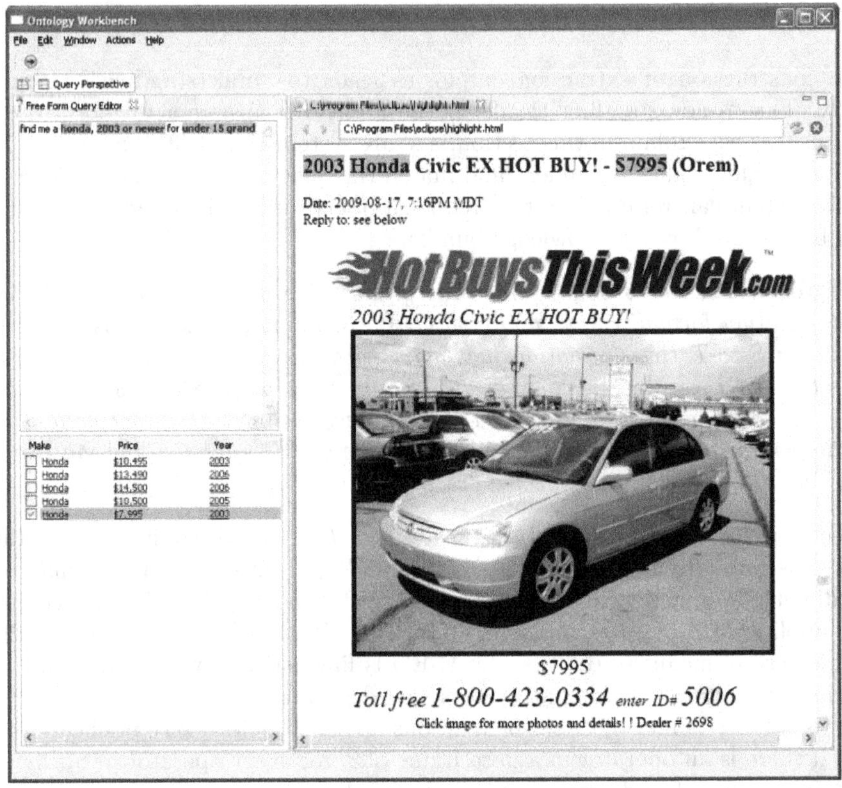

Fig. 6. Screenshot of WoK Prototype Showing Free-Form Query Processing

Example 7. Advanced Form-Query Processing: The form in Figure 7 is for an alerter system which we have implemented for craigslist.org. We use it as feedback to the user that the query has been understood. As such, it illustrates not only the ability of an OSM-EO to "read" and "understand," but also its ability to "write." Note, for example the conversion of "15 grand" to "$15,000" as well as the mnemonic names for predicates and operations. Besides providing feedback, this writing ability also lets the user know what else the OSM-EO knows about. A user U then has the opportunity to adjust the query or add additional constraints. For example, U may wish to also know if Toyotas are listed so long as they are not Camrys. Clicking on OR for $Make$ and adding $Toyota$ and then clicking on NOT for $Model$ and adding $Camry$ makes this possible. The plus icons show that more operators are available; clicking on the plus displays them. For example, the user might wish to limit prices with $Between(Car.Price, \$11K, \$16K)$. Since the OSM-EO has general recognizers for prices, U can enter them in any recognizable format.

Example 8. Fact Finding for Research Studies: In addition to "understanding" queries, it should be clear that "understanding" is also about fact finding. The fundamental intent of linguistically grounding extraction ontologies is for them to be able to recognize facts in structured, semi-structured, and unstructured text. As an example, we can exploit this fact-finding ability to gather facts for a bio-research study and store them as a knowledge bundle for further analysis [19]. Gathering tasks for these research studies often take trained bio-researchers several man-months of work. So, any significant speed-up extraction ontologies can provide would be of great benefit in bio-medical research.

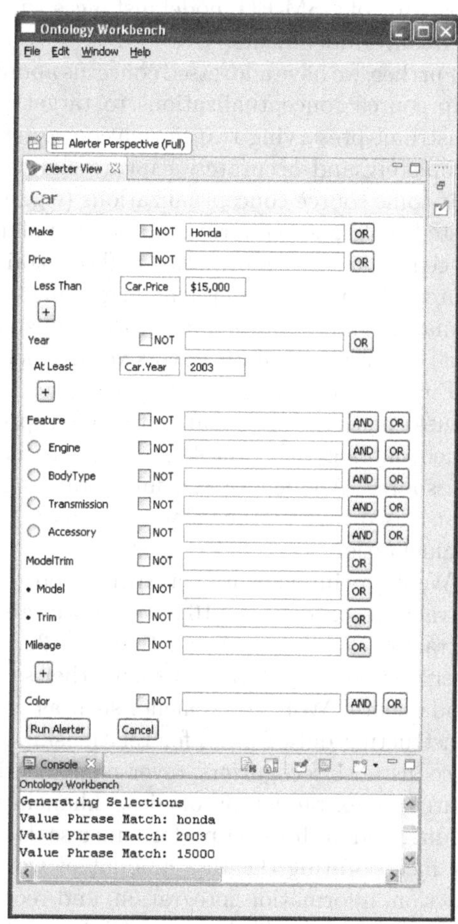

Fig. 7. Generated Form

Example 9. Knowledge Augmentation: Partial "understanding" is also useful. If two OSM-EO model instances, M_1 and M_2, can partially "understand" each other, they can be merged based on their common "understanding." M_1 can be seen as having been augmented by the part of M_2 that it did not already "understand" and vice versa. We can exploit "partial understanding" along with reverse engineering of semi-structured sources to grow ontologies. In our TANGO project, for example, we take a collection of ordinary tables with overlapping information, reverse engineer them one at a time creating an OSM-EO model instance for each and then merge each into a growing ontology that represents the entire collection [34].

6 Conclusion

We have defined a formal foundation for a web of knowledge (a WoK). We have also formalized fundamental WoK components: ontologies (OSM-O) in terms of decidable first-order logic and extraction ontologies (OSM-EO) linguistically grounded via data-frame recognizers. In addition, we have formalized a computational view of a WoK as a collection of interconnected knowledge bundles consisting of OSM-EO model instances with valid interpretations super-imposed over source documents.

Further, we have addressed concerns about WoK construction. Transformations map source conceptualizations to target conceptualizations. Information and constraint-preserving transformations guarantee that target conceptualizations completely and accurately capture source conceptualizations. We have shown that some source conceptualizations (e.g., specialized, but commonly occurring, nested tables, nested forms, and relational databases) have transformations guaranteed to preserve information and constraints. We conclude, however, that many source conceptualizations (ranging from semi-structured sources such as ordinary human-readable tables and forms to unstructured sources such as free-running text) require best-effort, pay-as-you-go methods.

Finally, we have formally addressed concerns about WoK usage. When transformations exist that map source predicates and operations to an established ontology, the ontology is said to have "understood" the information in the source. Applications of understanding include free-form query processing, form-based query processing, fact finding for research studies, and knowledge augmentation.

We have implemented a WoK prototype [18] including some prototypical extraction ontologies [16]. We have also done some work on automated extraction-ontology construction [24, 32, 33, 34] and some work on free-form query processing [2, 36]. We nevertheless still have much work to do, even on fundamental WoK components such as creating a sharable data-frame library, constructing data frames for relationship sets, finding ways to more easily produce instance recognizers, reverse-engineering of many genres of semi-structured sources to extraction ontologies, enhancing query processing, incorporating reasoning, and addressing performance scalability. We also see many opportunities for incorporating the vast amount of work done by others on information extraction, information integration, and record linkage. We cite as relevant examples: KnowItAll [20], best-effort information extraction [30], C-PANKOW [11], Q/A systems [28], bootstrapping pay-as-you-go data integration [29], large-scale deduplication [3], and OpenDMAP [23].

These collective efforts will eventually lead to a WoK—a realization of ideas of visionaries from Bush [9] to Berners-Lee [7] and Weikum [39]. Establishing a framework for a WoK by formalizing its basic components and establishing a firm theoretical foundation as we have done here can help enable this new kind of information system—a web of knowledge.

Acknowledgements. We would like to thank Cui Tao and Yihong Ding for coding the transformations of nested tables to OSM-EO model instances, Oliver Nina and Meher Shaikh for coding the craigslist.org alerter, Stephen W. Liddle for coding the SPARQL query augmenter and source highlighter, and many former and current students who have worked on our extraction-ontology engine. We also thank Andrea Cali for stimulating discussions about decidability and complexity issues.

References

1. Al-Kamha, R.: Conceptual XML for Systems Analysis. PhD dissertation, Brigham Young University, Department of Computer Science (June 2007)
2. Al-Muhammed, M., Embley, D.W.: Ontology-based constraint recognition for free-form service requests. In: Proceedings of the 23rd International Conference on Data Engineering (ICDE 2007), Istanbul, Turkey, April 2007, pp. 366–375 (2007)
3. Arasu, A., Re, C., Suciu, D.: Large-scale deduplication with constraints using Dedupalog. In: Proceedings of the 25th International Conference on Data Engineering (ICDE 2009), Shanghi, China, March/April 2009, pp. 952–963 (2009)
4. Aristotle: Metaphysics. Oxford University Press, New York; about 350 BC (1993 translation)
5. Astrova, I.: Reverse engineering of relational databases to ontologies. In: Proceedings of the First European Semantic Web Symposium, Heraklion, Crete, Greece, May 2004, pp. 327–341 (2004)
6. Atoum, J., Bader, D., Awajan, A.: Mining functional dependency from relational databases using equivalent classes and minimal cover. Journal of Computer Science 4(6), 421–426 (2008)
7. Berners-Lee, T., Hendler, J., Lassila, O.: The semantic web. Scientific American 36(25), 34–43 (2001)
8. Buitelaar, P., Cimiano, P., Haase, P., Sintek, M.: Towards linguistically grounded ontologies. In: Aroyo, L., Traverso, P., Ciravegna, F., Cimiano, P., Heath, T., Hyvönen, E., Mizoguchi, R., Oren, E., Sabou, M., Simperl, E. (eds.) ESWC 2009. LNCS, vol. 5554, pp. 111–125. Springer, Heidelberg (2009)
9. Bush, V.: As we think. The Atlantic Monthly 176(1), 101–108 (1945)
10. Cimiano, P.: Ontology Learning and Population from Text: Algorithm, Evaluation and Applications. Springer, New York (2006)
11. Cimiano, P., Ladwig, G., Staab, S.: Gimme' the context: Context-driven automatic semantic annotation with C-PANKOW. In: Proceedings of the 14th International World Wide Web Conference (WWW 2005), Chiba, Japan, May 2005, pp. 332–341 (2005)
12. Dalvi, N., Ré, C., Suciu, D.: Probabilistic databases: Diamonds in the dirt. Communications of the ACM 52(7), 86–94 (2009)
13. Ding, Y., Embley, D.W., Liddle, S.W.: Automatic creation and simplified querying of semantic web content: An approach based on information-extraction ontologies. In: Mizoguchi, R., Shi, Z.-Z., Giunchiglia, F. (eds.) ASWC 2006. LNCS, vol. 4185, pp. 400–414. Springer, Heidelberg (2006)
14. Elmagarmid, A.K., Ipeirotis, P.G., Verykios, V.S.: Duplicate record detection: A survey. IEEE Transactions on Knowledge and Data Engineering 18(1), 1–16 (2007)
15. Embley, D.W.: Object Database Development: Concepts and Principles. Addison-Wesley, Reading (1998)
16. Embley, D.W., Campbell, D.M., Jiang, Y.S., Liddle, S.W., Lonsdale, D.W., Ng, Y.-K., Smith, R.D.: Conceptual-model-based data extraction from multiple-record web pages. Data & Knowledge Engineering 31(3), 227–251 (1999)
17. Embley, D.W., Kurtz, B.D., Woodfield, S.N.: Object-oriented Systems Analysis: A Model-Driven Approach. Prentice Hall, Englewood Cliffs (1992)

18. Embley, D.W., Liddle, S.W., Lonsdale, D., Nagy, G., Tijerino, Y., Clawson, R., Crabtree, J., Ding, Y., Jha, P., Lian, Z., Lynn, S., Padmanabhan, R.K., Peters, J., Tao, C., Watts, R., Woodbury, C., Zitzelberger, A.: A conceptual-model-based computational alembic for a web of knowledge. In: Proceedings of the 27th International Conference on Conceptual Modeling, Barcelona, Spain, October 2008, pp. 532–533 (2008)
19. Embley, D.W., Liddle, S.W., Lonsdale, D.W., Stewart, A., Tao, C.: KBB: A knowledge-bundle builder for research studies. In: Proceedings of 2nd International Workshop on Active Conceptual Modeling of Learning (ACM-L 2009), Gramado, Brazil (November 2009) (to appear)
20. Etzioni, O., Cafarella, M., Downey, D., Kok, S., Popescu, A., Shaked, T., Soderland, S., Weld, D., Yates, A.: Unsupervised named-entity extraction from the web: An experimental study. Artificial Intelligence 165(1), 91–134 (2005)
21. Euzenat, J., Shvaiko, P.: Ontology Matching. Springer, Heidelberg (2007)
22. Gal, A., Modica, G.A., Jamil, H.M.: Ontobuilder: Fully automatic extraction and consolidation of ontologies from web sources. In: Proceedings of the 20th International Conference on Data Engineering, Boston, Massachusetts, March/April 2004, p. 853 (2004)
23. Hunter, L., Lu, Z., Firby, J., Baumgartner Jr., W.A., Johnson, H.L., Ogren, P.V., Cohen, K.B.: OpenDMAP: An open source, ontology-driven, concept analysis engine, with applications to capturing knowledge regarding protein transport, protein interactions and cell-type-specific gene expression. BMC Bioinformatics 9(8) (2008)
24. Lynn, S., Embley, D.W.: Semantically conceptualizing and annotating tables. In: Domingue, J., Anutariya, C. (eds.) ASWC 2008. LNCS, vol. 5367, pp. 345–359. Springer, Heidelberg (2008)
25. Pivk, A., Sure, Y., Cimiano, P., Gams, M., Rajkovič, V., Studer, R.: Transforming arbitrary tables into logical form with TARTAR. Data & Knowledge Engineering 60, 567–595 (2007)
26. Plato: Theaetetus. BiblioBazaar, LLC, Charleston, South Carolina, about 360BC (translated by Benjamin Jowett)
27. Rosati, R.: On the decidability and complexity of integrating ontologies and rules. Journal of Web Semantics 3(1), 61–73 (2005)
28. Roussinov, D., Fan, W., Robles-Flores, J.: Beyond keywords: Automated question answering on the web. Communications of the ACM 51(9) (September 2008)
29. Sarma, A.D., Dong, X., Halevy, A.: Bootstrapping pay-as-you-go data integration systems. In: Proceedings of SIGMOD 2008, Vancouver, British Columbia, Canada, June 2008, pp. 861–874 (2008)
30. Shen, W., DeRose, P., McCann, R., Doan, A., Ramakrishnan, R.: Toward best-effort information extraction. In: Proceedings of the ACM SIGMOD International Conference on Management of Data, Vancouver, British Columbia, Canada, June 2008, pp. 1031–1042 (2008)
31. Su, W., Wang, J., Lochovsky, F.: ODE: Ontology-assisted data extraction. ACM Transactions on Database Systems 12, 1–12 (2009)
32. Tao, C., Embley, D.W.: Automatic hidden-web table interpretation, conceptualization, and semantic annotation. Data & Knowledge Engineering 68(7), 683–703 (2009)
33. Tao, C., Embley, D.W., Liddle, S.W.: FOCIH: Form-based ontology creation and information harvesting. In: Laender, A.H.F., et al. (eds.) ER 2009. LNCS, vol. 5829, pp. 346–359. Springer, Heidelberg (2009)

34. Tijerino, Y.A., Embley, D.W., Lonsdale, D.W., Ding, Y., Nagy, G.: Toward ontology generation from tables. World Wide Web: Internet and Web Information Systems 8(3), 261–285 (2005)
35. Tirmizi, S., Sequeda, J., Miranker, D.: Translating SQL applications to the semantic web. In: Bhowmick, S.S., Küng, J., Wagner, R. (eds.) DEXA 2008. LNCS, vol. 5181, pp. 450–464. Springer, Heidelberg (2008)
36. Vickers, M.: Ontology-based free-form query processing for the semantic web. Master's thesis, Brigham Young University, Provo, Utah (June 2006)
37. W3C (World Wide Web Consortium) Semantic Web Activity Page, `http://www.w3.org/2001/sw/`
38. Weidong, Y., Ning, G., Baile, S.: Reverse engineering XML. In: Proceedings of the First International Multi-Symposiums on Computer and Computational Sciences (IMSCCS 2006), Hangzhou, Zhejiang, China, June 2006, vol. 2, pp. 447–454 (2006)
39. Weikum, G., Kasneci, G., Ramanath, M., Suchanek, F.: Database and information-retrieval methods for knowledge discovery. Communications of the ACM 52(4), 56–64 (2009)

Towards Controlled Query Evaluation for Incomplete First-Order Databases

Joachim Biskup, Cornelia Tadros, and Lena Wiese

Technische Universität Dortmund, Germany
{biskup,tadros,wiese}@ls6.cs.uni-dortmund.de

Abstract. Controlled Query Evaluation (CQE) protects confidential information, stored in an information system. It prevents harmful inferences due to a user's knowledge and reasoning. In this article we extend CQE to incomplete first-order databases, a data model which suits a broader range of applications than a previously studied propositional incomplete data model. Because of the complexity of the underlying implication problem, which describes the user's reasoning, the representation of the user's knowledge is the main obstacle to effective inference control. For knowledge representation, we introduce first-order modal logic to CQE. Especially, we deal with knowledge about a restricted data model in first-order logic. The restricted data model considered gives rise to a new problem: if the user is aware of the data model, his reasoning must be modeled appropriately. In the analysis of this "reasoning" model we consider both confidentiality and availability. Finally we show, how the considered data model can be reduced to the propositional case and analyze confidentiality properties of the resulting implementation.

1 Introduction

Database users can infer sensitive data even without being permitted direct access to it. A user with medical expertise, for instance, can deduce a clinical picture of some patient from the medication. The prevention of such harmful inferences is called *inference control*. Controlled Query Evaluation (CQE) is a framework of inference control in information systems.

Controlled Query Evaluation has been studied broadly for several parameters, e.g., the logic for data representation, the query language and enforcement methods, and is especially well understood in the case of complete logical databases [1,2,3,4,5] including optimization issues [6] and updates in the early stages [7,8]. A first approach to incomplete databases has already been made in [9,10] with more details on both papers in [11]. It yields a decidable framework for Controlled Query Evaluation in incomplete *propositional* databases. Yet for some applications, a fragment of *first-order logic* (abbr. FOL) is necessary to model the data; hence in this paper, we introduce a restricted data model using FOL sentences.

An important aspect of Controlled Query Evaluation is the representation of the user knowledge to keep track of the information revealed to him in addition

S. Link and H. Prade (Eds.): FoIKS 2010, LNCS 5956, pp. 230–247, 2010.

to his a priori knowledge. In the incomplete propositional case, modal logic is used for this purpose. Extending the data model of [11] to first-order databases, the user knowledge should be expressed in first-order modal logic (FOM). Thus, we investigate, how to exploit the richer semantics of FOM. Although a lot of fragments of first-order modal logic are known to be undecidable [12], using our restricted data model and FOM as a language for the user knowledge, CQE turns out to be reducible to the propositional case.

In [11] the complete data model was embedded into the CQE framework for incomplete propositional databases; the control mechanism has to account for the user's reasoning, who knows, that the data model is complete. With our restricted data model we face a similar situation and the user's reasoning due to these restrictions must be modeled; the proposed model is analyzed considering soundness and completeness.

In particular, this paper makes the following contributions:

1. We introduce a restricted first-order data model (called GFFD-databases) for which we can protect data by inference control with Controlled Query Evaluation.
2. We show that additional inference capabilities (due to the data model) can be modelled soundly and completely by a finite set of formulas in the user knowledge.
3. We show decidability of the problem of Controlled Query Evaluation for the restricted data model as well as FOM as a representation language for the user knowledge (and a confidentiality policy).
4. We present a reduction of Controlled Query Evaluation in our restricted data model to the propositional case and hence achieve an operationalization of the first-order problem.

The remainder of the introductory section recapitulates the basics of Controlled Query Evaluation and defines the data model, *GFFD-databases*, that is used throughout this article. Section 2 deals with the specification language and its semantics to represent knowledge about GFFD-databases. In the last sections we explain how Controlled Query Evaluation for incomplete propositional databases can be used for GFFD-databases: In Section 3 the needed transformations from the specification language to the corresponding propositional language are performed; Section 4 is devoted to the user's awareness of the data model, which must be included in a correct reduction; The reduction is finished in Section 5; Section 6 completes our analysis with confidentiality considerations.

1.1 Controlled Query Evaluation

First of all, the system of Controlled Query Evaluation needs a *logical frame* to represent the user's inferences by logical implication. In our case it will be first-order modal logic. The system of Controlled Query Evaluation has the following core components:

- **confidentiality policy**[1] $pot\_sec_{ps}^{GFFD}$**:** a set of sentences in a *specification language* defined as a fragment of the logical frame (such a sentence is called a **potential secret**)
- **user knowledge** $prior_{ps}^{GFFD} \subseteq log$**:** a set of sentences in the specification language expressing the user's a priori knowledge. The a priori knowledge is contained in a *log* in which any knowledge revealed to the user by answers to database queries is recorded additionally
- **censor:** a function that controls the answers given to the user
- **database instance** *db***:** a consistent set of sentences in a logic, which is described in detail in the next section

The user is not allowed to know that any of the *potential secrets* in the confidentiality policy *holds* in the database instance *db*. The actual semantics of a potential secret with respect to a database will be defined in Section 2.1. The *censor* determines the knowledge that would be revealed to the user by an answer to a database query. With the means of logical implication the censor finds out whether this knowledge and the knowledge stored in the logfile imply any potential secret. The latter would result in a violation of the confidentiality policy, so that the censor distorts the answer by *lying* or *refusal.* Lying is the modification of the given answer by returning a harmless answer from the possible answers *true*, *false* or *undef*, whereas refusal means that *refuse* is returned. In the case that data modification is used ("lying"), even the disjunction of all potential secrets must not be implied, see e.g. [2]. In the case that data restriction is used ("refusal"), the user might infer sensitive information by combining his a priori knowledge, the knowledge gained from the query answer, and also knowledge about the CQE algorithm. This kind of inference is called *meta inference*, see e.g. [11](Chapter 5).

Whereas the primary goal of CQE is confidentiality, *availability*, i.e. to provide all information required by a user, is also a security goal, since holding back or modifying information may even endanger lives (like in the medical example from the beginning). Considering availability to a maximum extent also prevents useless solutions, like always returning *refuse*.

CQE can be varied by several parameters. In the context of this paper we will assume that the user is aware of the confidentiality policy, the CQE algorithm and the data model, and has unlimited computing power. Harmful answers are refused by the CQE system, hence only data restriction is used.

1.2 The Data Model

Generally speaking an incomplete database instance is a set of sentences in FOL ([13]). Apart from the database instance a database administrator at design time can declare schema information, i.e., the used predicate symbols (and their arity) and values (objects), or integrity constraints. We now introduce our data model

---

[1] The notation $pot\_sec_{ps}^{GFFD}$ annotates the data model $GFFD$ and the type of policy *ps* (potential secrets) and distinguishes it from the propositional counterpart pot_sec_{ps}.

GFFD-databases. As for the schema, a *finite* alphabet of predicate symbols $\mathcal{P}$ and their arity and some finite set $Const := \{c_1, \ldots, c_n\}$ of constant symbols are fixed. $\mathcal{P}$ may not contain the equality predicate $=$. A *GFFD-database instance* *db* is a finite, consistent set of (negative and positive) ground facts; that is, ground facts are negated or non-negated closed predicates from $\mathcal{P}$ with constant symbols taken from $Const$. The set of all ground facts shall be denoted by $\mathcal{GF}$ and the set of all positive ground facts by $\mathcal{GF}^+$. Additionally it is required that *db* contains the finite domain axiom $FIN \equiv (\forall x : (x = c_1) \vee \ldots \vee (x = c_n)) \wedge EQU$, where EQU is a shorthand for the equality axioms. (Note that $=$ is only allowed in the axiom FIN.) The set $\{flu(Alice), \neg medication(MedA, Alice), FIN\}$, for instance, is a GFFD-database instance. The constants $Alice$ and $MedA$ are taken from $Const$.

A GFFD-database can be addressed by closed queries, that is, closed sentences ϕ in first-order logic. Open queries are omitted for simplicity, but might be emulated by a sequence of closed queries like in [4]. Query evaluation on an instance *db* is defined by the following function based on the operator $\models_{FOL}$ of logical implication in first-order logic:

$$eval(\phi)(db) := \begin{cases} true & \text{if } db \models_{FOL} \phi \\ false & \text{if } db \models_{FOL} \neg\phi \\ undef & \text{else} \end{cases} \tag{1}$$

Due to the restricted data model, the operator $\models_{FOL}$, which is undecidable in general, can be implemented by logical implication in propositional logic. In the

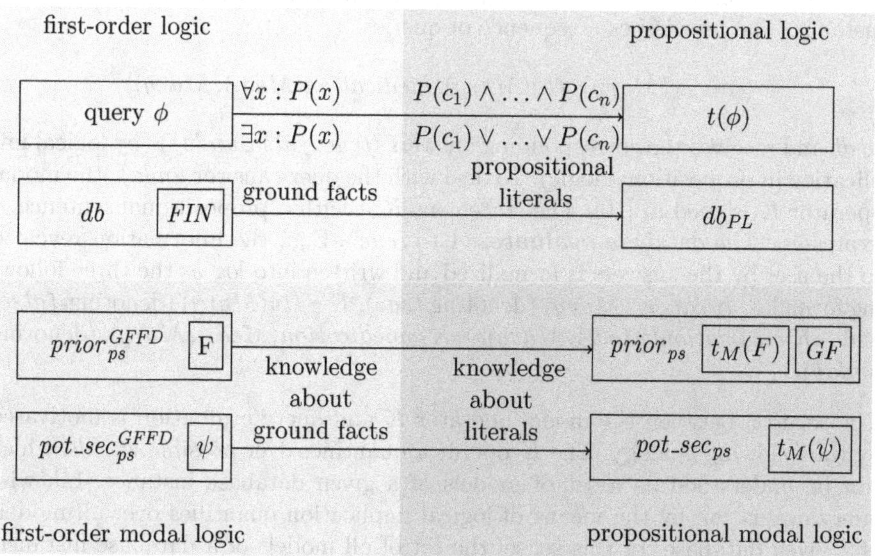

Fig. 1. Applying propositional CQE for GFFD-databases

context of GFFD-databases all queries will be transformed to propositional sentences which can be dealt with by a propositional database. A GFFD-database corresponds to a propositional database, interpreting ground facts as propositional literals and removing the axiom FIN. This interpretation of a GFFD-database is the motivation for applying the propositional CQE after some transformations. This approach is illustrated in Figure 1. We will explain the representation of the user's knowledge and the related transformations. The analogous transformation of queries will not be described for space reasons.

2 Specification Language for User Knowledge and Policy

In order to keep track of the user knowledge, all database answers are written into a logfile *log*. In addition to the logging, a security administrator can specify the user's a priori knowledge ($prior_{ps}^{GFFD}$) and any forbidden knowledge in the confidentiality policy ($pot\_sec_{ps}^{GFFD}$). Therefore a specification language ought to be comprehensible and handy to avoid errors at specification time. Expressiveness certainly is a design goal to suit all security requirements of an application. In [13] Reiter argued for KFOPCE [14] (a first-order modal logic) as an appropriate specification language for integrity constraints for incomplete first order databases. Similarly, we use a first-order modal logic as a specification language for user knowledge and confidentiality policy.

Before moving on to defining the specification language $\mathscr{L}_{ps}^{GFFD}$, the next example reviews, how user knowledge is tracked in the previous CQE approach [10] to account for incomplete information:

Example 1. Let $db = \{cancer(Mary), \neg flu(Mary)\}$ be a propositional database instance. The user issues a sequence of queries

$$\langle cancer(Mary), flu(Mary), medication(MedA, Mary)\rangle$$

to db and receives the corresponding answers $\langle true, false, undef\rangle$ (by logical implication in propositional logic). To deal with the query answer $undef$, the modal operator K is used in [10]. The statement $K\, A$ with a propositional sentence A expresses "*The database* **evaluates** *A to true*". E.g., the information revealed to the user by the answers is formalized and written into *log* as the three following formulas: $K\, cancer(Mary)$ (denoting $true$), $K\, \neg flu(Mary)$ (denoting $false$) and $\neg K\, medication(MedA, Mary) \wedge \neg K\, \neg medication(MedA, Mary)$ (denoting $undef$).

The relation between the modal operator K and query evaluation is motivated by the following analogy: The K-operator quantifies over *possible worlds*, which can be understood as a set of models of a given database instance. Likewise query evaluation by the means of logical implication quantifies over all models of a given database. In this sense, the set of all models of a database instance, i.e., its *information theoretic view*, represents the information contained in the database instance. Further, [10] assumes that the database only comprises correct information about the world. Thus the K-operator in [10] is interpreted in Kripke structures with one equivalence class for the models of a database with the actual world being among these models, which is an S5 system [15].

We now define the specification language $\mathscr{L}_{ps}^{GFFD}$ as a fragment of first-order modal logic S5 and use it for expressing the user knowledge as well as the potential secrets in the confidentiality policy. As the user is assumed to be aware of the data model, all his reasoning takes place in terms of reasoning about ground facts and the confidentiality policy is just concerned with knowledge about ground facts as well. The specification language $\mathscr{L}_{ps}^{GFFD}$ is defined formally as follows:

Definition 1 (Specification language for knowledge and policy). *The language $\mathscr{L}_{ps}^{GFFD}$ is defined inductively:*

- *ground facts:*
  Let $P(c_{i_1},\ldots,c_{i_k}) \in \mathcal{GF}^+$ be a ground fact. Each formula of the form $K\,P(c_{i_1},\ldots,c_{i_k})$ and $K\,\neg P(c_{i_1},\ldots,c_{i_k})$ is a sentence in $\mathscr{L}_{ps}^{GFFD}$.[2]
- $\neg,\vee,\wedge$*:*
  For sentences F and G in $\mathscr{L}_{ps}^{GFFD}$, the formulas $\neg F$, $F \vee G$ und $F \wedge G$ are also sentences in $\mathscr{L}_{ps}^{GFFD}$.
- $\exists,\forall$*:*
  For a sentence F in $\mathscr{L}_{ps}^{GFFD}$ containing a constant $c \in Const$ (i.e., F is not only built of zero-ary predicate symbols) the formulas $\exists x : F[x/c]$ and $\forall x : F[x/c]$ are also sentences in $\mathscr{L}_{ps}^{GFFD}$. The expression $F[x/c]$ stands for the substitution of c by x, wherever c occurs in F.

A short example gives an idea of what kind of knowledge can be expressed in this specification language.

Example 2. The following $\mathscr{L}_{ps}^{GFFD}$ sentences may be written in the user's logfile (expressing his knowledge) or in the confidentiality policy (expressing what the user should not know):

$K\,cancer(Alice) \wedge K\,\neg aids(Alice)$

- $cancer(Alice)$ and $\neg aids(Alice)$ are recorded in the database

$K\,cancer(Alice) \vee K\,cancer(Sue)$

- $cancer(Alice)$ or $cancer(Sue)$ or both are recorded in the database

$\forall k : \exists x : K\,hasDisease(x, k)$

- for each value c_k in the domain $\mathcal{U} := Const$ there is a ground fact $hasDisease(c_p, c_k)$ for some value c_p recorded in the database. It makes sense to introduce types, which is omitted here for simplicity.

The last formula can be built according to Definition 1 from a formula $K\,hasDisease(c_p, c_k)$ for arbitrary constants c_p and c_k.

[2] In full first-order modal logic syntax $K\,P(c_{i_1},\ldots,c_{i_k})$ should be written as $K < \lambda x_1,\ldots,x_k.P(x_1,\ldots,x_k) > (c_{i_1},\ldots,c_{i_k})$, because in general the syntax $K\,P(c_{i_1},\ldots,c_{i_k})$ is ambiguous in first-order modal logic. A formal definition of the syntax and semantics of the λ-operator can be found in [16].

2.1 A Formal Definition of the Semantics

So far we have stated the meaning of $\mathscr{L}_{ps}^{GFFD}$ sentences in natural language. Nevertheless a sentence should be evaluated with respect to an actual database instance, e.g., to verify, whether a secret holds in the instance or whether the instance is compatible with certain user knowledge. Following the propositional approach in [11](Chapter 4), the semantics will be defined on an S5 Kripke model as discussed after Example 1. The possible worlds represent first-order models of the actual database instance. We only consider Kripke models with *constant domain* and *rigid designators*[3], so that the intended semantics of the quantifiers is obtained (more details can be found in [17]). These constraints also appear in the definition of the semantics of KFOPCE [14].

For a constant domain we use the *constant domain semantics* of first-order modal logic with the domain $\mathcal{U}_{db} = Const$. A rigid designator is a constant that designates the same value in every possible world; hence, every constant $c_j \in Const$ is mapped to itself by the assignment function π_{db} of the Kripke model M_{db}^{GFFD} defined below. The set of possible worlds $\mathcal{S}_{db}^{+}$ represents a set of models of *db*. Consequently, the assignment function for predicates of a possible world s_i are defined according to the associated model M_i of *db*. The accessibility relation $\mathcal{K}_{db}$ is an equivalence relation, so that the model is an S5 Kripke model.

Definition 2 (The S5 Kripke model $M_{db}^{GFFD} = (\mathcal{S}_{db}^{+}, \mathcal{K}_{db}, \mathcal{U}_{db}, \pi_{db})$). *Let db be a GFFD-database instance. Then the model M_{db}^{GFFD} consists of the following four components:*

- *The set $\mathcal{S}_{db}^{+}$* [4] *represents the set of models $M_i = (\mathcal{U}_i, \varphi_i, \psi_i, \zeta_i)$ of db fulfilling the constraints $\mathcal{U}_i = Const$ and $\varphi_i(c_j) = c_j$ for every $c_j \in Const$. The world associated with M_i is denoted by s_i.*
- *The accessibility relation $\mathcal{K}_{db}$ is given by $\mathcal{K}_{db} := \mathcal{S}_{db}^{+} \times \mathcal{S}_{db}^{+}$.*
- *The domain $\mathcal{U}_{db}$ is given by $\mathcal{U}_{db} = Const$.*
- *The assignment function for variables $\pi_{db}(x)$ is arbitrary.*
  The assignment function for constants is $\pi_{db}(c_j, s_i) = c_j (= \varphi_i(c_j))$ for each $c_j \in Const$ and each $s_i \in \mathcal{S}_{db}^{+}$,
  and for predicate symbols P, $\pi_{db}(P, s_i) = \psi_i(P)$ for each $s_i \in \mathcal{S}_{db}^{+}$.

Based on the S5 Kripke model M_{db}^{GFFD} we will introduce a formal semantics of the specification language in the remainder of this section. No information, contained in a GFFD-database, must be lost in the representation M_{db}^{GFFD}. For this purpose, it is necessary that for any closed query ϕ the same answer is received by ordinary query evaluation, i.e., $db \models_{FOL} \phi$, and by a restricted query evaluation in the following sense: ϕ is evaluated to *true* w.r.t. *db* iff ϕ holds in all models of *db* in $\mathcal{S}_{db}^{+}$. Note that the axiom *FIN* is essential for this property, which becomes clear in the following example:

---

[3] More details on different semantics of first-order modal logic and rigid designators can be found in [16].

[4] We use the plus to emphasize that the worlds in M_{db}^{GFFD} represent only first-order structures, that are models of *db*.

Example 3. We have the simple setting $Const = \{c_1\}$, $\mathcal{P} = \{P\}$ and $db = \{P(c_1)\}$, deliberately omitting the axiom $FIN \equiv \forall x : (x = c_1) \land EQU$.

The model $M_1 := (\{c_1\}, \varphi_1(c_1) = c_1, \psi(P) = \{(c_1)\}, \zeta)$ is the only[5] model of db, where the constraints according to Definition 2 of M_{db}^{GFFD} hold. Also $\phi \equiv \forall x : P(x)$ holds in M_1, but the query ϕ is evaluated to $false$ w.r.t. db. In contrast, adding the axiom FIN, the query ϕ is evaluated to $true$ w.r.t. $db \cup \{FIN\}$.

The evaluation of $\mathscr{L}_{ps}^{GFFD}$ sentences regarding a database can now get a formal definition:

Definition 3 (The evaluation function $eval_{ps}^{GFFD}$). *Let db be a GFFD-database and F an $\mathscr{L}_{ps}^{GFFD}$ sentence. The function $eval_{ps}^{GFFD}$ evaluates F w.r.t. db either to true or $false$ and is defined as*

$$eval_{ps}^{GFFD}(F)(db) := (M_{db}^{GFFD}, s) \models F \text{ for an arbitrary } s \in S_{db}^{+}.$$

The proof that the model operator in the expression $(M_{db}^{GFFD}, s) \models F$ is independent of the choice of the world s in M_{db}^{GFFD} will not be shown here; it is of mere technical nature as it follows directly from the S5 properties. The semantics given by $eval_{ps}^{GFFD}$ is exactly what was described informally in Example 2 above. Due to the complexity of the formalization it will not be discussed in this article; the formalization can be found in [17].

3 Preparing the Reduction to the Propositional Case

At this point the specification language $\mathscr{L}_{ps}^{GFFD}$ was introduced together with a clear specification of the semantics, which can be used by a security administrator to write a confidentiality policy and the user's a priori knowledge.

Now CQE for propositional incomplete databases shall be applied. Hence the next step will be to transform the closed first-order queries, the confidentiality policy and the a priori knowledge to the propositional counterparts like in [11]. Certainly the semantics have to be preserved. We will end this section discussing this matter. Sentences of the propositional counterpart $\mathcal{L}_{PS}$ of $\mathscr{L}_{ps}^{GFFD}$ (see [11]) can be built applying Definition 1, but you have to replace ground facts with arbitrary propositional formulas over the alphabet $\mathcal{GF}^{+}$ in the first rule and skip the third rule. Figure 1 in the introductory section had visualized the idea of interpreting ground facts as propositional literals as the basis of the transformation. This idea is applied also to knowledge about ground facts in the following definition:

Definition 4 (The transformation t_M). *Let $\mathcal{L}_{PS}$ be the propositional counterpart of $\mathscr{L}_{ps}^{GFFD}$ (see [11]). Then the transformation $t_M : \mathscr{L}_{ps}^{GFFD} \to \mathcal{L}_{PS}$ is inductively defined as follows:*

[5] The function ζ is neglected since it is of no importance in our case.

- *ground facts:*
  Let $P(c_{i_1},\ldots,c_{i_k}) \in \mathcal{GF}^+$ *be a ground fact.*
  $t_M : K\,P(c_{i_1},\ldots,c_{i_k}) \mapsto K\,P(c_{i_1},\ldots,c_{i_k})$
  $t_M : K\,\neg P(c_{i_1},\ldots,c_{i_k}) \mapsto K\,\neg P(c_{i_1},\ldots,c_{i_k})$

Let F *and* G *be* $\mathscr{L}_{ps}^{GFFD}$ *sentences:*

- *propositional connectives:*
  $t_M : \neg F \mapsto \neg t_M(F)$
  $t_M : F \wedge G \mapsto t_M(F) \wedge t_M(G)$
  $t_M : F \vee G \mapsto t_M(F) \vee t_M(G)$
- *quantifiers:*
  $t_M : \exists x : F \mapsto t_M(F[c_1/x]) \vee \ldots \vee t_M(F[c_n/x])$
  $t_M : \forall x : F \mapsto t_M(F[c_1/x]) \wedge \ldots \wedge t_M(F[c_n/x])$
  where x *is not bound in* F

The transformation of closed query formulas to propositional query formulas is analogously done, except for the base case, in which predicates are mapped to atomic propositions. The only critical point of the transformation t_M is the semantics of the quantifiers to be preserved. We will take a closer look at the existential quantifier in the proof sketch of the following proposition, and we explain the role of the axiom FIN of the GFFD data model from Section 1.2. The counterparts of M_{db}^{GFFD} and $eval_{ps}^{GFFD}$ are M_{db} and $eval_{ps}$ in the propositional case in [11]. The Kripke structure M_{db}, similar to the first-order Kripke structure M_{db}^{GFFD}, represents the models of a propositional database db as an S5 Kripke model in (propositional) modal logic.

Proposition 1 (Semantics preservation under t_M). *Let a GFFD-database* db *be interpreted as a propositional database by* $db_{PL} := db \setminus \{FIN\}$*. Then it holds that*

$$eval_{ps}^{GFFD}(F)(db) = true \text{ iff } eval_{ps}(t_M(F))(db_{PL}) = true.$$

Sketch of Proof. By structural induction on $F \in \mathscr{L}_{ps}^{GFFD}$:

Base case: Let $P(c_{i_1},\ldots,c_{i_k}) \in \mathcal{GF}^+$ be a ground fact. Then it holds that $eval_{ps}^{GFFD}(K\,P(c_{i_1},\ldots,c_{i_k}))(db) = true$ if and only if it holds that $eval_{ps}(K\,P(c_{i_1},\ldots,c_{i_k}))(db_{PL}) = true$. Analogously for $\neg P(c_{i_1},\ldots,c_{i_k})$. See [17] for more details.

Inductive case: We demonstrate just the case of existential quantification to show the influences of the restricted data model "GFFD-database".

Let $F \equiv \exists x : G$ be a $\mathscr{L}_{ps}^{GFFD}$ sentence; then

$$eval_{ps}^{GFFD}(F)(db) = (M_{db}^{GFFD}, s) \models F \text{ for an arbitrary } s \in S_{db}^{+}$$

(Definition 3). This holds if and only if

$$(M_{db}^{GFFD}, s) \models G[c_j/x] \text{ for a } c_j \in Const.$$

Recall that the notation $G[c_j/x]$ denotes the substitution of all occurrences of x by c_j where x is not bound.

This equivalence is crucial, so we justify it in brief: For the interpretation of $\exists x : G$, a value $c \in \mathcal{U}_{db} = Const$ for x is chosen once and x is evaluated to c in *all worlds*[6] of $\mathcal{S}^+_{db}$ in the interpretation of G. In the interpretation of $G[c_j/x]$ the value of x is determined by the choice of c_j. As the identifiers $Const$ are *rigid*, in the interpretation of $G[c_j/x]$ the value of c_j, i.e., c_j, is assigned to x in *all worlds*. Choosing the constant c_j from $Const$ to substitute x in G thus corresponds to choosing the value for x in the interpretation of $\exists x : G$ in our setting.

By the induction hypothesis[7], $(M^{GFFD}_{db}, s) \models G[c_j/x]$ holds if and only if $eval_{ps}(t_M(G[c_j/x])) = true$ for a $c_j \in Const$.

This is equivalent to $eval_{ps}(t_M(G[c_1/x])) = true$ or ... or $eval_{ps}(t_M(G[c_n/x])) = true$. And this finally is shown in [11](Lemma 4.15) to be equivalent to $eval_{ps}(t_M(G[c_1/x]) \vee \ldots \vee t_M(G[c_n/x])) = true$ which is $eval_{ps}(t_M(F)) = true$ by Definition 4. □

The essential confinement of the data model is the axiom FIN in a GFFD-database, which enables us to set the universe $\mathcal{U}_{db}$ to $Const$ and to confine the interpretations to rigid designators. Hence the transformation t_M provides a means to transform a first-order modal $\mathscr{L}^{GFFD}_{ps}$ sentence into a propositional modal sentence and evaluate it in propositional modal logic.

4 Awareness of the Data Model

That the user is aware of the data model, means in the context of GFFD-databases, that he knows the schema specification, and the restrictions of an GFFD-database instance, as defined above. Further, the user knows the query evaluation function *eval*. Our assumptions are partly a precaution, because the user might be an insider as for example a system administrator. A security violation can arise if one does not take care of the user's awareness of the data model, i.e., that the user knows about the general restrictions, posed on GFFD-databases in Section 1.2. This becomes clear in the following scenario:

Example 4. A potential secret shall be: $\psi \equiv K\, cancer(Alice) \vee K\, cancer(Sue)$ (in propositional modal logic as this will be used later on). The query $\phi \equiv cancer(Alice) \vee cancer(Sue)$ shall be answered by *true*, which finally results in the logfile: $log = \{K\,\phi\}$. There is a Kripke model M in which ψ is not disclosed according to the logfile: $(M, w_1) \models log$ and $(M, w_1) \not\models \psi$. The model M is shown in the following figure:

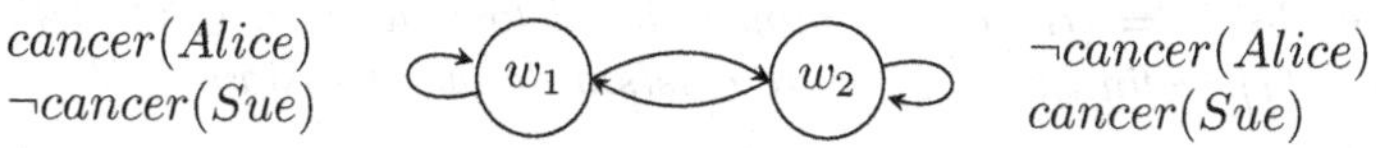

[6] All worlds of $\mathcal{S}^+_{db}$ are reachable from s for each $s \in \mathcal{S}^+_{db}$ by Definition 2.

[7] Note that $G[c_j/x]$ is a $\mathscr{L}^{GFFD}_{ps}$ sentence whereas G is not, because x is free in G.

Nevertheless the user knows, that ψ holds, since the database may only contain ground facts. To be precise the user will conclude from this restriction, that ϕ can only hold if one of the ground facts also holds. Therefore he believes ψ.

This observation can be generalized, saying: if a disjunction over a set Γ of ground facts holds in the database than at least one of the ground facts must hold. This generalization is expressed by the rule of pushing the K-operator into a disjunction of ground facts: $(K \bigvee_{G\in\Gamma} G) \rightarrow (\bigvee_{G\in\Gamma} K\, G)$. There is one exceptional case which is the disjunction being a tautology (compare with Example 5 below). Then certainly the user is not able to conclude any additional information. This can easily be checked: The disjunction over Γ is a tautology if and only if a ground fact and its negation are in Γ. The rule above is incorporated into the formula GF for all appropriate sets of ground facts:

Definition 5 (User knowledge GF of the $GFFD$-data model). *Given all appropriate sets of ground facts*

$$\mathcal{P}^*(\mathcal{GF}) := \{\Gamma \subset \mathcal{GF} \mid \bigvee_{G\in\Gamma} G \text{ is not a tautology}\},$$

let GF denote the conjunction of the rule above for all these sets, i.e.,

$$GF \equiv \bigwedge_{\Gamma\in\mathcal{P}^*(\mathcal{GF})} K(\bigvee_{G\in\Gamma} G) \rightarrow (\bigvee_{G\in\Gamma} K\, G).$$

The formula GF finitely represents the user's knowledge about the restriction "only ground facts" on GFFD-databases, since the set $\mathcal{GF}$ of all ground facts is finite according to the assumptions made in Section 1.2. The formula GF will be written into the user's a priori knowledge $prior_{ps}$ of the propositional CQE system (see Figure 1).

The knowledge expressed by GF can also be understood as the set of databases that the user considers possible when only knowing GF. This set must surely be the set of all GFFD-databases. We will give an analysis of GF regarding this matter: At first, each database instance that the user considers possible when only knowing GF must *possibly* be a GFFD-database from the user's point of view: i.e., before its transformation to propositional logic and viewed as a black box, which responds to queries. This is expressed by the property of *soundness* of GF:

Theorem 1 (Soundness). *For every (propositional) S5 structure $M = \langle\mathcal{S},\mathcal{K},\pi\rangle$ over the propositional symbols $\mathcal{GF}^+$ with $(M,s) \models GF$ for a world $s \in \mathcal{S}$ and $db_{M,s} := \{\alpha \mid \alpha$ is a propositional formula over $\mathcal{GF}^+$ with $(M,s) \models K\,\alpha\}$ (See [11] Definition 5.3) there exists a database instance db', such that*

1. *$db' \subseteq \mathcal{GF}$, and*
2. *$eval(A)(db') = eval(A)(db_{M,s})$ for each propositional formula A over $\mathcal{GF}^+$.*

The construction of $db_{M,s}$ can be viewed as an *inverse* construction to the propositional Kripke structure M_{db}, i.e., db can be retrieved from M_{db} by constructing $db_{M,s}$. In saying so, we neglect that $db_{M,s}$ is closed under logical implication (see [11]) whereas db might not. The database instance $db' \cup \{FIN\}$ can be interpreted as a GFFD-database. So the theorem above says in other words the following: after you invert the construction of M_{db} on a model satisfying GF, you obtain a database instance, which can be described using only propositional literals. This will be needed in the proof of confidentiality preservation.

Secondly, the set of databases, that the user considers possible at the moment according to GF must include all GFFD-databases. This is expressed by the property of *completeness* of GF:

Theorem 2 (Completeness). *Let db be GFFD-database and* $db_{PL} := db \setminus \{FIN\}$ *its propositional counterpart. Then its representation* $M_{db_{PL}}$ *over the propositional symbols* $\mathcal{GF}^+$ *is a model of* GF *in each world in* $\mathcal{S}_{db_{PL}}$.

The proof essentially uses the fact that the rules constituting GF do not apply to tautologies. The proofs of Theorem 1 and 2 can be found in [17].

Note that soundness is important for confidentiality and completeness important for availability. The following example shall explain, why the completeness property is important for availability:

Example 5. Let GF' denote an incomplete variant of GF where tautologies are not excluded and $pot\_sec_{ps} = \{K\, cancer(Alice)\}$ the confidentiality policy in the propositional CQE system [11]. The GFFD-database db in its propositional interpretation is $db_{PL} = \{flu(Alice), \neg medication(MedA, Alice)\}$.

The user's knowledge was recorded during previous system execution and together with $prior_{ps}$ stored in

$$log = \{K\, flu(Alice) \wedge K\, \neg cancer(Alice) \rightarrow K\, medication(MedA, Alice),\\ K\, flu(Alice), GF'\}.$$

Now the user issues further queries:

1. The user queries the database for the tautology $cancer(Alice) \vee \neg cancer(Alice)$.
2. CQE updates the logfile with $K\,(cancer(Alice) \vee \neg cancer(Alice))$.
3. The user makes the query $medication(MedA, Alice)$.
4. CQE refuses the correct answer *false*.

The refusal is necessary, since the logfile and the knowledge $K\, \neg medication(MedA, Alice)$ gained from a correct answer would imply $K\, cancer(Alice)$, which has to be protected. Nevertheless $cancer(Alice)$ does not even hold in the database, so that more answers than necessary are refused. With GF instead of GF' in the log, $K\, cancer(Alice)$ is not implied, when the correct answer is given in step 4.

5 Reduction to the Propositional Case

We summarize the results of the previous sections by introducing the implementation of a CQE-function cqe_{ps}^{GFFD} for GFFD-databases. The implementation invokes a propositional CQE implementation cqe_{ps}. The specification of the confidentiality policy and the users's a priori knowledge are transformed to the propositional specification language by t_M. The user's awareness of the data model is modeled in the a priori knowledge $t_M(prior_{ps}^{GFFD}) \cup \{GF\}$. Each query will be transformed to a propositional query by t. All these steps are visualized in Figure 1.

Definition 6 (Implementation of cqe_{ps}^{GFFD}). *For*

- *Q: a finite sequence of closed first-order queries*
- *db: a GFFD-database and* $db_{PL} := db \setminus \{FIN\}$ *its propositional interpretation*
- $prior_{ps}^{GFFD}$*: the user's a priori knowledge*
- $pot\_sec_{ps}^{GFFD}$*: the confidentiality policy*
- *t: the transformation from closed first-order queries to propositional queries*
- t_M*: the transformation between the specification languages*

the function cqe_{ps}^{GFFD} *is implemented by*

$$cqe_{ps}^{GFFD}(Q, db, prior_{ps}^{GFFD}, pot\_sec_{ps}^{GFFD}) := \\ cqe_{ps}(t(Q), db_{PL}, t_M(prior_{ps}^{GFFD}) \cup \{GF\}, t_M(pot\_sec_{ps}^{GFFD}))$$

The implementation cqe_{ps}^{GFFD} accepts an input *db*, *prior* and *pot_sec* only under additional *preconditions*: at first *db* must be a GFFD-database, so that the reduction works properly. Secondly, db_{PL}, $t_M(prior_{ps}^{GFFD})$ and $t_M(pot\_sec_{ps}^{GFFD})$ must fulfill the preconditions of the underlying propositional implementation cqe_{ps} [11].

6 Confidentiality Preservation

We now move to an analysis of confidentiality preservation of the implementation cqe_{ps}^{GFFD}. Recall that a potential secret ψ is not disclosed if from the user's point of view ψ might not hold in the database. The concept of an *alternative database instance* makes this more precise:

Definition 7 (Alternative database instance). *Let db be database instance and Q a finite query sequence. Further let*

$$cqe(Q, db, prior, pot\_sec) = \langle ans_1, \ldots, ans_m \rangle$$

denote the sequence of previous answers of CQE to Q, where $ans_i \in$ *{true,false,undef,refuse} is the answer of the CQE-function to the i-th query. A database instance db′ is called an alternative database instance of db if*

1. *db and db′ both satify the specified preconditions w.r.t. pot_sec and prior.*
2. $cqe(Q, db', prior, pot\_sec) = cqe(Q, db, prior, pot\_sec)$

In other words, the user cannot distinguish db and db' with the help of his query sequence Q and the CQE responses. If a potential secret ψ does not hold in an alternative database instance, ψ might not hold in the actual database instance db from the user's point of view.

A CQE-function must ensure that no potential secret is disclosed for all proper system configurations and all possible system executions:

Definition 8 (Confidentiality preservation). *A CQE-function preserves confidentiality, if for all database instances, finite query sequences, specifications of a user's a priori knowledge and a confidentiality policy, satisfying the specified preconditions, for each potential secret there is an alternative database instance in which the secret does not hold.*

Confidentiality preservation for cqe_{ps}^{GFFD} essentially depends on the preservation of the semantics in the transformations, the correct modeling of user's awareness of the data model and, of course, on the confidentiality preservation of the underlying propositional CQE system. A sketch of the proof makes these dependencies obvious. The key to the proof is, that the secure refusal censor of the propositional CQE enforces the invariant $log \not\models_{S5} \psi$ for every potential secret ψ in the confidentiality policy pot_sec_{ps}. This invariant guarantees, that there is a database instance, which is consistent with the user's knowledge log and falsifies the potential secret ψ.

Theorem 3 (Confidentiality preservation). *Function cqe_{ps}^{GFFD} preserves confidentiality if the function cqe_{ps} called by it is implemented with a secure refusal censor as in [11] (Chapter 6.2).*

Sketch of Proof. Let $\psi \in pot\_sec_{ps}^{GFFD}$ be a potential secret, db be a $GFFD$-database instance and $prior_{ps}^{GFFD}$ the specified a priori knowledge, such that this input satisfies the preconditions of cqe_{ps}^{GFFD}. Further let Q be a sequence of queries, i.e., closed formulas in first-order logic.

$t_M(\psi)$ is not disclosed by cqe_{ps}. In [11] (Chapter 6.2) an alternative database instance $db_{M,s}$ of the propositional interpretation db_{PL} of db is constructed from a model (M, s) of the logfile, such that ψ does not hold in $db_{M,s}$. Also by Definition 6 the knowledge GF about the data model has been written into the logfile initially, so that Theorem 1 now yields a database db' with $db' \subseteq \mathcal{GF}$ giving the same answers as $db_{M,s}$ to all propositional queries.

The algorithm cqe_{ps} has the same output on $db_{M,s}$ and db' since it only accesses them by queries. Therefore db' must also be an alternative database.

And ψ does not hold in db': The statement $K\,A$ means "*The database* ***evaluates*** *A to true*"(Section 2), so that the propositional specification language essentially expresses combined knowledge from queries. As db' and $db_{M,s}$ always give the same answers to queries and ψ does not hold in $db_{M,s}$, it does not hold in db' as well. The formal proofs will not be given here, but can be found in [17].

The database instance $\tilde{db} := db' \cup \{FIN\}$ is a GFFD-database instance, with $\tilde{db}, pot\_sec_{ps}^{GFFD}$ fulfilling the preconditions of cqe_{ps}^{GFFD}, by construction. Also $\tilde{db}$ gives the same answers as db to the issued query sequence Q, because cqe_{ps}^{GFFD}

just uses the propositional interpretation db', which is an alternative database instance of db_{PL}. Thus $\tilde{db}$ is an alternative database instance of the GFFD-database instance db.

Finally, because the transformation t_M preserves semantics by Proposition 1, it holds that $eval_{ps}^{GFFD}(\psi)(\tilde{db}) = eval_{ps}(t_M(\psi))(db') = false$. □

7 Related Work

The *propositionalization* of first-order theories has already been dealt with considering the efficiency of the transformation (e.g. see [18]). Yet we are not aware of similar results for first-order modal logic.

As for usage of modal logic in database theory, Levesque in [14] introduced the logic KFOPCE (modal **F**irst **O**rder **P**redicate **C**alculus with **E**quality) to allow for queries not only about the state of the world (as usual expressed in FOL) but also about the *knowledge* of the database. This is a desirable feature e.g. in the context of expert systems. As for the semantics of KFOPCE, constants define the (infinite) domain of discourse and are interpreted by themselves, i.e., are rigid in the FOM terminology, the same as with our semantics M_{db}^{GFFD}. Example 3 showed that a database must neccesarily contain the domain closure axiom FIN, so that the semantics M_{db}^{GFFD} preserves the information contained in the database. For KFOPCE the (infinite) domain closure assumption is implicit (as pointed out by [13]). An obvious difference is, that KFOPCE is said to be "weak S5", i.e., KD45 (e.g., in [13]). KD45 weakens S5 by dropping the axiom "$K\,p \rightarrow p$ " and adding the axiom "$\neg K\, false$"([15]). This means, that a database might have knowledge, which is false in the real world. In contrast, we assume (like in the CQE to date), that the database is accurate. Another difference is, that our data model is restricted to finite domains and ground facts, which led us to constrain the expressiveness of $\mathscr{L}_{ps}^{GFFD}$ likewise.

Inference control in incomplete databases has first been studied by [19] in the context of *propositional deductive databases*. The user's *awareness of the data model* is dealt with in [10,11] to embed the complete data model in the incomplete model. The user's a priori knowledge is adjusted, but this adjustment is not analyzed formally w.r.t. soundness and completeness.

The authors of [20] handle inferences, which the user can draw from a published XML document by applying XML constraints. In their context, also incomplete information occurs: A XML document is viewed as a tree structure. Enforcing the security policy, formulated as XQuery Patterns, certain subtrees are hidden. A subtree can represent either the *existence* of a data item or the *content* of a data item, depending on the semantics of its root node. By his inferences the user can also gain partial information, i.e., the existence of a subtree but not its exact position or content. Compared to our work, this approach employs pattern matching whereas we employ logical implication, so these results cannot be applied in our settings.

The authors of [21] evaluate their algorithms for inference control also by "the properties of *soundness* (i.e., everything that is generated by the algorithm is disclosed) and *completeness* (i.e., everything that can be disclosed is produced by the algorithm)". They relate these notions to availability and confidentiality respectively. The authors of [21] choose MLS databases – a complete data model – as the underlying data model and restrict database constraints to Horn-clauses and query-statements to selection and projection. A history of the answers is also maintained. User inference is modeled by the operator $\models_{\mathcal{D}}$ with a set $\mathcal{D}$ of Horn-constraints. This operator confines logical implication (in FOL) to the models of $\mathcal{D}$. The key algorithm computes from a history $\mathcal{P}$ the disclosed information $\mathcal{S} = \{(\textit{answer tuples,query}) = (PF, \phi) \mid \mathcal{P} \models_{\mathcal{D}} (PF, \phi)\}$ and has the desired properties. In our setting we use the implication operator of modal logic, so we can utilize dedicated solvers, whereas the chase algorithm in [21] is specially designed for Horn-constraints.

8 Conclusion and Future Work

We considered a restricted, incomplete data model in FOL, GFFD-databases, and the representation of knowledge about a GFFD-database. Firstly, this data model is covered by propositional logic, while still preserving the intended semantics of the represented knowledge about a GFFD-database instance in first-order modal logic. Especially, the logical implication in first-order modal logic, i.e., the user's reasoning, which suffers from the undecidability of the satisfiability problem in a lot of first-order modal fragments, is not needed because of the transformation into propositional modal logic. Secondly, ground facts will appear in other data models in first-order logic and thus we might use $\mathscr{L}_{ps}^{GFFD}$ as a basis for other specification languages for enhanced data models. Additionally, we dealt with the problem of inferences resulting from the user's awareness of the data model. These inferences were modeled as rules in the user's a priori knowledge. The model was analyzed in terms of soundness and completeness.

In future work we would like to expand the data model to more expressive ones. A useful enhancement is to allow *nulls*. Data models, containing nulls, benefit from the expressiveness of first-order modal logic: First-order modal logic offers additional expressiveness to first-order logic and propositional modal logic by the interaction of quantifiers and the K-operator. First we look at the interaction of the K-operator with quantifiers in detail:

Example 6. Consider the following scenario where a medical record contains first- and surname of the patient, his blood group and health impairment. Such a record is stored in the GFFD-database

$db_1 = \{\textit{medicalrecord}(Frank, Smith, A{+}, \textit{cardiac insufficiency}), FIN\}$
and the similar database

$db_2 = \{\exists x : \textit{medicalrecord}(Frank, Smith, A{+}, x), FIN\}$

with a null x. The statement $\exists x : \textit{medicalrecord}(Frank, Smith, A{+}, x)$ holds in both db_1 and db_2. It says: There is a medical record for Frank Smith with blood group A+ but with a *possibly unknown value* in the column health impairment. In first-order modal logic such knowledge can be expressed by

$K\, \exists x : \textit{medicalrecord}(Frank, Smith, A{+}, x).$

In contrast,

$\exists x : K\, medicalrecord(Frank, Smith, A+, x)$

already appeared in the specification language $\mathscr{L}_{ps}^{GFFD}$, stating: "There is a value from *Const* in the column health impairment such that 'Frank Smith suffers from it'".

The sentence $P \equiv \exists x : K\, medicalrecord(Frank, Smith, A+, x)$ taken as a potential secret protects the value of the column *health impairments*. The answer *true* to a query $\phi \equiv \exists x : medicalrecord(Frank, Smith, A+, x)$ need not reveal the secret P, since a null might be in the relevant column. So the blood group of Frank Smith can be asked for, whereas in GFFD-databases it cannot: The information "ϕ is *true*" results in the logfile entry P as suggested by the semantics of $\mathscr{L}_{ps}^{GFFD}$. For future work and extended data models, we might also adopt the results from the line of research related to KFOPCE [14], e.g., for nulls.

Reiter proposed KFOPCE as a language to specify integrity constraints ([13]). Likewise, we could use $\mathscr{L}_{ps}^{GFFD}$ to define integrity constraints for *GFFD*-databases. To check for integrity preservation the function $eval_{ps}^{GFFD}$ can be applied. Commonly in CQE, the user is assumed to know the integrity constraints. Because these constraints would be specified in $\mathscr{L}_{ps}^{GFFD}$, they can be written into $prior_{ps}^{GFFD}$.

As another enhancement the *equality* predicate should be allowed in ground facts. In this case, one also has to translate the equality axioms which should be contained in the propositional interpretation db_{PL} of a GFFD-database. This issue is dealt with in [22], which is also concerned about efficiency. Since the optimized transformation in [22] is just satisfiability preserving, one has to express the logical implication, during query evaluation, as a satisfiability problem ϕ in the natural way and transform ϕ afterwards (abandoning our strategy of generating db_{PL}). Equality needs special care in first-order modal logics semantics (see [16] for further details) and one also has to investigate if the results from [22] can be adopted to modal logic.

References

1. Biskup, J., Bonatti, P.: Controlled query evaluation for enforcing confidentiality in complete information systems. International Journal of Information Security 3, 14–27 (2004)
2. Biskup, J., Bonatti, P.: Lying versus refusal for known potential secrets. Data & Knowledge Engineering 38, 199–222 (2001)
3. Biskup, J., Bonatti, P.: Controlled query evaluation for known policies by combining lying and refusal. Annals of Mathematics and Artificial Intelligence 40, 37–62 (2004)
4. Biskup, J., Bonatti, P.: Controlled query evaluation with open queries for a decidable relational submodel. Annals of Mathematics and Artificial Intelligence 50, 39–77 (2007)
5. Biskup, J., Wiese, L.: Preprocessing for controlled query evaluation with availability policy. Journal of Computer Security 16(4), 477–494 (2008)

6. Biskup, J., Lochner, J.H.: Enforcing confidentiality in relational databases by reducing inference control to access control. In: Garay, J.A., Lenstra, A.K., Mambo, M., Peralta, R. (eds.) ISC 2007. LNCS, vol. 4779, pp. 407–422. Springer, Heidelberg (2007)
7. Biskup, J., Gogolin, C., Seiler, J., Weibert, T.: Requirements and protocols for inference-proof interactions in information systems. In: Backes, M., Ning, P. (eds.) Computer Security – ESORICS 2009. LNCS, vol. 5789, pp. 285–302. Springer, Heidelberg (2009)
8. Biskup, J., Seiler, J., Weibert, T.: Controlled query evaluation and inference-free view updates. In: Gudes, E., Vaidya, J. (eds.) Data and Applications Security XXIII. LNCS, vol. 5645, pp. 1–16. Springer, Heidelberg (2009)
9. Biskup, J., Weibert, T.: Confidentiality policies for controlled query evaluation. In: Barker, S., Ahn, G.-J. (eds.) Data and Applications Security 2007. LNCS, vol. 4602, pp. 1–13. Springer, Heidelberg (2007)
10. Biskup, J., Weibert, T.: Keeping secrets in incomplete databases. International Journal of Information Security 7(3), 199–217 (2008)
11. Weibert, T.: A Framework for Inference Control in Incomplete Logic Databases. PhD thesis, Technische Universität Dortmund (2008), `http://hdl.handle.net/2003/25116`
12. Wolter, F., Zakharyaschev, M.: Decidable fragments of first-order modal logics. The Journal of Symbolic Logic 66(3), 1415–1438 (2001)
13. Reiter, R.: What should a database know? Logic Programming 14, 127–153 (1992)
14. Levesque, H.L.: Foundations of a functional approach to knowledge representation. Artificial Intelligence 23, 155–212 (1984)
15. Halpern, J., Moses, Y.: A guide to the modal logics of knowledge and belief: Preliminary draft. In: Ninth International Joint Conference on Artificial Intelligence, pp. 480–490 (1985)
16. Fitting, M., Mendelsohn, R.L.: First-Order Modal Logic. Synthese Library, vol. 277. Kluwer Academic Publishers, Dordrecht (1998)
17. Tadros, C.: Kontrollierte Anfrageauswertung in unvollständigen prädikatenlogischen Datenbanken (in German). Diplomarbeit, Technische Universität Dortmund (2008), `http://ls6-www.cs.uni-dortmund.de/uploads/tx_ls6ext/Tadros2008Kontrollierte.pdf`
18. Ramachandran, D., Amir, E.: Compact propositional encodings of first-order theories. In: Proceedings of the Nineteenth International Joint Conference on Artificial Intelligence, IJCAI 2005, pp. 1579–1580 (2005)
19. Bonatti, P.A., Kraus, S., Subrahmanian, V.S.: Foundations of secure deductive databases. IEEE Transactions on Knowledge and Data Engineering 7, 406–422 (1995)
20. Yang, X., Li, C.: Secure XML publishing without information leakage in the presence of data inference. In: Proceedings of the Thirtieth International Conference on Very Large Data Bases, VLDB 2004, pp. 96–107 (2004)
21. Brodsky, A., Farkas, C., Jajodia, S.: Secure databases: Constraints, inference channels and monitoring disclosures. IEEE Transactions on Knowledge and Data Engineering 12(6), 900–919 (2000)
22. Gammer, I., Amir, E.: Solving satisfiability in ground logic with equality by efficient conversion to propositional logic. In: Miguel, I., Ruml, W. (eds.) SARA 2007. LNCS (LNAI), vol. 4612, pp. 169–183. Springer, Heidelberg (2007)

Bagging Decision Trees on Data Sets with Classification Noise*

Joaquín Abellán and Andrés R. Masegosa

Department of Computer Science and
Artificial Intelligence,
University of Granada, Spain
{jabellan,andrew}@decsai.ugr.es

Abstract. In many of the real applications of supervised classification techniques, the data sets employed to learn the models contains classification noise (some instances of the data set have wrong assignations of the class label), principally due to deficiencies in the data capture process. Bagging ensembles of decision trees are considered to be one of the most outperforming supervised classification models in these situations. In this paper, we propose Bagging ensemble of credal decision trees, which are based on imprecise probabilities, via the Imprecise Dirichlet model, and information based uncertainty measures, via the maximum of entropy function. We remark that our method can be applied on data sets with continuous variables and missing data. With an experimental study, we prove that Bagging credal decision trees outperforms more complex Bagging approaches in data sets with classification noise. Furthermore, using a bias-variance error decomposition analysis, we also justify the performance of our approach showing that it achieves a stronger and more robust reduction of the variance error component.

Keywords: Imprecise probabilities, Imprecise Dirichlet model, information based uncertainty measures, ensemble decision trees, classification noise.

1 Introduction

Supervised classification is an important field of data mining and machine learning research. It offers a wide range of different approaches to the problem of predicting the class of an object based on some indirect description of this object (for example, automatically decide whether an email is spam or not, by analyzing the words it contains). The variable to be predicted is normally called *class variable* and the rest *predictive attributes* or *features*. The applications of classification are important and distinguished in fields such as medicine, bioinformatics, physics, pattern recognition, economics, etc., and are used for disease

* This work has been jointly supported by Spanish Ministry of Education and Science under project TIN2007-67418-C03-03, by European Regional Development Fund (FEDER) and by the Spanish research programme Consolider Ingenio 2010: MIPRCV (CSD2007-00018).

S. Link and H. Prade (Eds.): FoIKS 2010, LNCS 5956, pp. 248–265, 2010.

diagnosis, meteorological forecasts, insurance, text classification, to name but a few.

Within a probabilistic approach, supervised classification problem is faced as an inference problem. The probability distribution of the class variable given the predictive attributes is estimated from a training data set and the quality of this estimation is then evaluated in an independent test data set. In order to estimate or learn this probability distribution, many different approaches can be employed.

Ensembles of classifiers are considered to be one of the most outperforming models in supervised classification, specially for the high classification accuracy performance they offer as well as the robustness to different issues which appear in real applications such as class imbalanced problems or training data sets with a very low size.

Among the different approaches to combine classification models, ensembles of decision trees are the most accepted and studied. Decision trees (or classification trees) are a special family of classifiers with a simple structure and very easy to interpret. But the important aspect of decision trees which make them very suitable to be employed in ensembles of classifiers is their inherent instability. This property causes that different training datasubsets from a given problem domain will produce very different trees. This characteristic was essential to consider them as suitable classifiers in ensemble schemes such as Bagging (Breiman [10]), Boosting (Freund and Schapire [15]) or Randomforest (Breiman [11]). It is proved that the techniques of combine multiple trees, or committees of trees, allow us to obtain better results than those that can be obtained from a single model. This approach is not just restricted to learning decision trees, it has been also applied to most other machine learning methods.

The performance of classifiers on data sets with classification noise is a very important issue for machine learning methods. Classification noise is named to those situations which appear when data sets have incorrect class labels in their training and/or test data sets. There are many relevant situations in which this problem can arise due to deficiencies in the data learning and/or test capture process (wrong disease diagnosis method, human errors in the class label assignation, etc).

Many studies have been concerned with the problems related to the performance of ensembles of decision trees in noisy data domains [13,15,18]. These studies showed as Boosting strongly deteriorates its performance while Bagging ensembles are the most robust and outperforming ensembles in these situations.

Noisy training data usually increases the variance in the predictions of the classifiers, therefore, Bagging ensembles based on variance-reducing methods work very well [10]. However, AdaBoost's performance is based on the exponential reweighting of incorrectly classified samples, so this approach invest too much effort on noisy samples that are incorrectly labelled.

In this study, we show as the employment of Bagging ensembles of decision trees, based on information/uncertainty measures and imprecise probabilities (see Klir [16]), can be a successful tool in classification problems with a high level of noise in the class variable.

This paper is organized as follows: in Section 2, we present previous knowledge necessary about decision trees and methods to ensemble decision trees; in Section 3, we focus on our method of Bagging credal decision trees; in Section 4, we present the results of experiments conducted to compare the performance of our method for Bagging credal decision trees with other Bagging scheme which uses the popular C4.5 method, on data sets with different levels of classification noise; and finally, Section 6 is devoted to the conclusions.

2 Previous Knowledge

2.1 Decision Trees

Decision trees are models based on a recursive partition method. The aim of which is to divide the data set using a single variable at each level. This variable is selected with a given criterion and it defines a set of set of data in which all the cases belong to the same value or state of that single variable.

The knowledge representation of a decision tree has a simple tree structure. It can be interpreted as a compact rule set in which each node of the tree is labelled with an predictive attribute that produces a ramification for each one of its values or states. The leaf nodes are labelled with a class label (i.e. a value or state of the class variable).

Many different approaches for inferring decision trees have been published. Quinlan's ID3 ([19]) and C4.5 ([21]) along with the CART approach of Breiman et al. [9] stand out among all of these.

C4.5 Tree Inducer

Here we will give a brief explanation of the most important aspects of this well known tree inducer which we will use as reference to compare our method. There are at least eight different releases of its concrete implementation, aiming to improve the efficiency, the handling of numeric attributes, missing values... and many of them provided different options and alternatives. We just highlight the main ideas behind all these approaches that were introduced by Quinlan in [21]:

Split Criteria: Information Gain [19] was firstly employed to select the split attribute at each branching node. But this measure was strongly affected but the number of states of the split attribute: attributes with a higher number of states were usually preferred. Quinlan's introduced for this new tree inducer the *Information Gain Ratio* (IGR) criterion which penalizes variables with many states. This score normalizes the information gain of an attribute X by its own entropy.

Handling Numeric Attributes: This tree inducer handles numeric attributes employing a very simple approach. Within this method, only binary split attributes are considered and each possible split point is evaluated and it is finally selected the one which induced a partition of the samples with the highest split score (i.e. the *Information Gain Ratio*).

Dealing with Missing Values: It is assumed that missing values are randomly distributed (*Missing at Random Hypothesis*). In order to compute the scores, the instances are split into pieces. The initial weight of an instance is equal to the unit, but when it goes going down a branch receives a weight equal to the proportion of instances that belongs to this branch (weights sum to 1). Information Gain based scores can work with this fractional instances using sum of weights instead of sum of counts.

When making predictions, C4.5 marginalizes the missing variable merging the predictions of all the possible branches that are consistent with the instance, using their previously computed weights.

Post-Pruning Process: Although there are many different proposals to carry out a post-pruning process of a decision tree, the technique employed by C4.5 is called *Pessimistic Error Pruning.* This method computes an upper bound of the estimated error rate of a given subtree employing a continuity correction of the Binomial distribution. When the upper bound of a subtree hanging from a given node is greater than the upper bound of the errors produced by the estimations of this node supposing it acts as a leaf, then this subtree is pruned.

2.2 Ensembles of Decision Trees

Ensembles of decision trees appear to present the best trade-off among performance, simplicity and theoretic bases in the family of classifier ensembles models. The basic idea consists of generating a set of different decision trees and combining them with a majority vote criteria. That is to say, when an unlabelled unclassified instance arises, each single decision tree makes a prediction and the instance is assigned to the class value with, normally, the highest number of votes. In this way, a diversity issue appears as a critical point when an ensemble is built [10]. If all decision trees are quite similar, the ensemble performance will not be much better than a single decision tree. However, if the ensemble is made up of a broad set of different decisions and exhibits good individual performance, the ensemble will become more robust, with a better prediction capacity.

There are many different approaches to this problem but Bagging [10], Random Forests [11] and AdaBoost [15] stand out as the best known and most competitive.

Breiman's Bagging (bootstrap aggregating) [10] is one of the first cases of an ensemble of decision trees. It is also the most intuitive and simple and performs very well. Diversity in Bagging is obtained by using bootstrapped replicas of the original training set: different training data sets are randomly drawn with replacement. And, subsequently, a single decision tree is built with each training data replica with the use of the standard approach [9]. Thus, each tree can be defined by a different set of variables, nodes and leaves. Finally, their predictions are combined by a majority vote.

Bagging approach can be employed with different decision tree inducers, although there is none considered as standard. There are studies were Bagging is employed with ID3, CART or C4.5 tree inducers and sometimes these inducers employed some post-pruning method. The very cited work of Diettrich [13] employs the C4.5 tree inducer (with and without post-pruning) to point out Bagging as an outperforming ensemble approach in data sets with classification noise, but there was not any definitive suggestion about the employment of pruning.

3 Bagging Credal Decision Trees

A new method which uses a split criterion based on uncertainty measures and imprecise probabilities (Imprecise Info-Gain criterion) to build simple decision trees was firstly presented in Abellán and Moral's method [3] and in a more complex procedure in Abellán and Moral [5]. In a similar way to ID3, this decision tree is only defined for discrete variables, it does not works with missing values, and it does not carry out a posterior pruning process.

In a recent work [7], these decision trees were introduced in a Bagging scheme and compared against similar ensembles which were built with several classic information split criteria based on frequentist approaches: Info-Gain [19], Info-Gain-Ratio [21] and Gini Index [9]; and a similar preprocessing step. The conclusions depicted in this study pointed out as the Bagging ensembles of single decision trees built with the Imprecise Info-Gain criteria outperformed these other classic split criteria on data sets with classification noise. These promising results encouraged the extension of the ensembles we presented in that paper to a broader class of data sets with continuous attributes and missing values as well as the introduction of a post-pruning method. Moreover, we aim to compare this new method with some of the state-of-the-art ensembles of classification trees in data sets with classification noise.

In the following subsections, we detail our method to build credal decision trees and its extension to be able to handle numeric attributes and deal with missing values. Moreover, we highlight the main differences respect to the current implementation of C4.5 release 8 with the aim of pointing out as our approach does not employ many of the free parameters introduced in this famous tree inducer.

3.1 Imprecise Information Gain

This is the split criteria employed to build credal decision trees. It is based on the application of uncertainty measures on convex sets of probability distributions. More specifically, probability intervals are extracted from the data set for each case of the class variable using Walley's imprecise Dirichlet model (IDM) [22], which represents a specific kind of convex sets of probability distributions (see Abellán [1]), and on these the entropy maximum is estimated. This is a total uncertainty measure which is well known for this type of set (see Abellán et al. [6]).

The IDM depends on a hyperparameter s and it estimates that (in a given data set) the probabilities for each value of the class variable are within the interval:

$$p(c_j) \in \left[\frac{n_{c_j}}{N+s}, \frac{n_{c_j}+s}{N+s}\right],$$

with n_{c_j} as the frequency of the set of values $(C = c_j)$, of the class variable, in the data set. The value of parameter s determines the speed with which the upper and lower probability values converge when the sample size increases. Higher values of s give a more cautious inference. Walley [22] does not give a definitive recommendation for the value of this parameter but he suggests values between $s = 1$ and $s = 2$. In Bernard [8], we can find reasons in favor of values greater than 1 for s.

Let $K(C)$ and $K(C|(X_i = x_t^i))$ be the following closed and convex sets of probability distributions, also called credal sets, q on Ω_C (the set of possible values or states of C):

$$K(C) = \left\{ q|\, q(c_j) \in \left[\frac{n_{c_j}}{N+s}, \frac{n_{c_j}+s}{N+s}\right] \right\},$$

$$K(C|(X_i = x_t^i)) = \left\{ q|\, q(c_j) \in \left[\frac{n_{\{c_j,x_t^i\}}}{N+s}, \frac{n_{\{c_j,x_t^i\}}+s}{N+s}\right] \right\},$$

with $n_{\{c_j,x_t^i\}}$ as the frequency of the set of values $\{C = c_j, X_i = x_t^i\}$ in the data set, with X_i a predictive attribute. We can define the Imprecise Info-Gain for each variable X_i as:

$$\mathbf{IIG}(C|X_i) = S(K(C)) - \sum_t p(x_t^i) S(K(C|(X_i = x_t^i))),$$

where $S()$ is the maximum entropy function of a credal set.

For the previously defined intervals and for a value of s between 1 and 2, it is very easy to obtain the maximum entropy using procedures of Abellán and Moral [2,4] or the specific one for the IDM of Abellán [1], which obtains its lower computational cost for $s = 1$.

We must remark that this new information measure can give us negative values. This is not a characteristic of classics scores, as the Information Gain or the Information Gain Ratio. If the data set is split, there is always a reduction of the entropy of the class variable with this scores but no with our new score.

3.2 Decision Tree Inducer

In this subsection we present the extension of the decision tree inducer presented in [3]. This extension aims to handle numeric attributes and deal with missing values. It is based on a direct adaptation of the C4.5 standard approaches to these situations. The main differences arise with strong simplification in the procedures with respect to the last implementation of C4.5 release 8.

Split Criteria: The attribute with the maximum Imprecise Info-Gain is selected as split attribute at each branching node.

This simple criteria contrast with the sophisticated conditions of C4.5 Release 8: it is selected the attribute with the highest Info-Gain Ratio score and whose Info-Gain score is higher than the average Info-Gain scores of the valid split attributes. These valid split attributes are those which are either numeric or whose number of values is smaller than the thirty percent of the number of instances which are in this branch.

Stop Criteria: The branching of the decision tree is stopped when there is no split attribute with a positive IIG score or there are 2 or less instances in a leaf.

In C4.5 release 8 is also established a minimum number of instances per leaf which is usually set to 2. But in addition to this, using the aforementioned condition in "Split Criteria" of valid split attributes, the branching of a decision tree is also stopped when there is not any valid split attribute.

Handling Numeric Attributes: Numeric attributes are handled using the same approach detailed in Section 2.1. Each possible split point is evaluated and the one which induces a binary partition with the highest Imprecise Info-Gain is selected.

C4.5 release 8 employs this approach but it establishes an additional requirement related to the minimum number of instances required for each partition. This minimum is set using the following heuristic: the ten per cent of the ratio between the number of instances which fall in this branch node and the number of values of the class variable.

It also corrects the information gain of the optimal partition subtracting to this final value the logarithm of the number of evaluated split points divided by the total number of instances in this branching node.

Dealing with Missing Values: We employ the same approach previously detailed in Section 2.1. The Imprecise Info-Gain can be also computed to those cases where there are fractional counts in the class variable.

Post-Pruning Process: Because the aim of this work is to propose a simple approach that exploits the robustness and performance of the Imprecise Info-Gain criterion, we introduce one the most simple pruning techniques: Reduced Error Pruning [20]. It is a bottom-up method that compares the error in a subtraining data set of one node with the error of its hanging subtree. If the error of the parent node is lower, then the subtree is pruned. In this implementation, the training data set was divided in 3 folds, two of them were employed to build the tree and the other one to estimate the errors.

4 Experimental Results

4.1 Experimental Set-Up

In our experimentation, we have used a wide and different set of 25 known data sets, obtained from the *UCI repository of machine learning databases* which can be directly downloaded from ftp://ftp.ics.uci.edu/machine-learning-databases. A brief description of these can be found in Table 1, where column "N" is the

Table 1. Data set Description

Data sets	N	Attrib	Num	Nom	k	Range
anneal	898	38	6	32	6	2-10
audiology	226	69	0	69	24	2-6
autos	205	25	15	10	7	2-22
breast-cancer	286	9	0	9	2	2-13
cmc	1473	9	2	7	3	2-4
colic	368	22	7	15	2	2-6
credit-german	1000	20	7	13	2	2-11
diabetes-pima	768	8	8	0	2	-
glass-2	163	9	9	0	2	-
hepatitis	155	19	4	15	2	2
hypothyroid	3772	29	7	22	4	2-4
ionosfere	351	35	35	0	2	-
kr-vs-kp	3196	36	0	36	2	2-3
labor	57	16	8	8	2	2-3
lymph	146	18	3	15	4	2-8
mushroom	8123	22	0	22	2	2-12
segment	2310	19	19	0	7	-
sick	3772	29	7	22	2	2
solar-flare1	323	12	0	12	2	2-6
sonar	208	60	60	0	2	-
soybean	683	35	0	35	19	2-7
sponge	76	44	0	44	3	2-9
vote	435	16	0	16	2	2
vowel	990	11	10	1	11	2
zoo	101	16	1	15	7	2

number of instances in the data sets, column "Attrib" is the number of predictive attributes, "Num" is the number of numerical variables, column "Nom" is the number of nominal variables, column "k" is the number of cases or states of the class variable (always a nominal variable) and column "Range" is the range of states of the nominal variables of each data set.

In the literature, ensembles of Bagging decision trees have been implemented using many different tree inducers such as CART [9], ID3 [19] or C4.5 [21]. We take as reference the work of Dietterich [13] where Bagging was evaluated with the last release of C4.5 (cited as Bagging-C4.5R8 or B-C4.5). More precisely, we took the implementation of this tree inducer included in the machine learning platform *Weka* [24].

Our Bagging ensembles of credal decision trees (cited as Bagging-CDT or B-CDT) was implemented using data structures of *Weka* and *Elvira* software[14] (other software platform for the evaluation of probabilistic graphical models). The parameter of the IDM was set to $s = 1$ (see Section 3.1).

Both Bagging ensembles were built with 100 decision trees. Although the number of trees can strongly affect the performance of the ensembles, this is a reasonable number of trees for the low-medium size of the data sets employed in this evaluation (see Table 1) and it has been used in other related works as in Freund and Schapire [15].

Using *Weka's* filters, we added the following percentages of random noise to the class variable: 0%, 5%, 10%, 20% and 30%, only in the training data set. The procedure to introduce noise was the following: a given percentage of instances of the training data set was randomly selected and, then, their current class values were randomly changed to other possible values. The instances belonging to the test data set were left unmodified. To estimate the classification accuracy of each classifier ensemble in each data set, we repeated 10 times a k-10 folds cross validation procedure and the average values were reported.

To compare both ensembles, we have used different statistical tests with the aim of having robust comparisons which, in other case, might be biased if only one statistical test is employed, because all of them are based on different assumptions (see Dempsar [12] and Witten and Frank [24] for a complete explanation and further references to these statistical tests).

Corrected Paired T-test: a corrected version of the Paired T-test implemented in *Weka*. It is used to avoid some problems of the original test with cross validation schemes. This test checks whether one classifier is better or worse than another on average, across all training and test datasubsets obtained from a given data set. We use this test on the training and test datasubsets obtained from a 10 times k-10 folds cross validation procedure on a original data set. The levels of significance used for this test is 0.05.
Counts of Wins, Losses and Ties: Sign Test: a binomial test that counts the number w of data sets on which an algorithm is the overall winner.
Wilcoxon Signed-Ranks Test: a non-parametric test which ranks the differences in performance of two classifiers of each data set, ignoring the sings, and compares the ranks for the positive and the negative differences.
Friedman Test: a non-parametric test which ranks the algorithms for each data set separately, the best performing algorithm getting the rank of 1, the second best rank 2,...The null hypothesis is that all the algorithms are equivalent. When the null-hypothesis is rejected, we can compare all the algorithms to each other using the average raking employing the **Nemenyi test.**

Except for the *Corrected Paired T-test*, we detail in bold face the name of the ensemble that has a statistically significant better classification accuracy (when the p-value is lower or equal than 0.05) and we write "=" when there is no statistical significant differences between both ensembles (p-values higher than 0.05).

4.2 Performance Evaluation without Tree Post-pruning

In this subsection, we compare our approach, B-CDT, with respect to B-C4.5 in terms of classification accuracy, both without post-pruning methods. We also give the average number of nodes of the trees in the different ensembles.

Table 2. Average results of B-C4.5 and B-CDT without post-pruning methods (full details per data set can be found in Appendix: Table 10)

	Classification Accuracy					Tree Size				
	0%	5%	10%	20%	30%	0%	5%	10%	20%	30%
Bagging-C45	87.5	86.8	86.0	82.8	77.2	93.7	225.9	328.7	484.4	591.0
Bagging-CDT	86.9	86.6	86.1	83.8	78.7	68.0	86.8	116.5	240.8	415.9

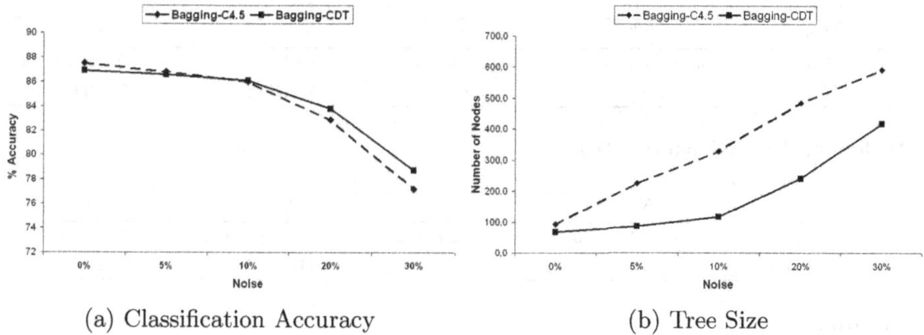

(a) Classification Accuracy (b) Tree Size

Fig. 1. Average results of B-C4.5 and B-CDT without post-pruning methods

In Table 2, we depict the average classification accuracy and the average size of the different trees for both ensembles and for the different noise levels. In Figure 1, these values are graphically represented in dashed lines for Bagging-C4.5R8 ensembles and in continuous lines for Bagging-CDT ensembles.

As can be seen, Bagging-C4.5R8 has a better average performance when no noise is added. And when random noise is introduced in the training data sets the performance of both ensembles degenerates. However, Bagging-CDT is more robust to the presence of noise and since the 20% of noise level it gets a better average classification accuracy.

Moreover, as can be seen Figure 1 (b), the size of the trees of Bagging-C4.5R8 lineally grows with the different noise levels in opposite to Bagging-CDT where the size of the trees is much lower and almost remains stable until the 10% of noise level.

In Table 3, we carry out an exhaustive comparison of the performance of both ensembles using the set of statistical tests detailed in Section 4.1. As can be seen, when no noise is introduced, the performance of Bagging-C4.5R8 is statistically better than Bagging-CDT, however when the different noise levels are introduced there is a shift in the results. For a low noise level, 5% and 10%, the advantage of Bagging-C4.5R8 disappears and there are no statistical differences between both ensembles. When the noise level is higher, 20% and 30%, Bagging-CDT outperforms Bagging-C4.5R8.

In Table 10 of the Appendix, we can see the comparative results about the accuracy of both methods on each data set for each level of noise. We can observe that with 0% of noise the Bagging-C4.5R8 method wins in 17 of the data sets,

Table 3. Tests results of B-C4.5 and B-CDT without post-pruning methods (see Section 4.1 for details)

Corrected Paired-T Test:

Notation: w-d-l, number of data sets where B-CDT respectively wins, draws and loses respect to B-C4.5.

0%	5%	10%	20%	30%
0-24-1	1-23-1	3-22-0	7-18-0	8-16-1

Sign Test:

Notation: w, number of data sets where an ensemble is the overall winner; p, the significance level.

0%	5%	10%	20%	30%
$w = 17, p < 0.1$	$w = 14, p > 0.1$	$w = 16, p > 0.1$	$w = 17, p < 0.1$	$w = 19, p < 0.05$
B-C4.5	=	=	**B-CDT**	**B-CDT**

Wilcoxon Signed-Ranks Test:

Notation: z, value of the normal two tailed distribution; p, the significance level.

0%	5%	10%	20%	30%
$z = -2.01, p < 0.05$	$z = -0.47, p < 0.1$	$z = -0.66, p > 0.1$	$z = -2.27, p < 0.05$	$z = -2.70, p < 0.01$
B-C4.5	=	=	**B-CDT**	**B-CDT**

Friedman Test:

Notation: F, value of the F-distribution; p, the significance level; Rank=(B-CDT, B-C4.5) the average rank.

0%	5%	10%	20%	30%
Rank=(1.7, 1.3)	Rank=(1.56, 1.44)	Rank=(1.36, 1.64)	Rank=(1.32,1.68)	Rank=(1.24,1.76)
$F = 4.57, p < 0.05$	$F = 0.35, p > 0.1$	$F = 2.04, p > 0.1$	$F = 3.57, p < 0.1$	$F = 8.89, p < 0.01$
B-C4.5	=	=	**B-CDT**	**B-CDT**

ties in 2, and loses in 6 with respect to the Bagging-CDT method. This situation changes when more level of noise is added to finish in the contrary situation when 30% of noise is added: Bagging-C4.5R8 wins in only 6 of the data sets and loses in 19 data sets with respect to the Bagging-CDT method.

4.3 Performance Evaluation with Tree Post-pruning

In this subsection, we analyze the performance of both ensembles where the decision trees now apply a post-pruning method (see Sections 2.1 and 3.2).

In Table 4, we show the average classification accuracy and the average size of the trees for both ensembles. In Figure 2 we also graphically show these values.

In [13], there were no conclusions about the suitability to introduce post-pruning methods for decision trees in Bagging ensembles. Our findings are similar, when no noise was introduced there were not statistical significant differences between Bagging-C4.5R8 ensembles with and without post-pruning methods (Wilconxon Signed-Ranks Test, z=-1.10, $p > 0.1$). We found the same conclusion for Bagging-CDT (Wilconxon Signed-Ranks Test, z=-1.48, $p > 0.1$). However when the noise rate is higher, the introduction of post-pruning methods is worthy and there is statistically significant differences for Bagging-C4.5R8 since 10% of noise level (Wilconxon Signed-Ranks Test, z=-2.65, $p < 0.01$). For Bagging-CDT, the introduction of post-pruning methods starts to be statistically significant at 20% of noise level (Wilconxon Signed-Ranks Test, z= -1.87,

Table 4. Average results of B-C4.5 and B-CDT with post-pruning methods (full details per data set can be found in Appendix: Table 11)

	Classification Accuracy					Tree Size				
	0%	5%	10%	20%	30%	0%	5%	10%	20%	30%
Bagging-C45	87.5	87.1	86.6	85.1	81.0	58.9	65.1	74.4	120.8	249.0
Bagging-CDT	86.0	85.9	85.7	84.6	82.0	32.4	34.6	38.6	61.8	109.7

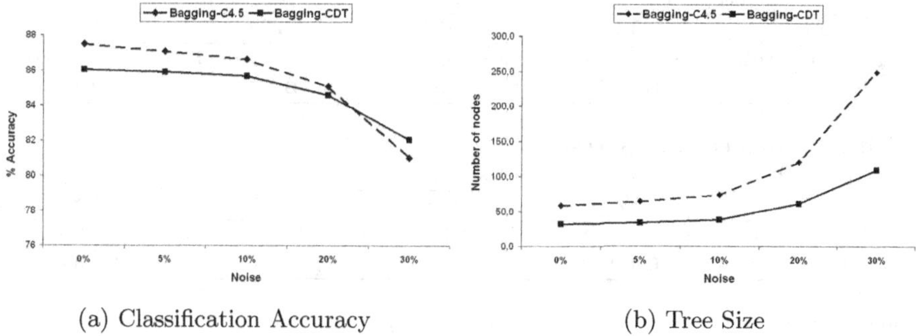

(a) Classification Accuracy (b) Tree Size

Fig. 2. Average results of B-C4.5 and B-CDT with post-pruning methods

$p < 0.1$) and it is clearly positive at 30% of noise level (Wilconxon Signed-Ranks Test, z= -1.87, $p < 0.01$). In this way, Bagging-CDT does not need post-pruning methods with low noise levels.

When we compare Bagging-C4.5R8 with Bagging-CDT the conclusions are mainly the same than in the previous Section 4.2. When no noise is added, Bagging-C4.5R8 performs better than Bagging-CDT but when the noise level increases there is a shift in the comparison and for 20% and 30% of noise levels, Bagging-CDT outperforms Bagging-C4.5R8.

As can be seen in Figure 2 (b), the size of the trees in both ensembles also grows with the noise level. However, the introduction of a post-pruning method helps to maintain a lower size for the trees at least with a noise level lower than 10%. When the noise level is higher or equal than 20%, the size of both ensembles quickly grows but for Bagging-CDT this increment is slower, as happened with the unpruned version.

In Table 5, we also carried out the comparison of both ensembles using our wide set of statistical tests. The conclusions are quite similar, Bagging-C4.5R8 outperforms Bagging-CDT with low noise levels (0% and 5%) and when the noise level increases there is a shift in the performances and Bagging-CDT statistically significantly outperforms Bagging-C4.5R8.

We can observe in Table 11 of the Appendix, similar situation that the one about the methods without pruning. In this table, we can see the results of the accuracy of both methods with pruning on each data set for each level of noise. We can observe that with 0% of noise the Bagging-C4.5R8 method wins in 18 of the data sets, ties in 3, and loses in 4 data sets with respect to Bagging-CDT

Table 5. Test results of B-C4.5 and B-CDT with post-pruning methods (see Section 4.1 for details)

Corrected Paired-T Test:

Notation: w-d-l, number of data sets where B-CDT respectively wins, draws and loses respect to B-C4.5.

0%	5%	10%	20%	30%
1-17-7	1-18-6	1-18-6	2-19-4	9-13-3

Sign Test:

Notation: w, number of data sets where an ensemble is the overall winner; p, the significance level.

0%	5%	10%	20%	30%
$w = 18, p < 0.05$	$w = 18, p < 0.05$	$w = 16, p > 0.1$	$w = 17, p < 0.1$	$w = 19, p < 0.05$
B-C4.5	**B-C4.5**	=	**B-CDT**	**B-CDT**

Wilcoxon Signed-Ranks Test:

Notation: z, value of the normal two tailed distribution; p, the significance level.

0%	5%	10%	20%	30%
$z = -3.04, p < 0.01$	$z = -2.12, p < 0.01$	$z = -1.39, p > 0.1$	$z = -0.65, p > 0.1$	$z = -1.87, p < 0.1$
B-C4.5	**B-C4.5**	=	=	**B-CDT**

Friedman Test:

Notation: F, value of the F-distribution; p, the significance level; Rank=(B-CDT, B-C4.5)the average rank.

0%	5%	10%	20%	30%
Rank=(1.76, 1.24)	Rank=(1.74, 1.26)	Rank=(1.66, 1.34)	Rank=(1.32,1.68)	Rank=(1.24,1.76)
$F = 8.89, p < 0.01$	$F = 7.18, p < 0.01$	$F = 2.73, p > 0.1$	$F = 3.07, p < 0.1$	$F = 8.89, p < 0.01$
B-C4.5	**B-C4.5**	=	**B-CDT**	**B-CDT**

method. Again, this situation changes when more level of noise is added to finish in the contrary situation when 30% of noise is added: Bagging-C4.5R8 method wins in only 6 of the data sets and loses in 19 data sets with respect to the Bagging-CDT method.

4.4 Bias-Variance Analysis

In this section, we attempt to analyze why Bagging-CDT performs better than Bagging-C4.5R8 in situations where there are high noise levels. To do that, we carry out a bias-variance decomposition of the percentage of the Error (Mean Squared Error) [17]:

$$Error = Bias^2 + Variance$$

Bias component represents the systematic component of the error resulting from the incapacity of the predictor to model the underlying distribution. However, variance represents the component of the error that stems from the particularities of the training sample (i.e. a measure of overfitting) and can be decreased by the increasing in the size of the data set. As both are added to the error, a bias-variance trade-off therefore takes place [17]: when we attempt to reduce bias by creating more complex models that fit better the underlying distribution of the

Table 6. Average Bias-Variance of B-C4.5 and B-CDT without post-pruning methods

	$Bias^2$					Variance				
	0%	5%	10%	20%	30%	0%	5%	10%	20%	30%
Bagging-C45	10.3	9.9	9.5	8.9	9.1	5.0	6.1	7.6	11.4	17.1
Bagging-CDT	11.1	10.8	10.3	9.6	10.0	4.8	5.6	7.0	10.1	15.5

Table 7. Average Bias-Variance of B-C4.5 and B-CDT with post-pruning methods

	$Bias^2$					Variance				
	0%	5%	10%	20%	30%	0%	5%	10%	20%	30%
Bagging-C45	10.7	10.4	10.1	9.4	9.3	4.5	5.3	6.2	9.0	13.9
Bagging-CDT	12.7	12.6	12.2	11.7	11.6	4.2	4.7	5.4	7.4	11.0

data, we take the risk of increasing the variance component due to overfitting of the learning data.

We detail the average percentage values of the decomposition of the classification error in bias and variance components for both ensembles without pruning, Table 6, and with post-pruning methods, Table 7. This decomposition was carried out using Weka's utilities and employing the methodology detailed in [23].

As expected, we can see that the introduction of post-pruning methods increases the bias component while reduces the variance, specially for high noise levels, in both ensembles. That is to say, post-pruning reduces the overfitting of the models, specially when the overfitting risk is higher as happens with high noise levels.

Bagging-CDT and Bagging-C4.5R8 employs different methods in order to approximate the underlying distribution of the data and make the classification predictions. The main difference is that Bagging-CDT are based on the Imprecise Info-Gain score which uses the maximum entropy of a credal set, in contradistinction to the classic information gain that measure the expected entropy. So, Imprecise Info-Gain is a more conservative measure than classic Info-Gain.

When no noise is added, the more conservative strategy of Bagging-CDT obtains a lower variance (there is an approximate difference of 0.2 in favor of B-CDT, Table 6) but a much higher bias component (there is an approximate difference of 0.8, Table 6), what results in a worse classification accuracy (the improvement in the variance in not compensated by the losses in the bias). However, when the data sets contains higher noise levels, this conservative strategy of Bagging-CDT is favored and the bad bias values it obtains are compensated by stronger improvements in the variance component. As can be seen in Tables 6 and 7, the difference in the bias between both ensembles remains stable across the different noise levels. However, the difference in the variance strongly increases with higher noise levels. As result, Bagging-CDT outperforms Bagging-C4.5R8 in data sets with medium-high noise levels.

Another indirect way to measure the overfitting of Bagging-C4.5R8 in data sets with classification noise was highlighted in Figures 1 (b) and 2 (b). As can be seen, the differences between the size of the trees of both ensembles increases with higher noise levels.

Table 8. Average Bias-Variance of B-C4.5 and B-CDT with post-pruning methods, using 100, 200, 300 and 500 trees in the Bagging scheme on data sets with high level of noise

	$Bias^2$ (20% noise)				Variance (20% noise)			
	100	200	300	500	100	200	300	500
Bagging-C45	9.4	9.5	9.4	9.4	9.0	8.9	8.9	8.9
Bagging-CDT	11.7	11.8	11.8	11.8	7.4	7.2	7.2	7.2

	$Bias^2$ (30% noise)				Variance (30% noise)			
	100	200	300	500	100	200	300	500
Bagging-C45	9.3	9.3	9.2	9.3	13.9	13.6	13.7	13.6
Bagging-CDT	11.6	11.7	11.8	11.6	11.0	10.7	10.5	10.6

Table 9. Average Bias-Variance of B-C4.5 and B-CDT without post-pruning methods, using 100, 200, 300 and 500 trees in the Bagging scheme on data sets with high level of noise

	$Bias^2$ (20% noise)				Variance (20% noise)			
	100	200	300	500	100	200	300	500
Bagging-C45	8.9	9.1	9.1	9.0	11.4	11.3	11.2	11.1
Bagging-CDT	9.6	9.7	9.7	9.8	10.1	10.0	9.9	9.9

	$Bias^2$ (30% noise)				Variance (30% noise)			
	100	200	300	500	100	200	300	500
Bagging-C45	9.1	9.1	9.0	9.1	17.1	16.9	16.9	16.8
Bagging-CDT	10.0	9.9	9.8	9.9	15.5	15.5	15.5	15.5

We can think that the increasing of the variance error for the Bagging-C4.5R8 method on data sets with high levels of classification noise, could be reduced taking an upper number of trees into the Bagging scheme. To analyze this possibility, we have repeated the experiments with 100, 200, 300 and 500 decision trees into the Bagging schemes on the data sets with 20% and 30% of classification noise. The results of the values of the bias and variance can be seen in Tables 8 and 9. As we can see, it exists a little bit decreasing in the variance values in both methods when we increase the number of trees, but the differences of the errors remain constant in favor of our method. Hence, we can say that the number of trees used is not an important parameter for the comparison of the methods.

5 Conclusions

In this paper, we have presented an interesting application of the information based uncertainty measures applied on credal sets. Using an experimental study, we have proved that our method of Bagging credal decision trees can reduce the percentage of error in classification when it is applied on data sets with medium-high level of classification noise.

We have compared our method with a similar scheme using decision trees built with the C4.5 procedure, which has a number of fix parameters to improve the accuracy. In the literature, we can see that Bagging classification methods using C4.5 procedure can obtain excellent results when it is applied on data sets with classification noise. However, our method based on a total uncertainty measure maintains the variance error component lower when the noise level is increased and, in consequence, obtains a better performance. In addition to this, our approach is more simple and requires a less number of parameters.

References

1. Abellán, J.: Uncertainty measures on probability intervals from Imprecise Dirichlet model. International Journal of General Systems 35(5), 509–528 (2006)
2. Abellán, J., Moral, S.: Maximum entropy for credal sets. International Journal of Uncertainty, Fuzziness and Knowledge-Based Systems 11(5), 587–597 (2003)
3. Abellán, J., Moral, S.: Building classification trees using the total uncertainty criterion. International Journal of Intelligent Systems 18(12), 1215–1225 (2003)
4. Abellán, J., Moral, S.: An algorithm that computes the upper entropy for order-2 capacities. International Journal of Uncertainty, Fuzziness and Knowledge-Based Systems 14(2), 141–154 (2006)
5. Abellán, J., Moral, S.: Upper entropy of credal sets. Applications to credal classification. International Journal of Approximate Reasoning 39(2-3), 235–255 (2005)
6. Abellán, J., Klir, G.J., Moral, S.: Disaggregated total uncertainty measure for credal sets. International Journal of General Systems 35(1), 29–44 (2006)
7. Abellán, J., Masegosa, A.R.: An experimental study about simple decision trees for bagging ensemble on data sets with classification noise. In: Sossai, C., Chemello, G. (eds.) ECSQARU 2009. LNCS, vol. 5590, pp. 446–456. Springer, Heidelberg (2009)
8. Bernard, J.M.: An introduction to the imprecise Dirichlet model for multinomial data. International Journal of Approximate Reasoning 39, 123–150 (2005)
9. Breiman, L., Friedman, J.H., Olshen, R.A.: Classification and Regression Trees. Wadsworth Statistics, Probability Series, Belmont (1984)
10. Breiman, L.: Bagging predictors. Machine Learning 24(2), 123–140 (1996)
11. Breiman, L.: Random Forests. Machine Learning 45(1), 5–32 (2001)
12. Demsar, J.: Statistical Comparison of Classifiers over Multiple Data Sets. Journal of Machine Learning Research 7, 1–30 (2006)
13. Dietterich, T.G.: An experimental comparison of three methods for constructing ensembles of decision trees: Bagging, boosting, and randomization. Machine Learning 40(2), 139–157 (2000)
14. Elvira: An Environment for Creating and Using Probabilistic Graphical Models. In: Proceedings of the First European Workshop on Probabilistic Graphical Models (PGM 2002), Cuenca, Spain, pp. 1–11 (2002)
15. Freund, Y., Schapire, R.E.: Experiments with a new boosting algorithm. In: Proceedings of the Thirteenth International Conference on Machine Learning, San Francisco, pp. 148–156 (1996)
16. Klir, G.J.: Uncertainty and Information: Foundations of Generalized Information Theory. John Wiley, Hoboken (2006)
17. Kohavi, R., Wolpert, D.: Bias plus variance decomposition for zero-one loss functions. In: Proceedings of the Thirteenth International Conference of Machine Learning, pp. 275–283 (1996)

18. Melville, P., Shah, N., Mihalkova, L., Mooney, R.J.: Experiments on ensembles with missing and noisy data. In: Roli, F., Kittler, J., Windeatt, T. (eds.) MCS 2004. LNCS, vol. 3077, pp. 293–302. Springer, Heidelberg (2004)
19. Quinlan, J.R.: Induction of decision trees. Machine Learning 1, 81–106 (1986)
20. Quinlan, J.: Simplifying decision trees. International Journal of Machine Learning Studies 27, 221–234 (1987)
21. Quinlan, J.R.: Programs for Machine Learning. Morgan Kaufmann series in Machine Learning (1993)
22. Walley, P.: Inferences from multinomial data: learning about a bag of marbles. Journal of the Royal Statistical Society, Series B 58, 3–57 (1996)
23. Webb, G., Conilione, P.: Estimating bias and variance from data, Technical Report (2005)
24. Witten, I.H., Frank, E.: Data Mining: Practical machine learning tools and techniques, 2nd edn. Morgan Kaufmann, San Francisco (2005)

Appendix

We present the results of each method, without and with a pruning procedure, on each data set and for each level of noise, in Tables 10 and 11 respectively.

Table 10. Classification accuracy of B-C4.5/B-CDT without pruning for the different noise levels

Data sets	0%	5%	10%	20 %	30%
anneal	98.9 / 98.89	98.83 / 98.78	98.05 / 98.5	95.34 / 97.42	89.44 / 93.97
audiology	81.83 / 80.41	81.32 / 80.36	80.84 / 79.28	76.25 / 75.57	73.37 / 71.51
autos	85.45 / 80.27	83.54 / 78.56	80.44 / 75.79	73.34 / 69.8	64.32 / 62.54
breast-cancer	70.43 / 70.35	69.03 / 70.63	67.17 / 69.87	63.4 / 66.2	59.83 / 61.24
cmc	52.19 / 53.21	51.17 / 52.5	50.12 / 51.82	48.38 / 50.14	45.41 / 47.51
horse-colic	85.51 / 84.91	85.21 / 84.15	84.55 / 83.71	81.46 / 80.73	76.04 / 75.16
german-credit	73.01 / 74.64	72.81 / 73.96	72.67 / 73.43	69.91 / 71.38	65.07 / 67.19
pima-diabetes	76.14 / 75.8	75.88 / 74.88	75.59 / 74.48	74.62 / 72.6	70.8 / 67.5
glass2	82.22 / 83	80.85 / 82.15	78.45 / 79.51	74.66 / 76.62	68.47 / 68.51
hepatitis	81.76 / 80.99	81.12 / 81.76	80.63 / 81.53	79.35 / 79.95	73.24 / 75.51
hypothyroid	99.62 / 99.59	99.53 / 99.55	99.3 / 99.48	98.34 / 99.37	95.9 / 98.82
ionosphere	92.57 / 91.23	92.25 / 91.54	91.8 / 90.58	87.84 / 86.7	79.86 / 78.38
kr-vs-kp	99.46 / 99.4	99.08 / 99.16	98.02 / 98.72	92.68 / 95.63	82.68 / 86.36
labor	86.43 / 86.53	83.67 / 84.27	84.2 / 84.77	79.47 / 79.53	76.33 / 76.73
lymphography	79.96 / 76.24	79.63 / 77.02	79.58 / 77.02	75.99 / 76	73.02 / 73.14
mushroom	100 / 100	99.67 / 99.99	98.82 / 99.93	94.15 / 98.67	84.24 / 89.06
segment	97.75 / 97.45	97.59 / 97.38	96.75 / 97.08	94.29 / 95.83	90.5 / 93.15
sick	98.97 / 98.97	98.68 / 98.66	98.08 / 98.47	96.14 / 97.87	90.34 / 94.64
solar-flare	97 / 97.31	96.29 / 97.22	94.96 / 97.07	90.17 / 94.93	81.55 / 86.29
sonar	80.07 / 80.78	79.43 / 80.35	77.45 / 79.47	74.77 / 76.27	69.52 / 71.75
soybean	92.28 / 90.47	91.95 / 90.5	91.22 / 90.25	88.07 / 87.7	83.45 / 81.65
sponge	93.91 / 92.63	92.75 / 92.57	91.39 / 92.68	87.89 / 90.57	77.45 / 84.05
vote	96.78 / 96.34	96.04 / 95.79	95.22 / 95.35	92.59 / 93.93	86.25 / 88.87
vowel	92.37 / 91.14	91.6 / 90.61	91.41 / 90.37	88.62 / 88.17	83.77 / 83.91
zoo	92.8 / 92.4	93.01 / 92.5	93.66 / 93.37	93.5 / 93.27	89.71 / 90.71

Table 11. Classification accuracy of B-C4.5/B-CDT with pruning for the different noise levels

Data sets	0%	5%	10%	20 %	30%
anneal	98.79 / 98.59	98.74 / 98.46	98.64 / 98.36	98.04 / 98.1	95.99 / 97.54
audiology	80.75 / 74.35	80.89 / 75.41	81.01 / 75.68	78.37 / 72.84	77.21 / 69.01
autos	84.39 / 72.65	82.71 / 70.5	79.99 / 67.43	73.88 / 63.51	64.91 / 57.87
breast-cancer	73.09 / 72.35	72.25 / 71.91	72.04 / 71.44	70.95 / 69.94	64.31 / 64.62
cmc	53.12 / 56.02	52.45 / 55.07	51.56 / 54.75	50.39 / 53.21	47.75 / 51.32
horse-colic	85.21 / 85.21	85.56 / 85.02	85.07 / 84.64	83.96 / 83.44	80.43 / 78.12
german-credit	74.73 / 75.26	74.36 / 74.88	73.92 / 74.66	71.9 / 73.85	67.27 / 70.65
pima-diabetes	76.17 / 75.92	75.83 / 76.34	75.53 / 75.84	74.76 / 75.3	71.09 / 71.53
glass2	81.97 / 80.44	80.98 / 79.83	78.2 / 79.79	75.22 / 77.62	69.01 / 72.54
hepatitis	81.37 / 81.57	81.45 / 83.03	82.06 / 82.64	80.63 / 81.38	75.18 / 80.33
hypothyroid	99.61 / 99.55	99.56 / 99.53	99.5 / 99.47	99.29 / 99.36	97.43 / 99.15
ionosphere	92.54 / 90.77	92.28 / 91.32	91.71 / 91.35	88.01 / 90.41	80.06 / 84.37
kr-vs-kp	99.44 / 98.92	99.29 / 98.92	99.17 / 98.79	97.5 / 97.99	88.91 / 95.14
labor	84.63 / 84.03	83.07 / 84.3	81.67 / 83.43	79.63 / 82.33	77.17 / 80.37
lymphography	79.69 / 77.51	78.15 / 77.78	78.63 / 78.1	77.49 / 77.44	75.82 / 76.53
mushroom	100 / 100	100 / 100	99.99 / 99.98	99.84 / 99.9	96.31 / 98.3
segment	97.64 / 96.74	97.58 / 96.54	97.14 / 96.49	95.81 / 96.28	92.33 / 95.99
sick	98.85 / 98.54	98.6 / 98.46	98.43 / 98.45	97.29 / 98.29	92.47 / 97.25
solar-flare	97.84 / 97.84	97.84 / 97.84	97.81 / 97.81	97.66 / 97.66	95.76 / 96.01
sonar	80.4 / 77.57	79.53 / 76.95	77.6 / 76.99	74.86 / 76.22	69.47 / 73.22
soybean	93.1 / 88.81	93.15 / 88.67	92.72 / 88.45	91.93 / 85.51	90.82 / 81.61
sponge	92.63 / 92.5	92.75 / 92.5	92.34 / 92.5	91.79 / 92.5	89.16 / 91.95
vote	96.69 / 95.52	96.16 / 95.49	95.91 / 95.56	95.17 / 95.49	91.54 / 94
vowel	92.14 / 87.54	91.47 / 86.38	91.38 / 85.83	88.76 / 84.87	84 / 83.11
zoo	92.5 / 92.61	92.71 / 92.7	93.77 / 93.77	93.99 / 91.39	91.31 / 90.53

Evolving Schemas for Streaming XML

Maryam Shoaran and Alex Thomo

University of Victoria, Victoria, Canada
{maryam,thomo}@cs.uvic.ca

Abstract. In this paper we model schema evolution for XML by defining formal language operators on Visibly Pushdown Languages (VPLs). Our goal is to provide a framework for efficient validation of streaming XML in the realistic setting where the schemas of the exchanging parties evolve and thus diverge from one another. We show that Visibly Pushdown Languages are closed under the defined language operators and this enables us to expand the schemas (for XML) in order to account for flexible or constrained evolution.

Keywords: XML schemas, Evolution, Streaming Data, Visibly Pushdown Languages.

1 Introduction

The ubiquitous theme in the modern theory of software systems is that evolution is unavoidable in real-world systems. The force of this fact is increasingly prominent today when software systems have numerous online interconnections with other systems and are more than ever under the user pressure for new changes and enhancements. It is often noted that no system can survive without being agile and open to change.

In this paper we propose ways to evolve schemas for XML in an online, streaming setting. As XML is by now the omnipresent standard for representing data and documents on the Web, there is a pressing need for having the ability to smoothly adapt schemas for XML to deal with changes to business requirements, and exchange standards.

One important use of schemas for XML is the validation of documents, which is checking whether or not a document conforms to a given schema. Notably, the validation is the basis of any application involving data-exchange between two or more parties.

Now in a scenario where the schemas of the exchanging parties possibly diverge from each other due to various changing business requirements, we need to expand the schemas making them more "tolerant" against incoming XML documents. To illustrate, suppose that there are parties exchanging patient records such as for example:

S. Link and H. Prade (Eds.): FoIKS 2010, LNCS 5956, pp. 266–285, 2010.

```
<record>
  <hospital> Victoria General </hospital>
  <patient>
    <name> Smith Brown </name>
    <address> 353 Douglas Str
      <phone> 250-234-5678 </phone>
    </address>
    <test> Complete blood count </test>
  </patient>
</record>
```

Suppose now that some sending party decides to suppress sending addresses containing the patient's phone number, apparently for privacy reasons. What we want is the system to adapt and continue to function in the face of this change. Namely, the schemas of the receiving parties need to evolve and be tolerant against the change. If we consider the XML documents as nested words (strings), and the schemas as languages of such words, then what we need is to expand the schemas with new words obtained from the original ones after deleting all the subwords of the form `<address>`w_1`<phone>`w_2`</phone>`w_3`</address>`, where w_1, w_2, and w_3 are properly nested words. Evidently, words `<address>`w_1`<phone>`w_2`</phone>`w_3`</address>` form a language, say D, and the problem becomes that of "deleting D from a schema language L." Similar arguments can also be made for expanding schemas by "inserting a language I into a schema language L."

Interestingly, language deletions and insertions have been studied as operators for regular languages in representing biological computations (cf.[8, 14]). In this paper, we investigate instead the deletion and insertion operators as means for evolving languages of nested words capturing the common schema formalism for XML.

We, also consider constrained variants of these language operators. Specifically, we provide means for specifying that we want to allow an operation to apply only at certain elements of the XML documents. For instance, in our example, we could specify that the above deletion is allowed to take place only inside the `patient` element and not anywhere else.

Schemas for XML. When it comes to XML schema specifications, the most popular ones are Document Type Definition (DTD), XML Schema ([21]) and Relax NG ([6]). Notably, all these schema formalisms can be captured by Extended Document Type Definitions (EDTDs) (cf. [7, 16, 17, 19]). It is well known that the tree languages specified by EDTDs coincide with (unranked) regular tree languages (cf. [7]).

In this paper, we will represent XML schemas by Visibly Pushdown Automata (VPAs) introduced in [3]. VPAs are in essence pushdown automata, whose push or pop mode can be determined by looking at the input only (hence their name). VPAs recognize Visibly Pushdown Languages (VPLs), which form a well-behaved and robust family of context-free languages. VPLs enjoy useful closure properties and several important problems for them are decidable. Furthermore, VPLs have been shown to coincide with the class of (word-encoded) regular tree languages, i.e. VPAs are equivalent in power with EDTDs. Recent work [15] has also shown that EDTDs can be directly compiled into VPAs.

Now, the validation problem reduces to the problem of accepting or rejecting the XML document (string) using a VPA built for the given schema. Notably, a VPA accepts or rejects an XML document without building a tree representation for it, and this is a clear advantage in a streaming setting, where transforming and storing the XML into a tree representation is a luxury we do not have.

Another reason for preferring VPAs over tree automata for XML is that VPAs are often more natural and exponentially more succinct than tree automata when it comes to "semi-formally" specify documents using pattern-based conditions on the global linear order of XML (cf. [4, 25]).

Also, considering the schemas for XML as word languages opens the way for a natural extension of deletion and insertion operations, thus making the schemas evolve in a similar spirit to biological computing artifacts.

We show that the deletion and insertion operations can be efficiently computed for VPLs, and furthermore they can be combined with useful constraints determining the scope of their applications.

Contributions. More specifically, our contributions in this paper are as follows.

1. We show that VPLs are closed under the language operations of deletion and insertion. This is in contrast to Context-Free Languages which are not closed under deletion, but only under insertion.
2. We introduce the extended operations of k-bounded deletion and insertion which allow the deletion and insertion, respectively, of k words in parallel. It is exactly these operations that are practical to use for evolving schemas for XML documents containing (or need to contain) multiple occurrences of words to be deleted (or inserted). For instance, a patient record might contain not only his/her own address, but the also the doctor's address, and all these addresses might need to be deleted. We show that the VPLs are closed under these extended operations as well.
3. We present an algorithm, which, given a schema VPL L, two sets $\mathcal{D}$ and $\mathcal{I}$ of allowed language deletions and insertions, respectively, and a positive integer k, produces a (succinctly represented) expanded language L' by applying not more than k operations in *parallel* from $\mathcal{D} \cup \mathcal{I}$ on L. Language L' contains *all* the possible "k-evolution" of L using operations from $\mathcal{D} \cup \mathcal{I}$. The difference from the k-bounded deletion and insertion is that now we allow these operations to be intermixed together.
4. We enhance the deletion and insertion operations by constraints that specify the allowed scope of the operations. We express these constraints by using XPath expressions which select the XML elements of interest. We present an algorithm which computes the "k-evolution" of a given schema VPL L under this constrained setting. The challenge is to be able to first mark non intrusively all the candidate spots for applying the operations, and then

apply them. This is because applying an operation could possibly change the structure of the words and thus harm matching of the other constraints.

Organization. The rest of the paper is organized as follows. In Section 2 we discuss related work. Section 3 reviews VPAs and VPTs (Visibly Pushdown Transducers). In Section 4.1 we study the deletion operation for VPLs. The k-bounded deletion for VPLs is also introduced there. In Section 4.2 we study the insertion operation for VPLs. The k-bounded insertion for VPLs is also introduced there. In Section 4.3 we present an algorithm for evolving a schema VPL by applying at most k language operations in parallel. In Section 5 we introduce constrained operations, and in Section 5.4 we present an algorithm to evolve schema VPLs using such operations. Finally, Section 6 concludes the paper.

2 Related Work

The first to propose using pushdown automata for validating streaming XML are Segoufin and Vianu in [20]. The notion of auxiliary space for validating streaming XML is also defined in this work. Auxiliary space is the stack space needed to validate an XML document and is proportional to the depth of the document.

VPLs and their recognizing devices, VPAs, are introduced in [3]. Logic-based characterizations are provided in [1, 2]. In [15], it is argued that VPAs are the apt device for the validation of streaming XML and a direct construction is given for going from EDTDs to equivalent VPAs.

The problem of error-tolerant validation has been studied in several works (cf. [5, 9, 23, 26]). These works use edit operations to modify the XML and possibly make it fit the schema. The difference of our work from these works is that we consider language operations rather than edit operations on XML trees. We note that performing edit operations might not correspond naturally to the user intention of changing an XML document or schema. For example to delete a complex `address` element we need several delete edit operations rather than just one language operation as in our setting. The latter, we believe, corresponds better to a user intention for deleting such an element in one-shot. Furthermore, with our language operations, the user is given the opportunity to specify the structure of the elements to be deleted or inserted, which as illustrated in the Introduction, is useful in practice.

Using edit operations on regular languages is studied in [11, 12, 13]. They revolve around the problem of finding paths in graph databases that approximately spell words in a given regular language.

The language operations of deletion and insertion are studied in [14] for regular and context free languages. As shown there, the regular languages are closed under deletion and insertion while context free languages are not closed under deletion, but closed under insertion.

Visibly Pushdown Transducers (VPTs) are introduced in [26] and [18]. The latter showed that VPLs are closed under transductions of VPTs which refrain from erasing open or close symbols. In this paper we will use this class of VPTs for some auxiliary marking operations on VPLs.

3 Visibly Pushdown Automata and Transducers

3.1 Visibly Pushdown Automata

VPAs were introduced in [3] and are a special case of pushdown automata. Their alphabet is partitioned into three disjoint sets of call, return and local symbols, and their push or pop behavior is determined by the consumed symbol. Specifically, while scanning the input, when a call symbol is read, the automaton pushes one stack symbol onto the stack; when a return symbol is read, the automaton pops off the top of the stack; and when a local symbol is read, the automaton only moves its control state.

Formally, a *visibly pushdown automaton* (VPA) A is a 6-tuple $(Q, (\Sigma, f), \Gamma, \tau, q_0, F)$, where

1. Q is a finite set of states.
2. – Σ is the alphabet partitioned into the (sub) alphabets Σ_c, Σ_l and Σ_r of call, local and return symbols respectively.
   – f is a one-to-one mapping $\Sigma_c \to \Sigma_r$. We denote $f(a)$, where $a \in \Sigma_c$, by $\bar{a}$, which is in Σ_r.[1]
3. Γ is a finite stack alphabet that (besides other symbols) contains a special "bottom-of-the-stack" symbol $\bot$.
4. q_0 is the initial state.
5. F is the set of final states.
6. $\tau = \tau_c \cup \tau_r \cup \tau_l \cup \tau_\epsilon$ is the transition relation and τ_c, τ_l, τ_r and τ_ϵ are as follows.
   – $\tau_c \subseteq Q \times \Sigma_c \times Q \times (\Gamma \setminus \bot)$
   – $\tau_r \subseteq Q \times \Sigma_r \times \Gamma \times Q$
   – $\tau_l \subseteq Q \times \Sigma_l \times Q$
   – $\tau_\epsilon \subseteq Q \times \{\epsilon\} \times Q$

We note that the ϵ-transitions do not affect the stack and they behave like in an NFA. They can be easily removed by an ϵ-removal procedure similar to the standard one for NFAs. However, we consider ϵ-transitions as they make expressing certain constructions more convenient.

A run of VPA A on word $w = x_0 \dots x_{k-1}$ is a sequence $\rho = (q_i, \sigma_i), \dots, (q_{i+k}, \sigma_{i+k})$, where $q_{i+j} \in Q$, $\sigma_{i+j} \in (\Gamma \setminus \{\bot\})^* \cdot \{\bot\}$, and for every $0 \le j \le k-1$, the following holds:

– If x_j is a call symbol, then for some $\gamma \in \Gamma$, $(q_{i+j}, x_j, q_{i+j+1}, \gamma) \in \tau_c$ and $\sigma_{i+j+1} = \gamma \cdot \sigma_{i+j}$ (Push γ).

[1] When referring to arbitrary elements of Σ_r, we will use $\bar{a}, \bar{b}, \dots$ in order to emphasize that these elements correspond to $a, b, \dots$ elements of Σ_c.

- If x_j is a return symbol, then for some $\gamma \in \Gamma$, $(q_{i+j}, x_j, \gamma, q_{i+j+1}) \in \tau_r$ and $\sigma_{i+j} = \gamma \cdot \sigma_{i+j+1}$ (Pop γ).
- If x_j is a local symbol, then $(q_{i+j}, x_j, q_{i+j+1}) \in \tau_l$ and $\sigma_{i+j+1} = \sigma_{i+j}$.
- If x_j is ϵ, then $(q_{i+j}, x_j, q_{i+j+1}) \in \tau_\epsilon$ and $\sigma_{i+j+1} = \sigma_{i+j}$.

A run is *accepting* if $q_i = q_0$, $q_{i+k} \in F$, and $\sigma_{i+k} = \perp$. A word w is accepted by a VPA if there is an accepting run in the VPA which spells w. A language L is a *visibly pushdown language* (VPL) if there exists a VPA that accepts all and only the words in L. The VPL accepted by a VPA A is denoted by $L(A)$.

When reasoning about XML structure and validity, the local symbols are not important, and thus, we consider the languages of XML schemas as VPLs on the alphabet $\Sigma_c \cup \Sigma_r$. Furthermore, we note that here, we are asking for an empty stack in the end of an accepting run because we are interested in VPLs of properly nested words.

Example 1. Suppose that we want to build a VPA accepting XML documents about movie collections. Such documents will have a *collection* element nesting any number of *movie* elements in them. Each *movie* element will nest a *title* element and any number of *star* elements. A VPA accepting well-formed documents of this structure is $A = (Q, (\Sigma, f), \Gamma, \tau, q_0, F)$, where

$$
\begin{aligned}
Q &= \{q_0, q_1, q_2, q_3, q_4, q_5, q_6, q_7\},\\
\Sigma &= \Sigma_c \cup \Sigma_r = \\
&\quad \{\mathit{collection}, \mathit{movie}, \mathit{title}, \mathit{star}\} \cup \{\overline{\mathit{collection}}, \overline{\mathit{movie}}, \overline{\mathit{title}}, \overline{\mathit{star}}\},\\
&\quad f \text{ maps the } \Sigma_c \text{ elements into their “bar”-ed counterparts in } \Sigma_r,\\
\Gamma &= \{\gamma_c, \gamma_m, \gamma_t, \gamma_s\} \cup \{\perp\},\\
F &= \{q_7\},\\
\tau &= \{(q_0, \mathit{collection}, q_1, \gamma_c), (q_1, \mathit{movie}, q_2, \gamma_m), (q_2, \mathit{title}, q_3, \gamma_t),\\
&\quad (q_3, \overline{\mathit{title}}, \gamma_t, q_4), (q_4, \mathit{star}, q_5, \gamma_s), (q_5, \overline{\mathit{star}}, \gamma_s, q_4),\\
&\quad (q_4, \overline{\mathit{movie}}, \gamma_m, q_6), (q_6, \overline{\mathit{collection}}, \gamma_c, q_7), (q_6, \epsilon, q_1)\}.
\end{aligned}
$$

We show this VPA in Fig. 1.

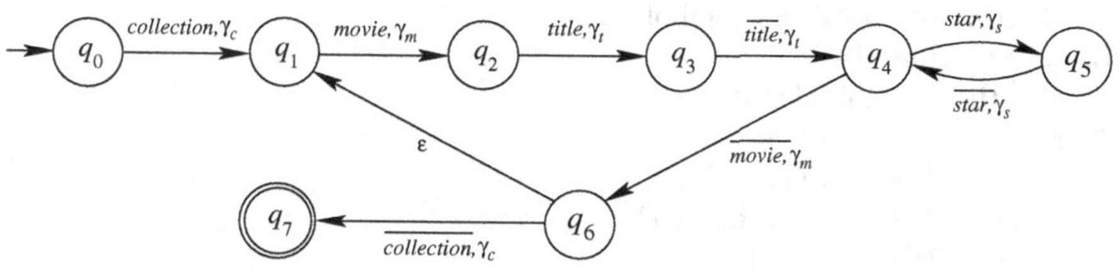

Fig. 1. Example of a VPA

Processing a Document with a VPA. As explained in [15], given a schema specification VPA $A = (Q, (\Sigma, f), \Gamma, \tau, q_0, F)$, the (exact) validation of an XML document (word) w amounts to accepting or rejecting w using A.

Intersection with Regular Languages. It can be shown that VPLs are closed under the intersection with regular languages. The construction is similar to the one showing closure of CFLs under intersection with regular languages. Formally we have

Theorem 1. *Let L be a VPL and R a regular language. Then, $L \cap R$ is a VPL.*

Proof Sketch. Let $A = (Q, (\Sigma, f), \Gamma, \tau^A, q_0, F)$ be an ϵ-free VPA for L, and $B = (P, \Sigma, \tau^B, p_0, G)$ an ϵ-free NFA for R. Now, language $L \cap R$ is accepted by the product VPA $C = (Q \times P, (\Sigma, f), \Gamma, \tau^C, q_0, F \times G)$, where

$$\begin{aligned}\tau^C = &\{((q,p), a, (q',p'), \gamma) : (q, a, q', \gamma) \in \tau^A \text{ and } (p, a, p') \in \tau^B\} \cup \\ &\{((q,p), \bar{a}, \gamma, (q',p')) : (q, \bar{a}, \gamma, q') \in \tau^A \text{ and } (p, \bar{a}, p') \in \tau^B\} \cup \\ &\{((q,p), l, (q',p')) : (q, l, q') \in \tau^A \text{ and } (p, l, p') \in \tau^B\}.\square\end{aligned}$$

The Language of All Properly Nested Words. In our constructions we will often use the language of all properly nested words which we denote by PN. All the VPLs of properly nested words are subsets of it. We consider the empty word ϵ to also be in PN.

3.2 Visibly Pushdown Transducers

A *visibly pushdown transducer* (VPT) T is a 7-tuple $(P, (I, f), (O, g), \Gamma, \tau, p_0, F)$, where

1. P is a finite set of states.
2. – I is the input alphabet partitioned into the (sub) alphabets I_c and I_r of input call and return symbols.
   – f is a one-to-one mapping $I_c \to I_r$. We denote $f(a)$, where $a \in I_c$, by $\bar{a}$.
3. – O is the output alphabet partitioned into the (sub) alphabets O_c and O_r of output call and return symbols respectively.
   – g is a one-to-one mapping $O_c \to O_r$. We denote $g(b)$, where $b \in O_c$, by $\bar{b}$.
4. Γ is a finite stack alphabet that (besides other symbols) contains a special "bottom-of-the-stack" symbol $\perp$.
5. p_0 is the initial state.
6. F is the set of final states.
7. $\tau = \tau_c \cup \tau_r \cup \tau_\epsilon$, where
   – $\tau_c \subseteq P \times I_c \times O_c \times P \times \Gamma$
   – $\tau_r \subseteq P \times I_r \times O_r \times \Gamma \times P$
   – $\tau_\epsilon \subseteq P \times \{\epsilon\} \times \{\epsilon\} \times P$.

We define an *accepting run* for T similarly as for VPAs. Now, given a word $u \in I^*$, we say that a word $w \in O^*$ is an *output of T for u* if there exists an accepting run in T spelling u as input and w as output.[2]

[2] In other words, we get u and w when concatenating the transitions' input and output components respectively.

A transducer T might produce more than one output for a given word u. We denote the set of all outputs of T for u by $T(u)$. For a language $L \subseteq I^*$, we define the *image of* L *through* T as $T(L) = \bigcup_{u \in L} T(u)$.

We note that in our definition of VPTs we disallow transitions which transduce a call or return symbol to ϵ. As [18] showed, VPLs are closed under the transductions of such non-erasing synchronized VPTs.

4 Language Deletion and Insertion

4.1 Language Deletion

In this section we present the language operation of deletion and show that VPLs are closed under this operation.

Let L and D be languages on Σ. The *deletion of* D *from* L, denoted by $L \longrightarrow D$, removes from the words of L one occurrence of some word in D. For example if $L = \{abcd, ab\}$ and $D = \{bc, cd, a\}$, then $L \longrightarrow D = \{ad, ab, bcd, b\}$.

Formally, the *deletion of* D *from* L is defined as:

$$L \longrightarrow D = \{w_1 w_2 : w_1 v w_2 \in L \text{ and } v \in D\}.$$

Kari (in [14]) showed that the regular languages are closed under deletion, whereas context-free languages are not. We show here that VPLs are closed under deletion.

Theorem 2. *If L and D are VPLs over Σ, then $L \longrightarrow D$ is a VPL as well.*

Proof
Construction. Let $A = (Q, (\Sigma, f), \Gamma, \tau, q_0, F)$, where $Q = \{q_0, \ldots, q_{n-1}\}$, be a VPA that accepts L. For every two states q_i and q_j in Q define the VPA

$$A_{ij} = (Q, (\Sigma, f), \Gamma, \tau, q_i, \{q_j\}),$$

which is the same as A, but with initial and final states being q_i and q_j, respectively. The language $L(A_{ij})$ (which we also denote by L_{ij}) is a VPL for each $q_i, q_j \in Q$. Consider now the VPA $A' = (Q, (\Sigma', f), \Gamma, \tau', q_0, F)$, with

1. $\Sigma' = \Sigma \cup \{\dagger\}$, where $\dagger$ is a fresh local symbol,
2. $\tau'_c = \tau_c$,
3. $\tau'_r = \tau_r$, and
4. $\tau'_l = \{(q_i, \dagger, q_j) : q_i, q_j \in Q \text{ and } L_{ij} \cap D \neq \emptyset\}$.

We then define language $L' = L(A') \cap (\Sigma^* \cdot \{\dagger\} \cdot \Sigma^*)$. This intersection extracts from $L(A')$ all the words marked by one $\dagger$. Such words are derived from the words of L containing some properly nested subword which is a word in D.

Now we obtain VPL L'' by substituting ϵ for $\dagger$ in L'. This is achieved by replacing the *local* $\dagger$ transitions, in the VPA for L', by ϵ transitions. We have that

Lemma 1. $L \longrightarrow D = L''$.

Proof. "$\subseteq$". Let $w \in L \longrightarrow D$. There exists $u \in L$, $v \in D$ such that $u = w_1 v w_2$ and $w = w_1 w_2$. Hence, there exists an accepting run of A for u:

$$\rho_u = (q_0, \sigma_0), \ldots, (q_i, \sigma_i), \ldots, (q_j, \sigma_j), \ldots, (q_f, \sigma_f),$$

where $(q_0, \sigma_0), \ldots, (q_i, \sigma_i)$ is a sub-run for w_1, $(q_i, \sigma_i), \ldots, (q_j, \sigma_j)$ is a sub-run for v, $(q_j, \sigma_j), \ldots, (q_f, \sigma_f)$ is a sub-run for w_2, and $q_f \in F$. As $v \in D$ is a properly nested word, we have that $\sigma_i = \sigma_j$.

Since $v \in L_{ij} \cap D$, the transition $(q_i, \dagger, q_j)$ exists in τ' and we have the following run in A':

$$\rho = (q_0, \sigma_0), \ldots, (q_i, \sigma_i), (q_j, \sigma_j), \ldots, (q_f, \sigma_f),$$

where the sub-run $(q_i, \sigma_i), (q_j, \sigma_j)$ reads symbol $\dagger$ and $\sigma_i = \sigma_j$ since $\dagger$ is a local symbol. It is a key point that $\sigma_i = \sigma_j$ in both of the sub-runs $(q_i, \sigma_i), \ldots, (q_j, \sigma_j)$ for v and $(q_i, \sigma_i), (q_j, \sigma_j)$ for $\dagger$. Thus, $w_1 \dagger w_2 \in L(A')$ and also $w_1 \dagger w_2 \in L' = L(A') \cap (\Sigma^* \cdot \{\dagger\} \cdot \Sigma^*)$. This proves that $w = w_1 w_2 \in L''$.

"$\supseteq$". Let $w \in L''$. As such, there exists a word $w' \in L' = L(A') \cap (\Sigma^*.\{\dagger\}.\Sigma^*)$ which is of the form $w_1 \dagger w_2$, where $w_1, w_2 \in \Sigma^*$. For w' there exists an accepting run in A':

$$\rho_{w'} = (q_0, \sigma_0), \ldots, (q_i, \sigma_i), (q_j, \sigma_j), \ldots, (q_f, \sigma_f),$$

where $(q_0, \sigma_0), \ldots, (q_i, \sigma_i)$ is a sub-run for w_1, $(q_j, \sigma_j), \ldots, (q_f, \sigma_f)$ is a sub-run for w_2, and $(q_i, \sigma_i), (q_j, \sigma_j)$ is a sub-run for $\dagger$ with $\sigma_i = \sigma_j$ (since $\dagger$ is a local symbol). This run indicates the existence of transition $(q_i, \dagger, q_j)$ in A', which means $L_{ij} \cap D \neq \emptyset$. This also implies that there exists some properly nested word $v \in D$ corresponding to a run $(q_i, \sigma_i), \ldots, (q_j, \sigma_j)$ in automaton A with $\sigma_i = \sigma_j$.

Since w' contains only one $\dagger$ symbol, there is only one transition from $\tau' \setminus \tau$ that has been traversed in run $\rho_{w'}$. Now it can be concluded that the following accepting run exists in A:

$$\rho = (q_0, \sigma_0), \ldots, (q_i, \sigma_i), \ldots, (q_j, \sigma_j), \ldots, (q_f, \sigma_f),$$

which indicates that $w_1 v w_2 \in L$. As $v \in D$, we have that $w = w_1 w_2 \in (w_1 v w_2 \longrightarrow v) \subseteq (L \longrightarrow D)$, i.e. $w \in L \longrightarrow D$. □

Finally, the claim of the theorem follows from the above lemma and the fact that the visibly pushdown languages are closed under the intersection with regular languages. □

Based on the construction of the above theorem we have that

Theorem 3. *Computing a VPA recognizing $L \longrightarrow D$ can be done in PTIME.*

***k*-Bounded Deletion.** We can also define a variant of deletion which allows for up to k deletions where k is a given positive integer. This is useful when we want to evolve a given schema language L by allowing the words of a given language

D to be deleted (in parallel) from k-locations in the words of L rather than just one.

Let L and D be languages on Σ. The *k-bounded deletion of D from L* is defined as:

$$L \xrightarrow{\leq k} D = \{w_1 w_2 \dots w_i w_{i+1} : w_1 v_1 w_2 \dots w_i v_i w_{i+1} \in L, \\ v_j \in D \text{ for } 1 \leq j \leq i, \text{ and } 1 \leq i \leq k\}.$$

Observe that $L \xrightarrow{\leq k} D$ cannot be obtained by iterative applications of k single deletions. For example, assume $L = \{abc\bar{c}\bar{b}\bar{a}\}$ and $D = \{b\bar{b}, c\bar{c}\}$. The language resulting from 2-bounded deletion of D from L is $\{ab\bar{b}\bar{a}\}$. On the other hand, by applying 2 successive deletions of D from L we have $(L \longrightarrow D) \longrightarrow D = \{a\bar{a}\}$.

We show now that VPLs are closed under the k-bounded deletion.

Theorem 4. *If L and D are VPLs over Σ, then $L \xrightarrow{\leq k} D$ is a VPL as well.*

Proof sketch
Construction. Let $A = (Q, (\Sigma, f), \Gamma, \tau, q_0, F)$, where $Q = \{q_0, \dots, q_{n-1}\}$, be a VPA that accepts L. We build a VPA A' as explained in the proof of Theorem 2. Now we set

$$L' = L(A') \cap \bigcup_{1 \leq h \leq k} (\Sigma^* \cdot \{\dagger\} \cdot \Sigma^*)^h.$$

This intersection extracts from $L(A')$ all the words marked by up to k †'s. Such words are derived from the words of L containing properly nested subwords which are words in D.

Now we obtain VPL L'' by substituting symbols † in L' by ϵ. This is achieved by replacing the local † transitions, in the VPA for L', by ϵ transitions. We can show that

Lemma 2. $L \xrightarrow{\leq k} D = L''$.

Based on this lemma the claim of the theorem follows. □

Based on the construction of the above theorem we have that

Theorem 5. *Computing a VPA recognizing $L \xrightarrow{\leq k} D$ can be done in PTIME if k is constant.*

On the other hand, if k is not constant, then the complexity is pseudo-polynomial in k. This is because of the intersection with $\bigcup_{1 \leq h \leq k} (\Sigma^* \cdot \{\dagger\} \cdot \Sigma^*)^h$ in the proof of Theorem 4.

4.2 Language Insertion

Before defining the insertion operation, we define the *Insertion-for-Local-Symbol (ILS)* operation. Let # be a local symbol and L and I be VPLs on $\Sigma_\# = \Sigma \cup \{\#\}$ and Σ, respectively. The *ILS of I into L* is defined as:

$$L \longleftarrow_\# I = \{w_1 v w_2 : w_1 \# w_2 \in L \text{ and } v \in I\}.$$

Here, we show that VPLs are closed under the ILS operation.

Theorem 6. *If L and I are VPLs over $\Sigma_\#$ and Σ, respectively, then $L \longleftarrow_\# I$ is a VPL over $\Sigma_\#$.*

Proof
Construction. Let $A = (Q, (\Sigma_\#, f), \Gamma, \tau, q_0, F)$ and $A^I = (P, (\Sigma, f), \Gamma, \tau^I, p_0, F^I)$ be VPA's accepting L and I, respectively. Starting with A, we construct VPA A' by adding for each (local) transition $(q_i, \#, q_j)$ a fresh copy of A^I connected with q_i and q_j with the following new transitions,

$$\{(q_i, \diamond, p_0)\} \cup \{(p_f, \epsilon, q_j) : p_f \in F^I\},$$

where $\diamond$ is a new local symbol.

In essence, A' will accept all the words accepted by A with an arbitrary number of # replaced by the words of language $\{\diamond\} \cdot I$. In order to have only one # replacement, we construct

$$L' = L(A') \cap (\Sigma_\#^* \cdot \{\diamond\} \cdot \Sigma_\#^*).$$

Then, we replace the single $\diamond$ by ϵ in L', achieved by changing the $(\_, \diamond, \_)$ transitions to $(\_, \epsilon, \_)$. Let L'' be the language obtained in this way. Now, we show that:

Lemma 3. $L \longleftarrow_\# I = L''$.

Proof. Let A'' be the VPA of L''.

"$\subseteq$". Let $w \in L \longleftarrow_\# I$. There exists $u \in L$, $v \in I$ such that $u = w_1 \# w_2$ and $w = w_1 v w_2$. Hence, there exists an accepting run:

$$\rho_u = (q_0, \sigma_0), \ldots, (q_i, \sigma_i), (q_j, \sigma_j), \ldots, (q_f, \sigma_f),$$

in A for u, where $(q_0, \sigma_0), \ldots, (q_i, \sigma_i)$ is a sub-run for w_1, $(q_i, \sigma_i), (q_j, \sigma_j)$ is a sub-run for symbol #, $(q_j, \sigma_j), \ldots, (q_f, \sigma_f)$ is a sub-run for w_2, and $q_f \in F$. As # is a local symbol, we have that $\sigma_i = \sigma_j$.

According to the construction in Theorem 6, we now have some accepting runs in A'' of the form:

$$\rho = (q_0, \sigma_0), \ldots, (q_i, \sigma_i), (p_0, \sigma_{i+1}), \ldots, (p_f, \sigma_{j-1}), (q_j, \sigma_j), \ldots, (q_f, \sigma_f),$$

where $p_f \in F^I$ and so $(p_0, \sigma_{i+1}), \ldots, (p_f, \sigma_{j-1})$ is a sub-run for some properly nested word $v \in I$. Since this sub-run is an accepting run in VPA A^I, we have that $\sigma_{i+1} = \sigma_{j-1}$. The (local) transitions (q_i, ϵ, p_0) and (p_f, ϵ, q_j) implies that $\sigma_i = \sigma_{i+1}$ and $\sigma_{j-1} = \sigma_j$, respectively. Therefore, we have that $\sigma_i = \sigma_j$ in the sub-run $(q_i, \sigma_i), (p_0, \sigma_{i+1}), \ldots, (p_f, \sigma_{j-1}), (q_j, \sigma_j)$ of ρ reading v. This is a key point that shows the sub-run $(q_j, \sigma_j), \ldots, (q_f, \sigma_f)$ in ρ can still read w_2 after reading v. Therefore, $w_1 v w_2 \in L''$.

"$\supseteq$". Let $w \in L''$. For w there exists an accepting run in A'':

$$\rho_w = (q_0, \sigma_0), \ldots, (q_i, \sigma_i), (p_0, \sigma_{i+1}), \ldots, (p_f, \sigma_{j-1}), (q_j, \sigma_j), \ldots, (q_f, \sigma_f),$$

where $q_f \in F$ and $p_f \in F^I$. The existence of sub-run $(p_0, \sigma_{i+1}), \ldots, (p_f, \sigma_{j-1})$ in ρ_w implies that w contains a subword, say v, belonging to I. Hence, w can be written as $w = w_1 v w_2$, where w_1 and w_2 correspond, respectively, to sub-runs $(q_0, \sigma_0), \ldots, (q_i, \sigma_i)$ and $(q_j, \sigma_j), \ldots, (q_f, \sigma_f)$. Since $(p_0, \sigma_{i+1}), \ldots, (p_f, \sigma_{j-1})$ in ρ_w is an accepting run in A^I, we have that $\sigma_{i+1} = \sigma_{j-1}$. Regarding the ϵ transitions (q_i, ϵ, p_0) and (p_f, ϵ, q_j) in A', we conclude $\sigma_i = \sigma_{i+1}$, $\sigma_{j-1} = \sigma_j$, and finally $\sigma_i = \sigma_j$. According to this equality and the way A'' is built, there exists a transition $(q_i, \#, q_j)$ in A, which indicates the existence of the following accepting run in A:

$$\rho = (q_0, \sigma_0), \ldots, (q_i, \sigma_i), (q_j, \sigma_j), \ldots, (q_f, \sigma_f),$$

Hence, $w_1 \# w_2 \in L$. As $v \in I$, we have that $w = w_1 v w_2 \in (w_1 \# w_2 \longleftarrow_{\#} v) \subseteq (L \longleftarrow_{\#} I)$, i.e. $w \in L \longleftarrow_{\#} I$. □

Finally, the claim of the theorem follows from the above lemma and the fact that the visibly pushdown languages are closed under the intersection with regular languages. □

Insertion. Let L and I be languages on Σ. The *insertion of I into L*, denoted by $L \longleftarrow I$, inserts into the words of L some word in I. For example if $L = \{ab, cd\}$ and $I = \{eg\}$, then $L \longleftarrow I = \{egab, aegb, abeg, egcd, cegd, cdeg\}$.

Formally, the *insertion of I into L* is defined as

$$L \longleftarrow I = \{w_1 v w_2 : w_1 w_2 \in L, \text{ and } v \in I\}.$$

Kari (in [14]) showed that the regular languages and context free languages are closed under insertion. We show here that VPLs are closed under insertion, too.

Theorem 7. *If L and I are VPLs over Σ, then $L \longleftarrow I$ is a VPL as well.*

Proof sketch
Construction. Let $A = (Q, (\Sigma, f), \Gamma, \tau, q_0, F)$ be a VPA that accepts L. Consider now the VPA $A' = (Q, (\Sigma', f), \Gamma, \tau', q_0, F)$, with

1. $\Sigma' = \Sigma \cup \{\#\}$, where $\#$ is a fresh local symbol,
2. $\tau'_c = \tau_c$,
3. $\tau'_r = \tau_r$, and
4. $\tau'_l = \{(q_i, \#, q_i) : q_i \in Q\}$.

We then define language $L' = L(A') \cap (\Sigma^* \cdot \{\#\} \cdot \Sigma^*)$. This intersection extracts from $L(A')$ all the words marked by only *one* $\#$. It can be verified now that

Lemma 4. $L \longleftarrow I = L' \longleftarrow_{\#} I$.

Finally the claim of the theorem follows from the above lemma, Theorem 6, and the fact that VPLs are closed under the intersection with regular languages. □

Based on the constructions of theorems 7 and 6, we have that

Theorem 8. *Computing a VPA recognizing $L \longleftarrow I$ can be done in PTIME.*

k-Bounded Insertion. We can also define a variant of insertion which allows for up to k insertions where k is a given positive integer. This is useful when we want to evolve a given schema language L by allowing the words of a given language I to be inserted (in parallel) into k-locations in the words of L rather than just one.

Before, we define the k-bounded ILS operation. Let L and I be languages on $\Sigma_{\#}$ and Σ, respectively.

The *k-bounded ILS of I into L* is defined as:

$$L \overset{\leq k}{\longleftarrow}_{\#} I = \{w_1 v_1 w_2 \ldots w_i v_i w_{i+1} : w_1 \# w_2 \ldots w_i \# w_{i+1} \in L, \\ v_j \in I \text{ for } 1 \leq j \leq i, \text{ and } 1 \leq i \leq k\}.$$

It can be verified that VPLs are closed under the k-bounded ILS operation. Namely, we have

Theorem 9. *If L and I are VPLs over $\Sigma_{\#}$ and Σ, respectively, then $L \overset{\leq k}{\longleftarrow}_{\#} I$ is a VPL over $\Sigma_{\#}$.*

Now, let L and I be languages on Σ. The *k-bounded insertion of I into L* is defined as:

$$L \overset{\leq k}{\longleftarrow} I = \{w_1 v_1 w_2 \ldots w_i v_i w_{i+1} : w_1 w_2 \ldots w_i w_{i+1} \in L, \\ v_j \in I \text{ for } 1 \leq j \leq i, \text{ and } 1 \leq i \leq k\}.$$

We show here that VPLs are closed under the operation of k-bounded insertion.

Theorem 10. *If L and I are VPLs over Σ, then $L \overset{\leq k}{\longleftarrow} I$ is a VPL as well.*

Proof sketch
Construction. Let $A = (Q, (\Sigma, f), \Gamma, \tau, q_0, F)$, where $Q = \{q_0, \ldots, q_{n-1}\}$, be a VPA that accepts L. We build a VPA A' as explained in the proof of Theorem 7. Then we define language

$$L' = L(A') \cap \bigcup_{1 \leq h \leq k} (\Sigma^* \cdot \{\#\} \cdot \Sigma^*)^h.$$

The above intersection extracts from $L(A')$ all the words marked by one up to k #'s. It can be verified now that

Lemma 5. $L \overset{\leq k}{\longleftarrow} I = L' \overset{\leq k}{\longleftarrow}_{\#} I$.

From this the claim of the theorem follows. □

Based on theorems 10 and 9, we have that

Theorem 11. *Computing a VPA recognizing $L \overset{\leq k}{\longleftarrow} I$ can be done in PTIME if k is constant.*

On the other hand, if k is not constant, then the complexity is pseudo-polynomial in k. This is because of the intersection with $\bigcup_{1 \leq h \leq k} (\Sigma^* \cdot \{\#\} \cdot \Sigma^*)^h$ in the proof of Theorem 10.

4.3 Transforming a VPL with Language Operations

In practice it is more useful to allow the schema transformation to be achieved by a set of deletion and insertion operations. For example, we can define a set $\mathcal{D} = \{D_1, \ldots, D_m\}$ and $\mathcal{I} = \{I_1, \ldots, I_n\}$ of allowed language deletions and insertions, respectively. With slight abuse of notation we will consider $D_1, \ldots, D_m$ and $I_1, \ldots, I_n$ to also denote their corresponding delete and insert operations, respectively. What we would like now is to apply (in parallel) up to k operations from $\mathcal{D} \cup \mathcal{I}$ on a given schema language L.

For this, given a VPA A for L, we extend the constructions described in the constructions of theorems 2 and 7. Specifically, let VPA $A = (Q, (\Sigma, f), \Gamma, \tau, q_0, F)$ have the states numbered as in Theorem 2. Also, for every two states q_i and q_j in Q define VPA A_{ij} and its accepted language L_{ij} as described in the same theorem. We construct now the VPA $A' = (Q, (\Sigma', f), \Gamma, \tau', q_0, F)$, with

1. $\Sigma' = \Sigma_c \cup \Sigma_r \cup \{\dagger_1, \ldots, \dagger_m\} \cup \{\#_1, \ldots, \#_n\}$
2. $\tau'_c = \tau_c$,
3. $\tau'_r = \tau_r$, and
4. $\tau'_l = \{(q_i, \dagger_x, q_j) : q_i, q_j \in Q \text{ and } L_{ij} \cap D_x \neq \emptyset, \text{ for } 1 \leq x \leq m\} \cup$
$\{(q_i, \#_y, q_i) : q_i \in Q \text{ and } 1 \leq y \leq n\}$.

VPA A' will accept language L' containing words with an arbitrary number of special local symbols. Each $\dagger_x$ represents a deletion corresponding to D_x, and each $\#_y$ represents an insertion corresponding to I_y. What we want, though, is to extract only those words of L', whose total number of the special symbols is not more than k. For this we construct the following intersection

$$L'' = L' \cap \bigcup_{1 \leq h \leq k} (\Sigma^* \cdot \{\dagger_1, \ldots, \dagger_m, \#_1, \ldots, \#_n\} \cdot \Sigma^*)^h.$$

Language L'' will contain all the words of L' with not more than k special symbols. Then, we obtain language L''' by replacing all the $\dagger_x$ for $1 \leq x \leq m$ by ϵ and performing n (one after the other) k-bounded ILS operations corresponding to each of $I_1, \ldots, I_n$ languages. It can be verified that

Theorem 12. *L''' is the result of applying from one up to k operations from $\mathcal{D} \cup \mathcal{I}$ on L.*

Taking the union $L \cup L'''$ gives us the new expanded schema language.

Based on the above construction, we have that

Theorem 13. *Computing a VPA recognizing L''' can be done in PTIME if k is constant.*

On the other hand, if k is not constant, then the complexity is pseudo-polynomial in k. This is because of the intersection with $\bigcup_{1 \leq h \leq k} (\Sigma^* \cdot \{\dagger_1, \ldots, \dagger_m, \#_1, \ldots, \#_n\} \cdot \Sigma^*)^h$ for obtaining L''.

5 Constrained Deletions and Insertions

Often we do not like the deletions and insertions to be performed in *unrestricted* places in the words of schemas for XML. Rather we would like them to apply only at certain parts of the words. For instance, taking the example given in the Introduction, one might want to apply operations only within a `patient` XML element.

For this, we assume that there is a given set of constraining rules that specify the conditions under which the operations can be applied on L.

We propose to express the conditions in the form of XPath expressions[3]. The alphabet of the XPath expressions is the set of XML elements corresponding to Σ_c (or Σ_r). Formally this alphabet is $\Sigma_e = \{\tilde{a} : a \in \Sigma_c\}$. As the validation problem considers the structure of XML only, we do not have data values in the XPath expressions.

Definition 1 *A deletion rule is a tuple (π, D), where π is an XPath expression, and D is a VPL.*

Such a rule implies that the words of nested language D can be deleted if they correspond to elements reached by XPath expression π. Of course, such elements need to further satisfy the structure imposed by D. An example of a deletion rule is $(/\tilde{a}/\tilde{b}/\tilde{c}, PN)$. Using this rule we can delete *all* the $\tilde{c}$ elements that are children of $\tilde{b}$ elements which in turn are children of $\tilde{a}$ elements. Surely, specifying $D = PN$ is a useful case in practice. However, we can further qualify the D language to allow only the deletion of those $\tilde{c}$ elements which contain some particular child, say an element $\tilde{d}$. In such a case, we set $D = \{c\} \cdot PN \cdot \{d\} \cdot PN \cdot \{\bar{d}\} \cdot PN \cdot \{\bar{c}\}$.

We denote by $\mathcal{D}$ the set of deletion rules.

Definition 2 *An insertion rule is a tuple (π, I), where π is an XPath expression, and I is a VPL.*

Such a rule implies that the words of nested language I can only be inserted as children of elements reached by XPath expression π. An example of an insertion rule is $(/\tilde{a}/\tilde{b}, PN)$.

We denote by $\mathcal{I}$ the set of insertion rules.

As shown in [10, 24], a unary (Core) XPath query can be represented by a VPA with a single output variable attached to some call transitions. The set of query answers consists of all the elements whose call (open) symbol binds to the variable during an accepting run.

Let π be an XPath expression and A_π a VPA for it. Let τ_c' be the subset of the A_π transitions with the output variable attached to them. It is easy to identify with the help of the stack the corresponding return transitions, τ_r'. Since we are not interested in outputting the result of the XPath queries, but rather just locate the elements of interest, we will assume that instead of having a variable attached to the transitions in τ_c', we have them simply marked by a dot ($\dot{}$). Also, we will similarly assume the transitions in τ_r' are marked as well.

---

[3] We consider in fact unary CoreXPath expressions.

We consider the alphabet of this VPA to be

$$(\Sigma_c \cup \dot{\Sigma}_c) \cup (\Sigma_r \cup \dot{\Sigma}_r),$$

where $\dot{\Sigma}_c$ and $\dot{\Sigma}_r$ are copies of Σ_c and Σ_r, respectively, with the symbols being marked by $(\dot{\ })$.

When a rule is applied on a given word (XML schema), we do not initially perform deletion or insertion on the word, but only color the related symbols. This is due to the fact that other rules can be further applied on the same document, and the XPath expressions for those rules were written considering the original version of that document in the given XML schema.

For example, suppose we have a set of deletion rules $\mathcal{D} = \{(/\tilde{a}/\tilde{b}, PN), (/\tilde{a}, \{ab\bar{b}\bar{a}\})\}$, and a given word $w = ab\bar{b}\bar{a}c\bar{c}$. Both of these rules can be applied on w. But if we apply the first rule and truly delete $b\bar{b}$, then the second rule can no longer be applied on the result word, which is $w' = a\bar{a}c\bar{c}$. On the other hand, only coloring b and $\bar{b}$, does not prevent the second rule from being applied on the word.

In the following we construct coloring VPTs based on the VPAs for the given rules. Then, we apply these VPTs on a schema language L.

5.1 Coloring VPTs for Deletion Rules

Let (π_x, D_x) be a deletion rule in $\mathcal{D}$. For each such rule we choose a distinct red color, r_x. In the following construction, we use alphabets Σ_c, Σ_r, $\Sigma_c^{r_x}$, $\Sigma_r^{r_x}$, where the last two are copies of the first two, respectively, having the symbols colored in r_x. For the sake of the discussion, we will consider the symbols of Σ_c, Σ_r as being colored in black.

Using A_{π_x} we construct the r_x-coloring VPT $T_{\pi_x}^{r_x}$, which has exactly the same states as A_{π_x}, and transitions:

1. (q, a, a, p, γ_a) for (q, a, p, γ_a) in A_{π_x},
2. $(q, \bar{a}, \bar{a}, \gamma_a, p)$ for $(q, \bar{a}, \gamma_a, p)$ in A_{π_x},
3. $(q, a, a^{r_x}, p, \gamma_a^{r_x})$ for $(q, \dot{a}, p, \gamma_{\dot{a}})$ in A_{π_x},
4. $(q, \bar{a}, \bar{a}^{r_x}, \gamma_a^{r_x}, p)$ for $(q, \dot{\bar{a}}, \gamma_{\dot{a}}, p)$ in A_{π_x}.

Intuitively, the "un-marked" transitions of A_{π_x} become "leave-unchanged" transitions in $T_{\pi_x}^{r_x}$, whereas the "marked" transitions of A_{π_x} become "black to red" transitions in $T_{\pi_x}^{r_x}$.

5.2 Coloring VPTs for Insertion Rules

Let (π_y, I_y) be an insertion rule in $\mathcal{I}$. For each such rule we choose a distinct green color, g_y. In the following construction, we use alphabets Σ_c, Σ_r, $\Sigma_c^{g_y}$, $\Sigma_r^{g_y}$, where the last two are copies of the first two, respectively, having the symbols colored in g_y.

Using A_{π_y} we construct the g_y-coloring VPT $T_{\pi_y}^{g_y}$, which has exactly the same states as A_{π_y}, and transitions:

1. (q, a, a, p, γ_a) for (q, a, p, γ_a) in A_{π_y},
2. $(q, \bar{a}, \bar{a}, \gamma_a, p)$ for $(q, \bar{a}, \gamma_a, p)$ in A_{π_y},
3. $(q, a, a^{g_y}, p, \gamma_a^{g_y})$ for $(q, \dot{a}, p, \gamma_{\dot{a}})$ in A_{π_y},
4. $(q, \bar{a}, \bar{a}^{g_y}, \gamma_a^{g_y}, p)$ for $(q, \dot{\bar{a}}, \gamma_{\dot{a}}, p)$ in A_{π_y}.

Intuitively, the "un-marked" transitions of A_{π_y} become "leave-unchanged" transitions in $T_{\pi_y}^{g_y}$, whereas the "marked" transitions of A_{π_x} become "black to green" transitions in $T_{\pi_y}^{g_y}$.

5.3 Color-Tolerant VPTs

Coloring VPTs presented in the two previous subsections can be applied only on black (normal) words. When we want to apply a coloring VPT on a word more than once or when we apply coloring VPTs for deletions and insertions one after the other, the VPTs have to be applicable also to words which have parts already colored. For example, suppose word $w = ab\bar{b}\bar{a}cd\bar{d}\bar{c}$ has the $b\bar{b}$ part already colored in a red (being so ready for deletion). Word w might be needed next to have $d\bar{d}$ colored in a green color (to become ready for an insertion). In order for a coloring VPT for insertion to be able to color $d\bar{d}$ in green, it has to be "color-tolerant" while reading the prefix $ab\bar{b}\bar{a}c$ of w.

Let T_{π_z} be a coloring VPT for a deletion as described in Subsection 5.1. Now we make T_{π_z} color-tolerant by adding the following colored copies of its transitions.

1. $(q, a^{r_x}, a^{r_x}, p, \gamma_a^{r_x})$ and $(q, a^{g_y}, a^{g_y}, p, \gamma_a^{g_y})$, for each transition (q, a, a, p, γ_a), and for every color r_x and g_y.
2. $(q, \bar{a}^{r_x}, \bar{a}^{r_x}, \gamma_a^{r_x}, p)$ and $(q, \bar{a}^{g_y}, \bar{a}^{g_y}, \gamma_a^{g_y}, p)$, for each transition $(q, \bar{a}, \bar{a}, \gamma_a, p)$, and for every color r_x and g_y.
3. $(q, a^{r_x}, a^{r_z}, p, \gamma_a^{r_z})$ and $(q, a^{g_y}, a^{r_z}, p, \gamma_a^{r_z})$, for each transition $(q, a, a^{r_z}, p, \gamma_a^{r_z})$, and for every color $r_x \neq r_z$ and g_y.
4. $(q, \bar{a}^{r_x}, \bar{a}^{r_z}, \gamma_a^{r_z}, p)$ and $(q, \bar{a}^{g_y}, \bar{a}^{r_z}, \gamma_a^{r_z}, p)$, for each transition $(q, \bar{a}, \bar{a}^{r_z}, \gamma_a^{r_z}, p)$, and for every color $r_x \neq r_z$ and g_y.

Finally, we mention that color-tolerant VPTs for insertions can be constructed in a similar way. Specifically, wherever there is an r_z superscript there will be a g_z one.

5.4 Transforming a VPL with Constrained Operations

Let $\mathcal{D} = \{(\pi_1, D_1), \ldots, (\pi_m, D_m)\}$ and $\mathcal{I} = \{(\pi'_1, I_1), \ldots, (\pi'_n, I_n)\}$ be the sets of rules for the allowed deletions and insertions, respectively. What we want is to apply up to k operations corresponding to the rules in $\mathcal{D} \cup \mathcal{I}$ on a given schema language L.

We start by constructing coloring VPT for each of the rules in $\mathcal{D} \cup \mathcal{I}$, as described in subsections 5.1, 5.2, and 5.3. Next, we transduce L by iteratively applying these VPTs one after the other (in no particular order) k times each. The result of this multiple transduction will be a language that is the same as L but with the words being colored to indicate the allowed places for deletions and insertions. For simplicity let us continue to use L for this colored version of the schema language.

Let A be a VPA for (the colored) L. Let A_{ij}, and its accepted language L_{ij}, be defined as in theorems 2 and 7. Also, let β be a transformation that uncolors words and languages. This transformation can be easily realized by a VPT. Now from A we build VPA A' keeping the same states and transitions, but adding the following transitions labeled by special local symbols.

$$\{(q_i, \dagger_x, q_j) : \beta(L_{ij} \cap (\Sigma_c^{r_x} \cdot PN \cdot \Sigma_r^{r_x})) \cap D_x \neq \emptyset, \text{ for } 1 \leq x \leq m\} \cup$$
$$\{(q_j, \#_y, q_j) : \text{ there exists a transition } (\_, \_^{g_y}, q_i, \_) \text{ and } L_{ij} \neq \emptyset\}.$$

The first set is for $\dagger$ transitions between the pairs of states connected by properly nested words in D_x. The words of interest in L_{ij} are those with a first and last symbol colored in red (r_x). We determine these words by intersecting with $(\Sigma_c^{r_x} \cdot PN \cdot \Sigma_r^{r_x})$. Then, we apply the β transformation in order to uncolor the resulting language and proceed with intersecting with D_x.

The second set indicates that if there is a call transition $(\_, \_^{g_y}, q_i, \_)$ colored by g_y (due to an insertion rule (π'_y, I_y)) in A, then in A' we have a transition labeled by $\#_y$ added in all of the states q_j reachable from q_i such that A_{ij} accepts a non-empty properly nested language including the empty word.

VPA A' will accept language L' containing words with at most k special symbols of each kind $(\dagger_1, \ldots, \dagger_m, \#_1, \ldots, \#_n)$. Each $\dagger_x$ represents a deletion corresponding to D_x, and each $\#_y$ represents an insertion corresponding to I_y. What we want, though, is to extract only those words of L', whose *total* number of the special symbols is not more than k. For this we construct the following intersection

$$L'' = L' \cap \bigcup_{1 \leq h \leq k} (\Sigma^* \cdot \{\dagger_1, \ldots, \dagger_m, \#_1, \ldots, \#_n\} \cdot \Sigma^*)^h.$$

Language L'' will contain all the words of L' with not more than k special symbols. Then, we obtain language L''' by replacing all the $\dagger_x$ for $1 \leq x \leq m$ by ϵ and performing m (one after the other) k-bounded ILS operations corresponding to each of $I_1, \ldots, I_n$ languages. Based on all the above it can be verified that

Theorem 14. *L''' is the result of applying from one up to k constrained operations from $\mathcal{D} \cup \mathcal{I}$ on L.*

Taking the union $L \cup L'''$ gives us the new expanded schema language.

Finally, regarding the complexity we have that

Theorem 15. *The colored VPA recognizing L can be computed in $O(\delta^{(m+n)k})$ time, where δ is an upper bound on the size of rule automata.*

Proof Sketch. The claim follows from the fact that each coloring VPT is applied k times, and we have $n + m$ such VPTs. □

We note that in practice, the numbers k, m, and n would typically be small. On the other hand, after having a colored L, obtaining L''' is polynomial.

6 Concluding Remarks

In this paper we proposed modeling the schema evolution for XML by using the language operations of deletion and insertion on VPLs. We showed that the VPLs are well-behaved under these operations and presented constructions for computing the result of the operations. Then, we introduced constrained operations which are arguably more useful in practice. In order to compute the results of constrained operations we developed special techniques (such as VPA coloring) achieving a compatible application of a set of different operations. Based on our techniques, the schema evolution operators can be applied in parallel without side-effect interactions.

References

1. Alur, R., Arenas, M., Etessami, K., Immerman, N., Libkin, L.: First-Order and Temporal Logics for Nested Words. Logical Methods in Computer Science 4(4) (2008)
2. Arenas, M., Barcelo, P., Libkin, L.: Regular Languages of Nested Words: Fixed Points, Automata, and Synchronization. In: Proc. 34th International Colloquium on Automata, Languages and Programming, Wroclaw, Poland, July 9-13, pp. 888–900 (2007)
3. Alur, R., Madhusudan, P.: Visibly Pushdown Languages. In: Proc. 36th ACM Symp. on Theory of Computing, Chicago, Illinois, June 13-15, pp. 202–211 (2004)
4. Alur, R.: Marrying Words and Trees. In: Proc. 26th ACM Symp. on Principles of Database Systems, Beijing, China, June 11-13, pp. 233–242 (2007)
5. Boobna, U., de Rougemont, M.: Correctors for XML Data. In: Proc. 2nd International XML Database Symposium, Toronto, Canada, August 29-30, pp. 97–111 (2004)
6. Clark, J., Murata, M.: RELAX NG Specification. OASIS (December 2001)
7. Comon, H., Dauchet, M., Gilleron, R., Jacquemard, F., Lugiez, D., Löding, C., Tison, S., Tommasi, M.: Tree Automata Techniques and Applications, October 12 (2007), `http://www.grappa.univ-lille3.fr/tata`
8. Daley, M., Ibarra, H.: Closure and decidability properties of some language classes with respect to ciliate bio-operations. Theor. Comput. Sci. 306(1-3), 19–38 (2003)
9. Flesca, S., Furfaro, F., Greco, S., Zumpano, E.: Querying and Repairing Inconsistent XML Data. In: Proc. 6th International Conference on Web Information Systems Engineering, New York, USA, November 20-22, pp. 175–188 (2005)
10. Gauwin, O., Caron, A.C., Niehren, J., Tison, S.: Complexity of Earliest Query Answering with Streaming Tree Automata. In: PLAN-X, San Francisco (January 2008)
11. Grahne, G., Thomo, A.: Approximate Reasoning in Semistructured Data. In: Proc. of the 8th International Workshop on Knowledge Representation meets Databases, Rome, Italy, September 15 (2001)
12. Grahne, G., Thomo, A.: Query Answering and Containment for Regular Path Queries under Distortions. In: Proc. of 3rd International Symposium on Foundations of Information and Knowledge Systems, Wilhelmminenburg Castle, Austria, February 17-20, pp. 98–115 (2004)
13. Grahne, G., Thomo, A.: Regular Path Queries under Approximate Semantics. Ann. Math. Artif. Intell. 46(1-2), 165–190 (2006)

14. Kari, L.: On Insertion and Deletion in Formal Languages. University of Turku, Department of Mathematics, Turku, Finland (1991)
15. Kumar, V., Madhusudan, P., Viswanathan, M.: Visibly Pushdown Automata for Streaming XML. In: Proc. of Int. Conf. on World Wide Web, Alberta, Canada, May 8-12, pp. 1053–1062 (2007)
16. Martens, W., Neven, F., Schwentick, T., Bex, G.J.: Expressiveness and complexity of XML Schema. ACM Trans. Database Syst. 31(3), 770–813 (2006)
17. Murata, M., Lee, D., Mani, M., Kawaguchi, K.: Taxonomy of XML schema languages using formal language theory. ACM Trans. Internet Techn. 5(4), 660–704 (2005)
18. Raskin, J.F., Servais, F.: Visibly Pushdown Transducers. In: Proc. 35th International Colloquium on Automata, Languages and Programming, Reykjavik, Iceland, July 7-11, pp. 386–397 (2008)
19. Schwentick, T.: Automata for XML - A survey. J. Comput. Syst. Sci. 73(3), 289–315 (2007)
20. Segoufin, L., Vianu, V.: Validating Streaming XML Documents. In: Proc. 21st ACM Symp. on Principles of Database Systems, Madison, Wisconsin, June 3-5, pp. 53–64 (2002)
21. Sperberg-McQueen, C.M., Thomson, H.: XML Schema 1.0 (2005), `http://www.w3.org/XML/Schema`
22. Staworko, S.: Personal Communication (2008)
23. Staworko, S., Chomicki, J.: Validity-Sensitive Querying of XML Data-bases. In: Proc. of 2nd International Workshop on Database Technologies for Handling XML Information on the Web, EDBT Workshops, Munich, Germany, March 26–31, pp. 164–177 (2006)
24. Staworko, S., Filiot, E., Chomicki, J.: Querying Regular Sets of XML Documents. In: Logic in Databases, Rome, Italy, May 19-20 (2008)
25. Thomo, A., Venkatesh, S.: Rewriting of Visibly Pushdown Languages for XML Data Integration. In: Proc. 17th ACM Conference on Information and Knowledge Management, Napa Valley, CA, October 26-30, pp. 521–530 (2008)
26. Thomo, A., Venkatesh, S., Ye, Y.Y.: Visibly Pushdown Transducers for Approximate Validation of Streaming XML. In: Proc. 5th International Symposium on Foundations of Information and Knowledge Systems, Pisa, Italy, February 11-15, pp. 219–238 (2008)

ONTO-EVO^{A}L an Ontology Evolution Approach Guided by Pattern Modeling and Quality Evaluation

Rim Djedidi[1] and Marie-Aude Aufaure[2]

[1] Computer Science Department, Supélec Campus de Gif
Plateau du Moulon – 3, rue Joliot Curie – 91192 Gif sur Yvette Cedex, France
rim.djedidi@supelec.fr
[2] MAS Laboratory, SAP Business Object Chair –Centrale Paris
Grande Voie des Vignes, F-92295 Châtenay-Malabry Cedex, France
marie-aude.aufaure@ecp.fr

Abstract. In this paper, we present a generic ontology evolution approach guided by a pattern-oriented process and a quality evaluation activity. Pattern modeling aims to drive and control change application while maintaining consistency of the evolved ontology. Evaluation activity –supported by an ontology quality model– is used to guide inconsistency resolution by assessing the impact of resolution alternatives –proposed by the evolution process- on ontology quality and selecting the resolution that preserves the quality of the evolved ontology.

Keywords: Ontology Evolution, Pattern Modeling and Application, Inconsistency Analysis and Resolution, Ontology Quality Evaluation.

1 Introduction

Ontology evolution regards the capability of managing the modification of an ontology in a consistent way. Ontology evolution is a complex problem: Besides identifying change requirements from several sources (modeled domain, usage environment, internal conceptualization, etc.), the management of a change –from a request to the final validation and application– needs to formally specify the required change, to analyze and resolve change effects on ontology, to implement the change, and to validate its final application. In a collaborative or distributed context, it is also necessary to propagate local changes to dependent artifacts and to globally validate changes. Moreover, to justify, explain or cancel a change, and manage ontology versions, change traceability has to be kept.

In our work[1], we focus on issues related to ontology change management in a local context and particularly on consistency maintenance. Driving change application while maintaining ontology consistency is a costly task in terms of time and complexity. An automated process is therefore essential. We have defined an

[1] This work is founded by the French National Research Agency ANR, as a part of the project DAFOE: Differential and Formal Ontology Editor.

S. Link and H. Prade (Eds.): FoIKS 2010, LNCS 5956, pp. 286–305, 2010.

ontology evolution approach ***ONTO-EVO^{A}L*** (Ontology-Evolution-Evaluation) guided by a pattern-oriented modeling and a quality evaluation activity. The patterns model the three dimensions: *Change*, *Inconsistency* and *Resolution Alternative*. Based on the modeled patterns and the conceptual links between them, we propose an automated process driving change application while maintaining consistency of the evolved ontology. In addition, the approach integrates an evaluation activity supported by an ontology quality model. The quality model is used to guide inconsistency resolution by assessing the impact of resolution alternatives –proposed by the evolution process– on ontology quality and selecting the alternative that preserves the quality of the evolved ontology.

The paper is organized as follows: in section 2, we detail ***ONTO-EVO^{A}L*** approach principles. Change Management Patterns (CMP) are described and illustrated in section 3. In section 4, we explain the role of ontology quality evaluation and describe the quality model. The implementation of the approach is presented in section 5. Before concluding and discussing further developments of this work, we report on related work in section 6.

2 *ONTO-EVO^{A}L* Approach Principles

The purpose of ***ONTO-EVO^{A}L*** approach is to drive in an automated way, the application of a change while preserving consistency of the evolved ontology. This implies the necessity of formally specifying a required change, detecting caused inconsistencies and explaining them, proposing solutions to resolve inconsistencies, guiding the ontology engineer to choose the suitable resolution and assisting him in the final validation.

To fulfill this purpose, we have defined *Change Management Patterns* (CMP) guiding the evolution process at three key phases: change specification, change

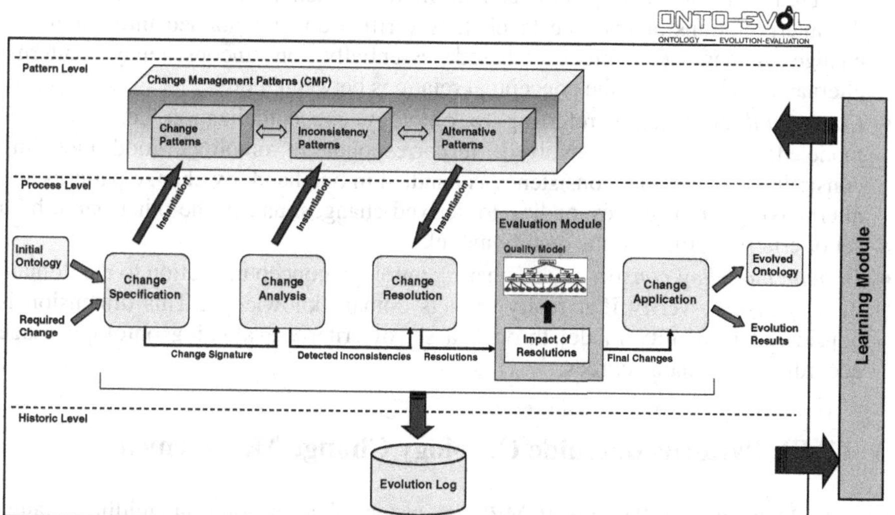

Fig. 1. ONTO-EVO^{A}L approach

analysis and change resolution (Fig. 1). Three categories of patterns are modeled: *Change Patterns* classifying types of changes, *Inconsistency Patterns* classifying types of logical inconsistencies and *Alternative Patterns* classifying types of inconsistency resolution alternatives. In addition, we have defined a *Quality Evaluation Module* guiding the evolution process in choosing the more suitable resolutions –if any– when resolving change impact. The evaluation is based on a *Quality Model* assessing the impact of proposed resolutions on ontology content and usage by means of quantitative metrics, to choose the resolution that preserve ontology quality.

The general architecture of the approach is organized through three levels (Fig. 1):

- *Process level* is the core of the approach. It includes its functional components: the four main change management phases, and the quality evaluation module. The process is detailed in [1];
- *Pattern level* represents the meta-model driving the process. It is described by CMP patterns (change patterns, inconsistency pattern, and alternative patterns) and the conceptual links between them. CMP patterns are modeled as an OWL DL ontology specifying the different change, inconsistency and alternative classes managed by ***ONTO-EVO^A^L*** approach and the semantic relations between them (section 3);
- *Historic level* insures the traceability of changes and treatments performed along change management process by saving all the results in the evolution log (section 5).

Based on this architecture, ***ONTO-EVO^A^L*** approach handles four kinds of consistency:

- *Structural consistency* referring to OWL DL language constraints and to the use of its constructors. It is handled by the formal specification of changes to apply by using change patterns which are based on OWL DL model;
- *Logical consistency* taking into account the formal semantics expressed in DL and the interpretations at instance level. It is verified and maintained throughout the change management process based essentially on inconsistency patterns, alternative patterns and the conceptual relations between them;
- *Conceptual consistency* referring to ontology conceptualization i.e. how it is modeled, modified, and evolved. It corresponds to ontology good modeling constraints. Conceptual consistency is handled in one hand, by change patterns and alternative patterns corresponding to derived changes, and in the other hand, by a set of criteria defined in the quality model;
- *Domain modeling consistency* comparing ontology conceptualization to the domain of discourse to verify if it really reflects domain knowledge. This dimension is considered in quality model through a set of criteria evaluating ontology usage according to domain sources.

3 CMP: Patterns to Guide Ontology Change Management

Change Management Patterns (CMP) are proposed as a solution guiding change application while maintaining consistency of the evolved ontology. They model categories of changes based on OWL meta-model, categories of logical inconsistencies that could be potentially caused by these changes considering OWL DL constraints, and categories of alternatives that possibly resolve these inconsistencies.

CMP are represented by two layers:

- *Presentation layer* as a catalogue of CMP patterns described in natural language and illustrated by UML diagram. This layer helps sharing and disseminating CMP patterns in particular for evaluation needs. Presentation layer formalism extends the representation of Ontology Design Patterns ODP [2] adopted from a common and accepted format describing design patterns in software engineering [3]. CMP are described by a general template containing the following slots:
  - General Properties: *General Information* about the pattern, and *Use Case* of the pattern;
  - Specific Properties related to CMP abstraction level including general properties of *Pattern Description* (some independent from pattern category and others extended to the different kind of CMP patterns), *Graphical Representation* of the pattern represented in UML format, and *Implementation* of the pattern in OWL DL language (not presented in this paper for page limit reason);
  - Properties about *Relationships* of the pattern with other patterns including *Relations to Other CMP*;
- *Formal specification layer* as an OWL DL ontology defining the semantics of CMP patterns and relations between them, and controlling their usage through restrictions expressed in DL. This layer facilitates change management process guidance. To facilitate the graphical illustration of CMP ontology, we present here (Fig. 2) an extract of the ontology presenting the three classes of patterns using UML formalism.

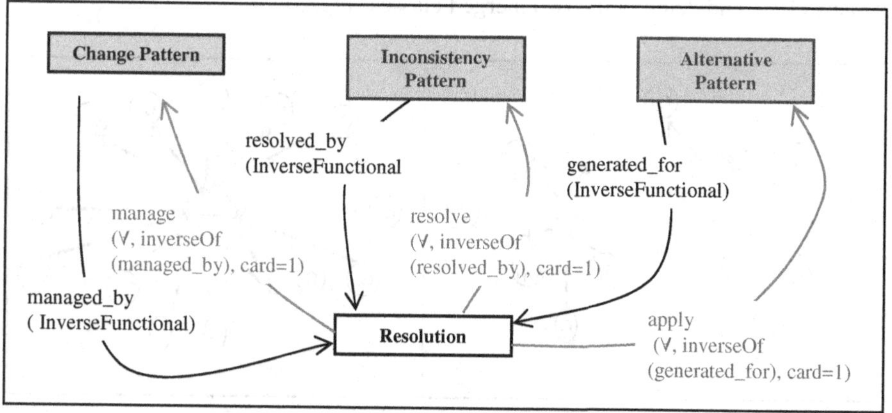

Fig. 2. Extract of CMP ontology illustrated in UML

3.1 Change Patterns

Change patterns aim to categorize changes, formally define their signification, their scope and their potential implications. Two categories of OWL changes are distinguished in the literature [4]: basic changes and complex changes. Change patterns cover all OWL basic changes and a first core of complex changes.

Basic change patterns describe all OWL basic changes derived from OWL meta-model. They model simple and indivisible OWL changes (e.g. add a new class or property, add a subsumption between two classes, define disjointness or equivalence between two classes, etc.).

However, often a change request corresponds to a set of changes that need to be applied together as one single logical sequence. In this case, it is important to apply the set of changes as a single transaction and to keep information that they meet together a same change requirement. This helps following ontology evolution and change rationales, and facilitates revoking applied changes as a transaction rollback operation in database management systems. Moreover, the effects of a set of changes applied in batch are not the same if these changes are applied separately. Thus, to extend the set of defined OWL basic operations with composite and specific change operations, we define complex change patterns. They correspond to composite OWL changes grouping logical sequences of basic and complex changes (e.g. enlarge the range of an object property to its super-class, merge two classes, etc.).

In this section, we describe an example of complex change. The reader can refer to [5] for a detailed description of a basic change example. The complex change example consists in merging two classes in a new one which includes adding a new class, adding subsumption relations between the new class and the super-classes of the merged classes, adding subsumption relations between the sub-classes of the merged classes and the new class, defining instantiation between the individuals of the merged classes and the new class, deleting instantiation relations between these individuals and the merged classes, aggregating properties of the merged classes to the new class, and deleting the two merged classes (Fig. 3).

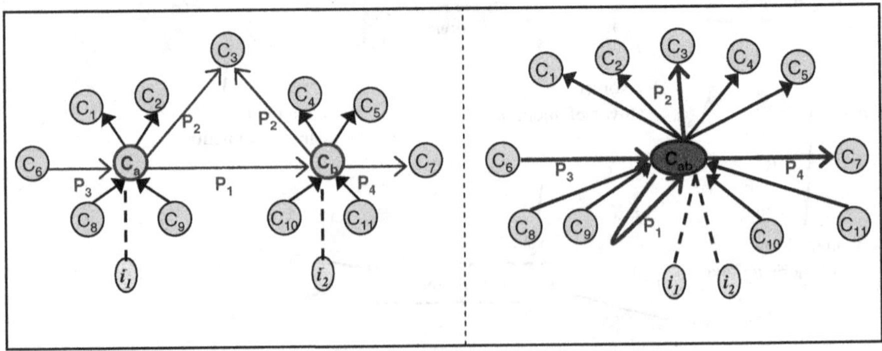

Fig. 3. Illustration of the complex change: merging two classes

The description of the corresponding pattern – in presentation layer template – is given by the following table (Tab. 1).

Table 1. Complex change pattern: Merge two classes

Complex change Pattern Example	
General Properties	
General Information	
Name*	Merge two classes.
Identifier*	CChP_MergeTwoClasses.
CMP Type*	Change Pattern.
Use Case	
Problem	Merge two classes.
Examples	How to merge two classes in a new one? Are there any constraints to consider? And how to compose this complex change?
CMP Abstraction Level	
Pattern Description	
Intent	The pattern models a complex change composed of basic and complex changes. It merges two classes and notifies if there is a constraint to check so that logical consistency will not be altered.
Consequences	The pattern merges two classes in a new one and aggregates all their properties, instances and sub-classes.
Scenarios	Define the class *Carnivorous-Plant* as a sub-class of the class *Animal.* Notify that the constraint: {class *Carnivorous-Plant* and class *Animal* should not be disjointed} has to be verified.
Change Abstraction Level	
Pattern Description	
Change Pattern Type*	Complex Change Pattern.
Object Type*	Class.
Involved Entity Type*	Class, Class
Arguments*	
Object*	ID of a first class (C1_ClassID), ID of a second class (C2_ClassID).
Referred Entities*	
First Class ID	ID of a first class (C1_ClassID)
Second Class ID	ID of a second class (C2_ClassID)
Merge Class ID	ID of merge class (C12_ClassID)
Intermediate Entities*	
ID(s) Super-Classes	ID(s) of super-classes of the two classes to merge {super_classID}
ID(s) Sub-Classes	ID(s) of sub-classes of the two classes to merge {sub_classID}
ID(s) Instances	ID(s) of instances of the two classes to merge {individualID}
ID(s) Properties	ID(s) of properties of the two classes to merge {propertyID}
Number of Sub-classes*	
Number of subclasses	Nb_s-cls
Number of Instances*	
Number of instances	Nb_i

Table 1. *(continued)*

Constraints	
Constraints	Ascendants of the two classes to merge (direct and indirect super-classes) are not disjoint ¬(Ascendants(C1_ClassID) disjointWith Ascendants(C2_ClassID)).
Complex Change Abstraction Level	
Pattern Description	
Sequence*	1) BChP_ AddClass (C12_ClassID) 2) CChP _AddSub-Class_Multi-Inheritance (C12_ClassID, {super_classID}) 3) Nb_s-cls times BChP _ AddSub-Class (sub_classID, C12_ClassID) 4) Nb_i times BChP _ExtendIndividualInstantiation (individualID, C12_ClassID) 5) Nb_i times BChP _DeleteIndividualInstantiation (individualID, C1_ClassID) BChP _ DeleteIndividualInstantiation (individualID, C2_ClassID) 6) CChP _AggregateProperties_TwoClasses ({propertyID}, C1_ClassID, C2_ClassID, C12_ClassID) 7) BChP _ Delete_Class (C1_ClassID) 8) BChP _ Delete_Class (C2_ClassID)

A graphical illustration of CMP ontology part (in UML) describing this complex change pattern is given by the following figure (Fig. 4).

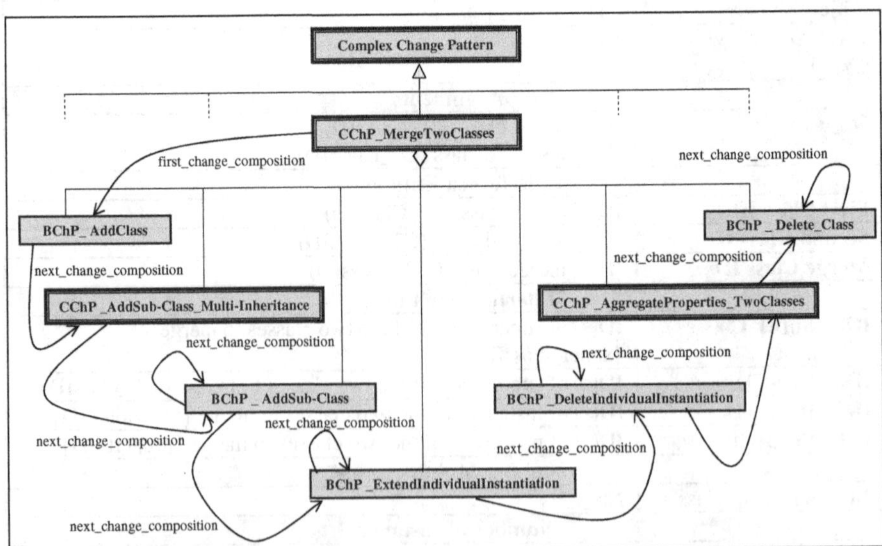

Fig. 4. Extract of CMP ontology part corresponding to the complex change pattern *merge two classes* illustrated in UML

3.2 Inconsistency Patterns

Logical inconsistency could be caused by a basic or a complex change. Ontology axioms need to be verified to detect consistency constraint violations. Change operation gives information about the entities referred by the change but does not necessarily provide information on the context application of the change (involved axioms, or axioms that could be altered by this change operation). It is not possible to intuitively anticipate all logical constraints related to its application. Modeling CMP patterns addresses this problem by defining: *i)* a set of constraints to verify for each class of change (in the properties of change patterns), and *ii)* a conceptual link between change and inconsistency patterns to predict possible inconsistency types that could be potentially caused by a class of changes. Thus, the analysis of change impact is guided by the explicit specification of the semantics of a requested change, the constraints to verify straightaway, and by indication about expected inconsistency types.

Inconsistency patterns categorize classes of logical inconsistencies and explain them to facilitate the interpretation of reasoner results, to guide the localization of axioms responsible of detected inconsistencies and, to prepare their resolution. They model a sub-set of OWL DL logical inconsistencies: disjointness inconsistencies related to subsumption and instantiation; inconsistencies related to the definition of a class as a collection of classes, inconsistencies related to class equivalence and complement, inconsistencies related to class equivalence and disjointness, inconsistencies related to value restrictions and inconsistencies related to cardinality restrictions.

To illustrate an example let's consider the inconsistency pattern "*Inapplicable Universal Value Restriction*". When a change adding a universal restriction value is required (corresponding to basic change pattern "*Add a Universal Property Restriction Value*"), we need to verify that the class involved by the change has not an existing universal restriction value – on the same property referred by this change – of which the set specifying the restriction value is disjoint to the set specifying the restriction value of the required change. This constraint can be described formally as follow:

For a change: $\mathtt{C} \sqsubseteq \forall \mathtt{P.C1}$ (C a class, and P a property)
Verify that: $\forall \mathit{C1} \in \mathtt{O}$, `if` $(\mathtt{C} \sqsubseteq \forall \mathtt{P.C1})$ `then` $\neg(\mathtt{C1}$ `disjointWith` $\mathtt{C2})$

This constraint aims to verify that the universal restriction value to add should not cause logical inconsistency and is applicable. It is handled by the conceptual relation between change pattern and inconsistency pattern mentioned above. A synthetic description of the inconsistency pattern "*Inapplicable Universal Value Restriction*" is given in the following table (Tab. 2).

Table 2. Synthetic description of the inconsistency pattern "*Inapplicable Universal Value Restriction*"

Inconsistency Pattern ***Inapplicable Universal Value Restriction***			
Property	**Description**	**Example**	
Implied Entities	- SourceClassID	C	
	- PropertyID	P	
	- TargetClassDisj1ID	C_1	
	- TargetClassDisj2ID	C_2	
Involved Entities	- SourceClassID	C	
	- PropertyID	P	
	- TargetClassDisj2ID	C_2	
Involved Axioms	- (TargetClassDisj1ID disjointWith TargetClassDisj2ID)	(C_1 disjointWith C_2)	
	- SourceClassID ⊑ ∀PropertyID. TargetClassDisj1ID	$C \sqsubseteq \forall P. C_1$	
Responsible Axioms	- SourceClassID ⊑ ∀PropertyID. TargetClassDisj2ID	$C \sqsubseteq \forall P. C_2$	

3.3 Alternative Patterns

Modeling alternative patterns aims to categorize classes of alternatives that can be generated to resolve types of logical inconsistencies supported by ***ONTO-EVO^{A}L*** approach. At the resolution phase of change management process (Fig. 1), conceptual relations defined (in CMP ontology) between the instantiated change pattern (corresponding to the required change), the instantiated inconsistency patterns (corresponding to logical inconsistencies detected by the reasoner), and the alternative patterns that could potentially resolve them, guide the generation of resolution alternative instances for each inconsistency caused by the change. This is based on the class *Resolution* corresponding to a change instance and an inconsistency instance as illustrated in (Fig. 2).

An alternative pattern represents an additional change (applied jointly to the required change) or a substitutive change to apply (replacing the required change) so that a logical inconsistency can be resolved. It is described as a change (basic or complex) and it inherits and extends change pattern properties.

Several resolution alternatives can be proposed for an inconsistency. Let's take an example (Fig. 5): if we need to apply a change adding a sub-class relation between two classes (instantiating the basic change pattern *BChP_AddSubClass*), and if this change causes a disjointness inconsistency (instantiating the inconsistency pattern IncP_DisjointnessSubsumption) related to the new defined subsumption because an existing super-class is disjoint with the super-class referred by this change, two alternative patterns are proposed by CMP ontology. The first one *AlP_DefineHybridClass* corresponds to a substitutive resolution extending a complex change. It resolves a disjointness inconsistency –related to a subsumption– by: 1) defining a hybrid class as a union of the definitions of the disjoint classes implicated in the inconsistency to be resolved; 2) defining a subsumption between the most specific common super-class of the disjoint classes implicated in the inconsistency and the hybrid class created; and then 3) defining a subsumption between the hybrid class and the sub-class involved in the inconsistency (Fig. 5). The second one *AlP_EnlargeClassDefinition* corresponds to a substitutive resolution extending a basic change. It resolves a disjointness inconsistency –related to a subsumption– by enlarging the definition of the sub-class involved in the inconsistency by adding –in its description– a union of the definitions of disjoint classes implicated in the inconsistency (Fig. 5).

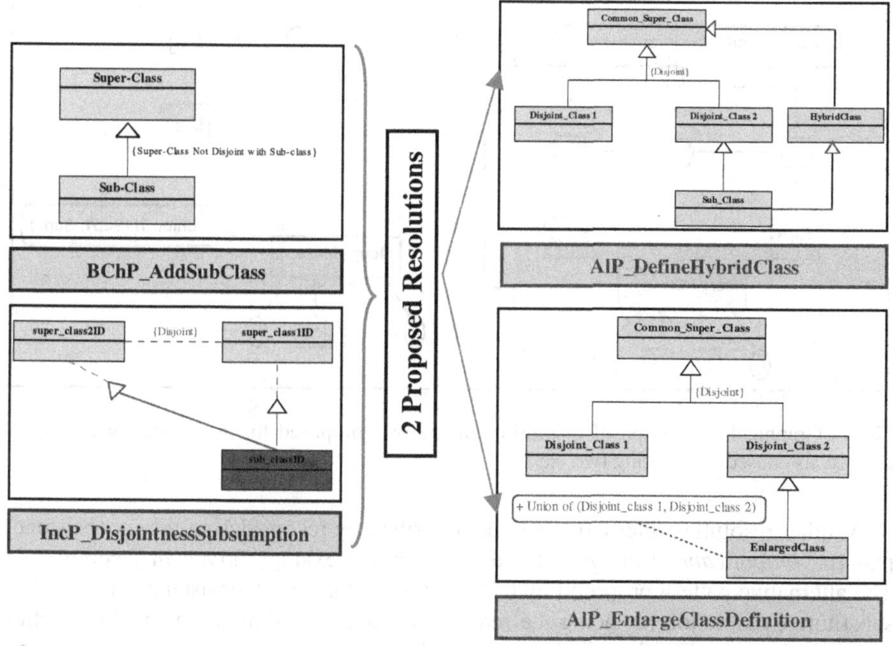

Fig. 5. Example of two proposed resolutions for a basic change instance and a logical inconsistency instance

Reconsidering the example of complex change pattern CChP_MergeTwoClasses given in (section 3.1), a logical inconsistency could be caused if the ascendants of the two classes to merge (direct or indirect super-classes) are disjoints. Two resolution alternatives are proposed by CMP ontology in this case (Fig. 6): the first one (*al1*) defines a generalized subsumption between the merge class and the most specific common super-class of the disjoint ascendants; the second one (*al2*) enlarges the definition of the merge class by adding –in its description– a union of the definitions of disjoint ascendants.

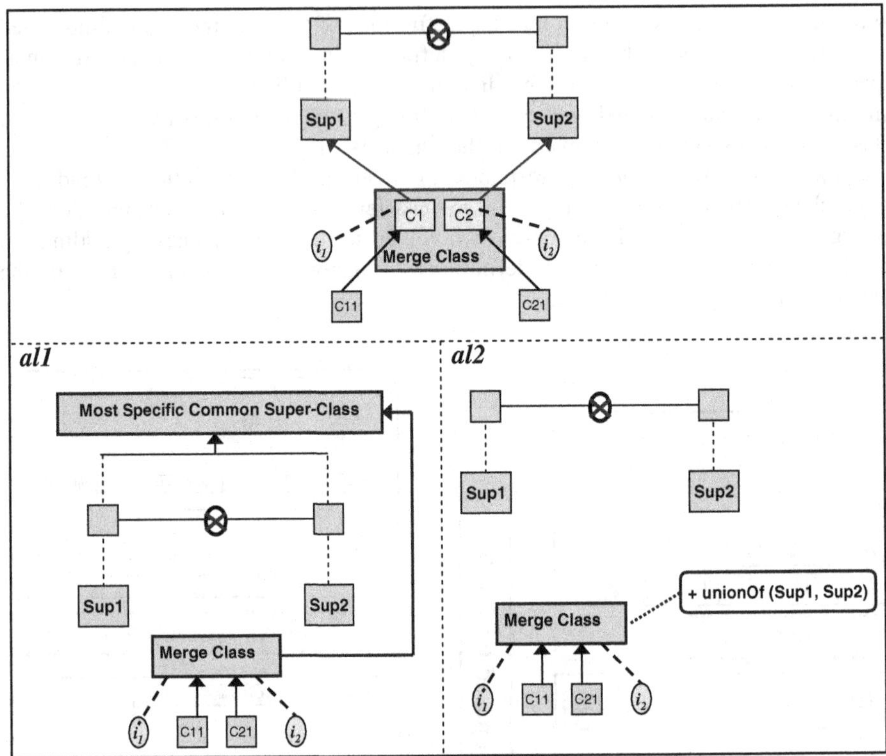

Fig. 6. Graphical illustration of resolution alternatives proposed for inconsistency of disjoint ascendants caused by merging two classes

Another resolution alternative can be described by reconsidering the inconsistency pattern "*Inapplicable Universal Value Restriction*" example given in (section 3.2). The alternative pattern proposed by CMP ontology for this inconsistency consists in a substitutive resolution replacing the required change of adding a universal restriction which is inapplicable, by another universal restriction value applied on the same property by specifying the two disjoint classes as a value of the restriction. Axioms implicated in the inconsistency (involved and responsible axioms) and the transformation proposed by the resolution alternative, are summarized by the following table (Tab. 3).

Table 3. Synthetic description of the resolution of the inconsistency "*Inapplicable Universal Value Restriction*"

Axioms implicated in an inconsistency of an Inapplicable Universal Value Restriction
$\underline{Ax_1}$: SourceClassID ⊑ ∀PropertyID. TargetClassDisj1ID $\underline{Ax_2}$: SourceClassID ⊑ ∀PropertyID. TargetClassDisj2ID
$\underline{Ax_3}$: (TargetClassDisj1ID disjointWith TargetClassDisj2ID)

Table 3. *(continued)*

Application of the alternative pattern resolving an inapplicable universal value restriction
⇨ Transformation : delete Ax_1, keep Ax_3 and add Ax Ax : SourceClassID ⊑ ∀PropertyID. (TargetClassDisj1ID ∪ TargetClassDisj2ID) Ax_3 : (TargetClassDisj1ID disjointWith TargetClassDisj2ID)
* axiom Ax_2 corresponding to the required change will not be applied as the resolution proposed for this change is defined as a substitutive resolution.

3.4 List of Modeled CMP Patterns

The list of CMP patterns modeled in CMP ontology is given by the following table (Tab. 4).

Table 4. List of modeled CMP patterns

Basic Change Patterns	
Class of patterns	**List of patterns**
Class	Add (class, class union, class intersection, class complement, enumeration), Extend Definition (by a class union, a class intersection, a class complement, an enumeration), Delete.
Sub-class	Add, Delete.
Class Equivalence	Add (to a class, a class union, class intersection, a class complement, an enumeration), Delete.
Class Disjointness	Add (to a class, a class union, class intersection, a class complement, an enumeration), Delete.
Value Restriction	Add (Equal, Universal, Existential), Delete.
Cardinality Restriction	Add (maximal, minimal, exactly), Delete.
Property	Add, Delete.
Domain Property	Add (class, class union, class intersection, class complement, enumeration, value restriction, cardinality restriction), Modify, Delete.
Range Property	Add (object property (class union, class intersection), data property), Modify (object property, data property), Delete (object property, data property).
Sub-Property	Add, Delete.
Property Equivalence	Add, Delete.
Functional Property	Set (object property), Unset.
Inverse Functional Property	Set, Unset.
Inverse Property	Add, Delete.
Symmetric Property	Set, Unset.
Transitive Property	Set, Unset.

Table 4. *(continued)*

Individual	Add, Extend Instantiation (class, property domain or range), Delete, Delete Instantiation (class, property domain or range)
Individual Difference	Add, Delete.
Individual Identity	Add, Delete.
Complex Change Patterns	
Sub-hierarchy	Add a class and its sub-classes.
Multi-Inheritance	Add a sub-class in multi-inheritance.
Merge	Merge two classes.
Generalization	Pull up a class.
Specialization	Pull down a class.
Deletion	Delete a class and attaching its properties to its sub-classes, and its instances and sub-classes to a super-class.
Property Domain	Extend property domain to a super-class, Restrain property domain to a sub-class.
Property Range	Extend property range to a super-class, Restrain property range to a sub-class.
Instantiation	Extend multiple instantiation of an individual (add several instantiations to an existing individual).
Property Aggregation	Aggregate properties of two classes.
Complex Value Restriction	Complex (class union, intersection or complement) universal restriction value, Complex existential restriction value, Complex exactly restriction value.
Inconsistency Patterns	
Disjointness	Subsumption, Instantiation, Ascendants.
Value Restriction	Inapplicable universal value restriction, Incompatible universal and existential value restrictions.
Class Deletion	Obsolete dependencies (properties, sub-classes and instances) of a deleted class.
Complex Definition of Property Domain or Range	Intersection of disjoint domains, Intersection of disjoint ranges.
Alternative Patterns	
Resolving Disjointness Inconsistency Related to Subsumption by a Hybrid Class	Define a hybrid class for a subsumption, Attach all subsumption relations to a hybrid class.
Resolving Disjointness Inconsistency Related to Instantiation by a Hybrid Class	Define a hybrid class for an instance, Attach all instances to a hybrid class.

Table 4. (*continued*)

Resolving Disjointness Inconsistency Based on Class Definitions	Extend definition of all subsumption relations, Extend definition of a sub-class, Extend definition of a sub-class by considering disjoint ascendants.
Resolving Subsumption Inconsistency by Generalization	Define a generalized subsumption.
Resolving Inconsistency of Value Restriction	Resolve inapplicable universal value restriction, resolve incompatible universal and existential value restriction, Resolve incompatible existential and universal value restriction.
Resolving Inconsistency of Obsolete Dependencies	Detach obsolete dependencies, Reclassify and redistribute obsolete dependencies.
Resolving Inconsistency of Disjoint Intersections of Property Domain or Range	Transform domain intersection into domain union, Transform range intersection into range union.

4 Ontology Quality Evaluation

To more automate the change management process, we have integrated in ***ONTO-EVO^{A}L*** approach a quality evaluation module. Thus, rather than presenting the proposed resolutions to the ontology engineer to choose suitable ones, they are sorted according to their impact on ontology quality and hence, if quality is preserved, required and/or derived changes (selected resolutions) are directly applied.

Quality evaluation has a larger significance than guiding the process. Change effects should not be considered only for consistency maintenance. It is also interesting to take into account the rationales behind the change and the specificity of the ontology usage. According to our purpose, we have focused on automatically measurable metrics giving a quantitative perspective of ontology quality and, we have defined a hierarchical quality model consisting of three levels: *Quality Features*, *Quality Criteria* (for each feature) and *Quality Metrics* (for each criterion). These features, criteria and metrics are selected from the literature and, adapted and enriched according to our needs. We have focused particularly on work described in [6] [7] [8] [9] [10].

4.1 Quality Features

Quality features considered are ontology *Content* and *Usage*. Ontology content includes structural and semantic characteristics of the ontology. Its evaluation is based on complexity, cohesion, conceptualization and abstraction criteria. Ontology usage is evaluated through ontology completeness and comprehension degree criteria. The assessment of these criteria is calculated through quantifiable metrics (Tab.5) that are defined independently of any ontology language and any modeled domain so that the quality model can be reused for other ontology evaluation purposes.

4.2 Quality Criteria

Complexity criterion assesses structural and semantic links between ontology entities and the navigability in ontology structure. *Cohesion* criterion takes into account connected ontology components (classes and instances). *Conceptualization* criterion corresponds to design richness of ontology content. *Abstraction* criterion indicates class abstraction level (generalization/specialization) by measuring the depth of subsumption hierarchies. *Completeness* evaluates if the ontology covers domain relevant properties. This criterion is based on class (and property) label conformity with the keywords of the domain. *Comprehension* criterion assesses the facility of understanding ontology through entity naming and annotations describing class and relation definitions and also the facility of explaining class differential with parents and siblings.

Some evaluation criteria can be contradictory: change that leads to a more complex ontology for example, can improve cohesion. To face this problem, each criterion is weighted (w_i) at the beginning of evolution process, by domain expert according to its relevance for the modeled domain and for the application using the ontology.

4.3 Quality Metrics

The different metrics evaluating these criteria are described bellow (Tab.5).

Table 5. Quality metrics description

Criteria		Metrics
Content		
Complexity	(w_1)	- Average number of paths to reach a class from root. (CP)
		- Average depth of the ontology. (D)
		- Average number of semantic relations (object properties assigned to classes) per class. (SRC)
Cohesion	(w_2)	- Average number of connected components (classes and instances). (CC)
Conceptualization	(w_3)	- Semantic Richness: Ratio of the total number of semantic relations assigned to classes, divided by the total number of ontology relations (object properties and subsumption relations). (SR)
		- Attribute Richness: Ratio of the total number of attributes (data properties describing ontology classes), divided by the total number of ontology classes. (AR)
		- Inheritance Richness: Average number of sub-classes per class. (IR)
Abstraction	(w_4)	- Average depth of the ontology. (D)

Table 5. *(continued)*

Usage		
Completeness	(w_5)	- Precision: percentage of class labels used in the ontology and appearing in a standard (domain representative data sources), compared to the total number of ontology class labels. (PREC)
		- Recall: percentage of terms used in a standard (domain representative data sources) and appearing in the ontology as class labels, compared to the total number of terms in the standard. (REC)
Comprehension	(w_6)	- Ratio of annotated classes divided by the total number of classes. (AC)
		- Ratio of annotated instances divided by the total number of instances. (AIR)
		- Ratio of annotated semantic relations (object properties) divided by the total number of semantic relations. (ASR)

5 Prototype of *ONTO-EVO^A L* System

ONTO-EVO^A L is implemented in a first prototype developed in Java language and based on OWLAPI[2]. It supports core functionalities of the approach and a first set of CMP patterns. The global technical architecture is described by the following figure (Fig. 7).

The main components are: *Pattern Descriptor* extracting sub-hierarchy corresponding to a CMP pattern from CMP Ontology, *Pattern Selector* verifying matching of a change pattern to a required change or inconsistency pattern to a caused inconsistency, *Pattern Generator* generating alternative patterns based on resolution relations with instantiated change and inconsistency patterns, *Pattern Instantiator* instantiating a CMP pattern and adapting it to the context of its application, *Change Specifier* formalizing a change request according to a change pattern instance, *Change Simulator* applying temporary a change to manage it, *Inconsistency Localizer* localizing and diagnosing a detected inconsistency, *Global Resolution Generator* combining the different alternatives proposed for inconsistencies caused by a change, and constituting all possible global change resolutions, *Quality Evaluator* evaluating the impact of all coherent global resolutions generated on ontology quality based on quality model, weights of the different criteria, and on the quality of the initial ontology (before the change), *Change Validator* deciding about change application according to quality evaluation results, *Archiver* saving traces of all change treatment results and ontology evolutions, and *Learning Module* that aims to exploit evolution log knowledge base to enrich and enhance CMP ontology.

Consistency analysis to detect caused inconsistencies is performed by employing *Pellet Reasoner* [11]. Pellet supports both terminological level *TBox* (classes and properties) and assertional level *ABox* (individuals) of OWL DL and provides entailment justifications. CMP application to resolve inconsistency is detailed in [5].

---

[2] http://owlapi.sourceforge.net/

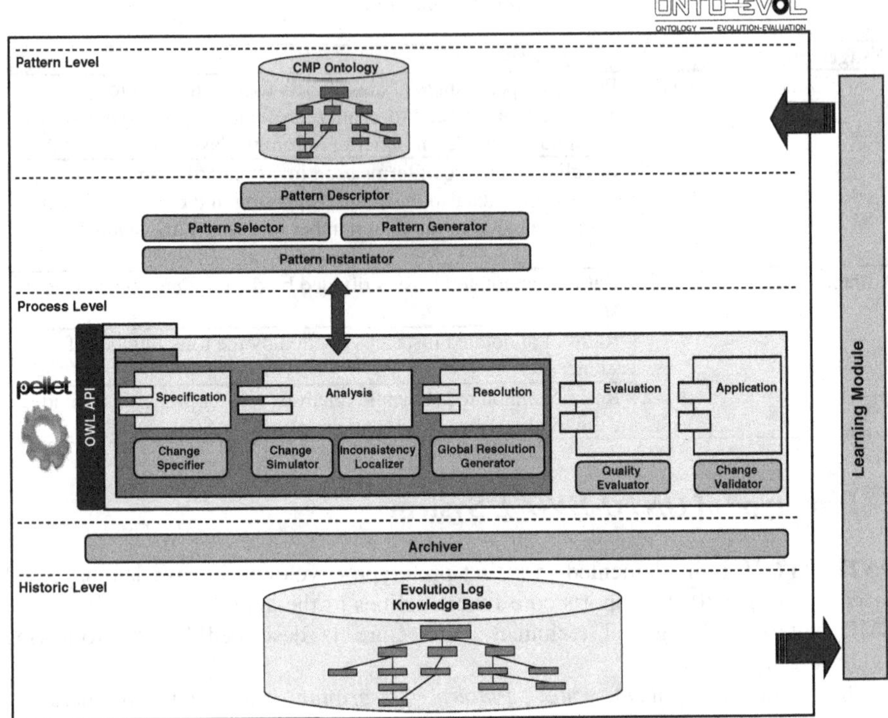

Fig. 7. Technical global architecture of ONTO-EVO^{A}L approach

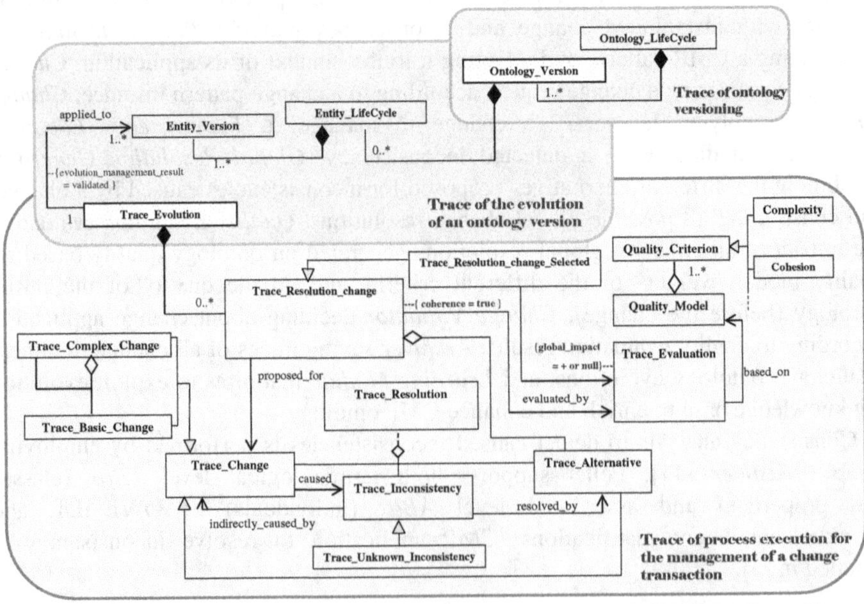

Fig. 8. Graphical representation in UML of evolution log ontology

Evolution log contains traces of modifications applied to ontology and the results of the management of their application. It facilitates controlling change management process and the continuous maintenance of an ontology and its different versions. It is modeled as an OWL DL ontology covering the three dimensions to be taken into account by an evolution historic: *i)* trace of process execution for the management of a change transaction, *ii)* trace of the evolution of an ontology version, and *iii)* trace of ontology versioning (Fig. 8).

6 Discussion and Related Work

Discussion with related work is tackled through three parts: pattern modeling, ontology evolution approaches and integrating evaluation aspects in ontology evolution process.

Pattern modeling was adopted in web ontology design to propose guidelines and provide reusable ontological component catalogue[3]. The notion of ontology design pattern has been introduced by [12] [13] [14] [15]. Change Management Patterns *CMP* are close to Ontology Design Patterns[4] *ODP*, particularly Logical Ontology Patterns *LOP* and Content Ontology Patterns *COP* [16]. *Change Patterns* of CMP can be considered as *COP* patterns for ontology domain i.e. ontology design patterns solving modeling problems of the domain 'ontology'. *Alternative Patterns* of CMP can be defined as *LOP* patterns resolving a problem of logical inconsistency. The alternative pattern "*Define Hybrid Class Resolving Disjointness due to Subsumption*" (Fig. 5) is indeed proposed as an *LOP* pattern on ODP portal[5] as described in [17]. In [18], a catalogue of anti-patterns for formal ontology debugging is presented. *Logical Anti-Patterns* –modeling errors detected by DL reasoners– are quite close to some Inconsistency *Patterns* of CMP.

Concerning OWL ontology evolution approaches, in [19], a pattern-driven approach was adopted for ontology evolution. The patterns determine the evolution operation to be performed: population (adding new instances) or enrichment (extension by new concepts and properties). In [20], authors have introduced resolution strategies based on OWL Lite model. The resolution is limited to the identification of axioms that should be removed to resolve inconsistencies and, their presentation to the user. In ***ONTO-EVO^{A}L*** approach, we tend to minimize axiom removing solutions by proposing alternatives that merge, divide, generalize or specialize classes and properties and redistribute instances to preserve existent knowledge. Alternative selection is also guided by quality evaluation.

In [21], a controlled evolution of ontologies through semiotic-based evaluation methods was presented. A tool *S-OntoEval* assessing ontology quality was described. Evaluation metrics are categorized into semiotic levels.

[3] Example: http://sourceforge.net/projects/odps/

[4] http://ontologydesignpatterns.org/

[5] http://ontologydesignpatterns.org/wiki/Submissions:Define_Hybrid_Class_Resolving_Disjointness_due_to_Subsumption

7 Conclusion and Future Work

In this paper, we present an ontology evolution approach ***ONTO-EVO^{A}L*** optimizing and automating change management while maintaining consistency and quality of the evolved ontology. The main contributions of the approach are: a pattern-oriented modeling driving and controlling change management process; and a quality evaluation activity guiding change resolution by selecting the more suitable alternatives.

Currently, we are developing a learning module enriching and enhancing *CMP* by considering evolution log information, new change compositions not yet supported by *Change Patterns*, detected inconsistencies not yet classified by *Inconsistency Patterns*, new potential resolution alternatives not yet modeled by *Alternative Patterns*, and also new possible relation instantiations between CMP.

References

1. Djedidi, R., Aufaure, M.-A.: Ontology Change Management. In: Paschke, A., Weigand, H., Behrendt, W., Tochtermann, K., Pellegrini, T. (eds.) I-Semantics 2009, Proceedings of I-KNOW 2009 and I-SEMANTICS 2009, pp. 611–621. Verlag der Technischen Universitt Graz (2009) ISBN 978-3-85125-060-2
2. Gangemi, A., Gomez-Perez, A., Presutti, V., Suarez-Figueroa, M.C.: Towards a Catalog of OWL-based Ontology Design Patterns. In: CAEPIA 2007. Neon project publications (2007), http://www.neon-project.org
3. Gamma, E., Helm, R., Johnson, R., Vlissides, J.: Design Patterns: Elements of Reusable Object-Oriented Software. Addison-Wesley, Reading (1995)
4. Klein, M.: Change Management for Distributed Ontologies. Ph.D. Thesis, Dutch Graduate School for Information and Knowledge Systems (2004)
5. Djedidi, R., Aufaure, M.-A.: Change Management Patterns (CMP) for Ontology Evolution Process. In: D'Aquin, M., Antoniou, G. (eds.) Proceedings of the 3rd International Workshop on Ontology Dynamics (IWOD 2009) in ISWC 2009. CEUR Workshop Proceedings, vol. 519 (2009)
6. Tartir, S., Budak Arpinar, I.: Ontology Evaluation and Ranking using OntoQA. In: The proceedings of the International Conference on Semantic Computing ICSC 2007, pp. 185–192. IEEE Computer Society, Los Alamitos (2007)
7. Vrandecic, D., Sure, Y.: How to Design better Ontology Metrics. In: Franconi, E., Kifer, M., May, W. (eds.) ESWC 2007. LNCS, vol. 4519, pp. 311–325. Springer, Heidelberg (2007)
8. Gangemi, A., Catenacci, C., Ciaramita, M., Lehmann, J.: Ontology Evaluation and Validation: an Integrated Model for the Quality Diagnostic Task. Technical Report , Laboratory of Applied Ontologies–CNR, Rome (2005), http://www.loa-cnr.it/Publications.html
9. Gangemi, A., Catenaccia, C., Ciaramita, M., Lehmann, J.: Modelling Ontology Evaluation and Validation. In: Sure, Y., Domingue, J. (eds.) ESWC 2006. LNCS, vol. 4011, pp. 140–154. Springer, Heidelberg (2006)
10. Gangemi, A., Catenaccia, C., Ciaramita, M., Lehmann, J.: Qood grid: A Meta-Ontology Based Framework for Ontology Evaluation and Selection. In: Vrandecic, D., Suarez-Figueroa, M.C., Gangemi, A., Sure, Y. (eds.) CEUR Workshop Proceedings of Evaluation of Ontologies for the Web EON 2006, vol. 179 (2006)

11. Sirin, E., Parsia, B., Cuenca Grau, B., Kalyanpur, A., Katz, Y.: Pellet: A practical OWL-DL reasoner. Journal of Web Semantics 5(2) (2007)
12. Gangemi, A., Catenacci, C., Battaglia, M.: Inflammation ontology design pattern: an exercise in building a core biomedical ontology with descriptions and situations. In: Pisanelli, D.M. (ed.) Ontologies in Medicine. IOS Press, Amsterdam (2004)
13. Rector, A., Rogers, J.: Patterns, properties and minimizing commitment: Reconstruction of the Galen upper ontology in OWL. In: Motta, E., Shadbolt, N.R., Stutt, A., Gibbins, N. (eds.) EKAW 2004. LNCS (LNAI), vol. 3257. Springer, Heidelberg (2004)
14. Svatek, V.: Design patterns for semantic web ontologies: Motivation and discussion. In: 7th conference on Business Information Systems, Poznan (2004)
15. Gangemi, A.: Ontology Design Patterns for Semantic Web Content. In: Gil, Y., Motta, E., Benjamins, V.R., Musen, M.A. (eds.) ISWC 2005. LNCS, vol. 3729, pp. 262–276. Springer, Heidelberg (2005)
16. Presutti, V., Gangemi, A., David, S., Aguado De Cea, G., Suarez-Figueroa, M., Montiel-Ponsoda, E., Poveda, M.: Library of design patterns for collaborative development of networked ontologies. Deliverable D2.5.1, NeOn project (2008)
17. Djedidi, R., Aufaure, M.-A.: Define Hybrid Class Resolving Disjointness Due to Subsumption, `http://ontologydesignpatterns.org/wiki/Submissions:Define_Hybrid_Class_`
18. Corcho, O., Roussey, C., Vilches Blazquez, L.M., Perez, I.: Pattern-based OWL Ontology Debugging Guidelines, In: Blomqvist, E., Sandkuhl, K., Scharffe, F., Svatek, V. (eds.) CEUR Workshop Proceedings, vol. 516, pp. 68–82 (2009)
19. Castano, S., Espinosa, S., Ferrara, A., Karkaletsis, V., Kaya, A., Melzer, S., Moller, R., Montanelli, S., Petasis, G.: Ontology Dynamics with Multimedia Information: The BOEMIE Evolution Methodology. In: Proceedings of International Workshop on Ontology Dynamics (IWOD 2007), ISWC 2007 Conference (2007)
20. Haase, P., Stojanovic, L.: Consistent Evolution of OWL Ontologies. In: Gomez-Perez, A., Euzenat, J. (eds.) ESWC 2005. LNCS, vol. 3532, pp. 182–197. Springer, Heidelberg (2005)
21. Dividino, R., Sonntag, D.: Controlled Ontology Evolution through Semiotic-based Ontology Evaluation. In: Proceedings of International Workshop on Ontology Dynamics (IWOD 2008), ISWC 2008 Conference, pp. 1–14 (2008)

Towards a Paradoxical Description Logic for the Semantic Web

Xiaowang Zhang[1,2], Zuoquan Lin[2], and Kewen Wang[1]

[1] Institute for Integrated and Intelligent Systems, Griffith University, Australia
[2] School of Mathematical Sciences, Peking University, China
{zxw,lzq}@is.pku.edu.cn, k.wang@griffith.edu.au

Abstract. As a vision for the future of the Web, the Semantic Web is an open, constantly changing and collaborative environment. Hence it is reasonable to expect that knowledge sources in the Semantic Web contain noise and inaccuracies. However, as the logical foundation of Ontology Web Language in the Semantic Web, description logics fail to tolerate inconsistent information. The study of inconsistency handling in description logics is an important issue in the Semantic Web. One major approach to inconsistency handling is based on so-called *paraconsistent reasoning*, in which standard semantics is refined so that inconsistencies can be tolerated. Four-valued description logics are not satisfactory for the Semantic Web in that its reasoning is a bit far from standard semantics. In this paper, we present a paraconsistent description logic called *paradoxical description logic*, which is based on a three-valued semantics. Compared to existing paraconsistent description logics, our approach is more suitable for dealing with inconsistent ontologies in that paraconsistent reasoning under our semantics provides a better approximation to the standard reasoning. An important result in this paper is that we propose a sound and complete tableau for paradoxical description logics.

1 Introduction

The Semantic Web[1], which is conceived as a future generation of the World Wide Web (WWW), is an evolving development of the Web by defining the meaning (semantics) of information and services on the Web in order to make it possible to process and use the Web content by computer programs [1]. In an open, constantly changing and collaborative environment like the Semantic Web, knowledge bases and data in a real world will rarely be perfect. Many reasons, such as modeling errors, migration from other formalisms, merging ontologies and ontology evolution, potentially bring inconsistent knowledge. Then it is unrealistic to expect that real knowledge bases and data are always logically consistent [2,3]. Unfortunately, inconsistencies cause undesired problems in reasoning with classical logic. As long as inconsistencies are involved in reasoning, inference can be described as exploding, or trivialised. That is to say, anything, no matter whether it is meaningful, can follow from an inconsistent set of assumptions. Let Γ be a theory and ϕ a formula. $\Gamma \vdash \phi$ denotes that ϕ is a consequence from Γ using classical

[1] http://semanticweb.org/

S. Link and H. Prade (Eds.): FoIKS 2010, LNCS 5956, pp. 306–325, 2010.

logic where $\vdash$ is denoted the classical consequence relation. If Γ is inconsistent, then $\Gamma \vdash \varphi$ and $\Gamma \vdash \neg\varphi$ for some φ in the language. According to the inference rule of classical logic *ex falso quodlibet*,

$$\frac{\phi \qquad \neg\phi}{\psi},$$

for any ψ in the language, $\Gamma \vdash \psi$. So inconsistencies cause classical logic to collapse. No useful inferences follow from an inconsistent theory, in the sense that inference is trivial [2]. Today, description logics (DLs) [4] have become a cornerstone of the Semantic Web for its use in the design of ontologies. For instance, the OWL-DL and OWL-Lite sub-languages of the W3C-endorsed Web Ontology Language (OWL)[2] are based on DLs. However, DLs, as a fragment of classical first-order logic (FOL), break down in the presence of inconsistent knowledge [2]. As a result, inconsistency handling in DLs has attracted much attention in the Semantic Web community and Artificial Intelligence (AI) community.

The study of inconsistency handling in AI has a long tradition and many works exist [5,6,7,8,9,10,11,12,13,14] etc. The issue of how to transfer these existing results in classical logic. As a result, several approaches to handle inconsistencies in DLs have been proposed, which can be roughly divided into two fundamentally different categories. The first one is based on the assumption that inconsistencies indicate erroneous data which are to be repaired in order to obtain a consistent ontology, e.g., by pinpointing the parts of an inconsistent ontology, which cause inconsistencies, and remove or weaken axioms in these parts to restore consistencies [15,16,17,18]. In these approaches, researchers hold a common view that knowledge bases should be completely free of inconsistencies, and thus try to eliminate inconsistencies from them immediately by any means possible. This view is regarded as "too simplistic for developing robust intelligent systems, and fails to use the benefits of inconsistent knowledge in intelligent activities, or to acknowledge the fact that living with inconsistencies seems to be unavoidable" in [2]. Inconsistencies in knowledge are the norm in the real world, and so should be formalized and used, rather than always rejected [2]. Another category of approaches, called *paraconsistent approaches*, is not to simply avoid inconsistencies but to tolerate them by applying a non-standard reasoning method to obtain meaningful answers [19,20,21,22,23]. In this kind of approaches, inconsistencies are treated as a natural phenomenon in realistic data and are tolerated in reasoning. The call for robust intelligent systems has led to an increased interest in inconsistency tolerance in AI. Compared with the former, the latter acknowledges and distinguishes the different epistemic statuses between "*the assertion is true*" and "*the assertion is true with conflict*". So far, paraconsistent approaches to inconsistency handling in DLs, such as [19,20,21], are based on Belnap's four-valued semantics [5]. The major idea is to use additional truth values *underdefined* (i.e. neither true nor false) and *overdefined* (i.e. *contradictory*, or, both true and false). To a certain extent *four-valued semantics* can help us to handle inconsistencies. However the four-valued semantics is relatively weak because some important inference rules are not valid in the four-valued semantics. These rules include: (1) the *law of excluded middle*: $\models A \sqcup \neg A(a)$ where A is a concept name

[2] `http://www.w3.org/TR/owl-features/`

and a is an individual. This rule captures the toutology in logics, and (2) some *intuitive equivalences* including: $\mathcal{O} \models C \sqsubseteq D$ if and only if $\mathcal{O} \models \neg C \sqcup D(a)$ for any individual a (where C and D are concepts). Such rules are employed to reduce reasoning tasks w.r.t. ABoxes and TBoxes to reasoning tasks w.r.t. only ABoxes [24]. These shortcomings are inherent limitations of four-valued logics in paraconsistent reasoning. Besnard and Hunter [8] proposed a paraconsistent approach for classical logic, known as *quasi-classical logic* (abbreviated *QC*). This paraconsistent logic has been introduced to DLs by Zhang et al [22,23], where two semantics *strong semantics* and *weak semantics* are introduced in DLs in order to make more useful inference rules valid. The *concept inclusions* under two semantics are defined differently in order to maintain some intuitive equivalences at the cost of effectiveness. However, the *excluded middle rule* is still disabled in QC DLs. Based on the *argumentation theory* in [9], Zhang et al [25] present a hybrid approach to deal with inconsistent ontologies in description logics. A drawback of the approach is that the complexity of the argumentative semantics is much higher than that of classical logic.

In short, existing approaches to paraconsistency for DLs employ a "nonstandard" semantics that is from "*relevance logics*"[3]. In such semantics, to tolerate inconsistencies, the cost is to weaken the inference power. For instance, their inference power even are weaker than that of classical logics in reasoning with consistent classical ontologies since they are relevant logics. However, DLs, as classical logics, are not relevant logics. Naturally, we expect that a paraconsistent approach to inconsistency handling in DLs can preserve properties of classical logics as many as possible. That is to say, paraconsistent DLs is expected to be more close to classical DLs. For instance, candidate paraconsistent DLs can satisfy as inference rules (or proof rules) of DLs as many as possible. The motivation of this paper is to present a "*brave-reasoning*"[4] version of DLs which is more close to DLs.

In this paper, we present a paraconsistent extension of DL $\mathcal{ALC}$, called *paradoxical DL* $\mathcal{ALC}$ (written by $\mathcal{ALC}_{LP}$), which is an extension of $\mathcal{ALC}$ with semantics of logics of paradox (LP) presented by Priest in [6,7]. We introduce a paraconsistent semantics for $\mathcal{ALC}$, which is closer to the standard semantics than the four-valued one. We prove that the paradoxical entailment satisfies the *excluded middle rule* and *intuitive equivalence*. A signed tableau for $\mathcal{ALC}_{LP}$ is proposed to implement paraconsistent reasoning in DL-based ontologies. Firstly, we introduce two symbols $\mathbf{T}$ (for *true*) and $\mathbf{F}$ (for *false*) in $\mathcal{ALC}_{LP}$ which are employed to mark the states of axioms in a given situation. Secondly, signed expansion rules are provided to construct signed tableaus for ABoxes. Finally, closed conditions are modified to tolerate inconsistencies and capture paraconsistency in the signed tableau. Furthermore, we prove that the signed tableau is sound and complete with respect to the paradoxical entailment in $\mathcal{ALC}_{LP}$.

The rest of the paper is organized as follows: in the next section, we gives a short introduction of the syntax, the semantics of LP and DL $\mathcal{ALC}$. Then Section 3 introduces the paradoxical semantics for DL $\mathcal{ALC}$ and two basic inference problems. Section 4

---

[3] http://plato.stanford.edu/entries/logic-relevance/

[4] In general, there are two reasoning mechanisms: "*cautious reasoning* " and "*brave reasoning*". Cautious reasoning is done w.r.t. a local ontology and its neighbours while brave reasoning is done w.r.t. the transitive closure.

presents a signed tableau algorithm for $\mathcal{ALC}_{LP}$ to implement paraconsistent reasoning. Section 5 discusses related works. Finally, in Section 6, we conclude this paper and discuss some future work.

2 Preliminaries

In this section, we recall some basics of logics of paradox (LP) and DL $\mathcal{ALC}$.

2.1 Logics of Paradox

A *logic of paradox* (LP) is a three-valued logic, in which one additional value is used to represent conflict besides the two standard truth values. We refer the reader to [6,7,26,27] for further details.

A *paradoxical interpretation* π assigns to each atomic sentence ϕ in $\mathcal{L}$ one of the following three values: f (false and false only), t (true and true only) and B (both true and false). A sentence ϕ is *true* under π if $\pi(\phi) = t$ or $\pi(\phi) = B$; and ϕ is *false* under π if $\pi(\phi) = f$ or $\pi(\phi) = B$. Under any paradoxical interpretation, truth values can be conventionally extended to non-atomic sentences as follows:

(1) – $\neg\phi$ is true if and only if ϕ is false;
 – $\neg\phi$ is false if and only if ϕ is true;
(2) – $\phi \wedge \psi$ is true if and only if both ϕ and ψ are true;
 – $\phi \wedge \psi$ is false if and only if either ϕ or ψ is false;
(3) – $\forall x.\phi(x)$ is true if and only if for all closed terms t, $\phi(t)$ is true;
 – $\forall x.\phi(x)$ is false if and only if for some closed terms t, $\phi(t)$ is false.

Note that $\phi \vee \psi \equiv \neg(\neg\phi \wedge \neg\psi)$, $\phi \rightarrow \psi \equiv \neg(\phi \wedge \neg\psi)$ and $\exists x.\phi(x) \equiv \neg\forall x.\neg\phi(x)$ where $\equiv$ is the *logical equivalence* (*relationship*).

For the sake of simplicity, we assume that every object in the domain is named by a closed term. Similarly, other connectives and qualifiers can be easily defined via $\neg$, $\wedge$ and $\forall$ as a standard way.

- $\pi(\phi) = t$ if ϕ is true but not false under π;
- $\pi(\phi) = f$ if ϕ is false but not true under π;
- $\pi(\phi) = B$ if ϕ is both true and false under π.

Let Γ be a set of sentences and ψ a sentence. We say Γ *paradoxically entails* ψ, denoted by $\Gamma \models_{LP} \psi$, if and only if ψ is true in all paradoxical models of Γ, that is, for any paradoxical interpretation π, if every member of Γ is true under π, then ψ is also true under π.

Example 1. *Let ϕ and ψ be two different atomic sentences. It is easy to see that $\phi \wedge \neg\phi \models \psi$, but $\phi \wedge \neg\phi \not\models_{LP} \psi$. This can be seen by noting that $\phi \wedge \neg\phi$ is true (actually it is both true and false) but ψ is not true, under the paradoxical interpretation π with $\pi(\phi) = B$ and $\pi(\psi) = f$.*

By the above example, LP gets rid of trivial problem of classical logic, and as a paraconsistent logic, inconsistencies are indeed localized in the sense. Furthermore, the *law of excluded middle* is valid in LP. For instance, let ϕ be an atomic sentence. It easily concludes that $\models_{LP} \phi \vee \neg\phi$ since $\phi \vee \neg\phi \equiv \neg(\neg\phi \wedge \phi)$.

2.2 Description Logic $\mathcal{ALC}$

Description logics (DLs) is a well-known family of knowledge representation formalisms. For more comprehensive background reasoning, we refer the reader to Chapter 2 of the DL Handbook [4]. DLs can be seen as fragments of *first-order predicate logic*. They are different from their predecessors such as semantic networks and frames in that they are equipped with a formal, logic-based semantics. In DLs, elementary descriptions are *concept names* (unary predicates) and *role names* (binary predicates). Complex descriptions are built from them inductively using concept and role constructors provided by the particular DLs under consideration.

In this paper, we consider $\mathcal{ALC}$ which is a simple yet relatively expressive DL. $\mathcal{AL}$ is the abbreviation of attributive language and $\mathcal{C}$ denotes "*complement*". Let N_C and N_R be pairwise disjoint and countably infinite sets of *concept names* and *role names* respectively. Let N_I be an infinite set of *individual names*. We use the letters A and B for concept names, the letter R for role names, and the letters C and D for concepts. $\top$ and $\bot$ denote the *universal concept* and the *bottom concept* respectively. The set of $\mathcal{ALC}$ concepts is the smallest set such that:

(1) every concept name is a concept;
(2) if C and D are concepts, R is a role name, then the following expressions are also concepts: $\neg C$ (*full negation*), $C \sqcap D$ (*concept conjunction*), $C \sqcup D$ (*concept disjunction*), $\forall R.C$ (*value restriction on role names*) and $\exists R.C$ (*existential restriction on role names*).

For instance, the concept description $Person \sqcap Female$ is an $\mathcal{ALC}$-concept describing those persons that are female. Suppose $hasChild$ is a role name, the concept description $Person \sqcap \forall hasChild.Female$ expresses those persons whose children are all female. The concept $\forall hasChild.\bot \sqcap Person$ describes those persons who have no children.

An interpretation $I = (\Delta^I, \cdot^I)$ consisting of a non-empty domain Δ^I and a mapping $\cdot^I$ which maps every concept to a subset of Δ^I and every role to a subset of $\Delta^I \times \Delta^I$, for all concepts C, D and a role$R$, satisfies conditions in the following table.

A *general concept inclusion axiom (GCI)* or *a terminology* is an inclusion statement of the forms $C \sqsubseteq D$, where C and D are two (possibly complex) $\mathcal{ALC}$ concepts

Table 1. Syntax and Semantics of $\mathcal{ALC}$

Constructor	Syntax	Semantics
concept	A	$A^I \subseteq \Delta^I$
role	R	$R^I \subseteq \Delta^I \times \Delta^I$
individual	o	$o^I \in \Delta^I$
top concept	$\top$	Δ^I
bottom concept	$\bot$	$\emptyset^I$
negation	$\neg C$	$\Delta^I \setminus C^I$
conjunction	$C_1 \sqcap C_2$	$C_1^I \cap C_2^I$
disjunction	$C_1 \sqcup C_2$	$C_1^I \cup C_2^I$
existential restriction	$\exists R.C$	$\{x \mid \exists y, (x, y) \in R^I$ and $y \in C^I\}$
value restriction	$\forall R.C$	$\{x \mid \forall y, (x, y) \in R^I$ implies $y \in C^I\}$

(concepts for short). It is the statement about how concepts are related to each other. We use $C \equiv D$ as an abbreviation for the symmetrical pair of GCIs $C \sqsubseteq D$ and $D \sqsubseteq C$, called a *concept definition*. An interpretation I satisfies a GCI $C \sqsubseteq D$ if and only if $C^I \subseteq D^I$, and it satisfies a GCI $C \equiv D$ if and only if $C^I = D^I$. A finite set of GCIs is called a *TBox*. An *acyclic TBox* is a finite set of concept definitions such that every concept name occurs at most once on the left-hand side of an axiom, and there is no cyclic dependency between the definitions.

We can also formulate statements about individuals. A *concept* (*role*) *assertion axiom* has the form $C(a)$ ($R(a, b)$), where C is a concept description, R is a role name, and a, b are individual names. An *ABox* contains a finite set of concept axioms and role axioms. In the ABox, one describes a specific state of affairs of an application domain in terms of concept and roles.

To give a semantics to ABoxes, we need to extend interpretations to individual names. For each individual name a, $\cdot^I$ maps it to an element $a^I \in \Delta^I$. We assume the mapping $\cdot^I$ satisfies the *unique name assumption* (UNA) in which $a^I \neq b^I$ for distinct names a and b.

An interpretation I satisfies a concept axiom $C(a)$ if and only if $a^I \in C^I$; I satisfies a role axiom $R(a, b)$ if and only if $(a^I, b^I) \in R^I$. An ontology $\mathcal{O}$ consists of a TBox and an ABox, i.e., it is a set of GCIs and assertion axioms.

An interpretation I is a *model* of a DL (TBox or ABox) axiom if and only if it satisfies this axiom; and it is a model of an ontology $\mathcal{O}$ if it satisfies every axiom in $\mathcal{O}$. A concept D *subsumes* a concept C w.r.t. a TBox $\mathcal{T}$ if and only if each model of $\mathcal{T}$ is a model of axiom $C \sqsubseteq D$. An ABox $\mathcal{A}$ is consistent w.r.t. a TBox $\mathcal{T}$ if and only if there exists a common model of $\mathcal{T}$ and $\mathcal{A}$.

Given an ontology $\mathcal{O}$ and a DL axiom ϕ, we say $\mathcal{O}$ *entails* ϕ, denoted $\mathcal{O} \models \phi$, if and only if every model of $\mathcal{O}$ is a model of ϕ. A concept C is *satisfiable* if and only if there exists an individual a such that the ABox $\{C(a)\}$ is consistent; and *unsatisfiable* otherwise. A concept C is *satisfiable* w.r.t. a TBox $\mathcal{T}$ if and only if there exists a model I of $\mathcal{T}$ such that $C^I \neq \emptyset$ is consistent; and *unsatisfiable* otherwise.

Two basic reasoning problems, namely, *instance checking* (checking whether an individual is an instance of a given concept) and *subsumption checking* (checking whether a concept subsumes a given concept) can be reduced to the problem of consistency by the following lemma.

Lemma 1 **([4]).** *Let $\mathcal{O}$ be an ontology, C, D concepts and a an individual in $\mathcal{ALC}$.*

(1) *$\mathcal{O} \models C(a)$ if and only if $\mathcal{O} \cup \{\neg C(a)\}$ is inconsistent.*
(2) *$\mathcal{O} \models C \sqsubseteq D$ if and only if $\mathcal{O} \cup \{C \sqcap \neg D(\iota)\}$ is inconsistent where ι is a new individual not occurring in $\mathcal{O}$.*

The following lemma by Horrocks et al [24] shows that satisfiability, unsatisfiability and consistency of a concept w.r.t. an acyclic TBox can be reduced to the corresponding reasoning task w.r.t the empty TBox. This result is obtained by introducing a "*universal*" role U, that is, if y is reachable from x via a role path, then $\langle x, y \rangle \in U^I$.

Lemma 2 **([24]).** *Let C, D be concepts, $\mathcal{A}$ an ABox and $\mathcal{T}$ an acyclic TBox in $\mathcal{ALC}$. Define*

$$C_{\mathcal{T}} := \bigsqcap_{C_i \sqsubseteq D_i \in \mathcal{T}} \neg C_i \sqcup D_i.$$

Then the following properties hold:

(1) *C is satisfiable w.r.t. $\mathcal{T}$ if and only if $C \sqcap C_\mathcal{T} \sqcap \forall U.C_\mathcal{T}$ is satisfiable;*
(2) *D subsumes C w.r.t. $\mathcal{T}$ if and only if $C \sqcap \neg D \sqcap C_\mathcal{T} \sqcap \forall U.C_\mathcal{T}$ is unsatisfiable;*
(3) *$\mathcal{A}$ is consistent w.r.t. $\mathcal{T}$ if and only if $\mathcal{A} \cup \{C_\mathcal{T} \sqcap \forall U.C_\mathcal{T}(a) \mid a \in U_\mathcal{A}\}$ is consistent, where $U_\mathcal{A}$ is a set of all individuals occurring in $\mathcal{A}$.*

Since a reasoning problem w.r.t an ABoxe and an acyclic TBoxe can be reduced to the same reasoning problem w.r.t. the empty TBox. So in this paper, we mainly consider reasoning problems w.r.t. ABoxes.

3 Paradoxical Description Logic $\mathcal{ALC}$

In this section, we present the syntax and semantics of the paradoxical description logic $\mathcal{ALC}$, named $\mathcal{ALC}_{LP}$.

3.1 Syntax and Semantics

Syntactically, $\mathcal{ALC}_{LP}$ hardly differs from $\mathcal{ALC}$. In $\mathcal{ALC}_{LP}$, complex concepts and assertions are defined exactly in the same way. In the following, let $\mathcal{L}$ be the language of $\mathcal{ALC}_{LP}$.

Semantically, *paradoxical interpretations* map individuals to elements of the domain of the interpretation, as usual. However, as for concepts, to allow for reasoning with inconsistencies, a paradoxical interpretation over domain Δ^I assigns to each concept C a pair $\langle +C, -C\rangle$ of (not necessarily disjoint) subsets of Δ^I and $+C \cup -C = \Delta^I$.

Intuitively, $+C$ is the set of elements which are known to belong to the extension of C, while $-C$ is the set of elements which are known to be not contained in the extension of C. $+C$ and $-C$ are not necessarily disjoint but mutual complemental w.r.t. the domain. In this case, we do not consider that *incomplete information* since it is not valuable for users but a statement for insufficient information. Under the semantics of $\mathcal{ALC}_{LP}$, three situations which may occur in terms of containment of an individual in a concept are considered as follows:

- we know it is contained;
- we know it is not contained;
- we have contradictory information, i.e., the individual is both contained in the concept and not contained in the concept.

There are several equivalent ways of how this intuition can be formalized, one of which is described in the following.

Formally, a paradoxical interpretation is a pair $I = (\Delta^I, \cdot^I)$ with Δ^I as domain, where $\cdot^I$ is a function assigning elements of Δ^I to individuals, and subsets of $(\Delta^I)^2$

Table 2. Semantics of $\mathcal{ALC}_{LP}$ Concepts

Constructor Syntax	Semantics
A	$A^I = \langle +A, -A \rangle$, where $+A, -A \subseteq \Delta^I$
R	$R^I = \langle +R, -R \rangle$, where $+R, -R \subseteq \Delta^I \times \Delta^I$
o	$o^I \in \Delta^I$
$\top$	$\langle \Delta^I, \emptyset \rangle$
$\bot$	$\langle \emptyset, \Delta^I \rangle$
$C_1 \sqcap C_2$	$\langle +C_1 \cap +C_2, -C_1 \cup -C_2 \rangle$, if $C_i^I = \langle +C_i, -C_i \rangle$ for $i = 1, 2$
$C_1 \sqcup C_2$	$\langle +C_1 \cup +C_2, -C_1 \cap -C_2 \rangle$, if $C_i^I = \langle +C_i, -C_i \rangle$ for $i = 1, 2$
$\neg C$	$(\neg C)^I = \langle -C, +C \rangle$, if $C^I = \langle +C, -C \rangle$
$\exists R.C$	$\langle \{x \mid \exists y, (x, y) \in \text{proj}^+(R^I) \text{ and } y \in \text{proj}^+(C^I)\}$, $\{x \mid \forall y, (x, y) \in \text{proj}^+(R^I) \text{ implies } y \in \text{proj}^-(C^I)\} \rangle$
$\forall R.C$	$\langle \{x \mid \forall y, (x, y) \in \text{proj}^+(R^I) \text{ implies } y \in \text{proj}^+(C^I)\}$, $\{x \mid \exists y, (x, y) \in \text{proj}^+(R^I) \text{ and } y \in \text{proj}^-(C^I)\} \rangle$

to concepts, so that the conditions in Table 2 are satisfied, where functions $\text{proj}^+(\cdot)$ and $\text{proj}^-(\cdot)$ are defined by $\text{proj}^+\langle +C, -C \rangle = +C$ *and* $\text{proj}^-\langle +C, -C \rangle = -C$. Table 2 for role restrictions are designed in such a way that the semantic equivalences $\neg(\forall R.C) = \exists R.(\neg C)$ and $\neg(\exists R.C) = \forall R.(\neg C)$ are retained – this is the most convenient way for us to handle role restrictions, as it will allow for a straightforward translation from $\mathcal{ALC}_{LP}$ to classical $\mathcal{ALC}$.

As for roles, actually we require only the positive part of the extension since $\neg R$ is not a constructor in $\mathcal{ALC}$, i.e., $\text{proj}^+(R^I) \cap \text{proj}^-(R^I) = \emptyset$ for any $\mathcal{ALC}$ role R. However, $\neg R$ is a constructor in OWL 2[5], such as $\mathcal{EL}$ [28]. Thus, inconsistencies may occur in roles in OWL 2. For instance, an $\mathcal{EL}$ ontology tells that $Mike$ does not like $Jane$, i.e., $\neg like(Mike, Jane)$ while it is inferred that $Mike$ likes $Jane$, i.e., $like(Mike, Jane)$. That is, $\text{proj}^+(R^I) \cap \text{proj}^-(R^I) = \{(Mike, Jane)\}$. This example shows that, to deal with inconsistency, an interpretation should assign a pair of sets to a role. For a given domain Δ^I and a concept C, under the constraints $+C \cap -C = \emptyset$, paradoxical interpretations just become standard two-valued interpretations.

The correspondence between truth values from $\{t, f, B\}$ and concept extensions can be observed easily: for an individual $a \in \Delta^I$ and a concept name C, we have that

- $C^I(a) = t$, if and only if $a^I \in \text{proj}^+(C^I)$ and $a^I \notin \text{proj}^-(C^I)$;
- $C^I(a) = f$, if and only if $a^I \notin \text{proj}^+(C^I)$ and $a^I \in \text{proj}^-(C^I)$;
- $C^I(a) = B$, if and only if $a^I \in \text{proj}^+(C^I)$ and $a^I \in \text{proj}^-(C^I)$.

The paraconsistent semantics defined above ensures that a number of useful equivalences from classical DLs are still valid, including *double negation law* and *De Morgan Law*. The following theorem summarizes these desirable characteristics.

[5] http://www.w3.org/TR/owl2-profiles/

Theorem 1. *For any paradoxical interpretation I, concepts C, D and a role R in $\mathcal{ALC}_{LP}$, the following claims hold.*

(1) $(C \sqcap \top)^I = C^I$	(2) $(C \sqcup \top)^I = \top^I$
(3) $(C \sqcap \bot)^I = \bot^I$	(4) $(C \sqcup \bot)^I = C^I$
(5) $(\neg\top)^I = \bot^I$	(6) $(\neg\bot)^I = \top^I$
(7) $(\neg(C \sqcup D))^I = (\neg C \sqcap \neg D)^I$	(8) $(\neg(C \sqcap D))^I = (\neg C \sqcup \neg D)^I$
(9) $(\neg(\forall R.C))^I = (\exists R.\neg C)^I$	(10) $(\neg(\exists R.C))^I = (\forall R.\neg C)^I$
(11) $(\neg\neg C)^I = C^I$	(12) $proj^+((C \sqcup \neg C)^I) = proj^+(\top^I)$

From (12) in Theorem 1, we can see that $(C\sqcup\neg C)^I$ may be different from $\top^I$ for some concepts C. For instance, if $\Delta^I = \{a, b, c\}$ and $C^I = \langle\{a, b\}, \{b, c\}\rangle$ then $(C\sqcup\neg C)^I = \langle\{a, b, c\}, \{b\}\rangle$ while $\top^I = \langle\{a, b, c\}, \emptyset\rangle$, i.e., $(C\sqcup\neg C)^I \neq \top^I$. Similarly, we conclude that $(C \sqcap \neg C)^I \neq \bot^I$.

By Theorem 1, we can equivocally transform the negation of a complex concept into its *negation normal form* (NNF) which negation ($\neg$) only occurs on front of concept names in it.

Having defined paradoxical interpretations, we now introduce the *paradoxical satisfaction* relation.

Definition 1. *The* ***paradoxical satisfaction****, denoted $\models_{LP}$, is a binary relation between a set of paradoxical interpretations and a set of axioms defined as follows:*

For any paradoxical interpretation I,

(1) *$I \models_{LP} C(a)$ if and only if $a^I \in +C$, where $C^I = \langle +C, -C\rangle$;*
(2) *$I \models_{LP} R(a, b)$ if and only if $(a^I, b^I) \in +R$, where $R^I = \langle +R, -R\rangle$;*
(3) *$I \models_{LP} C_1 \sqsubseteq C_2$ if and only if $(\Delta^I \backslash -C_1) \subseteq +C_2$, for $i = 1, 2$, $C_i^I = \langle +C_i, -C_i\rangle$.*

where C, C_1, C_2 are concepts, R a role and a an individual in $\mathcal{ALC}$.

Note that I paradoxically satisfies a concept assertion $C(a)$ if and only if a^I is known to be an member of C^I; I paradoxically satisfies a role assertion $R(a, b)$ if and only if it is known that a^I is related to b by the binary relation R^I; and, I paradoxically satisfies a concept inclusion $C \sqsubseteq D$ if and only if every member of C^I is a member of D^I. Thus, concept C and its negation $\neg C$ are taken as two different concepts under paradoxical satisfaction. Based on the paradoxical satisfaction of concept inclusions, when there is an inconsistency with one concept, it will not be propagated to other concepts through axioms. For instance, let $\mathcal{T} = \{A \sqsubseteq B, A \sqsubseteq \neg B, B \sqsubseteq C\}$ be a TBox, I a paradoxical interpretation such that $I \models_{LP} A \sqsubseteq B$, $I \models_{LP} A \sqsubseteq \neg B$ and $I \models_{LP} B \sqsubseteq C$. Thus $\text{proj}^+(B^I) \cap \text{proj}^-(B^I) = \text{proj}^+(A^I)$. Then $\Delta^I \backslash \text{proj}^-(A^I) \not\subseteq \text{proj}^+(C^I)$. Therefore, $I \not\models_{LP} A \sqsubseteq C$. Intuitively, because concept B is unsatisfiable w.r.t. $\mathcal{T}$, i.e., inconsistencies would occur in any ontologies by integrating $\mathcal{T}$ with some ABox. Fortunately, inconsistencies are not transmitted into more general concepts.

In addition, the following result asserts that the *intuitive equivalence* is valid in $\mathcal{ALC}_{LP}$.

Theorem 2. *Let C_1, C_2 be concepts and I a paradoxical interpretation in $\mathcal{ALC}_{LP}$.*

$$I \models_{LP} C_1 \sqsubseteq C_2 \text{ if and only if } I \models_{LP} \neg C_1 \sqcup C_2(a) \text{ for every individual } a.$$

Theorem 2 in company with Lemma 1 ensures that a reasoning problem w.r.t. an ABox and a TBox can be equivalently transformed into the same reasoning problem w.r.t. an ABox (i.e. the TBox is empty).

In $\mathcal{ALC}_{LP}$, a paradoxical interpretation I *paradoxically satisfies* an ABox $\mathcal{A}$ (i.e., a paradoxical model of it), denoted by $I \models_{LP} \mathcal{A}$, if and only if it satisfies each assertion in $\mathcal{A}$; I *paradoxically satisfies* a TBox $\mathcal{T}$, denoted by $I \models_{LP} \mathcal{T}$, if and only if it satisfies each inclusion axiom in $\mathcal{T}$; I *paradoxically satisfies* an ontology $\mathcal{O}$, denoted by $I \models_{LP} \mathcal{O}$, if and only if it satisfies its ABox and its TBox. An ontology $\mathcal{O}$ is *paradoxically consistent (inconsistent)* if and only if there exists (does not exist) such a paradoxical model. In the following, we cut out "*paradoxically*" or "*paradoxical*" from terms for simplicity if no confusion occurs.

The following theorem shows that any classical ontology is paradoxically consistent, that is, contradictions are tolerated under the semantics of $\mathcal{ALC}_{LP}$.

Theorem 3. *There exists a paradoxical model of $\mathcal{O}$ for any ontology $\mathcal{O}$ in $\mathcal{ALC}$.*

Next, we introduce the *paradoxical entailment* relation between ontologies and axioms based on paradoxical models.

Definition 2. *Let $\mathcal{O}$ be an ontology and ϕ be an axiom in $\mathcal{ALC}_{LP}$. We say $\mathcal{O}$* ***paradoxically entails*** *ϕ, denoted $\mathcal{O} \models_{LP} \phi$, if and only if for every paradoxical model I of $\mathcal{O}$, $I \models_{LP} \phi$.*

The following example shows that the paradoxical entailment does not satisfy the *ex contradictione quodlibet*, i.e., $\mathcal{ALC}_{LP}$ is paraconsistent.

Example 2. *Let A be a concept name and a an individual in $\mathcal{ALC}_{LP}$. Then $\{A(a), \neg A(a)\} \models_{LP} A(a)$ and $\{A(a), \neg A(a)\} \models_{LP} \neg A(a)$. Let I be a paradoxical interpretation such that $A^I(a) = B$ and $A'^I(a) = f$ where b is an individual and A' a concept name. That is, I is a paradoxical model of $\{A(a), \neg A(a)\}$ but I is not a paradoxical model of $A'(b)$. Then $\{A(a), \neg A(a)\} \not\models_{LP} A'(b)$.*

In the following, we build the relationship between paradoxical interpretations and classical interpretations in $\mathcal{ALC}_{LP}$ by introducing the definition of *isomorphic map*.

Definition 3. *Let I be a paradoxical interpretation and I_c be a classical interpretation. The* ***isomorphic map*** *transforming I into I_c is defined as follows: for each axiom ϕ,*

$$\begin{cases} I_c \models \phi, \text{ if and only if } I \models_{LP} \phi; \\ I_c \not\models \phi, \text{ if and only if } I \not\models_{LP} \phi. \end{cases}$$

By Definition 3, we conclude that for any paradoxical interpretation I, there exists a classical interpretation I_c such that the isomorphic map (defined in Definition 3) can transform I into I_c.

It is a pleasing property that reasoning in $\mathcal{ALC}_{LP}$ can be reduced easily into reasoning in the standard $\mathcal{ALC}$. Thus, reasoning algorithms in $\mathcal{ALC}$ can be used to implement paraconsistent reasoning in $\mathcal{ALC}_{LP}$.

In general, the reasoning power of $\mathcal{ALC}_{LP}$ is weaker than that of $\mathcal{ALC}$.

Theorem 4. *Let $\mathcal{O}$ be an ontology and ϕ an axiom in $\mathcal{ALC}$.*

$$\text{If } \mathcal{O} \models_{LP} \phi \text{ then } \mathcal{O} \models \phi\text{; but not vice versa.}$$

However, when ontologies are classically consistent, $\mathcal{ALC}_{LP}$ has the same inference power as $\mathcal{ALC}$.

Theorem 5. *Let $\mathcal{O}$ be a classically consistent ontology and ϕ an axiom in $\mathcal{ALC}$.*

$$\mathcal{O} \models_{LP} \phi \text{ if and only if } \mathcal{O} \models \phi$$

The next result shows that the set of tautologies in $\mathcal{ALC}_{LP}$ is exactly the set of tautologies in $\mathcal{ALC}$.

Theorem 6. *Let ϕ be an axiom in $\mathcal{ALC}$.*

$$\models_{LP} \phi \text{ if and only if } \models \phi$$

The *excluded middle rule* is also valid in $\mathcal{ALC}_{LP}$.

Corollary 1. $\models_{LP} C \sqcup \neg C(a)$ *for any concept C and individual a in $\mathcal{ALC}$.*

By Corollary 1, we conclude that the semantics of $\mathcal{ALC}_{LP}$ is different from the semantics of relevance logic.

3.2 Inference Problems

The *instance checking* and *subsumption checking* are two important inference problems in $\mathcal{ALC}$. In this section, we look at these two problem in the setting of $\mathcal{ALC}_{LP}$.

First, we give their definition as follows:

- **Instance Checking:** an individual a is a *paradoxical instance* of a concept C w.r.t. an ontology $\mathcal{O}$ if $\mathcal{O} \models_{LP} C(a)$;
- **Subsumption Checking:** a concept C is *paradoxically subsumed* by a concept D w.r.t. a TBox $\mathcal{T}$ if $\mathcal{T} \models_{LP} C \sqsubseteq D$.

An important result is that Lemma 2 still holds in $\mathcal{ALC}_{LP}$.

Corollary 2. *Let C, D be concepts, $\mathcal{A}$ an ABox and $\mathcal{T}$ an acyclic TBox in $\mathcal{ALC}_{LP}$. Then we have*

(1) *C is paradoxically satisfiable w.r.t. $\mathcal{T}$ if and only if $C \sqcap C_{\mathcal{T}} \sqcap \forall U.C_{\mathcal{T}}$ is paradoxically satisfiable;*

(2) *D paradoxically subsumes C w.r.t. $\mathcal{T}$ if and only if $C \sqcap \neg D \sqcap C_{\mathcal{T}} \sqcap \forall U.C_{\mathcal{T}}$ is paradoxically unsatisfiable;*

(3) *$\mathcal{A}$ is paradoxically consistent w.r.t. $\mathcal{T}$ if and only if $\mathcal{A} \cup \{C_{\mathcal{T}} \sqcap \forall U.C_{\mathcal{T}}(a) \mid a \in U_{\mathcal{A}}\}$ is paradoxically consistent, where $U_{\mathcal{A}}$ is a set of all individuals occurring in $\mathcal{A}$.*

By Corollary 2, the problem of subsumption checking can be reduced to the problem of instance checking and the problem of instance checking w.r.t. ABoxes and TBoxes can be reduced to the problem of instance checking w.r.t. only ABoxes. Thus, in the rest of the paper we will mainly consider these reasoning problems w.r.t. ABoxes.

Example 3. *Let $\mathcal{T}$ be a TBox and let C, D, and E be concepts in $\mathcal{ALC}_{LP}$. By Definition 1 and Definition 2, if D is classical satisfiable w.r.t. $\mathcal{T}$, that is, for any paradoxical interpretation I of $\mathcal{T}$, $proj^+(D^I) \cap proj^-(D^I) = \emptyset$, then $\{C \sqsubseteq D, D \sqsubseteq E\} \models_{LP} C \sqsubseteq E$; if D is classical unsatisfiable w.r.t. $\mathcal{T}$, we have $\{C \sqsubseteq D, D \sqsubseteq E\} \not\models_{LP} C \sqsubseteq E$.*

From Example 3, we can see that concept inclusions do not satisfy the transitivity in $\mathcal{ALC}_{LP}$ when TBoxes contain unsatisfiable concepts. This implies that inconsistencies are not transited in $\mathcal{ALC}_{LP}$. Because of this, the principle of reasoning with inconsistent ontologies here is reasonable in the real world since the negative influence arising from inconsistencies is minimized or localized to some extent.

4 Signed Tableau Algorithm

In this section, we discuss the problem of instance checking and provide a sound and complete tableau for $\mathcal{ALC}_{LP}$ by extending the signed tableau for LP developed by Lin [27] to $\mathcal{ALC}_{LP}$. As explained before, we assume that the TBoxe of an ontology is empty.

Let $\mathcal{A}$ be an ABox and ϕ a concept assertion in $\mathcal{ALC}_{LP}$. Under the paradoxical semantics of $\mathcal{ALC}_{LP}$, an axiom can be both true and false and triviality of classical $\mathcal{ALC}$ is destroyed. There are two approaches to destroying the triviality of proofs. One approach is to restrict the rule of *reductio ad absurdum*: $\mathcal{A}$ paradoxically entails ϕ if and only if we can infer a contradiction from $\mathcal{A}$ and $\neg\phi$ by using relevant information from ϕ. With this idea of using relevant information from ϕ, we can formulate the proof procedure by retaining the tableau rules for classical logic but modifying closedness conditions of the tableau to get rid of the triviality. The other approach to formulating the restricted *reduction* to *absurdity* is using *signed tableau* as we will present in the following.

We first introduce two symbols $\mathbf{T}$ (for *true*) and $\mathbf{F}$ (for *false*) in $\mathcal{L}$. A *signed axiom* has the form $\mathbf{T}\phi$ (called $\mathbf{T}$-axiom) or $\mathbf{F}\phi$ (called $\mathbf{F}$-axiom), where $\phi \in \mathcal{L}$. A paradoxical interpretation I is extended to a signed axiom as follows:

- $I \models_{LP} \mathbf{T}\phi$ if and only if $I \models_{LP} \phi$;
- $I \models_{LP} \mathbf{F}\phi$ if and only if $I \models_{LP} \neg\phi$.

Let $\mathcal{L}^* = \{\mathbf{T}\phi, \mathbf{F}\phi \mid \phi \in \mathcal{L}\}$. In this section, we mainly consider our tableaus based on $\mathcal{L}^*$. Let $\mathbf{T}\mathcal{A} = \{\mathbf{T}\psi \mid \psi \in \mathcal{A}\}$. A *signed literal* is a signed axiom in which the objective axiom of symbol $\mathbf{T}$ or $\mathbf{F}$ is a literal. A *signed ABox* is a set of signed concept assertions and signed role assertions. A *signed tableau* is a tree whose branches are actually sequences of signed axioms. A signed axiom is in NNF if negation ($\neg$) only occurs in front of concept names occurring it.

Definition 4. *Let $\mathcal{A}$ be a signed ABox, C_1, C_2, C concepts, R a role and x, y, z individuals in $\mathcal{ALC}_{LP}$. The* ***signed expansion rules*** *are defined in Table 3, where* $\mathbf{S}$ *is the place-holder of* $\mathbf{T}$ *and* $\mathbf{F}$.

In Table 3, nine signed expansion rules are introduced to construct a signed tableau for a given ABox. The first four rules similar to expansion rules of classical tableau

Table 3. Signed Expansion Rules in $\mathcal{ALC}$

$\sqcap_S$-rule	Condition: $\mathcal{A}$ contains $\mathbf{S}C_1 \sqcap C_2(x)$, but not both $\mathbf{S}C_1(x)$ and $\mathbf{S}C_2(x)$. Action: $\mathcal{A}' := \mathcal{A} \cup \{\mathbf{S}C_1(x), \mathbf{S}C_2(x)\}$.
$\sqcup_S$-rule	Condition: $\mathcal{A}$ contains $\mathbf{S}C_1 \sqcup C_2(x)$, but neither $\mathbf{S}C_i(x)$ for $i = 1, 2$. Action: $\mathcal{A}' := \mathcal{A} \cup \{\mathbf{S}C_1(x)\}$, $\mathcal{A}'' := \mathcal{A} \cup \{\mathbf{S}C_2(x)\}$.
$\exists_S$-rule	Condition: $\mathcal{A}$ contains $\mathbf{S}\exists R.C(x)$, but there is no individual name z such that $\mathbf{S}C(z)$ and $\mathbf{T}R(x, z)$ are in $\mathcal{A}$. Action: $\mathcal{A}' := \mathcal{A} \cup \{\mathbf{S}C(y), \mathbf{T}R(x, y)\}$ where y is an individual name not occurring in $\mathcal{A}$.
$\forall_S$-rule	Condition: $\mathcal{A}$ contains $\mathbf{S}\forall R.C(x)$ and $\mathbf{T}R(x, y)$, but is does not contain $\mathbf{S}C(y)$. Action: $\mathcal{A}' := \mathcal{A} \cup \{\mathbf{S}C(y)\}$.
$\neg\neg_S$-rule	Condition: $\mathcal{A}$ contains $\mathbf{S}\neg\neg C(x)$, but not $\mathbf{S}C(x)$. Action: $\mathcal{A}' := \mathcal{A} \cup \{\mathbf{S}C(x)\}$.
$\neg\sqcap_S$-rule	Condition: $\mathcal{A}$ contains $\mathbf{S}\neg(C_1 \sqcap C_2)(x)$, but neither $\mathbf{S}\neg C_i(x)$ for $i = 1, 2.$. Action: $\mathcal{A}' := \mathcal{A} \cup \{\mathbf{S}\neg C_1(x)\}$, $\mathcal{A}'' := \mathcal{A} \cup \{\mathbf{S}\neg C_2(x)\}$.
$\neg\sqcup_S$-rule	Condition: $\mathcal{A}$ contains $\mathbf{S}\neg(C_1 \sqcup C_2)(x)$, but not both $\mathbf{S}\neg C_1(x)$ and $\mathbf{S}\neg C_2(x)$. Action: $\mathcal{A}' := \mathcal{A} \cup \{\mathbf{S}C_1(x), \mathbf{S}C_2(x)\}$.
$\neg\exists_S$-rule	Condition: $\mathcal{A}$ contains $\mathbf{S}\neg\exists R.C(x)$ and $\mathbf{T}R(x, y)$, but is does not contain $\mathbf{S}\neg C(y)$. Action: $\mathcal{A}' := \mathcal{A} \cup \{\mathbf{S}\neg C(y)\}$.
$\neg\forall_S$-rule	Condition: $\mathcal{A}$ contains $\mathbf{S}\neg\forall R.C(x)$, but there is no individual name z such that $\mathbf{S}\neg C(z)$ and $\mathbf{T}R(x, z)$ are in $\mathcal{A}$. Action: $\mathcal{A}' := \mathcal{A} \cup \{\mathbf{S}\neg C(y), \mathbf{T}R(x, y)\}$ where y is an individual name not occurring in $\mathcal{A}$.

for $\mathcal{ALC}$-ABoxes are applied to expand the tableau. Furthermore, the last five rules are applied to transform axioms into equivalent ones in NNF by pushing negations inwards using a combination of *De Morgan's laws* and some equivalences in Theorem 1.

Definition 5. *Let $\mathcal{A}$ be a signed ABox. A* ***signed tableau*** *for $\mathcal{A}$ is defined inductively according to the signed expansion rules in Table 3.*

A signed tableau can be obtained from an initial signed ABox by applying the nine rules in Table 3 repeatedly. We say that a signed axiom is *marked* when a signed expansion rule is used up in it; otherwise, a signed axiom is called *non-marked*. In the following, we define the complete signed tableau. Initially, we transform a given $\mathcal{ALC}$ ABox $\mathcal{A}$ and a query ϕ into a signed $\mathcal{ALC}$ ABox $\mathcal{A}'$ by computing the union of $\{\mathbf{T}\psi \mid \psi \in \mathcal{A}\}$ and $\{\mathbf{F}(\phi)\}$.

Definition 6. *A branch* b *of a signed tableau $\mathbb{T}$ for $\mathcal{A}$ is* ***complete*** *if and only if all non-marked axioms in* b *are signed literals, that is to say, every signed expansion rule that can be used to extend* b *has been applied at least once. A signed tableau $\mathbb{T}$ for $\mathcal{A}$ is a* ***complete tableau*** *if and only if every branch of $\mathbb{T}$ is complete.*

In the following, we define the closed signed tableau by modifying the closed conditions.

Definition 7. *A branch* b *of a signed tableau $\mathbb{T}$ is* ***closed*** *if and only if one of the following two conditions is satisfied:*

(1) $\mathbf{T}A(a) \in \mathbf{b}$ *and* $\mathbf{F}A(a) \in \mathbf{b}$*; or*
(2) $\mathbf{F}A(a) \in \mathbf{b}$ *and* $\mathbf{F}\neg A(a) \in \mathbf{b}$*;*

where A is a concept name and a is an individual.

A signed tableau $\mathbb{T}$ is ***closed*** *if and only if every branch of $\mathbb{T}$ is closed.*

By Definition 7, the condition $\{\mathbf{T}A(a), \mathbf{T}\neg A(a)\} \subseteq \mathbf{b}$ of closedness in signed tableaux is no longer taken as one of closed conditions since the kind of closedness is brought about because of inconsistencies in ABoxes. In short, the closed condition of signed tableaux for $\mathcal{ALC}$-ABoxes is different from that of classical tableau for $\mathcal{ALC}$-ABoxes.

Compared to classical tableau for DLs, if there exists a branch containing $\mathbf{T}A(a)$ and $\mathbf{T}\neg A(a)$ is closed, then the tableau is *refutation complete* for classical $\mathcal{ALC}$ in the sense that for any signed ABox $\mathcal{A}$, $\mathcal{A}$ is inconsistent if and only if there exists a signed tableau for $\mathcal{A}$ such that every branch of the tableau is closed. Therefore, according to the rule of *reductio ad absurdum*, $\mathcal{A}$ paradoxically entails ϕ if and only if every branch of the signed tableau for $\mathbf{T}\mathcal{A} \cup \{\mathbf{F}\phi\}$ is closed.

Definition 8. *$\mathbb{T}$ is a* ***signed tableau*** *for $\mathbf{T}\mathcal{A}$ and $\mathbf{F}\phi$ if the root node of $\mathbb{T}$ contains only the set $\mathbf{T}\mathcal{A}\cup\{\mathbf{F}\phi\}$. We denote $\mathcal{A} \vdash_{LP} \phi$ if and only if the signed tableau for $\mathbf{T}\mathcal{A}\cup\{\mathbf{F}\phi\}$ are closed.*

As mentioned earlier, the first approach to obtaining a paraconsistent logic is by restricting the rule of *reductio ad absurdum*. In $\mathcal{ALC}_{LP}$, we only need to weaken the closure conditions as defined above. This leads to the following soundness and completeness theorem of signed tableau w.r.t. the semantics of $\mathcal{ALC}_{LP}$.

Theorem 7. *Let $\mathcal{A}$ be an ABox, C a concept and a an individual in $\mathcal{ALC}_{LP}$.*

$$\mathcal{A} \vdash_{LP} C(a) \text{ if and only if } \mathcal{A} \models_{LP} C(a).$$

The following example illustrates the soundness and completeness in Theorem 7.

Example 4. *In Example 2, let $\mathcal{A} = \{A(a), \neg A(a)\}$. It can be easily checked that $\{A(a), \neg A(a)\} \vdash_{LP} A(a)$, since the signed tableau for $\mathbf{T}\mathcal{A} \cup \{\mathbf{F}(\neg A(a))\}$ has only one branch $\{\mathbf{T}A(a),\ \mathbf{T}\neg A(a),\ \mathbf{F}\neg A(a)\}$ which is closed. However, $\{A(a), \neg A(a)\} \nvdash_{LP} B(a)$, since the signed tableau for $\mathbf{T}\mathcal{A} \cup \{\mathbf{F}\neg B(a)\}$ has only a branch $\{\mathbf{T}A(a), \mathbf{T}\neg A(a),\ \mathbf{F}\neg B(a)\}$ which is not closed.*

Unfortunately, the following example shows the fact that *disjunctive syllogism* is invalid in $\mathcal{ALC}_{LP}$.

Example 5. *Let A, B be concepts and a an individual. It is easy to check that $\{A(a), \neg A \sqcup B(a)\} \nvdash_{LP} B(a)$, since the signed tableau for $\{\mathbf{T}A(a), \mathbf{T}\neg A \sqcup B(a), \mathbf{F}\neg B(a)\}$ has only a branch $\{\mathbf{T}A(a), \mathbf{T}B(a), \mathbf{F}\neg B(a)\}$ which is closed, while $\{\mathbf{T}A(a), \mathbf{T}\neg A(a),\ \mathbf{F}\neg B(a)\}$ is not closed. Then, $\{A(a), \neg A \sqcup B(a)\} \not\models_{LP} B(a)$ by Theorem 7.*

By Theorem 5, *disjunctive syllogism* is still valid in classical consistent ontologies. No new inference can be drawn from those possible inconsistent knowledge by using *disjunctive syllogism*. Therefore, in this paper, the paraconsistent reasoning mechanism in

$\mathcal{ALC}$ is "*brave*" under certain conditions. In particular, when we do not know which knowledge causes inconsistencies in an inconsistent ontologies, our reasoning mechanism becomes "*cautious*". Intuitively, inferences following from an ontology, no matter whether it is inconsistent, are rather rational.

5 Related Work

In this section, we mainly compare $\mathcal{ALC}_{LP}$ with $\mathcal{ALC}4$ presented in [21] which is paraconsistent by $\mathcal{ALC}$ integrating with Belanp's four-valued logics [5]. The similarities and differences between LP [6], three-valued logics [11] and four-valued logics still exist between $\mathcal{ALC}_{LP}$ and $\mathcal{ALC}4$.

In $\mathcal{ALC}4$, a 4-interpretation over domain Δ^I assigns a pair $\langle P, N\rangle$ each concept C, where P and N are of (neither necessarily disjoint nor necessarily mutual complement) subsets of Δ^I. The correspondence between truth values from $\{t, f, B, U\}$ and concept extensions is the obvious one: for instances $a \in \Delta^I$ and concept name C we have

- $C^I(a) = t$, if and only if $a^I \in \text{proj}^+(C^I)$ and $a^I \notin \text{proj}^-(C^I)$;
- $C^I(a) = f$, if and only if $a^I \notin \text{proj}^+(C^I)$ and $a^I \in \text{proj}^-(C^I)$;
- $C^I(a) = B$, if and only if $a^I \in \text{proj}^+(C^I)$ and $a^I \in \text{proj}^-(C^I)$;
- $C^I(a) = U$, if and only if $a^I \notin \text{proj}^+(C^I)$ and $a^I \notin \text{proj}^-(C^I)$.

In $\mathcal{ALC}4$, a 4-model of an axiom is defined as follows.

Definition 9 ([21]). *Let I be a 4-interpretation. $\models_4$ denotes the four-valued satisfaction relationship in $\mathcal{ALC}4$.*

- concept assertion*: $I \models_4 C(a)$ iff $a^I \in proj^+(C^I)$;*
- material inclusion*: $I \models_4 C \mapsto D$ iff $\overline{proj^-(C^I)} \subseteq proj^+(D^I)$;*
- internal inclusion*: $I \models_4 C \sqsubset D$ iff $proj^+(C^I) \subseteq proj^+(D^I)$;*
- strong inclusion*: $I \models_4 C \rightarrow D$ iff $proj^+(C^I) \subseteq proj^+(D^I)$ and $proj^-(D^I) \subseteq proj^-(C^I)$.*

where C, D are concepts and a is an individual in $\mathcal{ALC}4$.

Definition 10 ([21]). *Let $\mathcal{O}$ be an ontology and ϕ be an axiom in $\mathcal{ALC}4$. ϕ is a* ***4-valued entailment*** *of $\mathcal{O}$, denoted by $\mathcal{O} \models_4 \phi$, if and only if for any 4-model I of $\mathcal{O}$, I is a 4-model of ϕ.*

The following theorem directly follows from Definition 1, Definition 2, Definition 9 and Definition 10.

Theorem 8. *Let $\mathcal{O}$ be an ontology, C, D concepts and a an individual in $\mathcal{ALC}$. Then*

(1) *If $\mathcal{O} \models_4 C(a)$ then $\mathcal{O} \models_{LP} C(a)$.*
(2) *If $\mathcal{O} \models_4 C \mapsto D$ then $\mathcal{O} \models_{LP} C \sqsubseteq D$*

Theorem 8 shows that $\mathcal{ALC}_{LP}$ has stronger inference power than $\mathcal{ALC}4$.

On the other hand, compared with the semantics of $\mathcal{ALC}4$, the incomplete information can also be expressed in $\mathcal{ALC}_{LP}$. For instance, let $\mathcal{A} = \{C(a), \neg C(a), C \sqcup$

$D(a), \neg C \sqcup \neg D(a)\}$. For any 4-model I_4 of $\mathcal{A}$, $D^{I_4}(a) = U$. In fact, $\mathcal{A} \not\models_{LP} D(a)$ and $\mathcal{A} \not\models_{LP} \neg D(a)$.

Another paraconsistent DL is QC $\mathcal{ALC}$, which is proposed in [22] by adapting Besnard and Hunter's quasi-classical logics [8] to $\mathcal{ALC}$. Because the QC semantics is based on the four-valued semantics, the differences between the paradoxical semantics and the four-valued semantics still retain between the paradoxical semantics and the QC semantics. Beside those differences, there are at least three differences between paradoxical DLs and QC DLs as follows.

- The QC semantics is characterized by two satisfaction relationships, namely, QC strong satisfaction and QC weak satisfaction. The QC weak satisfaction is tolerating inconsistencies while the QC strong satisfaction is enhancing the inference power. Furthermore, the concept inclusions (subsumptions) are differently defined under the two satisfaction relationships in order to capture the intuitive equivalence between concept inclusions and disjunction of concepts in QC $\mathcal{ALC}$. Compared with the QC semantics, it is intuitive that the paradoxical semantics is characterized by only one satisfaction (paradoxical satisfaction) and the concept inclusions is defined by the paradoxical satisfaction.
- In QC DLs [23], the tableau algorithm for ABoxes is obtained by modifying the disjunction expansion rules and changing closed conditions however the tableau algorithm for ABoxes (signed tableau algorithm) is obtained by only modifying the closed conditions in paradoxical DL $\mathcal{ALC}$. Thus, the signed tableau algorithm is more suitable for implementing paraconsistent reasoning.
- The *excluded middle rule* is valid in $\mathcal{ALC}_{LP}$ by Corollary 1. However, since QC $\mathcal{ALC}$ inherits properties of $\mathcal{ALC}_4$, the *excluded middle law* is invalid QC $\mathcal{ALC}$, that is, $\not\models_Q A \sqcup A(a)$ where A is a concept name and a is an individual (see Example 2 in [22]).

In summary, the paradoxical semantics is more suitable for dealing with inconsistent ontologies in that paraconsistent reasoning under our semantics provides a better approximation to the standard reasoning.

6 Conclusion and Further Work

In this paper, we have presented a non-standard DL $\mathcal{ALC}_{LP}$, which is a paraconsistent variant of $\mathcal{ALC}$ by adopting the semantics of LP. $\mathcal{ALC}_{LP}$ can be applied to formalize reasoning in the presence of inconsistency while still preserving some important inference rules such as the *excluded middle rule*. Our new logic $\mathcal{ALC}_{LP}$ is different from the 4-valued DL $\mathcal{ALC}4$ in that the *excluded middle rule* is not valid in $\mathcal{ALC}4$. Several useful properties of our logic is shown. We have also introduced a tableau for $\mathcal{ALC}_{LP}$ and shown that it is sound and complete. This tableau algorithm has been provided to implement instance checking in ABoxes. In the future, we will develop efficient implementation of $\mathcal{ALC}_{LP}$ by using existing efficient reasoners. It is also interesting to investigate the extension of our approach to other DLs, such as expressive DL $\mathcal{SHION}(\mathcal{D})$, which is the logical foundation of the Web Ontology Language (OWL) DL in Semantic Web, and tractable DLs, such as $\mathcal{EL}$++, Horn-DLs, and DL-Lite.

Acknowledgment

The authors would like to gratefully acknowledge the helpful comments and suggestions of three anonymous reviewers, which have improved the presentation. This work was supported by the Australia Research Council (ARC) Discovery Projects DP0666107.

References

1. Berners-Lee, T., Hendler, J., Lassila, O.: The semantic web. Scientific American (2001)
2. Bertossi, L., Hunter, A., Schaub, T. (eds.): Inconsistency Tolerance. LNCS, vol. 3300, pp. 237–269. Springer, Heidelberg (2005)
3. Schaffert, S., Bry, F., Besnard, P., Decker, H., Decker, S., Enguix, C.F., Herzig, A.: Paraconsistent reasoning for the semantic web. In: Proc. of the 4th International Semantic Web Conf. Workshop on Uncertainty Reasoning for the Semantic Web (ISWC-URSW 2005), Ireland, pp. 104–105 (2005)
4. Baader, F., Calvanese, D., McGuinness, D.L., Nardi, D., Patel-Schneider, P.F. (eds.): The Description Logic Handbook: Theory, Implementation, and Applications. Cambridge University Press, Cambridge (2003)
5. Belnap, N.D.: A useful four-valued logic. Modern uses of multiple-valued logics, 7–73 (1977)
6. Priest, G.: Logic of paradox. J. of Philosophical Logic 8, 219–241 (1979)
7. Priest, G.: Reasoning about truth. Artif. Intell. 39(2), 231–244 (1989)
8. Besnard, P., Hunter, A.: Quasi-classical logic: Non-trivializable classical reasoning from incosistent information. In: Froidevaux, C., Kohlas, J. (eds.) ECSQARU 1995. LNCS, vol. 946, pp. 44–51. Springer, Heidelberg (1995)
9. Elvang-Gøransson, M., Hunter, A.: Argumentative logics: Reasoning with classically inconsistent information. Data Knowl. Eng. 16(2), 125–145 (1995)
10. Benferhat, S., Dubois, D., Lang, J., Prade, H., Saffiotti, A., Smets, P.: A general approach for inconsistency handling and merging information in prioritized knowledge bases. In: Proc. of the 6th International Conf. on Principles of Knowledge Representation and Reasoning (KR 1998), Italy, pp. 466–477. Morgan Kaufmann, San Francisco (1998)
11. Konieczny, S., Marquis, P.: Three-valued logics for inconsistency handling. In: Flesca, S., Greco, S., Leone, N., Ianni, G. (eds.) JELIA 2002. LNCS (LNAI), vol. 2424, pp. 332–344. Springer, Heidelberg (2002)
12. Lang, J., Marquis, P.: Resolving inconsistencies by variable forgetting. In: Proc. of the 8th International Conf. on Principles of Knowledge Representation and Reasoning (KR 2002), France, pp. 239–250. Morgan Kaufmann, San Francisco (2002)
13. Coste-Marquis, S., Marquis, P.: Recovering consistency by forgetting inconsistency. In: Hölldobler, S., Lutz, C., Wansing, H. (eds.) JELIA 2008. LNCS (LNAI), vol. 5293, pp. 113–125. Springer, Heidelberg (2008)
14. Arieli, O.: Distance-based paraconsistent logics. Int. J. Approx. Reasoning 48(3), 766–783 (2008)
15. Huang, Z., van Harmelen, F., ten Teije, A.: Reasoning with inconsistent ontologies. In: Proc. of the 19th International Joint Conf. on Artificial Intelligence (IJCAI 2005), UK, pp. 454–459. Professional Book Center (2005)
16. Haase, P., van Harmelen, F., Huang, Z., Stuckenschmidt, H., Sure, Y.: A framework for handling inconsistency in changing ontologies. In: Gil, Y., Motta, E., Benjamins, V.R., Musen, M.A. (eds.) ISWC 2005. LNCS, vol. 3729, pp. 353–367. Springer, Heidelberg (2005)

17. Qi, G., Liu, W., Bell, D.A.: A revision-based approach to handling inconsistency in description logics. Artif. Intell. Rev. 26(1-2), 115–128 (2006)
18. Qi, G., Du, J.: Model-based revision operators for terminologies in description logics. In: Proc. of the 21st International Joint Conf. on Artificial Intelligence (IJCAI 2009), USA, pp. 891–897 (2009)
19. Patel-Schneider, P.F.: A four-valued semantics for terminological logics. Artif. Intell. 38(3), 319–351 (1989)
20. Schlobach, S., Cornet, R.: Non-standard reasoning services for the debugging of description logic terminologies. In: Proc. of the 18th International Joint Conf. on Artificial Intelligence (IJCAI 2003), Mexico, pp. 355–362. Morgan Kaufmann, San Francisco (2003)
21. Ma, Y., Hitzler, P., Lin, Z.: Algorithms for paraconsistent reasoning with OWL. In: Franconi, E., Kifer, M., May, W. (eds.) ESWC 2007. LNCS, vol. 4519, pp. 399–413. Springer, Heidelberg (2007)
22. Zhang, X., Lin, Z.: Paraconsistent reasoning with quasi-classical semantic in ALC. In: Calvanese, D., Lausen, G. (eds.) RR 2008. LNCS, vol. 5341, pp. 222–229. Springer, Heidelberg (2008)
23. Zhang, X., Xiao, G., Lin, Z.: A tableau algorithm for handling inconsistency in OWL. In: Aroyo, L., Traverso, P., Ciravegna, F., Cimiano, P., Heath, T., Hyvönen, E., Mizoguchi, R., Oren, E., Sabou, M., Simperl, E. (eds.) ESWC 2009, Greece. LNCS, vol. 5554, pp. 399–413. Springer, Heidelberg (2009)
24. Horrocks, I., Sattler, U., Tobies, S.: Reasoning with individuals for the description logic SHIQ. CoRR cs.LO/0005017 (2000)
25. Zhang, X., Zhang, Z., Lin, Z.: An argumentative semantics for paraconsistent reasoning in description logic ALC. In: Proc. of the 22nd International Workshop on Description Logics (DL 2009), UK. CEUR Workshop Proceedings 477, CEUR-WS.org (2009)
26. Lin, Z.: Three-valued nonmonotonic logic. In: Proc. of the 23rd IEEE International Symposium on Multiple-Valued Logic (ISMVL 1993), USA, pp. 42–47. IEEE Computer Society, Los Alamitos (1993)
27. Lin, Z.: Tableau systems for paraconsistency and minimal inconsistency. J. of Computer Sci. and Technol. 13(2), 174–188 (1998)
28. Baader, F., Peñaloza, R., Suntisrivaraporn, B.: Pinpointing in the description logic EL. In: Proc. of the 20th International Workshop on Description Logics (DL 2007), Italy. CEUR Workshop Proceedings 250, CEUR-WS.org (2007)
29. Baader, F., Horrocks, I., Sattler, U.: Description Logics. In: van Harmelen, F., Lifschitz, V., Porter, B. (eds.) Handbook of Knowledge Representation. Elsevier, Amsterdam (2007)
30. Schmidt-Schauß, M., Smolka, G.: Attributive concept descriptions with complements. Artif. Intell. 48(1), 1–26 (1991)

Appendix: Proof Sketches

Proof of Theorem 1
It is easy to check these properties by the definitions of complex concepts in Table 2.

Proof of Theorem 2
For any individual a in $\mathcal{ALC}_{LP}$, $a^I \in (\neg C_1 \sqcup C_2)^I$ if and only if $a^I \in \text{proj}^+((\neg C_1)^I)$ or $a^I \in \text{proj}^+(C_2^I)$ by Definition 1 if and only if $a^I \notin \text{proj}^+((\neg C_1)^I)$ implies $a^I \in \text{proj}^+(C_2^I)$, that is, $a^I \in \Delta^I \setminus \text{proj}^-(C_1^I)$ implies $a^I \in \text{proj}^+(C_2^I)$ if and only if $I \models_{LP} \neg C_1 \sqcup C_2(a)$.

Proof of Theorem 3

Let I be a paradoxical interpretation of $\mathcal{O}$ such that $\text{proj}^+(C^I) \cap \text{proj}^-(C^I) = \Delta^I$ and $R^I = \langle \Delta^I \times \Delta^I, \emptyset \times \emptyset \rangle$ for any concept C and role R occurring in $\mathcal{O}$. It easily shows that for any axiom $C(a)$, $R(a,b)$ and $C \sqsubseteq D$ of $\mathcal{O}$, $I \models_{LP} C(a)$, $I \models_{LP} R(a,b)$ and $I \models_{LP} C \sqsubseteq D$ respectively where C, D are concepts, R is a role and a, b are individual in $\mathcal{ALC}_{LP}$. Therefore, I is a paradoxical model of $\mathcal{O}$.

Proof of Theorem 4

If $\mathcal{O} \models_{LP} \phi$ then there exists a paradoxical interpretation I of $\mathcal{O} \cup \{\phi\}$. A classical interpretation I_c is obtained by constructing isomorphically a map $\nu : I \to I_c$ by Definition 3. It easily shows that $I_c \models \mathcal{O} \cup \{\phi\}$ by induction on the numbers of connectives occurring in ϕ and the definition of satisfiability. Therefore, $\mathcal{O} \models \phi$. On the other hand, let $\mathcal{O} = \{A(a), \neg A(a)\}$. $\mathcal{O} \models B(a)$ while $\mathcal{O} \not\models_{LP} B(a)$ by Example 2.

Proof of Theorem 5

For any paradoxical interpretation I of $\mathcal{O}$, $I \models_{LP} \phi$ if and only if $I \models \neg\phi$ since $\mathcal{O}$ is classical consistent, that is, a classical interpretation I_c, which is obtained by constructing isomorphically a map $\nu : I \to I_c$ by Definition 3, is I in fact. Therefore, $\mathcal{O} \models_{LPm} \phi$ if and only if $\mathcal{O} \models \phi$.

Proof of Theorem 6

($\Rightarrow$) If $\models_{LP} \phi$ then for any paradoxical interpretation I of ϕ. A classical interpretation I_c is obtained by constructing isomorphically a map $\nu : I \to I_c$ by Definition 3. It easily shows that $I_c \models \phi$ by induction on the numbers of connectives occurring in ϕ and the definition of satisfiability. Therefore, $\models \phi$.

($\Leftarrow$) If $\models \phi$ then for any classical interpretation I_c of ϕ. A paradoxical interpretation I is obtained by constructing isomorphically a map $\nu : I_c \to I$ by Definition 3. It easily shows that $I \models_{LP} \phi$ by induction on the numbers of connectives occurring in ϕ and the definition of satisfiability. Therefore, $\models_{LP} \phi$.

Proof of Corollary 1

$\models_{LP} C \sqcup \neg C(a)$ by Theorem 6 since $\models C \sqcup \neg C(a)$.

Proof of Corollary 2

The proof is similar to that of Lemma 2.

Proof of Theorem 7

Let $\mathbb{T}$ be a signed tableau for $\mathbf{T}\mathcal{A} \cup \{\mathbf{F}C(a)\}$. We prove this property by induction of the structure of C.

($\Rightarrow$) Assume that $\mathcal{A} \vdash_{LP} C(a)$. We need to prove that $\mathcal{A} \models_{LP} C(a)$.

In basic step, we consider two cases, namely, case 1. C is a concept name and case 2. C a complex concept.

Case 1: let C be the form A where A is a concept name. We assume that $\mathcal{A} \vdash_{LP} A(a)$, that is, $\mathbf{T}\mathcal{A} \cup \{\mathbf{F}A(a)\}$ is closed. Supposed that I is a paradoxical model of $\mathcal{A}$ and

$I \not\models_{LP} A(a)$, i.e., $\mathbf{T}A^I(a) = f$. It easily shows that by induction on the structure of $\mathbb{T}$ there exists a branch $\mathbf{b}$ with respect to I in $\mathbb{T}$. Thus, $\{\mathbf{T}A(a), \mathbf{F}A(a)\} \not\subseteq \mathbf{b}$ since $\mathbf{T}A^I(a) = f$ and $\{\mathbf{F}A(a), \mathbf{F}\neg A(a)\} \not\subseteq \mathbf{b}$ since $\mathbf{F}\neg A(a) \notin \mathbf{b}$. Thus, $\mathbf{b}$ isn't closed, that is, $\mathbf{T}\mathcal{A} \cup \{\mathbf{F}A(a)\}$ aren't closed which contradicts our assumption. Therefore, $\mathcal{A} \models_{LP} A(a)$.

Case 2: let C has one of forms of $\neg A, A \sqcap B, A \sqcup B, \forall R.A$ and $\exists R.A$ where A and B are concept names. It directly follows that $\mathcal{A} \vdash_{LP} C(a)$ by the proof of Case 1 since $\mathbf{F}\neg A^I(a) = t \Leftrightarrow \mathbf{F}A^I(a) = f$, $\mathbf{F}(A \sqcap B)^I(a) = t \Leftrightarrow \mathbf{F}\neg A^I(a) = f$ or $\mathbf{F}\neg B^I(a) = f$, $\mathbf{F}(A \sqcup B)^I(a) = t \Leftrightarrow \mathbf{F}\neg A^I(a) = f$ and $\mathbf{F}\neg B^I(a) = f$, $\mathbf{F}(\forall R.A)^I(a) = t$ and $\mathbf{T}R^I(a,b) = t \Leftrightarrow \mathbf{F}\neg A^I(b) = f$ and $\mathbf{F}(\exists R.A)^I(a) = t \Leftrightarrow \mathbf{F}\neg A^I(\iota) = f$ and $\mathbf{T}R^I(a,\iota) = t$ where ι is a new individual no occurring in $\mathbb{T}$ before by applying paradoxical expansion rules in Table 3.

In induction step, we assume that this property holds when C_1 and C_2 general concepts. It analogously shows that $\mathcal{A} \models_{LP} C(a)$ when C has the form of $\neg C_1, C_1 \sqcap C_2, C_1 \sqcup C_2, \forall R.C_1$ and $\exists R.C_1$ by the proof of Case 2.

Therefore, if $\mathcal{A} \vdash_{LP} C(a)$ then $\mathcal{A} \models_{LP} C(a)$.

($\Leftarrow$) Assume that $\mathcal{A} \models_{LP} C(a)$. We need to prove that $\mathcal{A} \vdash_{LP} C(a)$.

In basic step, we consider two case, namely, case 1. C is a concept name and case 2. C a complex concept.

Case 1: let C be the form A where A is a concept name. We assume that $\mathcal{A} \models_{LP} A(a)$, that is, for any paradoxical model I of $\mathcal{A}$, $I \models_{LP} A(a)$. Supposed that $\mathbf{T}\mathcal{A} \cup \{\mathbf{F}A(a)\}$ isn't closed, that is, there exists a branch $\mathbf{b}$ in $\mathbb{T}$ such that $\{\mathbf{T}A(a), \mathbf{F}A(a)\} \not\subseteq \mathbf{b}$ and $\{\mathbf{F}A(a), \mathbf{F}\neg A(a)\} \not\subseteq \mathbf{b}$. Since $\mathbf{F}\neg A(a) \notin \mathbf{b}$, $\mathbf{T}A(a) \notin \mathbf{b}$. Thus there exists a paradoxical model I' with respect to $\mathbf{b}$ of $\mathbb{T}$ such that $\mathbf{T}A^{I'}(a) = f$, then $A^{I'}(a) = f$. It easily shows that by induction on the structure of $\mathbb{T}$ since $A^{I'}(a) = f$. Thus, I' isn't a paradoxical model of $A(a)$ which contradicts our assumption. Therefore, $\mathcal{A} \vdash_{LP} A(a)$.

Case 2: let C be of forms $\neg A, A \sqcap B, A \sqcup B, \forall R.A$ and $\exists R.A$ where A and B are concept names. It directly follows that $\mathcal{A} \models_{LP} C(a)$ by the proof of Case 1 since $\mathbf{F}\neg\neg A(a) \in \mathbf{b} \Leftrightarrow \mathbf{F}A(a) \in \mathbf{b}$, $\mathbf{F}(A \sqcap B)(a) \in \mathbf{b} \Leftrightarrow \mathbf{F}A(a) \in \mathbf{b}$ or $\mathbf{F}B(a) \in \mathbf{b}$, $\mathbf{F}(A \sqcup B)(a) \in \mathbf{b} \Leftrightarrow \mathbf{F}A(a) \in \mathbf{b}$ and $\mathbf{F}B(a) \in \mathbf{b}$, $\mathbf{F}(\exists R.A)(a) \in \mathbf{b}$ and $\mathbf{T}R(a,b) \in \mathbf{b} \Leftrightarrow \mathbf{F}A(b) \in \mathbf{b}$ and $\mathbf{F}(\forall R.A)(a) \in \mathbf{b} \Leftrightarrow \mathbf{F}A(\iota) \in \mathbf{b}$ and $\mathbf{T}R(a,\iota) \in \mathbf{b}$ where ι is a new individual no occurring in $\mathbb{T}$ before by applying paradoxical expansion rules in Table 3.

In induction step, we assume that this property holds when C_1 and C_2 general concepts. It analogously shows that $\mathcal{A} \vdash_{LP} C(a)$ when C has the form of $\neg C_1, C_1 \sqcap C_2, C_1 \sqcup C_2, \forall R.C_1$ and $\exists R.C_1$ by the proof of Case 2.

Therefore, if $\mathcal{A} \models_{LP} C(a)$ then $\mathcal{A} \vdash_{LP} C(a)$.

Towards a Unified Model of Preference-Based Argumentation

Jean-Rémi Bourguet[1,3], Leila Amgoud[2], and Rallou Thomopoulos[1,3]

[1] INRA, UMR1208, Montpellier, France
{bourgujr, rallou}@supagro.inra.fr
[2] Institut de Recherche en Informatique (IRIT), Toulouse, France
amgoud@irit.fr
[3] CNRS and Université Montpellier II, LIRMM, Montpellier, France

Abstract. Argumentation is a reasoning model based on the construction and the evaluation of arguments. In his seminal paper, Dung has proposed the most abstract argumentation framework. In that framework, arguments are assumed to have the same strength. This assumption is unfortunately strong and often unsatisfied. Consequently, three extensions of the framework have been proposed in the literature. The first one assumes that an argumentation framework should be equipped with a (partial or total) preorder representing a preference relation between arguments, and capturing a difference of strengths of the arguments. The source of this preference relation is not specified, thus it can be instantiated in different manners. The second extension claims that the strength of an argument depends on the value(s) promoted by this argument. The third extension states that the set of arguments is equipped with several preorders; each of them expresses preferences between arguments in a given context.

The contribution of this paper is two-fold: first, it proposes a comparative study of these extensions of Dung's framework. It clearly shows under which conditions two proposals are equivalent. The second contribution of the paper consists in integrating the three extensions into a common more expressive framework.

1 Introduction

Argumentation is a reasoning model based on the construction and the evaluation of interacting arguments. It has been applied to nonmonotonic reasoning (e.g. [7]), decision making (e.g. [3,5,8]), and for modeling different types of dialogues including negotiation (e.g. [10,12]). Most of the models developed for the above applications are grounded on the abstract argumentation framework proposed by Dung in [7]. That framework consists of a set of arguments and a binary relation on that set, expressing conflicts among arguments. An argument gives a reason for believing a claim, for doing an action. It is worth mentioning that in this framework arguments are assumed to have the same strength. This assumption is quite strong since it is natural to consider an argument built from certain information stronger than another grounded on defeasible information.

S. Link and H. Prade (Eds.): FoIKS 2010, LNCS 5956, pp. 326–344, 2010.

Consequently, three different extensions of the framework have been proposed in the literature. The first one, proposed in [1], assumes that in addition to the conflict relation among arguments, another binary relation (called preference relation) on the set of arguments is available. This relation captures the differences in strengths of the arguments. The source of this relation is not specified, thus it can be instantiated in different manners. The second extension is proposed in [4] and extended in [9]. It claims that the strength of an argument depends on the value promoted by this argument. Each argument is assumed to promote a value. The values may not have the same importance. Thus, the argument promoting the most important value is considered as stronger than the others. The third extension, proposed in [2], states that the set of arguments is equipped with several preorders; each of them expresses preferences between arguments in a given context. It may be the case that for two arguments α and β, α is preferred to β in a given context and β is preferred to α in another context. This extension aims at generalizing the preference-based model defined in [1]. It is important to compare the three extensions and to highlight the similarities and the differences between them.

The contribution of this paper is two-fold: first, it proposes a comparative study of these extensions of Dung's framework. It clearly shows under which conditions two proposals are equivalent. The second contribution of the paper consists in integrating the three extensions into a common more expressive framework, whose properties are investigated.

The rest of the paper is organized as follows: Section 2 briefly recalls Dung's framework as well as its three extensions. Section 3 presents a comparative study of the three extensions. Section 4 proposes a unifying framework that captures the features of the three extensions.

2 Recalling Abstract Argumentation Frameworks

This section briefly recalls Dung's abstract argumentation framework as well as its three extensions. The three frameworks are illustrated by a running example that shows the power and the weaknesses of each of them.

2.1 Dung's Abstract Framework

An argumentation process follows three main steps: 1) constructing *arguments* and counter-arguments, 2) evaluating the *acceptability* of the different arguments, and 3) concluding or defining the *justified conclusions*. In [7], an argumentation framework is defined as follows:

Definition 1 (Dung's Argumentation Framework). *An* argumentation framework *is a pair* $\mathtt{AF} = \langle \mathcal{A}, \mathcal{R} \rangle$ *where* $\mathcal{A}$ *is a set of arguments and* $\mathcal{R} \subseteq \mathcal{A} \times \mathcal{A}$ *is an* attack *relation. An argument* α *attacks an argument* β *iff* $(\alpha, \beta) \in \mathcal{R}$.

In the above definition, arguments are abstract entities. Their origin and structure are left unknown. Note that we can associate each argumentation system with a directed graph whose nodes are the different arguments, and the edges represent the attack relation between them.

Among all the conflicting arguments, one has to define which arguments to keep for inferring conclusions or for making decisions. In [7], different semantics for the notion of acceptability have been proposed. For the purpose of this paper, we only recall admissible semantics.

Definition 2 (Conflict-Free, Defense, Admissible Semantics). *Let* $\mathcal{B} \subseteq \mathcal{A}$.

- $\mathcal{B}$ *is* conflict-free *iff* $\nexists\ \alpha_i, \alpha_j \in \mathcal{B}$ *such that* $(\alpha_i, \alpha_j) \in \mathcal{R}$.
- $\mathcal{B}$ defends *an argument* $\alpha_i \in \mathcal{B}$ *iff for each argument* $\alpha_j \in \mathcal{A}$, *if* $(\alpha_j, \alpha_i) \in \mathcal{R}$, *then* $\exists\ \alpha_k \in \mathcal{B}$ *such that* $(\alpha_k, \alpha_j) \in \mathcal{R}$.
- *A conflict-free set* $\mathcal{B}$ *of arguments is an* admissible extension *iff* $\mathcal{B}$ *defends all its elements.*

Let us illustrate the abstract framework through a simple example, describing a multi-criteria decision making situation. A French national nutritional health programme (PNNS), launched in 2001, aims at improving the state of health of the whole population by acting on several major determinants of citizens life, especially bread consumption. A primary objective of this programme is to increase the fraction of complex carbohydrates in the diet, and to reduce the fraction of simple carbohydrates. The part of simple carbohydrates on total carbohydrates is denoted SCP (Simple Carbohydrates Proportion). An action proposed by the decision makers is then to change the type of flour, labeled according to its ash value (mineral content), used in bread. The following table summarizes the performances obtained for two actions (bread type $\mathbb{T}_{65}$ and $\mathbb{T}_{80}$) and for several criteria (ash value, fibers and SCP) [6]. The objective is to choose between two breads, bread obtained with $\mathbb{T}_{65}$ and bread obtained with $\mathbb{T}_{80}$, on the basis of their performance in these criteria.

Example 1. Table below summarizes the performances in the different criteria.

	Ash value (av) (%)	Fibers (fb) (g/100g)	SCP (sc) (%)
Bread $\mathbb{T}_{80}$	0.80	4.2	3.85
Bread $\mathbb{T}_{65}$	0.65	3.8	4.11

In this application, an argument gives an information and consequently a reason for choosing a given bread. Note that all the performances are supposed to be in favor of a choice. The following six arguments are thus built:

- $\mathbb{T}_{80}^{av}$ ➡ Bread $\mathbb{T}_{80}$ should be chosen since its ash value is 0.80 %,
- $\mathbb{T}_{80}^{fb}$ ➡ Bread $\mathbb{T}_{80}$ should be chosen since its fibers content is 4.2 g/100g,
- $\mathbb{T}_{80}^{sc}$ ➡ Bread $\mathbb{T}_{80}$ should be chosen since its SCP content is 3.85 %,
- $\mathbb{T}_{65}^{av}$ ➡ Bread $\mathbb{T}_{65}$ should be chosen since its ash value is 0.65 %,
- $\mathbb{T}_{65}^{fb}$ ➡ Bread $\mathbb{T}_{65}$ should be chosen since its fibers content is 3.8 g/100g,
- $\mathbb{T}_{65}^{sc}$ ➡ Bread $\mathbb{T}_{65}$ should be chosen since its SCP content is 4.11 %.

Since only one bread type will be chosen, any pair of arguments that do not support the same option is considered as conflicting (i.e. $\in \mathcal{R}_{ex}$[1]). The following figure summarizes the different conflicts between arguments.

The system has two maximal (for set inclusion) admissible extensions $\{\mathbb{T}_{80}^{av}, \mathbb{T}_{80}^{fb}, \mathbb{T}_{80}^{sc}\}$ and $\{\mathbb{T}_{65}^{av}, \mathbb{T}_{65}^{fb}, \mathbb{T}_{65}^{sc}\}$ each of them supports a bread. Thus, the two breads $\mathbb{T}_{65}$ and $\mathbb{T}_{80}$ are equally preferred in Dung's system.

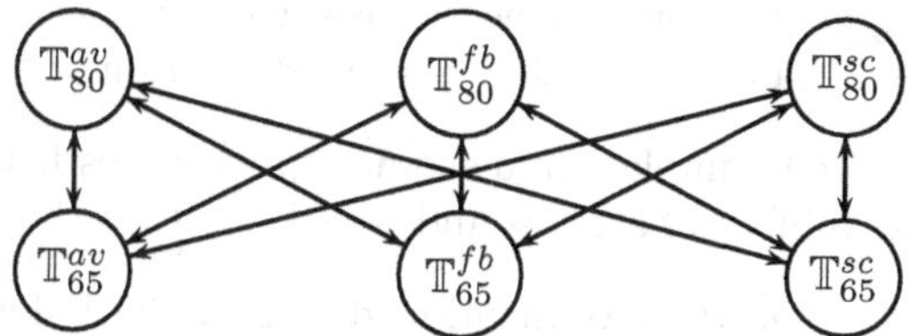

The above example shows that this framework is not powerful enough for making decisions. Indeed, for some criteria, it is possible to conclude that a bread is better than the other one (e.g. for SCP, bread $\mathbb{T}_{80}$ shows a better performance than bread $\mathbb{T}_{65}$). However, since the framework does not take into account the strengths of arguments, it has only solved the conflicts between arguments and concluded that both options are acceptable.

For comparison purposes, we will define a notion of equivalent frameworks. Two argumentation frameworks are said equivalent if they return exactly the same extensions under a given semantics.

Definition 3 (Equivalent frameworks). *Let* $\mathtt{AF}_1$, $\mathtt{AF}_2$ *be two argumentation frameworks.* $\mathtt{AF}_1$ *and* $\mathtt{AF}_2$ *are equivalent iff* $Ext(\mathtt{AF}_1) = Ext(\mathtt{AF}_2)$, *where* $Ext(\mathtt{AF}_i)$ *is the set of all extensions of* $\mathtt{AF}_i$ *under a given semantics.*[2]

2.2 Preference Based Argumentation Framework

In [1], it has been argued that arguments may have different strengths. In the previous example, it is clear that the argument $\mathbb{T}_{80}^{sc}$ is stronger than $\mathbb{T}_{65}^{sc}$. This information should be exploited in the argumentation framework. It allows to reduce the number of attacks among arguments. The idea is that an attack may fail if the attacked argument is stronger than its attacker.

Definition 4 (Preference-Based Argumentation Framework (PAF)). *A* preference-based argumentation framework *is a tuple* $\mathtt{PAF} = \langle \mathcal{A}, \mathcal{R}, \succeq \rangle$ *where* $\mathcal{A}$ *is a set of arguments,* $\mathcal{R} \subseteq \mathcal{A} \times \mathcal{A}$ *is an attack relation, and* $\succeq \subseteq \mathcal{A} \times \mathcal{A}$ *is a (partial or total) preorder*[3]. *For* $\alpha, \beta \in \mathcal{A}$, $(\alpha, \beta) \in \succeq$ *(or* $\alpha \succeq \beta$*) means that* α *is at least as strong as* β.

The relation $\succeq$ is general and may be instantiated in different manners. In order to evaluate the acceptability of arguments in a preference-based argumentation framework (PAF), a Dung style framework is associated to this PAF. Dung's semantics are then applied to the new framework.

[1] $\mathcal{R}_{ex}$ denotes the attack relation used in the example.

[2] In proofs, admissible semantics is used to establish equivalence between framework.

[3] A preorder is a binary relation that is *reflexive* and *transitive*.

Definition 5. *Let* PAF $= \langle \mathcal{A}, \mathcal{R}, \succeq \rangle$ *be a preference-based argumentation framework. The* AF *associated with* PAF *is the pair* $\langle \mathcal{A}, \texttt{Def} \rangle$ *where* $\texttt{Def} \subseteq \mathcal{A} \times \mathcal{A}$ *such that* $(\alpha, \beta) \in \texttt{Def}$ *iff* $(\alpha, \beta) \in \mathcal{R}$ *and* $(\beta, \alpha) \notin \succ$[4].

Dung's semantics are applied to the framework $\langle \mathcal{A}, \texttt{Def} \rangle$ in order to evaluate arguments of PAF $= \langle \mathcal{A}, \mathcal{R}, \succeq \rangle$.

Property 1. The argumentation frameworks associated respectively with $\texttt{PAF}_1 = \langle \mathcal{A}, \mathcal{R}, \succ \rangle$ and $\texttt{PAF}_2 = \langle \mathcal{A}, \mathcal{R}, \succeq \rangle$ (with $\succ$ the strict relation of $\succeq$) are equivalent.

Let us now re-consider Example 1 and see how preferences between arguments will help reduce the number of attacks and possibly return the expected result.

Example 2 (Example 1 cont.). As mentioned above, the following preferences hold between arguments: $\mathbb{T}^{sc}_{80} \succ \mathbb{T}^{sc}_{65}$ ($\mathbb{T}^{av}_{80}$, $\mathbb{T}^{av}_{65}$, $\mathbb{T}^{fb}_{80}$, $\mathbb{T}^{fb}_{65}$ are indifferent).

Another source of preferences between arguments is the importance of the criteria. Let us, for instance, assume that SCP is more important than fibers and ash value content, and that fibers and ash value are equally important. Thus any argument referring to SCP is stronger than any argument referring to ash value or fibers. The graph of $\langle \mathcal{A}, \texttt{Def} \rangle$ is summarized below.

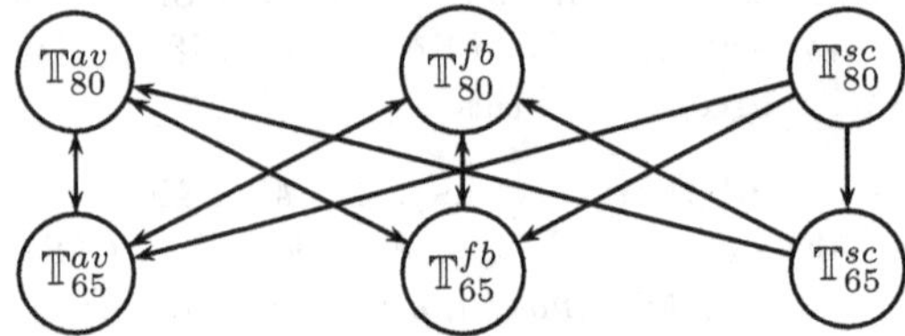

The framework has only one maximal (for set inclusion) admissible extension which is $\{\mathbb{T}^{av}_{80}, \mathbb{T}^{fb}_{80}, \mathbb{T}^{sc}_{80}\}$. The bread obtained with $\mathbb{T}_{80}$ is preferred to bread obtained with $\mathbb{T}_{65}$ in this PAF. Note that if we change the importance of the criteria and assume that SCP and fibers are equally important and both are more important than ash value, then two extensions $\{\mathbb{T}^{av}_{80}, \mathbb{T}^{fb}_{80}, \mathbb{T}^{sc}_{80}\}$ and $\{\mathbb{T}^{av}_{65}, \mathbb{T}^{fb}_{65}, \mathbb{T}^{sc}_{65}\}$ are obtained.

2.3 Value-Based Argumentation Framework

In [4], Bench Capon tried to formalize ideas of Perelman [11]. The latter emphasizes the importance of promoting values through arguments. In other terms, an argument may promote a value like, for instance, health, economy, etc. A value-based argumentation framework (VAF) is defined as follows:

Definition 6 (Value-Based Argumentation Framework). *A* value-based argumentation framework *is a tuple* VAF $= \langle \mathcal{A}, \mathcal{R}, \mathcal{V}, \texttt{val}, \texttt{Pref} \rangle$ *where* $\mathcal{A}$ *is a set of arguments,* $\mathcal{R} \subseteq \mathcal{A} \times \mathcal{A}$ *is an attack relation,* $\mathcal{V}$ *is a set of values,* $\texttt{val}: \mathcal{A} \mapsto \mathcal{V}$*, and* $\texttt{Pref} \subseteq \mathcal{V} \times \mathcal{V}$ *is an irreflexive, asymmetric and transitive strict relation.*

[4] We recall that $(\alpha, \beta) \in \succ$ iff $(\alpha, \beta) \in \succeq$ and $(\beta, \alpha) \notin \succeq$.

Like in [1], an argumentation framework à la Dung is associated to each VAF as follows:

Definition 7. *Let* $\mathtt{VAF} = \langle \mathcal{A}, \mathcal{R}, \mathcal{V}, \mathtt{val}, \mathtt{Pref} \rangle$ *be a VAF. The AF associated with* $\mathtt{VAF}$ *is* $\langle \mathcal{A}, \mathtt{defeats} \rangle$ *where* $\mathtt{defeats} \subseteq \mathcal{A} \times \mathcal{A}$ *such that* $(\alpha, \beta) \in \mathtt{defeats}$ *iff* $(\alpha, \beta) \in \mathcal{R}$ *and* $(\mathtt{val}(\beta), \mathtt{val}(\alpha)) \notin \mathtt{Pref}$.

As for PAFs, Dung's acceptability semantics are applied to the AF for evaluating the different arguments. Let us now illustrate this framework through the running example. For that purpose, one needs to define what the values will be as well as the preference relation between those values. There are several possibilities: the first one consists in considering the different criteria as values. The second solution considers each performance as a possible value. In what follows, we will mix the first and second solutions, and we will show that considering only criteria (or performances) as possible values is not powerful enough to get a meaningful result.

Example 3 (Example 1 cont.). Assume that $\mathcal{V} = \{v^{av}, v^{fb}, v^{sc}_{-}, v^{sc}_{+}\}$ such that $(v^{sc}_{-}, v^{sc}_{+}) \in \mathtt{Pref}$ and the two values v^{sc}_{-} and v^{sc}_{+} are preferred to the others (i.e $(v^{sc}_{+}, v^{fb}), (v^{sc}_{+}, v^{av}), (v^{sc}_{-}, v^{fb}), (v^{sc}_{-}, v^{av}) \in \mathtt{Pref}$. The function $\mathtt{val}$ is defined as follows: $\mathtt{val}(\mathbb{T}^{av}_{80}) = \mathtt{val}(\mathbb{T}^{av}_{65}) = v^{av}$, $\mathtt{val}(\mathbb{T}^{fb}_{80}) = \mathtt{val}(\mathbb{T}^{fb}_{80}) = v^{fb}$, $\mathtt{val}(\mathbb{T}^{sc}_{80}) = v^{sc}_{-}$ and $\mathtt{val}(\mathbb{T}^{sc}_{65}) = v^{sc}_{+}$. The graph associated with the framework $\langle \mathcal{A}, \mathtt{defeats} \rangle$ is depicted below:

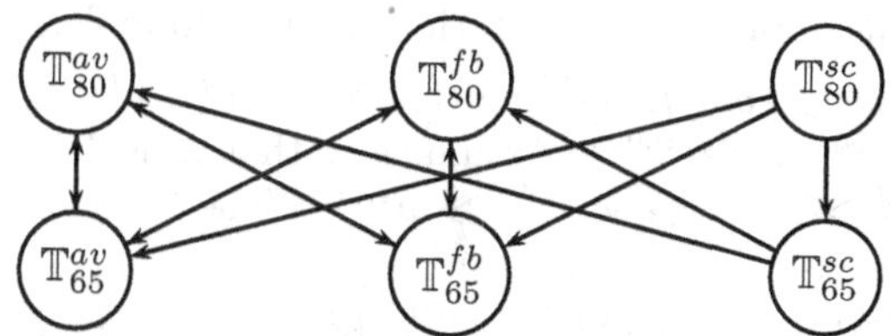

The framework has only one maximal (for set inclusion) admissible extension which is $\{\mathbb{T}^{av}_{80}, \mathbb{T}^{fb}_{80}, \mathbb{T}^{sc}_{80}\}$. Note that this result is that obtained by the PAF. If we assign the same value v^{sc} for the two arguments $\mathbb{T}^{sc}_{80}$ and $\mathbb{T}^{sc}_{65}$ and assume that this value is preferred to the others, then two extensions $\{\mathbb{T}^{av}_{80}, \mathbb{T}^{fb}_{80}, \mathbb{T}^{sc}_{80}\}$ and $\{\mathbb{T}^{av}_{65}, \mathbb{T}^{fb}_{65}, \mathbb{T}^{sc}_{65}\}$ are obtained.

It seems necessary to distinguish between several values (expressing respectively the considered criteria and its performance) which are here combined for v^{sc}_{-} and v^{sc}_{+}. This implies that it is necessary to allow an argument to support several values, enforcing the expressivity of the framework.

2.4 New Value-Based Argumentation Framework

This kind of VAFs introduced in [9] accounts for an extension of the classical VAF introduced by Bench-Capon in [4]. An argument in this framework may promote several values. There are then many ways for comparing pairs of arguments.

Definition 8 (Extended Valued-Based Framework). *An* extended value-based argumentation framework *(VSAF) is a tuple* $\langle \mathcal{A}, \mathcal{R}, \mathcal{V}, \arg, \gg \rangle$ *where* $\mathcal{A}$

is a set of arguments, $\mathcal{R} \subseteq \mathcal{A} \times \mathcal{A}$ *is an attack relation,* $\mathcal{V}$ *is a set of values,* $\arg : \mathcal{V} \mapsto 2^{\mathcal{A}}$ *such that* $\arg(v)$ *is .the set of arguments promoting value* v*, and* *gg is a partial order on* $\mathcal{V}$.

Since an argument may promote several values, then there are several ways for comparing pairs of arguments. Examples of this relation, denoted by $Pref_{\nabla}$, are given below.

Definition 9 (Preference Relations). *Let* $\langle \mathcal{A}, \mathcal{R}, \mathcal{V}, \arg, \gg \rangle$ *be a VSAF. Let* $\alpha, \beta \in \mathcal{A}$.

- $(\alpha,\ \beta) \in \mathtt{Pref}_M$ *iff* $|\arg^{-1}(\{\alpha\})| > |\arg^{-1}(\{\beta\})|$
- $(\alpha, \beta) \in \mathtt{Pref}_{Bc}$ *iff* $\exists v \in \mathcal{V}$ *such that* $\alpha \in \arg(v)$ *and* $\forall v' \in \mathcal{V}$ *with* $\beta \in \arg(v')$, $(v, v') \in \gg$.

The first relation prefers the argument that promotes most values while the second one privileges the argument that promotes the most important value.

In order to evaluate arguments in a VSAF, a Dung style framework is associated to this extended framework, and thus acceptability semantics are applied.

Definition 10. *Let* $\mathtt{VSAF} = \langle \mathcal{A}, \mathcal{R}, \mathcal{V}, \arg, \gg \rangle$ *be a VSAF and* $\mathtt{Pref}_{\nabla}$ *be particular preference relation between arguments. The argumentation framework associated with* $\mathtt{VSAF}$ *is* $\langle \mathcal{A}, \mathtt{defeats}_{\nabla} \rangle$ *where* $\mathtt{defeats}_{\nabla} \subseteq \mathcal{A} \times \mathcal{A}$ *and for* $\alpha, \beta \in \mathcal{A}$, $(\alpha, \beta) \in \mathtt{defeats}_{\nabla}$ *iff* $(\alpha,\ \beta) \in \mathcal{R}$ *and* $(\beta,\ \alpha) \notin Pref_{\nabla}$.

Let us illustrate this approach through the running example.

Example 4 (Example 1 cont.). Assume that $\mathcal{V} = \{v^{av},\ v^{fb},\ v^{sc},\ v^{3.85\%},\ v^{4.11\%},\ v^{2.0\%},\ v^{1.8\%},\ v^{4.2g/100g},\ v^{3.8g/100g}\}$. Assume also that $\gg$ is defined as follows: $v^{sc} \gg v^{av}, v^{fb} \gg v^{1.8\%} \gg v^{2.0\%} \gg v^{3.85\%} \gg v^{4.11\%}$. The function arg is defined as follows:

$\arg(v^{av}) = \{\mathbb{T}^{av}_{80},\ \mathbb{T}^{av}_{65}\}$ $\quad \arg(v^{fb}) = \{\mathbb{T}^{fb}_{80},\ \mathbb{T}^{fb}_{65}\}$ $\quad \arg(v^{sc}) = \{\mathbb{T}^{sc}_{80},\ \mathbb{T}^{sc}_{65}\}$
$\arg(v^{2.0\%}) = \mathbb{T}^{av}_{80}$ $\quad \arg(v^{1.8\%}) = \mathbb{T}^{av}_{65}$ $\quad \arg(v^{4.2g/100g}) = \mathbb{T}^{fb}_{80}$
$\arg(v^{3.8g/100g}) = \mathbb{T}^{fb}_{65}$ $\quad \arg(v^{3.85\%}) = \mathbb{T}^{sc}_{80}$ $\quad \arg(v^{4.11\%}) = \mathbb{T}^{sc}_{65}$

The graph associated with the framework $\mathtt{VSAF} = \langle \mathcal{A}, \mathtt{defeats}_{Bc} \rangle$ is depicted below:

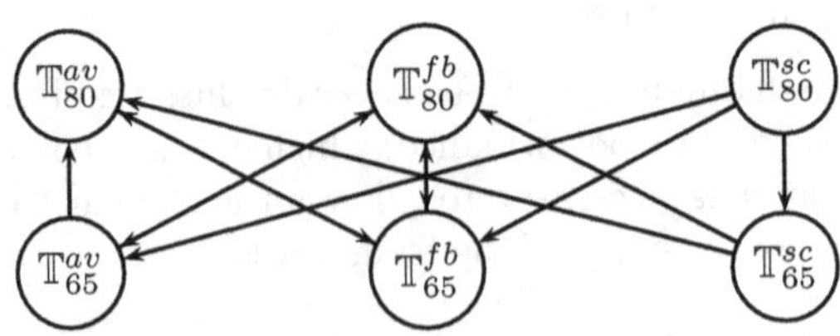

The framework has only one maximal (for set inclusion) admissible extension which is $\{\mathbb{T}^{av}_{80},\ \mathbb{T}^{fb}_{80},\ \mathbb{T}^{sc}_{80}\}$, the same result is obtained by using $\mathtt{VAF}$ system. Note that if we consider contexts, improving the expressivity of the models, preferences between values (especially $v^{1.8\%} \gg v^{2.0\%}$) can have a contextual validity (v^{sc}) which, for this model, would not necessarily return exactly the same graph depicted with $\mathtt{VAF}$ (see example 3).

2.5 Argumentation Framework Based on Contextual Preferences

In works on PAFs and VAFs, preferences between arguments are assumed to be not conflicting. However, in real applications this is not always true, for instance when we consider multiple points of view. Let us consider the case of the running example. Assume that two points of view can express different preferences about fibers content: the baker's point of view can be to prefer bread with a lower fibers content (preventing consumer satiety) whereas the miller's point of view can be to prefer a higher fibers content in flour improving yield. Thus, in baker context $\mathbb{T}_{65}^{fb}$ is stronger than $\mathbb{T}_{80}^{sc}$, while in miller context $\mathbb{T}_{80}^{sc}$ is stronger than $\mathbb{T}_{65}^{fb}$. In [2], an extension of PAF has been proposed. The idea is to assume that the set $\mathcal{A}$ of arguments is equipped with several preference relations $\succeq_1, \dots \succeq_n$, each of them expressing non-conflicting preferences between arguments in a particular *context*. Contexts (e.g. agents, points of view, criteria to be taken into account in a decision choice, etc.) are assumed to be ordered by a complete and strict relation denoted by $\rhd$. Note that for two arguments α and β, it may be the case that α is preferred to β in a given context and β is preferred to α in another one.

Definition 11 (CPAF). *A contextual preference-based argumentation framework (CPAF) is a tuple* $\mathtt{CPAF} = \langle \mathcal{A}, \mathcal{R}, \mathcal{C}, \rhd, \succeq_1, \dots \succeq_n \rangle$ *where* $\mathcal{A}$ *is a set of arguments,* $\mathcal{R}$ *is an attack relation,* $\mathcal{C}$ *is a finite set of contexts s.t.* $|\mathcal{C}| = n$, $\rhd$ *is a strict total order on the contexts, and* $\succeq_i$ *is a (partial or total) preorder associated with context* c_i.

In order to evaluate arguments in a CPAF, again an argumentation framework is associated to this CPAF. For that purpose, the different preference relations $\succeq_i$ are aggregated into a unique relation denoted by $\otimes^{\rhd}(\succeq_1, \dots \succeq_n)$. An example of such aggregation consists of keeping all the preferences of the strongest context, then to add the preferences of the next important context that are not conflicting with those of the first one. The same process is repeated until there is no remaining context. Note that there are several ways for aggregating preferences. For the purpose of our paper, we keep this aggregation abstract and can thus be instantiated in different manners.

Definition 12. *Let* $\mathtt{CPAF} = \langle \mathcal{A}, \mathcal{R}, \mathcal{C}, \rhd, \succeq_1, \dots \succeq_n \rangle$ *and* $\otimes^{\rhd}(\succeq_1, \dots \succeq_n)$ *be an aggregated preference relation between arguments. The argumentation framework associated with* CPAF *is* $\langle \mathcal{A}, \mathtt{Def} \rangle$ *where* $\forall \alpha, \beta \in \mathcal{A}$, $(\alpha, \beta) \in \mathtt{Def}$ *iff* $(\alpha, \beta) \in \mathcal{R}$ *and* $(\beta, \alpha) \notin \otimes^{\rhd}(\succeq_1, \dots, \succeq_n)$.

Dung's acceptability semantics are then applied on the framework $\langle \mathcal{A}, \mathtt{Def} \rangle$ for evaluating the arguments of the set $\mathcal{A}$.

Example 5 (Example 1. cont.). Let $\mathcal{C} = \{\mathfrak{Pnns}, \mathfrak{Baker}, \mathfrak{Miller}\}$ with $\mathfrak{Pnns} \rhd \mathfrak{Baker} \rhd \mathfrak{Miller}$. The set of arguments is equipped with three preference relations, respectively denoted $\succeq_{\mathfrak{B}}$, $\succeq_{\mathfrak{M}}$ and $\succeq_{\mathfrak{P}}$ ($\mathfrak{B}$ stands for $\mathfrak{Baker}$, $\mathfrak{M}$ for $\mathfrak{Miller}$ and $\mathfrak{P}$ for $\mathfrak{Pnns}$). These relations are defined as follows: $\mathbb{T}_{65}^{fb} \succeq_{\mathfrak{B}} \mathbb{T}_{80}^{fb}$; $\mathbb{T}_{80}^{fb} \succeq_{\mathfrak{M}} \mathbb{T}_{65}^{fb}$ and $\mathbb{T}_{80}^{sc} \succeq_{\mathfrak{P}} \mathbb{T}_{65}^{sc} \succeq_{\mathfrak{P}} \{\mathbb{T}_{80}^{fb}, \mathbb{T}_{65}^{fb}, \mathbb{T}_{80}^{av}, \mathbb{T}_{65}^{av}\}$. The aggregated relation is not in this case the union of the three relations because there is a contradiction between preferences (on arguments related to fibers) expressed in contexts $\mathfrak{Miller}$ and $\mathfrak{Baker}$. The order on the contexts induces the aggregated preference $\otimes^{\triangleright}(\succeq_{\mathfrak{P}}, \succeq_{\mathfrak{B}}, \succeq_{\mathfrak{M}})$ and the graph associated with this CPAF and depicted in figure below:

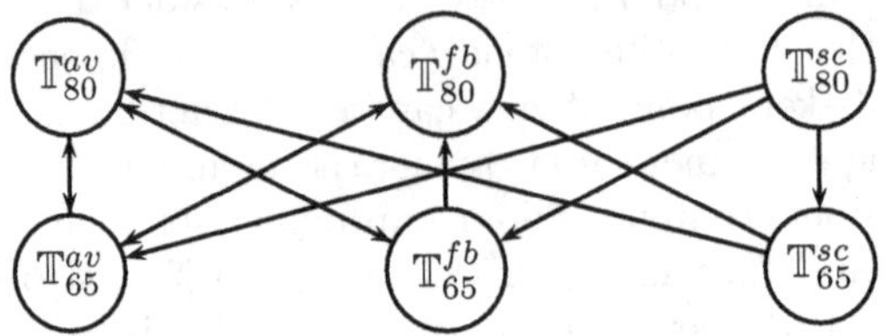

The framework has only one maximal (for set inclusion) admissible extension which is $\{\mathbb{T}_{80}^{av}, \mathbb{T}_{80}^{fb}\mathbb{T}_{80}^{sc}\}$. Thus, bread obtained with $\mathbb{T}_{80}$ is preferred to bread obtained with $\mathbb{T}_{65}$.

3 Comparing Different Abstract Argumentation Frameworks

This section compares the different argumentation frameworks previously presented in terms of equivalence, on the basis of Definition 3.

3.1 Comparing Dung's Framework and PAF

Dung's argumentation framework can be seen as a particular case of a preference-based argumentation framework. Several situations in which PAF and AF are equivalent can be highlighted, in particular when there is no strict preference between arguments, and when all the attacks between arguments succeed (i.e if an argument α attacks an argument β then β is not preferred to α).

Property 2. The argumentation framework $\mathtt{AF} = \langle \mathcal{A}, \mathcal{R} \rangle$ is equivalent to the argumentation framework associated with $\mathtt{PAF} = \langle \mathcal{A}, \mathcal{R}, \succeq \rangle$ iff:

- $\nexists \alpha, \beta \in \mathcal{A}$ such that $(\alpha, \beta) \in \succ$, or
- $\nexists \alpha, \beta \in \mathcal{A}$ such that $(\alpha, \beta) \in \mathcal{R}$ and $(\beta, \alpha) \in \succ$.

3.2 Comparing Bench Capon's Framework and Extension

Bench Capon's framework can be seen as a particular case of the extension proposed in [9] and that assumes that an argument may promote more than one value. The following properties describes the situations under which a VAF is equivalent to a VSAF.

Property 3.

- The two argumentation frameworks $\langle \mathcal{A}, \mathtt{defeats} \rangle$ and $\langle \mathcal{A}, \mathtt{defeats}_{Bc} \rangle$ associated respectively with $\mathtt{VAF} = \langle \mathcal{A}, \mathcal{R}, \mathcal{V}, \mathtt{val}, \mathtt{Pref} \rangle$[5] and $\mathtt{VSAF} = \langle \mathcal{A}, \mathcal{R}, \mathcal{V}, \mathrm{arg}, \gg \rangle$ are equivalent iff $\mathtt{val} = \mathrm{arg}^{-1}$ and $\mathtt{Pref} = \gg$.

[5] $\mathcal{V}=\{v_1, \ldots, v_n\}$ such that for i<j $(v_i, v_j) \in \mathtt{Pref}$.

- The two argumentation frameworks $\langle\mathcal{A}, \texttt{defeats}\rangle$ and $\langle\mathcal{A}, \texttt{defeats}_M\rangle$ associated respectively with $\texttt{VAF} = \langle\mathcal{A}, \mathcal{R}, \mathcal{V}, \texttt{val}, \texttt{Pref}\rangle$ and $\texttt{VSAF} = \langle\mathcal{A}, \mathcal{R}, \mathcal{V}', \text{arg}, \gg\rangle$ are equivalent iff $|\text{arg}^{-1}(\texttt{val}^{-1}(v_i))| = \text{i}$.

3.3 Comparison between the VAF and the PAF

In this section we show that several VAFs can be associated to the same PAF while a unique PAF is associated to a VAF.

Equivalent PAF^V Built from VAF

Definition 13. *Let* $\texttt{VAF} = \langle\mathcal{A}, \mathcal{R}, \mathcal{V}, \texttt{val}, \texttt{Pref}\rangle$ *be a value-based argumentation framework. From a* VAF, *a preference-based argumentation framework can be defined by* $\texttt{PAF}^V = \langle\mathcal{A}, \mathcal{R}, \succeq^V\rangle$, *with the preference relation* $\succeq^V \subseteq \mathcal{A}\times\mathcal{A}$ *defined as follows:* $\forall\alpha, \beta \in \mathcal{A}$ $(\alpha, \beta) \in \succ^V$ *iff* $(\texttt{val}(\alpha), \texttt{val}(\beta)) \in \texttt{Pref}$.

It is easy to show that the relation $\succ^V$ has the same properties as the relation Pref.

Property 4. The relation $\succ^V$ is irreflexive, asymmetric and transitive.

Property 5. The argumentation frameworks associated with VAF and PAF^V are equivalent.

Definition 13 and Property 5 are illustrated through the following example.

Example 6 (Example 3. cont.). The preference relation extracted from VAF is as follows: $\mathbb{T}^{sc}_{80} \succ^V \mathbb{T}^{sc}_{65} \succ^V \{\mathbb{T}^{av}_{80}, \mathbb{T}^{av}_{65}, \mathbb{T}^{fb}_{80}, \mathbb{T}^{fb}_{65}\}$. The system PAF^V, built with $\succeq^V$, has only one preferred extensions $\{\mathbb{T}^{av}_{80}, \mathbb{T}^{fb}_{80}, \mathbb{T}^{sc}_{80}\}$, similarly to VAF.

VAF's Equivalence Classes Built from PAF

Bijective construction: A value-based argumentation framework can be intuitively built from a PAF by assigning to each argument a distinct value, and exactly transferring the prioritization of arguments to their corresponding values. This framework, denoted $\texttt{VAF}^P_b$, is defined as follows:

Definition 14. *Let* $\texttt{PAF} = \langle\mathcal{A}, \mathcal{R}, \succeq\rangle$ *be a preference-based system. A* $\texttt{VAF}^P_b$ *defined from a* PAF *is a tuple* $\langle\mathcal{A}, \mathcal{R}, \mathcal{V}_b, \texttt{val}_b, \texttt{Pref}^P_b\rangle$ *such that:*
$\mathcal{V}_b$ *is a set of values with the same cardinality as* $\mathcal{A}$ $(|\mathcal{V}_b|=|\mathcal{A}|)$, $\texttt{val}_b$ *is a bijection from* $\mathcal{A}$ *to* $\mathcal{V}_b$, *and* $\texttt{Pref}^P_b \subseteq \mathcal{V}_b \times \mathcal{V}_b$ *is defined by:* $\forall\ v_\alpha, v_\beta \in \mathcal{V}_b$, $(v_\alpha, v_\beta) \in \texttt{Pref}^P_b$ *iff* $(\texttt{val}_b^{-1}(v_\alpha), \texttt{val}_b^{-1}(v_\beta)) \in \succ$, *where* $\texttt{val}_b^{-1}$ *denotes the inverse function of* $\texttt{val}_b$.

Example 7 (Example 2. cont.). For instance, from the PAF presented in Example 2, $\texttt{VAF}^P_b$ can be built by associating each argument ($\mathbb{T}^{av}_{65}$, $\mathbb{T}^{av}_{80}$, ...) with a value, e.g. its name ("$\mathbb{T}^{av}_{65}$", "$\mathbb{T}^{av}_{80}$", ...), keeping the same preferences for the names as for the underlying arguments.

The following property shows that the relation $\texttt{Pref}^P_b$ has the same properties as the relation $\succ$.

Property 6. The relation $\texttt{Pref}_b^P$ is irreflexive, asymmetric and transitive.

Property 7. The argumentation frameworks associated with PAF and $\texttt{VAF}_b^P$ are equivalent.

In order to show that different VAFs can be mapped from one PAF, we first define a relation between values in the target set $\mathcal{V}_b$ of a value-based argumentation framework $\texttt{VAF}_b^P$. This relation called typologic equivalence and denoted Te is defined as follows:

Definition 15. *Two values v_α, $v_\beta \in \mathcal{V}_b$ belongs to the typologic equivalence relation* Te, *i.e., $(v_\alpha, v_\beta) \in \texttt{Te}$ iff:*

- $\forall\, v_\gamma \in \mathcal{V}_b$, $(\texttt{val}_b^{-1}(v_\gamma), \texttt{val}_b^{-1}(v_\alpha)) \in \succ$ *iff* $(\texttt{val}_b^{-1}(v_\gamma), \texttt{val}_b^{-1}(v_\beta)) \in \succ$,
- $\forall\, v_\delta \in \mathcal{V}_b$, $(\texttt{val}_b^{-1}(v_\alpha), \texttt{val}_b^{-1}(v_\delta)) \in \succ$ *iff* $(\texttt{val}_b^{-1}(v_\beta), \texttt{val}_b^{-1}(v_\delta)) \in \succ$.

The following properties for the typologic equivalence Te are satisfied.

Property 8. Te is reflexive, symmetric and transitive.

Considering properties of this relation, Te defines an equivalence relation on the set $\mathcal{V}_b$. It can be also defined equivalence classes partitioning the set $\mathcal{V}_b$ into several disjoint subsets, all the elements in a given equivalence class being equivalent among themselves.

Definition 16. *The equivalence class of an element v_α in $\mathcal{V}_b$ equipped by the equivalence relation* Te, *denoted* $\texttt{Te}(v_\alpha)$, *is the subset of all images of v_α by* Te: $\texttt{Te}(v_\alpha) = \{v_\beta \in \mathcal{V}_b \mid (v_\alpha, v_\beta) \in \texttt{Te}\}$.

The set of all equivalence classes in $\mathcal{V}_b$ given by the equivalence relation Te is called quotient set of $\mathcal{V}_b$ by Te.

Definition 17. *The quotient set of $\mathcal{V}_b$ by* Te, *denoted $V_b/\texttt{Te}$ is the set of all equivalence classes of $\mathcal{V}_b$ according to* Te. *It is defined as follows:* $\mathcal{V}_b/\texttt{Te} = \{\texttt{Te}(v) \mid v \in \mathcal{V}_b\}$.

Surjective Construction: Another way to represent a PAF with a value-based argumentation framework can be to assign a same value for the set of arguments being themselves indifferent according to the preference relation $\succeq$. This framework, denoted $\texttt{VAF}_s^P$, is defined as follows:

Definition 18. *Let* $\texttt{PAF} = \langle \mathcal{A}, \mathcal{R}, \succeq \rangle$ *be a PAF. A* $\texttt{VAF}_s^P$ *defined from a* PAF *is a tuple* $\langle \mathcal{A}, \mathcal{R}, \mathcal{V}_s, \texttt{val}_s, \texttt{Pref}_s^P \rangle$ *such that:*
$\mathcal{V}_s$ is a set of values with a cardinality at most equal to the cardinality of $\mathcal{A}$, $\texttt{val}_s$ *is a function from $\mathcal{A}$ to $\mathcal{V}_s$.*
$\texttt{Pref}_s^P \subseteq \mathcal{V}_s \times \mathcal{V}_s$ *is such that:*

1. $\forall\, \alpha, \beta \in \mathcal{A}$, *if* $\texttt{val}_s(\alpha) = \texttt{val}_s(\beta)$ *then* $\forall\, \gamma \in \mathcal{A}$ *it holds that* $(\alpha, \gamma) \in \succ$ *iff* $(\beta, \gamma) \in \succ$ *and it holds that* $(\gamma, \alpha) \in \succ$ *iff* $(\gamma, \beta) \in \succ$,
2. $(v_\alpha, v_\beta) \in \texttt{Pref}_s^P$ *iff* $\forall\, \alpha \in \texttt{val}_s^{-1}(v_\alpha)$ *and* $\forall\, \beta \in \texttt{val}_s^{-1}(v_\beta)$ *it holds that* $(\alpha, \beta) \in \succ$, *where* $\texttt{val}_s^{-1}$ *denotes the inverse function of* $\texttt{val}_s$.

Example 8 (Example 2. cont.). For instance, from the PAF presented in Example 2, a $\mathtt{VAF}_s^P$ can be built by associating with the arguments $\mathbb{T}_{65}^{av}$ and $\mathbb{T}_{80}^{av}$ a value v_α, with the arguments $\mathbb{T}_{65}^{fb}$ and $\mathbb{T}_{80}^{fb}$ a value v_β, with the argument $\mathbb{T}_{80}^{sc}$ a value v_γ and with the argument $\mathbb{T}_{65}^{sc}$ a value v_δ, such that $(v_\gamma, v_\delta) \in \mathtt{Pref}_s^P$, $(v_\delta, v_\alpha) \in \mathtt{Pref}_s^P$ and $(v_\delta, v_\beta) \in \mathtt{Pref}_s^P$.

The relation $\mathtt{Pref}_s^P$ has the same properties as the relation $\succ$.

Property 9. The relation $\mathtt{Pref}_s^P$ is irreflexive, asymmetric and transitive.

Property 10. PAF and $\mathtt{VAF}_s^P$ are equivalent.

There is one surjective construction denoted $\mathtt{VAF}_{s(min)}^P$, giving a minimal target set of value denoted $\mathcal{V}_s^{min}$ (in the sense of cardinality) for which the related $\mathtt{VAF}_s^P$ holds in Definition 18 and a $\mathtt{VAF}_b^P$ satisfies Definition 14 for a same PAF.

Example 9 (Example 2. cont.). From the PAF presented in Example 2, the $\mathtt{VAF}_{s(min)}^P$ can be built by associating with the arguments $\mathbb{T}_{65}^{av}$, $\mathbb{T}_{80}^{av}$, $\mathbb{T}_{65}^{fb}$ and $\mathbb{T}_{80}^{fb}$ a common value, (e.g. v_α), with the argument $\mathbb{T}_{80}^{sc}$ a value v_γ and with the argument $\mathbb{T}_{65}^{sc}$ a value v_δ, such that $(v_\gamma, v_\delta) \in \mathtt{Pref}_s^P$ and $(v_\delta, v_\alpha) \in \mathtt{Pref}_s^P$.

Property 11. $|\mathcal{V}_s^{min}| = |\mathcal{V}_b \ / \ \mathtt{Te}|$

It is worth mentioning that this mapping does not necessarily distinguish indifferent arguments and incomparable arguments.

3.4 Related Work

Previous works, in particular [9], have focused on the comparison of the argumentation frameworks AF, PAF, VAF, and VSAF. However this comparison is purely syntactical, since no formal definition of the equivalence between frameworks was proposed.

Comparison of PAF and VAF. Moreover, [9] indicates that each PAF can be represented by various VAFs, which is correct, but then claims that all of these VAFs have the same topology, that is, each of them is a renaming of the others. Although stated by the Lemma 4 of [9], this statement is not correct. Indeed, a counter-example is the following: Let a PAF be defined by $\langle \mathcal{A}, \mathcal{R}, \succeq \rangle$ with $\mathcal{A} = \{\alpha, \beta, \gamma\}$ and $\succeq = \{(\alpha, \beta), (\alpha,\gamma)\}$. Consider the two following VAFs. $\mathtt{VAF}_1 = \langle \mathcal{A}, \mathcal{R}, \mathcal{V}_1, \mathtt{val}_1, \mathtt{Pref}_1 \rangle$, with $\mathcal{V}_1 = \{v_1, v_2, v_3\}$, $\mathtt{val}_1(\alpha) = v_3$, $\mathtt{val}_1(\beta) = v_2$ and $\mathtt{val}_1(\gamma) = v_1$, and $\mathtt{Pref}_1 = \{(v_3, v_2), (v_3, v_1)\}$. $\mathtt{VAF}_2 = \langle \mathcal{A}, \mathcal{R}, \mathcal{V}_2, \mathtt{val}_2, \mathtt{Pref}_2 \rangle$, with $\mathcal{V}_2 = \{v_1, v_2\}$, $\mathtt{val}_2(\alpha) = v_2$, $\mathtt{val}_2(\beta) = v_1$ and $\mathtt{val}_2(\gamma) = v_1$, and $\mathtt{Pref}_1 = \{(v_2, v_1)\}$. Although $\mathtt{VAF}_1$ and $\mathtt{VAF}_2$ are both equivalent to the PAF, they do not have the same topology. Actually, the above statement would stand if $\mathtt{Pref}_i$ were total orders, which is not assumed.

Comparison of PAF and VSAF. For each value specification argumentation framework, there is at most one preference-based argumentation framework it represents. From definition, $\langle \mathcal{A}, \mathcal{R}, \mathcal{V}, \arg, \gg, MM\rangle$ represents $\langle \mathcal{A}, \mathcal{R}, \mathtt{Pref}_{MM}\rangle$ if and only if $\mathtt{Pref}_{MM}$ is the least specific relation among the $\mathtt{Pref}'_{MM}$ such that $\langle \mathcal{A}, \mathcal{R}, \mathtt{Pref}'_{MM}\rangle$ satisfies $\langle \mathcal{A}, \mathcal{R}, \mathcal{V}, \arg, \gg, MM\rangle$. Let v_1, $v_2 \in \mathcal{V}$ such that $\arg(v_1) = \{\alpha, \beta, \gamma\}$ and $\arg(v_2) = \{\gamma, \delta\}$, the two systems $\langle \mathcal{A}, \mathcal{R}, \mathtt{Pref}^1_{MM}\rangle$ and $\langle \mathcal{A}, \mathcal{R}, \mathtt{Pref}^2_{MM}\rangle$ satisfy $\langle \mathcal{A}, \mathcal{R}, \mathcal{V}, \arg, \gg, MM\rangle$ with $\mathtt{Pref}^1_{MM} = \{(\alpha,\gamma), (\gamma,\delta), (\delta,\gamma)\}$ and $\mathtt{Pref}^2_{MM} = \{(\beta,\gamma), (\gamma,\delta), (\delta,\gamma)\}$. There is no more specific relation between $\mathtt{Pref}^1_{MM}$ and $\mathtt{Pref}^2_{MM}$ according to Definition presented in [13], and contradicting Theorem 5 presented in [9].

3.5 Comparing the PAF and the CPAF

A CPAF can be viewed as several PAFs completely ordered with a relation r_{PAF}, and aggregated using the operator denoted $\otimes^{r_{PAF}}$ and defined as follows:

Definition 19. *Let* $\{\mathtt{PAF}_1 = \langle \mathcal{A}, \mathcal{R}, \succeq_1\rangle, \ldots, \mathtt{PAF}_n = \langle \mathcal{A}, \mathcal{R}, \succeq_n\rangle\}$ *be a set of preference-based argumentation frameworks, totally ordered by a relation denoted* r_{PAF}. $\otimes^{r_{PAF}}(\mathtt{PAF}_1, \ldots, \mathtt{PAF}_n)$ *is a* CPAF $= \langle \mathcal{A}, \mathcal{R}, \mathcal{C}, \rhd, \succeq_{c_1}, \ldots \succeq_{c_n}\rangle$ *such that:*

- $\mathcal{C}$ *is a set of* n *contexts, each* c_i *associated with* $\mathtt{PAF}_i$,
- $\rhd$ *is a total preorder on* $\mathcal{C} \times \mathcal{C}$, *such that* $(c_i, c_j) \in \rhd$ *iff* $(\mathtt{PAF}_i, \mathtt{PAF}_j) \in r_{PAF}$,
- $\succeq_{c_i} = \succeq_i$.

It is clear according to this definition that PAF can be viewed as a particular case of CPAF with $n = 1$.

On the other hand, the evaluation of a CPAF relies on an aggregation function (see Definition 11), in order to provide a unique defeat relation, which leads to the computation of a PAF.

Definition 20. *Given a* CPAF $= \langle \mathcal{A}, \mathcal{R}, \mathcal{C}, \rhd, \succeq_1, \ldots \succeq_n\rangle$, *an aggregated preference-based argumentation framework can be defined as follows:* $\mathtt{PAF}_{ag} = \langle \mathcal{A}, \mathcal{R}, \otimes^{\rhd}(\succeq_1, \ldots \succeq_n)\rangle$.

4 The Unifying Framework

The aim of this section is to propose an argumentation framework generalizing the previous preference-based argumentation frameworks and improving their expressivity.

4.1 An Argument Can Be Expressed in One or Several Contexts

The extensions of Dung's framework are integrated into a common more expressive framework that can be used in a multicriteria decision situation.

Definition 21. *An* argumentation framework based on multi-contextual preferences *(*MCPAF*) is a pair* $\langle \mathcal{A}, \mathtt{Def}\rangle$, *where* Def *is defined as follows:* $\forall \alpha, \beta \in \mathcal{A}$, $(\alpha, \beta) \in \mathtt{Def}$ *iff* $(\alpha, \beta) \in \oplus^{\unrhd}(\mathcal{R}_1, \ldots, \mathcal{R}_n)$ *and* $(\beta, \alpha) \notin \otimes^{\unrhd}(\succeq_1, \ldots \succeq_n)$ *such that:*

- $\mathcal{C} = c_1, \ldots, c_n$ *is the set of contexts,*
- $\trianglerighteq$ *is a complete preordering on* $\mathcal{C} \times \mathcal{C}$,
- $\mathcal{A}_1, \ldots, \mathcal{A}_n$ *are sets of arguments,* $\mathcal{A}_i \subseteq \mathcal{A}$ *(with* $\mathcal{A} = \cup_{i \in [1,n]} \mathcal{A}_i$*) is the set of arguments which are expressed in the context* c_i,
- $\mathcal{R}_1, \ldots, \mathcal{R}_n$ *are binary relations representing contextual attacks,* $\mathcal{R}_i \subseteq \mathcal{A}_i \times \mathcal{A}_i$ *concerns the attack of arguments expressed in context* c_i,
- $\succeq_1, \ldots, \succeq_n$ *is the set of contextual preferences,* $\succeq_i \subseteq \mathcal{A}_i \times \mathcal{A}_i$ *is a partial preordering and concerns preferences of argument expressed in context* c_i,
- $\oplus^{\trianglerighteq}$[6] *(resp.* $\otimes^{\trianglerighteq}$*) is an aggregation operator of contextual attacks (resp. preferences).*

MCPAF *will be represented as a tuple:* $\langle \mathcal{A}_1, \ldots, \mathcal{A}_n, \mathcal{R}_1, \ldots, \mathcal{R}_n, \mathcal{C}, \trianglerighteq, \succeq_1, \ldots, \succeq_n \rangle$.

This system allows one to relate that an argument can be expressed in one or several contexts, to compare two arguments in the same context with a preference relation or to express an attack between two arguments in a given context.

4.2 CPAF Is a Particular Case of a MCPAF

CPAF is an argumentation system which can be seen as a particular case of a MCPAF reduced to a strict order between contexts and where all sets of arguments $\mathcal{A}_i$ and contextual attacks $\mathcal{R}_i$ are similar. In the following definition a multi-contextual preferences argumentation framework is built from a CPAF (denoted MCPAF^C).

Definition 22. *Let a* CPAF *be a tuple such that* $\text{CPAF} = \langle \mathcal{A}, \mathcal{R}, \mathcal{C}, \triangleright, \succeq_1, \ldots \succeq_n \rangle$, MCPAF^C *is also a tuple built from* CPAF *such that*
$\text{MCPAF}^C = \langle \mathcal{A}_1^C, \ldots, \mathcal{A}_n^C, \mathcal{R}_1^C, \ldots, \mathcal{R}_n^C, \mathcal{C}, \trianglerighteq^C, \succeq_1, \ldots, \succeq_n \rangle$ *with:*
$\mathcal{A}_1^C = \ldots = \mathcal{A}_n^C = \mathcal{A}$, $\mathcal{R}_1^C = \ldots = \mathcal{R}_n^C = \mathcal{R}$, $\trianglerighteq^C = \triangleright$.

Property 12. MCPAF^C and CPAF are equivalent.

4.3 Aggregation Operator of Contextual Preferences

Since the set of contexts in a MCPAF is equipped with a complete preordering, there are indifferent contexts w.r.t $\trianglerighteq$. It thus possible to stratify the set $\mathcal{C}$ of contexts $\mathcal{C}^1, \ldots, \mathcal{C}^m$ such that for all $c, c' \in \mathcal{C}^i$, $(c, c') \in \trianglerighteq$ and $(c', c) \in \trianglerighteq$. Moreover, for any $c \in \mathcal{C}^i$ and $c' \in \mathcal{C}^j$ with $j > i$, it holds that $(c', c) \in \triangleright$ (meaning that $(c', c) \in \trianglerighteq$ and $(c, c') \notin \trianglerighteq$).

In each subset of preferences $\bigcup_i \succeq^i$, there may be contradictory preferences whose set is denoted CP^i.

Definition 23. $\text{CP}^i \subseteq \mathcal{A} \times \mathcal{A}$ *is the set of contradictory preferences between arguments expressed in equivalent contexts of level at most equal to i.*
$\text{CP}^1 = \{(\alpha, \beta) \mid \exists\, c_k, c_l \in \mathcal{C}^1$ *s.t.* $(\alpha, \beta) \in \succeq_k$ *and* $(\beta, \alpha) \in \succeq_l\}$
$\text{CP}^i = \{(\alpha, \beta) \mid \exists\, c_k, c_l \in \mathcal{C}^i$ *s.t.* $(\alpha, \beta) \in \succeq_k$ *and* $(\beta, \alpha) \in \succeq_l\} \cup_{r \in [1, i-1]} \text{CP}^r$

---

[6] For comparison purpose, $\oplus^{\trianglerighteq}$ can be axiomatized as follows: $\oplus^{\trianglerighteq}(\mathcal{R}_1, \ldots, \mathcal{R}_n) = \mathcal{R}_1$ if $\mathcal{R}_2 = \ldots = \mathcal{R}_n$ and $\oplus^{\trianglerighteq}(\mathcal{R}_c, \ldots, \mathcal{R}_c) = \mathcal{R}_c$.

Definition 24. *An aggregation operator of contextual preferences for* MCPAF *can be defined as follows:* $\otimes^{\trianglerighteq}(\succeq_1, \ldots \succeq_n) = \Pi_n$*:*

$$\Pi_1 = \{(\alpha, \beta) \in \succeq^1 \text{ and } (\alpha, \beta) \notin \mathtt{CP}^1\}$$
$$\Pi_{k+1} = \Pi_k \cup \{(\alpha, \beta) \in \succeq^{k+1} \text{ and } (\beta, \alpha) \notin \mathtt{CP}^{k+1} \cup \Pi_k\}$$

In CPAF, the set of contexts is equipped with a total order, therefore after discretization of preferences about the fiber criterion within two audiences (represented as the contexts $\mathfrak{Miller}$ and $\mathfrak{Baker}$), this framework doesn't allow to consider contexts as indifferent. The benefit generated by enforced expressivity is counterbalanced by a limitation in the ordering of contexts, involving an important difference for Def relation. Moreover, even though a preference can be expressed in a given context, arguments lose the expressivity obtained in VSAF (where an argument can promote one or several values). MCPAF allows to encompass advantages included in VSAF and CPAF without a loss of PAF generality.

Example 10 (Example 1 cont.). Table below summarizes contexts, sets of arguments expressed in these contexts, contextual preferences and attacks, represented in MCPAF through the running example. The ordering of the set of contexts is described as follows:

$\mathfrak{Pnns} \triangleright \mathfrak{Baker} \sim \mathfrak{Miller} \triangleright \mathfrak{Ash\ Value} \sim \mathfrak{Fibers} \sim \mathfrak{SCP} \sim \% \sim \mathrm{g/100g} \sim 0.80 \sim \ldots \sim 42.$

c_i	$\mathfrak{Pnns}$	$\mathfrak{Baker}$	$\mathfrak{Miller}$	$\mathfrak{Ash\ Value}$	...	%	...	42
$\succeq_i$	$\mathbb{T}^{sc}_{80} \succ_{\mathfrak{P}} \mathbb{T}^{sc}_{65} \succ_{\mathfrak{P}} \{\mathbb{T}^{fb}_{80}, \ldots, \mathbb{T}^{av}_{65}\}$	$\mathbb{T}^{fb}_{65} \succeq_{\mathfrak{B}} \mathbb{T}^{fb}_{80}$	$\mathbb{T}^{fb}_{80} \succeq_{\mathfrak{M}} \mathbb{T}^{fb}_{65}$	$\emptyset$	...	$\emptyset$	...	$\emptyset$
$\mathcal{A}_i$	$\mathbb{T}^{sc}_{80}, \mathbb{T}^{sc}_{65}, \mathbb{T}^{fb}_{80}, \mathbb{T}^{fb}_{65}, \mathbb{T}^{av}_{80}, \mathbb{T}^{av}_{65}$	$\mathbb{T}^{fb}_{65}, \mathbb{T}^{fb}_{80}$	$\mathbb{T}^{fb}_{80}, \mathbb{T}^{fb}_{65}$	$\mathbb{T}^{av}_{80}, \mathbb{T}^{av}_{65}$	...	$\mathbb{T}^{sc}_{80}, \mathbb{T}^{sc}_{65}, \mathbb{T}^{av}_{80}, \mathbb{T}^{av}_{65}$	...	$\mathbb{T}^{fb}_{80}$
$\mathcal{R}_i$	$\mathcal{R}_{ex}$	$\emptyset$	$\emptyset$	$\emptyset$	...	$\emptyset$	...	$\emptyset$

MCPAF framework has only one maximal (for set inclusion) admissible extension which is $\{\mathbb{T}^{av}_{80}, \mathbb{T}^{fb}_{80}, \mathbb{T}^{sc}_{80}\}$. The bread obtained with $\mathbb{T}_{80}$ is preferred to bread obtained with $\mathbb{T}_{65}$ in this system on the basis of the aim of the nutritional program and no preferences between actors point of views (e.g. bakers and millers).

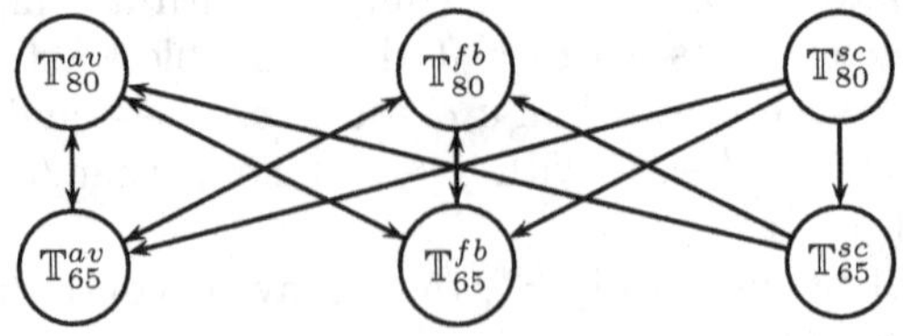

5 Conclusion

Comparing different argumentation framework can be a hard task, especially since there are few propositions in literature on the ways to achieve this task. In this paper, we have proposed to compare frameworks on the basis of their extensions under a given semantics. We have considered two argumentation frameworks as equivalent if they return exactly the same extensions. Then, we have compared well-known frameworks (AF, VAF, VSAF, PAF, CPAF) under the light of this comparison method. It is also clearly shown that these frameworks can be considered as equivalent under particular conditions.

We have then proposed a more general framework (MCPAF) generalizing the others as special cases, and allowing for fine representations of contextual specificities. Although its benefits have to be evaluated on more complex real-world problems, we think that it will prove useful in multiple criteria decision problems in presence of multiple actors. We therefore plan to apply it to agronomical issues.

Indeed a case study recently investigated covers questions related to policy decisions concerning public health. In a first step, area describes knowledge base from bread nutritional formulation using different types of flour, then it combines arguments (coming from the actors points of view) put into a decisional system. Finally, the ambition is to refine a consensus decision that satisfies both public authorities and consumers through all the actors involved in the bread transformation process.

References

1. Amgoud, L., Cayrol, C.: A reasoning model based on the production of acceptable arguments. Annals of Mathematics and Artificial Intelligence 34, 197–216 (2002)
2. Amgoud, L., Parsons, S., Perrussel, L.: An argumentation framework based on contextual preferences. In: Proceedings of the International Conference on Formal and Applied and Practical Reasoning (FAPR 2000), pp. 59–67 (2000)
3. Amgoud, L., Prade, H.: Using arguments for making and explaining decisions. Artif. Intell. 173(3-4), 413–436 (2009)
4. Bench-Capon, T.J.M.: Persuasion in practical argument using value-based argumentation frameworks. Journal of Logic and Computation 13(3), 429–448 (2003)
5. Bonet, B., Geffner, H.: Arguing for decisions: A qualitative model of decision making. In: Proceedings of the 12th Conference on Uncertainty in Artificial Intelligence (UAI 1996), pp. 98–105 (1996)
6. Bourre, J.-M., Bégat, A., Leroux, M.-C., Mousques-Cami, V., Pérandel, N., Souply, F.: Valeur nutritionnelle (macro et micro-nutriments) de farines et pains français. Médecine et Nutrition 44(2), 49–76 (2008)
7. Dung, P.M.: On the acceptability of arguments and its fundamental role in nonmonotonic reasoning, logic programming and n-person games. Artificial Intelligence Journal 77, 321–357 (1995)
8. Fox, J., Das, S.: Safe and Sound. Artificial Intelligence in Hazardous Applications. AAAI Press, The MIT Press (2000)
9. Kaci, S., van der Torre, L.: Preference-based argumentation: Arguments supporting multiple values. Int. J. Approx. Reasoning 48(3), 730–751 (2008)
10. Kraus, S., Sycara, K., Evenchik, A.: Reaching agreements through argumentation: a logical model and implementation 104, 1–69 (1998)
11. Perelman, C.: Justice, Law and Argument. Reidel, Dordrecht (1980)
12. Sycara, K.: Persuasive argumentation in negotiation. Theory and Decision 28, 203–242 (1990)
13. Yager, R.R.: Entropy and specificity in a mathematical theory of evidence. In: Classic Works of the Dempster-Shafer Theory of Belief Functions, pp. 291–310 (2008)

Appendix

Proof. of Property 1. Assume that Ext($\mathtt{PAF}_1$), Ext($\mathtt{PAF}_2$) be the sets of admissible extensions in these two abstract frameworks $\mathtt{PAF}_1$ and $\mathtt{PAF}_2$. Let us show that $Ext(\mathtt{PAF}_2) \subseteq Ext(\mathtt{PAF}_1)$ and $Ext(\mathtt{PAF}_1) \subseteq Ext(\mathtt{PAF}_2)$:

1. Ext($\mathtt{PAF}_2$) $\subseteq$ Ext($\mathtt{PAF}_1$). Let $\mathcal{E} \in$ Ext($\mathtt{PAF}_1$). Assume that $\mathcal{E} \notin$ Ext($\mathtt{PAF}_2$). This means that $\mathcal{E}$ is not an admissible extension in $\mathtt{PAF}_2$. According to Definition 2, there are two possibilities:

   ∗Case 1: $\mathcal{E}$ is not conflict-free in $\mathtt{PAF}_2$. $\exists\ \alpha,\ \beta \in \mathcal{E}$ such that $(\alpha, \beta) \in \mathtt{Def}_2$ (where $\mathtt{Def}_2$ is built from $\mathcal{R}$ and $\succ$, the strict relation of $\succeq$, by Definition 4). This means that $(\alpha, \beta) \in \mathcal{R}$ and $(\beta, \alpha) \notin \succ$. It holds also that $(\alpha, \beta) \in \mathtt{Def}_1$ (where $\mathtt{Def}_1$ is defined from $\mathcal{R}$ and $\succ$). Thus, $\mathcal{E}$ is not conflict-free in $\mathtt{PAF}_1$. This contradicts the fact that $\mathcal{E}$ is an admissible extension in $\mathtt{PAF}_1$.

   ∗Case 2: $\mathcal{E}$ does not defend its elements in $\mathtt{PAF}_2$. This means that $\exists\ \alpha \in \mathcal{E}$, $\exists\ \beta \in \mathcal{A}$ such that $(\beta, \alpha) \in \mathtt{Def}_2$ and $\nexists\ \gamma \in \mathcal{E}$ such that $(\gamma, \beta) \in \mathtt{Def}_2$. According to Definition 4, since $(\beta, \alpha) \in \mathtt{Def}_2$ then $(\beta, \alpha) \in \mathcal{R}$ and $(\alpha, \beta) \notin \succ$, i.e $(\beta, \alpha) \in \mathtt{Def}_1$. But, $\mathcal{E}$ is an admissible extension in $\mathtt{PAF}_1$, i.e $\exists\ \gamma \in \mathcal{E}$ such that $(\gamma, \beta) \in \mathtt{Def}_1$, i.e $(\gamma, \beta) \in$ R and $(\beta, \gamma) \notin \succ$, i.e $(\gamma, \beta) \in \mathtt{Def}_2$. This conclusion contradicts the fact that $\mathcal{E}$ does not defend its arguments in $\mathtt{PAF}_2$.
2. Ext($\mathtt{PAF}_1$) $\subseteq$ Ext($\mathtt{PAF}_2$). Let $\mathcal{E} \in$ Ext($\mathtt{PAF}_2$). Assume that $\mathcal{E} \notin$ Ext($\mathtt{PAF}_1$). This means that $\mathcal{E}$ is not an admissible extension in $\mathtt{PAF}_1$. According to Definition 2, there are two possibilities:

   ∗Case 1: $\mathcal{E}$ is not conflict-free in $\mathtt{PAF}_1$. $\exists\ \alpha,\ \beta \in \mathcal{E}$ such that $(\alpha, \beta) \in \mathtt{Def}_1$ ($\mathtt{Def}_1$ is built from $\mathcal{R}$ and $\succ$). This means that $(\alpha, \beta) \in \mathcal{R}$ and $(\beta, \alpha) \notin \succ$. By Definition 4, it holds also that $(\alpha, \beta) \in \mathtt{Def}_2$ (where $\mathtt{Def}_2$ is defined from $\mathcal{R}$ and $\succ$, the strict relation of $\succeq$). Thus, $\mathcal{E}$ is not conflict-free in $\mathtt{PAF}_2$. This contradicts the fact that $\mathcal{E}$ is an admissible extension in $\mathtt{PAF}_2$.

   ∗Case 2: $\mathcal{E}$ does not defend its elements in $\mathtt{PAF}_1$. This means that $\exists\ \alpha \in \mathcal{E}$, $\exists\ \beta \in \mathcal{A}$ such that $(\beta, \alpha) \in \mathtt{Def}_1$ and $\nexists\ \gamma \in \mathcal{E}$ such that $(\gamma, \beta) \in \mathtt{Def}_1$. However, since $(\beta, \alpha) \in \mathtt{Def}_1$ then $(\beta, \alpha) \in \mathcal{R}$ and $(\alpha, \beta) \notin \succ$. By Definition 4, $\mathtt{Def}_2$ is built from $\mathcal{R}$ and $\succ$, the strict relation of $\succeq$, i.e $(\beta, \alpha) \in \mathtt{Def}_2$. But, $\mathcal{E}$ is an admissible extension in $\mathtt{PAF}_2$, i.e $\exists\ \gamma \in \mathcal{E}$ such that $(\gamma, \beta) \in \mathtt{Def}_2$, i.e $(\gamma, \beta) \in \mathcal{R}$ and $(\beta, \gamma) \notin \succ$, i.e $(\gamma, \beta) \in \mathtt{Def}_1$. This conclusion contradicts the fact that $\mathcal{E}$ does not defend its arguments in $\mathtt{PAF}_1$.

Proof. of Property 2. Assume that Ext(AF), Ext(PAF) be the sets of admissible extensions in these two abstract frameworks AF and PAF. Let us show that $Ext(\mathtt{AF}) \subseteq Ext(\mathtt{PAF})$ and $Ext(\mathtt{PAF}) \subseteq Ext(\mathtt{AF})$:

1. Ext(AF) $\subseteq$ Ext(PAF). Let $\mathcal{E} \in$ Ext(PAF). Assume that $\mathcal{E} \notin$ Ext(AF). This means that $\mathcal{E}$ is not an admissible extension in AF. According to Definition 2, there are two possibilities:

   ∗Case 1: $\mathcal{E}$ is not conflict-free in AF. $\exists\ \alpha, \beta \in \mathcal{E}$ such that $(\alpha, \beta) \in \mathcal{R}$. But $(\beta, \alpha) \notin \succ$ since $\nexists\ (\beta, \alpha) \in \succ$, it holds also that $(\alpha, \beta) \in$ Def (Def is built from $\mathcal{R}$ and $\succ$ according to Definition 4). This contradicts the fact that $\mathcal{E}$ is conflict-free in PAF.

   ∗Case 2: $\mathcal{E}$ does not defend its elements in AF. This means that $\exists\ \alpha \in \mathcal{E}$, $\exists\ \beta \in \mathcal{A}$ such that $(\beta, \alpha) \in \mathcal{R}$ and $\nexists\ \gamma \in \mathcal{E}$ such that $(\gamma, \beta) \in \mathcal{R}$. However, since $(\beta, \alpha) \in \mathcal{R}$ and $(\alpha, \beta) \notin \succ$, i.e $(\beta, \alpha) \in$ Def. But, $\mathcal{E}$ is an admissible extension in PAF, i.e $\exists\ \gamma \in \mathcal{E}$ such that $(\gamma, \beta) \in$ Def, i.e $(\gamma, \beta) \in \mathcal{R}$ and $(\beta, \gamma) \notin \succ$ (by property), i.e $(\gamma, \beta) \in \mathcal{R}$. This conclusion contradicts the fact that $\mathcal{E}$ does not defend its arguments in AF.
2. Ext(PAF) $\subseteq$ Ext(AF). Let $\mathcal{E} \in$ Ext(AF). Assume that $\mathcal{E} \notin$ Ext(PAF). This means that $\mathcal{E}$ is not an admissible extension in PAF. According to Definition 2, there are two possibilities:

   ∗Case 1: $\mathcal{E}$ is not conflict-free under PAF. On the other hand, $\mathcal{E}$ is conflict-free in AF, i.e $\nexists\ \alpha, \beta \in \mathcal{E}$ such that $(\alpha, \beta) \in \mathcal{R}$. It holds also that $(\alpha, \beta) \notin$ Def. (Def is built from $\mathcal{R}$ and $\succ$ according to Definition 4). This contradicts the fact that $\mathcal{E}$ is not conflict-free in PAF.

   ∗Case 2: $\mathcal{E}$ does not defend its elements in PAF. $\mathcal{E}$ is an admissible extension in AF, i.e if $\exists\ \alpha \in \mathcal{E}$, $\exists\ \beta \in \mathcal{A}$ such that $(\beta, \alpha) \in \mathcal{R}$ then $\exists\ \gamma \in \mathcal{E}$ such that $(\gamma, \beta) \in \mathcal{R}$. By property $(\alpha, \beta) \notin \succ$ and $(\beta, \gamma) \notin \succ$. This means that $\exists\ \alpha \in \mathcal{E}$, $\exists\ \beta \in \mathcal{A}$ such that $(\beta, \alpha) \in$ Def and $\exists\ \gamma \in \mathcal{E}$ such that $(\gamma, \beta) \in$ Def (Def is built from $\mathcal{R}$ and $\succ$ according to Definition 4). This conclusion contradicts the fact that $\mathcal{E}$ does not defend its arguments in PAF.

- If $\nexists\ \alpha, \beta \in \mathcal{A}$, such that $(\alpha,\beta) \in \mathcal{R}$ and $(\beta, \alpha) \in \succ$.

Assume that Ext(AF), Ext(PAF) be the sets of admissible extensions in these two abstract frameworks AF and PAF. Let us show that Ext(AF) $\subseteq$ Ext(PAF) and Ext(PAF) $\subseteq$ Ext(AF):

1. Ext(AF) $\subseteq$ Ext(PAF). Let $\mathcal{E} \in$ Ext(PAF). Assume that $\mathcal{E} \notin$ Ext(AF). This means that $\mathcal{E}$ is not an admissible extension in AF. According to Definition 2, there are two possibilities:

   ∗Case 1: $\mathcal{E}$ is not conflict-free under AF. $\exists\ \alpha, \beta \in \mathcal{E}$ such that $(\alpha, \beta) \in \mathcal{R}$. But $(\beta, \alpha) \notin \succ$ since $\nexists\ (\beta, \alpha)$ such that $(\beta, \alpha) \in \succ$ and $(\alpha, \beta) \in \mathcal{R}$, it holds also that $(\alpha, \beta) \in$ Def (Def is built from $\mathcal{R}$ and $\succ$ according to Definition 4). This contradicts the fact that $\mathcal{E}$ is conflict-free in PAF.

$*$Case 2: $\mathcal{E}$ does not defend its elements in AF. This means that $\exists\ \alpha \in \mathcal{E}$, $\exists\ \beta \in \mathcal{A}$ such that $(\beta, \alpha) \in \mathcal{R}$ and $\nexists\ \gamma \in \mathcal{E}$ such that $(\gamma, \beta) \in \mathcal{R}$. However, since $(\beta, \alpha) \in \mathcal{R}$ and by property $(\alpha, \beta) \notin \succ$, i.e $(\beta, \alpha) \in$ Def (Def is built from $\mathcal{R}$ and $\succ$ according to Definition 4). But, $\mathcal{E}$ is an admissible extension in PAF, i.e $\exists\ \gamma \in \mathcal{E}$ such that $(\gamma, \beta) \in$ Def, i.e $(\gamma, \beta) \in \mathcal{R}$ and $(\beta, \gamma) \notin \succ$ (according to Definition 4), i.e $(\gamma, \beta) \in \mathcal{R}$. This conclusion contradicts the fact that $\mathcal{E}$ does not defend its arguments in AF.

2. Ext(PAF) $\subseteq$ Ext(AF). Let $\mathcal{E} \in$ Ext(AF). Assume that $\mathcal{E} \notin$ Ext(PAF). This means that $\mathcal{E}$ is not an admissible extension in PAF. According to Definition 2, there are two possibilities:

   $*$Case 1: $\mathcal{E}$ is not conflict-free in PAF. On the other hand, $\mathcal{E}$ is conflict-free in AF, i.e $\nexists\ \alpha, \beta \in \mathcal{E}$ such that $(\alpha, \beta) \in \mathcal{R}$. It holds also that $(\alpha, \beta) \notin$ Def. (Def is built from $\mathcal{R}$ and $\succ$ according to Definition 4). This contradicts the fact that $\mathcal{E}$ is not conflict-free in PAF.

   $*$Case 2: $\mathcal{E}$ does not defend its elements in PAF. $\mathcal{E}$ is an admissible extension in AF, i.e if $\exists\ \alpha \in \mathcal{E}$, $\exists\ \beta \in \mathcal{A}$ such that $(\beta, \alpha) \in \mathcal{R}$ then $\exists\ \gamma \in \mathcal{E}$ such that $(\gamma, \beta) \in \mathcal{R}$. By property if $\exists\ \alpha, \beta$ such that $(\beta, \alpha) \in \mathcal{R}$ then $(\alpha, \beta) \notin \succ$ and if $\exists\ \gamma, \beta$ such that $(\gamma, \beta) \in \mathcal{R}$ then $(\beta, \gamma) \notin \succ$. This means that $\exists\ \alpha \in \mathcal{E}$, $\exists\ \beta \in \mathcal{A}$ such that $(\beta, \alpha) \in$ Def and $\exists\ \gamma \in \mathcal{E}$ such that $(\gamma, \beta) \in$ Def (Def is built from $\mathcal{R}$ and $\succ$ according to Definition 4). This conclusion contradicts the fact that $\mathcal{E}$ does not defend its arguments in PAF.

Proof. of Property 4.

1. $\succ^V$ is irreflexive,
   Assume that $\succ^V$ is reflexive: $\forall\ \alpha \in \mathcal{A}$, $(\alpha, \alpha) \in \succ^V$. This means that $(\mathtt{val}(\alpha), \mathtt{val}(\alpha)) \in$ Pref, but this is not possible since Pref is irreflexive, showing a contradiction.

2. $\succ^V$ is asymmetric,
   Assume that $\alpha, \beta \in \mathcal{A}$ such that $(\alpha, \beta) \in \succ^V$ and $(\beta, \alpha) \in \succ^V$:
   - $(\alpha, \beta) \in \succ^V \Rightarrow (\mathtt{val}(\alpha), \mathtt{val}(\beta)) \in$ Pref,
   - $(\beta, \alpha) \in \succ^V \Rightarrow (\mathtt{val}(\beta), \mathtt{val}(\alpha)) \in$ Pref,

   By Definition 6, this is impossible since Pref is asymmetric, thus $\succ^V$ is also asymmetric.
3. $\succ^V$ is transitive,
   Let $\alpha, \beta, \gamma \in \mathcal{A}$, assume that $(\alpha, \beta) \in \succ^V$, $(\beta, \gamma) \in \succ^V$ and $(\alpha, \gamma) \notin \succ^V$:
   - $(\alpha, \beta) \in \succ^V \Rightarrow (\mathtt{val}(\alpha), \mathtt{val}(\beta)) \in$ Pref (1),
   - $(\beta, \gamma) \in \succ^V \Rightarrow (\mathtt{val}(\beta), \mathtt{val}(\gamma)) \in$ Pref (2),

   Since $(\alpha, \gamma) \notin \succ^V \Rightarrow (\mathtt{val}(\alpha), \mathtt{val}(\gamma)) \notin$ Pref. However, since Pref is transitive, from (1) and (2) it follows that $\succ^V$ is transitive, showing a contradiction.

Two Complementary Classification Methods for Designing a Concept Lattice from Interval Data

Mehdi Kaytoue[1], Zainab Assaghir[1,2], Nizar Messai[1], and Amedeo Napoli[1]

[1] Laboratoire Lorrain de Recherche en Informatique et ses Applications (LORIA) Campus Scientifique, B.P. 235, 54500 Vandœuvre-lès-Nancy, France
kaytouem@loria.fr, zassaghir@colmar.inra.fr, messai@loria.fr, napoli@loria.fr

[2] Institut National de la Recherche Agronomique (INRA)
28, rue de Herrlisheim, B.P. 20507, 68021 Colmar, France

Abstract. This paper holds on the application of two classification methods based on formal concept analysis (FCA) to interval data. The first method uses a similarity between objects while the second considers so-called pattern structures. We deeply detail these methods in order to show their close links. This parallel study helps understanding complex data with concept lattices. We explain how the second method obtains same results and how to handle missing values. Most importantly, this is achieved in full compliance with the FCA-framework, and thus benefits from existing and efficient tools such as algorithms. Finally, an experiment on real-world data in agronomy has been carried out for decision helping in agricultural practices.

1 Introduction

Many classification problems can be formalized by means of *formal contexts*. A context materializes a set of objects, a set of attributes, and a binary relation between the two sets indicating whether an object has or not an attribute [1,2]. It is generally represented by a binary table, where a cross (×) denotes the fact that the object in line possesses the attribute in column, e.g. Table 2. According to the so-called *Galois connection*, one may classify within *formal concepts* a set of objects sharing a same maximal set of attributes, and vice-versa. Concepts are ordered within a lattice structure called *concept lattice* within the Formal Concept Analysis (FCA) framework [2]. This mathematical structure supports potential knowledge discovery in databases that benefits of an important "FCA-toolbox" for building, visualizing and interpreting concept lattices [2,3,4,5].

In many cases, data are not binary, but rather *complex* and *heterogeneous*: an object is no more linked or not with an attribute, but takes on values for this attribute (possibly none or several). These values can either be number, intervals, graphs, ... Table 1 is an example of such data. Then, a concept lattice can not be naturally built from these data: an operation called *conceptual scaling*, i.e. binarization, turns them into binary tables [2]. In this paper, we consider numerical data where objects are described by numbers or intervals of numbers.

S. Link and H. Prade (Eds.): FoIKS 2010, LNCS 5956, pp. 345–362, 2010.

A classical scaling operation is given by the following rule: "an object is linked with an attribute iff its value for this attribute is greater than a given threshold", e.g. [6]. This kind of rules enables for example to build Table 2 from Table 1. Also, attribute domains may be split into intervals, e.g. [7]. However, the scaling operation is not always reliable and easy to perform. Most of the time, such an operation is dependent on a domain expert (when available) and leads to a certain loss of information. Moreover, the partition of domain or interval of values is arbitrary and crisp: a value is either in an interval or in another.

Classifying objects having similar attribute values within same concepts may be thought as a more natural way. In that sense, authors of [8,9] defined *FCA guided by Similarity* denoted by FCAS in this paper. They propose to consider a similarity relation between "numerical" objects to directly build the concept lattice, i.e. without scaling. Intuitively, two objects are similar if the difference of their value does not exceed a given parameter for each attribute, e.g. $[2,4] \simeq_\theta [4,8]$ means that both values are similar with a parameter $\theta = 6$. Quite naturally, this similarity relation is not transitive and raises a problem for ordering concepts. The authors propose to consider a pairwise similarity of objects instead, and give applications to biological resource retrieval on the web. However, the associated theory is complex and provides no efficient algorithm at present.

On another hand, some works have been proposed to adapt the FCA process to complex data such as graph data with so-called *pattern structures* [10,11] and *logical contexts* where objects have propositional descriptions [12,13] also *without scaling*. In [14], pattern structures have been used to build a concept lattice from numerical data. So-called *Interval-based pattern structures* (IPS) relies on a theory in full compliance with FCA and thus benefits of its "toolbox" including efficient algorithms. However, the notion of similarity of objects is complex and different from the intuitive one used in FCAS: it relies on a subsumption relation between object descriptions, e.g. $[2,8] \sqsubseteq [4,8]$.

This research work holds on a study of the relations between two classification methods based on FCA and extending the classification ability of FCA for dealing with objects with many-valued attributes in an original way. Actually, the parallel study of FCAS and IPS helps to understand how these two methods are interrelated and how they can be applied to complex data for building concept lattices. IPS uses a framework in full compliance with FCA with efficient and scalable algorithms. In turn, FCAS brings an intuitive notion of similarity and helps understanding the resulting concept lattices.

The paper is organized as follows. After recalling main FCA definitions in Section 2, we detail both FCAS and IPS methods (Section 3 and 4). This is important to better highlight the links between these two methods in Section 5 where we discuss the notion of similarity and show how IPS obtains same results as FCAS in full compliance with FCA. We also present in this section how to consider missing values when building the lattice. Finally, an experiment with real-world agronomic data supports the notions discussed in this paper and addresses the problem of decision helping in agricultural practices (Section 6).

2 Formal Concept Analysis

The following definitions are taken from [2]. Let G be a set of objects, M a set of attributes, and I a binary relation defined on the Cartesian product $G \times M$, i.e. $I \subseteq G \times M$. The triple (G, M, I) is called a formal context. For an object $g \in G$ and an attribute $m \in M$, $(g, m) \in I$ or simply gIm means that "g has attribute m". The tabular representation of a formal context lies in Figure 2. Each row (resp. column) corresponds to an object in G (resp. attribute in M). A table entry contains a cross ($\times$) iff the object in row has the attribute in column. For example the object g_1 has the attribute $m_1 \leq 8$, i.e. $(g_1, m_1 \leq 8) \in I$. Given a set of objects $A \subseteq G$ and a set of attributes $B \subseteq M$, the two following derivation operators $(\cdot)'$ define a Galois connection between 2^G and 2^M:

$$A' = \{m \in M \mid \forall g \in A : gIm\} \qquad B' = \{g \in G \mid \forall m \in B : gIm\}$$

The first operator associates to any set A of objects the maximal set of attributes common to all the objects in A. For example $\{g_4\}' = \{m_2 \leq 19, m_3 \geq 0.5\}$. Dually, the second operator associates to any set B of attributes the maximal set of objects having all the attributes in B. For example $\{m_2 \leq 19, m_3 \geq 0.5\}' = \{g_4, g_6\}$. A pair (A, B) such that $A' = B$ and $B' = A$, is called a *(formal) concept*, and A and B are respectively called the *extent* and the *intent* of the concept. $(\{g_4, g_6\}, \{m_2 \leq 19, m_3 \geq 0.5\})$ is an example of concept. Concepts are partially ordered by the following subsumption relation

$$(A_1, B_1) \leq (A_2, B_2) \Leftrightarrow A_1 \subseteq A_2 \ (\Leftrightarrow B_2 \subseteq B_1).$$

For example $(\{g_2\}, \{m_1 \leq 8, m_2 \leq 19\}) \leq (\{g_1, g_2\}, \{m_1 \leq 8\})$. With respect to this partial order, the set of all formal concepts forms a complete lattice called *concept lattice*. The concept lattice associated to Table 2 is given in Figure 1 with *reduced labelling* : each node is a concept while a line denotes an order relation between two concepts. The extent of a concept is composed of objects labelling its sub-concepts w.r.t $\leq$. Dually, its intent is composed of attributes labelling its super-concepts.

Many data, e.g. in life sciences, are not given directly by binary relations but rather consist in complex objects described by numbers-, intervals- or graphs- valued attributes. These data can be described by many-valued contexts (G, M, W, I) where W is a set of attribute values, such as $(g, m, w) \in I$, written $m(g) = w$, means that the attibute m takes the value w for the object g. Many-valued contexts require to be turned into formal contexts with so-called conceptual scaling. The choice of a scale is done w.r.t. goals of the data-analysis and directly affects the size and interpretation of resulting concept lattice. Many scales are proposed in [2], but they all face the same problem: avoiding loss of information produces large and dense contexts hard to process with existing algorithms while loss of information produces contexts easier to process but leads to incomplete representation of many-valued contexts [4,14]. An example of scaling turns interval data of Table 1 to formal context of Table 2. There, each attribute of the interval data is scaled separately. Consider

Table 1. Interval data

	m_1	m_2	m_3
g_1	$[2, 4]$	$[25, 29]$	0.3
g_2	$[4, 8]$	19	0.1
g_3	$[10, 15]$	29	0.5
g_4	$[9, 13]$	17	0.5
g_5	$[8, 13]$	$[17, 19]$	0.3
g_6	$[9, 15]$	$[14, 19]$	$[0.5, 0.7]$

Table 2. A formal context

I	$m_1 \leq 8$	$m_2 \leq 19$	$m_3 \geq 0.5$
g_1	×		
g_2	×	×	
g_3			×
g_4		×	×
g_5		×	
g_6		×	×

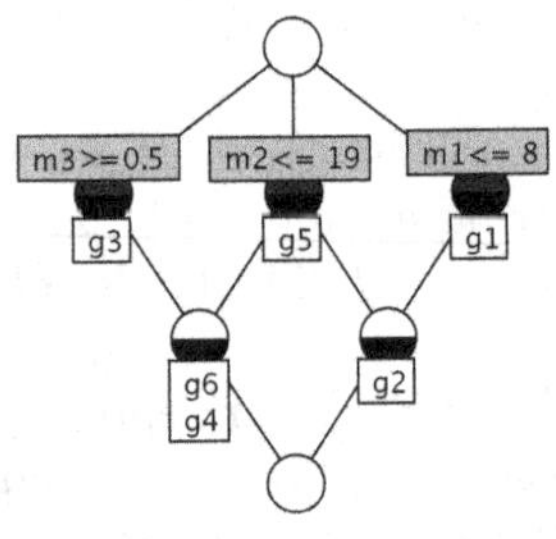

Fig. 1. A concept lattice

the first attribute, m_1. The scale can be interpreted as a "rule" stating in which conditions should an object possesses the attribute in the formal context. This rule is here simple and is actually given by the derived attribute label, $m_1 \leq 8$. This means that an object possesses the attribute m_1 in the formal context iff its value for this attribute is lower or equal than 8, and if the value is an interval, its right border must be lower or equals that 8. Scale of the other attributes follows the same idea. Table 2 is naturally an incomplete representation of Table 1. To control this loss of information, one should use other scales creating more binary attributes, generally leading to contexts hard to process [4].

3 FCAS: An FCA-Based Method with Object Similarity

FCAS is an FCA based method allowing to build a complete lattice from complex data without scaling and considering similarity between objects from a many-valued context [8,9]. Table 1 shows the kind of contexts we are interested in: contexts (G, M, W, I) such as attribute values in W are intervals of numbers or simply numbers. Firstly we recall an intuitive similarity between intervals and the problem it sets. Then, pairwise similarity is shown to be a interesting solution and is used to define the Galois connection to build a concept lattice.

3.1 Similarity between Intervals

In FCA, a set of objects A possesses an attribute m iff any single object of A possesses m. When objects are described by numbers or intervals, the sharing is not straightforward and requires scaling procedure to obtain a formal context. By contrast, usual intuition calls for a classical similarity between numbers or intervals: a set of objects possesses an attribute iff all their values are similar for this attribute. In other words, two values are similar if their difference is not significant. Formally, given $[\alpha_i, \beta_i]$ and $[\alpha_j, \beta_j]$ two intervals of real numbers, and θ a similarity threshold, then w_i and w_j are said to be similar iff:

$$[\alpha_i, \beta_i] \simeq_\theta [\alpha_j, \beta_j] \Leftrightarrow max(\beta_i, \beta_j) - min(\alpha_i, \alpha_j) \leq \theta$$

The similarity threshold θ expresses the maximal variation allowed between two similar intervals and reflects the precision requirements to be considered during the analysis of data. For example, with $\theta = 6$, $[2,4] \simeq_\theta [4,8]$ but $[2,4] \not\simeq_\theta [9,13]$ whereas for $\theta = 11$ the three intervals are similar. It is important to notice that the similarity operator $\simeq_\theta$ is not transitive: with $\theta = 9$, $[2,4] \simeq_\theta [4,8]$, $[4,8] \simeq_\theta [9,13]$ but $[2,4] \not\simeq_\theta [9,13]$.

3.2 Similarity between Objects

FCAS introduces the notion of object similarity as follows.

- Two objects g_1 and g_2 share an attribute m iff $m(g_1) \simeq_\theta m(g_2)$. θ may be different for each attribute, as attributes may have a different domain of values.
- A set of objects $A \subseteq G$ shares an attribute m whenever *any pair* of objects in A shares m. This is why it is called a pairwise similarity of objects.
- A set of objects $A \subseteq G$ shares a set $B \subseteq M$ of attributes whenever any pair of objects in A shares all attributes in B. Then A is said to be *valid* w.r.t. B.

When a set of objects shares a set of attributes, objects are pairwise similar w.r.t. this set of attributes. For example, if we consider $\theta = 6$ for attribute m_1, $\theta = 4$ for attribute m_2, and $\theta = 0.2$ for attribute m_3, then objects in $\{g_3, g_4, g_6\}$ are pairwise similar w.r.t. m_1 and m_3: they share the attributes m_1 and m_3. This means that each pair of objects has similar values for attributes m_1 and m_3. For the attribute m_1 this means that $m_1(g) \simeq_\theta m_1(h)$ for any $g, h \in \{g_3, g_4, g_6\}$, e.g. $m_1(g_3) \simeq_\theta m_1(g_6)$.

From these statements, a Galois connection can be defined. A first operator associates to a set of objects the set of attributes they share and for each of these attributes, the interval of values containing all of them (this is required to order attributes). As a result, this operator gives a set of pairs (attribute,interval). Dually, the second operator associates to a set of pairs, the maximal set of objects that share attributes from pairs in this set. These operators are detailed later.

3.3 Maximal Sets of Pairwise Similar Objects

In spirit of FCA, it is important to determine maximal sets of pairwise similar objects. This corresponds to the notion of closed sets (on which relies the definition of a concept). As in classical FCA, one has to characterize maximal sets of objects sharing maximal sets of attributes. For example, $\{g_3, g_6\}$ is valid, as well as $\{g_3, g_4, g_6\}$ for the same attributes m_1 and m_3. This very last set only will determine a formal concept.

Starting from a set of objects, the idea to obtain its maximal set of pairwise similar objects is the following. Given a set of objects A, one should (i) search for all objects similar with all objects in A, (ii) remove all pairs of objects that are not pairwise similar, and finally (iii) build the description of remaining objects, i.e. an interval needed for the Galois connection. (i) and (ii) can be seen as a closure in mathematical morphology, consisting in (i) a dilatation and (ii) an erosion by a structuring element characterizing θ [15].

(i) Set of Reachable Objects. Given an interval context (G, M, W, I), $g_i \in G$ *reaches* $g_j \in G$ w.r.t. $m \in M$ whenever $m(g_i) \simeq_\theta m(g_j)$. The set of all *reachable* objects from a valid set of objects $A \subseteq G$ w.r.t. m is defined as follows:

$$\mathfrak{R}(A, m) = \{g_i \in G \mid m(g_i) \simeq_\theta m(g), \forall g \in A\}$$

The set of reachable objects from A w.r.t. $B \subseteq M$ is: $\mathfrak{R}(A, B) = \bigcap_{m \in B} \mathfrak{R}(A, m)$. Considering the interval context in Table 1 and a threshold $\theta = 0.2$ for attribute m_3, then $\mathfrak{R}(\{g_1\}, m_3) = \{g_1, g_2, g_3, g_4, g_5\}$. This set of objects is not valid with respect to m_3 because $m(g_2) \not\simeq_\theta m(g_3)$ and $m(g_2) \not\simeq_\theta m(g_4)$. Actually, this is due to the fact that in the general case, the set of objects $\mathfrak{R}(A, m)$ may not be valid w.r.t. m because of the non transitivity of "$\simeq_\theta$".

(ii) Maximal Valid Set of Reachable Objects. The maximal valid set of objects containing A is the subset of $\mathfrak{R}(A, m)$ obtained by removing from $\mathfrak{R}(A, m)$ all pairs of objects which do not share m (i.e. g_i, g_j such that $m(g_i) \not\simeq_\theta m(g_j)$). Formally this set is defined as follows:

$$\mathfrak{R}_v(A, m) = \mathfrak{R}(A, m) \setminus \{g_i, g_j \mid m(g_i) \not\simeq_\theta m(g_j)\}.$$

The maximal valid set containing A w.r.t. $B \subseteq M$ is: $\mathfrak{R}_v(A, B) = \bigcap_{m \in B} \mathfrak{R}_v(A, m)$. In the example, $\mathfrak{R}_v(\{g_1\}, m_3) = \{g_1, g_5\}$ (i.e. obtained from $\mathfrak{R}(\{g_1\}, m_3)$ by removing g_2, g_3, and g_4).

(iii) Description of a Maximal Valid Set of Objects. When $A \subseteq G$ shares an attribute $m \in M$ ($\mathfrak{R}(A, m) \neq \varnothing$) then $A \subseteq \mathfrak{R}_v(A, m)$ and $\mathfrak{R}_v(A, m)$ shares m. The interval describing the set $\mathfrak{R}_v(A, m)$ is given by:

$$\gamma(A, m) = [min(\alpha_i), max(\beta_i)] \text{ for } [\alpha_i, \beta_i] = m(g_i), g_i \in \mathfrak{R}_v(A, m)$$

When a set of objects A shares an attribute m for a threshold θ, then we say that A shares $(m, \gamma(A, m))$. For example, $\{g_1, g_2\}$ shares $(m_1, [2, 8])$ for a threshold $\theta = 6$. When A is not valid w.r.t. m then $\gamma(A, m) = \varnothing$. Indeed, consider $\theta = 6$ and the attribute m_1. The objects g_1 and g_2 share m_1. The objects g_3 and g_4 share m_1. However g_1, g_2, g_3, and g_4 do not share m_1. This means that an object description, has to be composed of pairs: the first value gives an attribute name while the second provides with its value.

3.4 Building the Concept Lattice

In [8], it is shown that the two following operators form a Galois connection between 2^G and the partially ordered set $(M \times \mathcal{I}_\Theta, \sqsubseteq)$. $\mathcal{I}_\Theta$ is the set of all intervals possibly returned by the function γ. $\sqsubseteq$ orders pairs (attribute,interval) by inclusion of intervals of same attributes. With $A \subseteq G$ and $B \subseteq M \times \mathcal{I}_\Theta$:

$$A^\uparrow = \{(m, \gamma(A, m)) \in M \times \mathcal{I}_\Theta \mid \gamma(A, m) \neq \varnothing\}$$
$$B^\downarrow = \mathfrak{R}_v(\{g \in G \mid \forall (m, [\alpha, \beta]) \in B,\ m(g) \simeq_\theta [\alpha, \beta]\}, B)$$

$A^\uparrow$ is the set of attributes shared by all the objects in A and $B^\downarrow$ is the set of objects sharing all attributes in B. We illustrate these operators on our example, with resp. $\theta = 6$, $\theta = 4$ and $\theta = 0.2$ for resp. attributes m_1, m_2 and m_3:

$$\{g_3, g_6\}^\uparrow = \{(m_1, [9, 15]), (m_3, [0.5, 0.7])\}$$
$$\{(m_1, [9, 15]), (m_3, [0.5, 0.7])\}^\downarrow = \{g_3, g_4, g_6\}.$$

The pair $(A, B) = (\{g_3, g_4, g_6\}, \{(m_1, [9, 15]), (m_3, [0.5, 0.7])\})$ is a concept as $A^{\uparrow} = B$ and $A = B^{\downarrow}$. The set of all concepts classically ordered by $(A_1, B_1) \leq (A_2, B_2) \Leftrightarrow A_1 \subseteq A_2 (\Leftrightarrow B_2 \sqsubseteq B_1)$ generates a complete lattice. A lattice built on real-wolrd data is given in last Section.

4 IPS: Pattern Structures with an Interval Algebra

This section presents another approach to build a concept lattice from interval data without scaling [14]. In classical FCA, the operators of the Galois connection put in correspondence elements of the lattices $(2^G, \subseteq)$ of objects and $(2^M, \subseteq)$ of attributes and vice-versa. These lattices are partially ordered sets. This means that if one needs to build concept lattices where objects are not described by binary attributes but by complex descriptions (graphs, intervals, ...), one has to define a partial ordering of object descriptions. This is the main idea of *pattern structures* formalizing objects from G and their descriptions called *patterns* from a set D where patterns are ordered in a meet-semi-lattice $(D, \sqcap)$ [10]. Indeed in classical FCA, if we consider the lattice of attributes $(2^M, \subseteq)$, it is straightforward that $\forall N, O \subseteq M$, then $N \subseteq O \Leftrightarrow N \cap O = N$, e.g. with $M = \{a, b, c\}$, $\{a, b\} \subseteq \{a, b, c\} \Leftrightarrow \{a, b\} \cap \{a, b, c\} = \{a, b\}$. The set-intersection operator $\cap$ has the properties of a meet operator in a semi-lattice. This is the underlying idea for ordering patterns with a subsumption relation $\sqsubseteq$: given two patterns $c, d \in D$, $c \sqsubseteq d \Leftrightarrow c \sqcap d = c$. Then, how to build the concept lattice is in full compliance with FCA theory.

4.1 Pattern Structures

The following definitions are taken from [10]. Let G be a finite set of objects, let $(D, \sqcap)$ be a meet-semi-lattice of potential object descriptions and let $\delta : G \longrightarrow D$ be a mapping giving a description $\delta(g)$ for any object $g \in G$. Then $(G, (D, \sqcap), \delta)$ is called a *pattern structure*. Elements of D are called *patterns* and are ordered by subsumption relation $\sqsubseteq$: given $c, d \in D$, $c \sqsubseteq d \Leftrightarrow c \sqcap d = c$. The following operators $(.)^{\square}$ form a Galois connection between the powerset of G and $(D, \sqsubseteq)$:

$$A^{\square} = \bigsqcap_{g \in A} \delta(g) \qquad for\ A \subseteq G,$$

$$d^{\square} = \{g \in G | d \sqsubseteq \delta(g)\} \qquad for\ d \in (D, \sqcap).$$

The meet operator $\sqcap$ is also called a *similarity operation*: it gives a description representing similarity of objects from any arbitrary set. For example, consider two objects $g, h \in G$, their "similarity" is represented by the meet of their descriptions $\delta(g) \sqcap \delta(h)$. Indeed, the first operator denoted by $(.)^{\square}$ takes a set of objects A and returns a description shared by all objects, i.e. "similar" to those of $A^{\square}$. On the other hand, the second derivation operator $(.)^{\square}$ takes a description d and returns the set of objects having this description, i.e. description of objects in $d^{\square}$ are "similar" to d. Concepts of $(G, (D, \sqcap), \delta)$ are pairs (A, d), where $A \subseteq G$,

$d \in (D, \sqcap)$, such that $A^{\square} = d$ and $A = d^{\square}$ and are classically ordered by $(A_1, d_1) \leq (A_2, d_2) \Leftrightarrow A_1 \subseteq A_2 \ (\Leftrightarrow d_2 \sqsubseteq d_1)$ to give rise to a concept lattice.

Finally, let us recall that the principle of FCA algorithms compared in [4] is to propose an efficient strategy to generate each concept the fewest times as possible. Indeed, it may often happen that several sets of objects have the same closure since they are not necessarily all closed sets. Processing a pattern structure as shown in [10,14] requires slight modifications of these algorithms and consists in implementing the operators $(.)^{\square}$ for the considered type of patterns. Thus processing a pattern structure takes advantages of the possibilities of all existing efficient FCA algorithms detailed in [4].

4.2 Similarity between Intervals

Intervals are patterns: they may be ordered within a meet-semi-lattice making them potential object descriptions. The meet $\sqcap$ of two intervals $[a_1, b_1]$ and $[a_2, b_2]$, with $a_1, b_1, a_2, b_2 \in \mathbb{R}$ is: $[a_1, b_1] \sqcap [a_2, b_2] = [min(a_1, a_2), max(b_1, b_2)]$, i.e. the largest interval containing them. Indeed, when c and d are intervals, $c \sqsubseteq d \Leftrightarrow c \sqcap d = c$ holds:

$$\begin{aligned}
[a_1, b_1] \sqsubseteq [a_2, b_2] \Leftrightarrow \quad & [a_1, b_1] \sqcap [a_2, b_2] &= [a_1, b_1] \\
\Leftrightarrow & [min(a_1, a_2), max(b_1, b_2)] &= [a_1, b_1] \\
\Leftrightarrow & \quad a_1 \leq a_2 \text{ and } b_1 \geq b_2 & \\
\Leftrightarrow & \quad [a_1, b_1] &\supseteq [a_2, b_2].
\end{aligned}$$

This definition means that, contrarily to intuition, smaller intervals subsume larger intervals containing them, and that the meet of n intervals is the smallest interval containing all of them. Figure 2 gives an example of meet-semi-lattice of intervals. The interval labelling a node is the meet of all intervals labelling its ascending nodes, e.g. $[0.1, 0.5] = [0.1, 0.3] \sqcap [0.3, 0.5]$, and is also subsumed by these intervals, e.g. $[0.1, 0.5] \sqsubseteq [0.3, 0.5]$. In other words, if $[a_2, b_2] \subseteq [a_1, b_1]$ then $[a_1, b_1] \sqsubseteq [a_2, b_2]$; but if $[a_2, b_2] \not\subseteq [a_1, b_1]$ then $[a_1, b_1] \sqcap [a_2, b_2]$ returns the largest interval containing both $[a_1, b_1]$ and $[a_2, b_2]$.

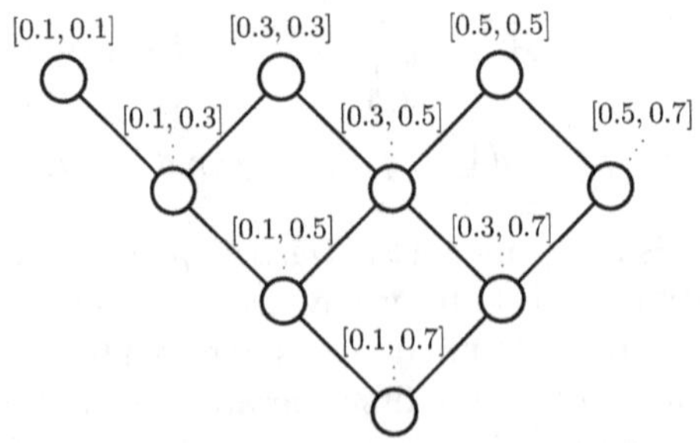

Fig. 2. A meet-semi-lattice of intervals

4.3 Similarity between Objects

As objects are generally described by several intervals, each one standing for a given attribute, *interval vectors* have been introduced as p-dimensional vector of intervals. When e and f are interval vectors, we write $e = \langle [a_i, b_i] \rangle_{i \in [1,p]}$ and $f = \langle [c_i, d_i] \rangle_{i \in [1,p]}$. Interval vectors are patterns: they may be partially ordered within a meet-semi-lattice. Indeed, the similarity operation $\sqcap$ and consequently subsomption relation $\sqsubseteq$ are given by:

$$\begin{aligned} e \sqcap f &= \langle [a_i, b_i] \rangle_{i \in [1,p]} \sqcap \langle [c_i, d_i] \rangle_{i \in [1,p]} \\ &= \langle [a_i, b_i] \sqcap [c_i, d_i] \rangle_{i \in [1,p]} \end{aligned} \qquad \begin{aligned} e \sqsubseteq f &\Leftrightarrow \langle [a_i, b_i] \rangle_{i \in [1,p]} \sqsubseteq \langle [c_i, d_i] \rangle_{i \in [1,p]} \\ &\Leftrightarrow [a_i, b_i] \sqsubseteq [c_i, d_i], \ \forall i \in [1,p] \end{aligned}$$

These definitions state that computing $\sqcap$ (resp. testing $\sqsubseteq$) for interval vectors results in computing $\sqcap$ (resp. testing $\sqsubseteq$) between intervals of each dimension, e.g. $\langle [9, 15], [14, 29] \rangle \sqsubseteq \langle [10, 15], [29, 29] \rangle$ as $[9, 15] \sqsubseteq [10, 15]$ and $[14, 29] \sqsubseteq [29, 29]$. Then, each dimension of a vector corresponds to one and only one attribute or column of a dataset and requires a canonical order of vector dimensions.

4.4 Building the Concept Lattice

As interval vectors are patterns, Table 1 shows a pattern structure $(G, (D, \sqcap), \delta)$ where $G = \{g_1, \dots, g_6\}$, D is a set of interval vectors or 3-dimensional vectors, where each component corresponds to an attribute or a column of the table. $(D, \sqcap)$ is composed of five interval vectors, i.e. a description for each object, plus all possible meets: by definition, any pair of elements (d, e) of a meet-semi-lattice admits a meet $d \sqcap e$. Description of g_3 is $\delta(g_3) = \langle [10, 15], [29, 29], [0.5, 0.5] \rangle$. Operators of the general Galois connection given in [10] are applied.

$$\begin{aligned} \{g_3, g_6\}^\square &= \textstyle\bigsqcap_{g \in \{g_3, g_6\}} \delta(g) \\ &= \delta(g_3) \sqcap \delta(g_6) \\ &= \langle [10, 15], [29, 29], [0.5, 0.5] \rangle \sqcap \langle [9, 15], [14, 19], [0.5, 0.7] \rangle \\ &= \langle [10, 15] \sqcap [9, 15], [29, 29] \sqcap [14, 19], [0.5, 0.5] \sqcap [0.5, 0.7] \rangle \\ &= \langle [9, 15], [14, 29], [0.5, 0.7] \rangle \end{aligned}$$

$$\begin{aligned} \langle [9, 15], [14, 29], [0.5, 0.7] \rangle^\square &= \{g \in G \mid \langle [9, 15], [14, 29], [0.5, 0.7] \rangle \sqsubseteq \delta(g)\} \\ &= \{g_3, g_4, g_6\} \end{aligned}$$

Obviously, g_3 and g_6 belongs to $\langle [9, 15], [14, 29], [0.5, 0.7] \rangle^\square$. g_4 also belongs to this set as its description is composed, for each dimension, of an interval that is included in the corresponding interval in $\langle [9, 15], [14, 29], [0.5, 0.7] \rangle$, i.e. $\langle [9, 15], [14, 29], [0.5, 0.7] \rangle \sqsubseteq \delta(g_4)$. Deriving the set $\{g_3, g_6\}$ with both Galois connection operators forming a closure operator makes the pair $(A, d) = (\{g_3, g_4, g_6\}, \langle [9, 15], [14, 29], [0.5, 0.7] \rangle)$ a pattern concept, i.e. $A^\square = d$ and $A = d^\square$. Partial ordering of all concepts is in full compliance with FCA and gives rise to a concept lattice. A particular example is given later for real-world data.

5 FCAS Formalized by Means of Pattern Structures

Previously, we have detailed two methods for building a concept lattice from interval data. This section presents our main contributions. Firstly, it highlights the links existing between both methods and shows how the general formalism of pattern structures obtains same results as FCAS on interval data. In other words, we show how to handle with patterns structures a similarity and a pairwise similarity like in FCAS, taking advantage of efficient algorithms. Another contribution, useful for real-world experiments, shows how handling missing values with patterns structures. Consequently, this section also shows how both methods benefit from each other.

5.1 First Statements

Both methods rely on a Galois connection between two partially ordered sets, i.e. $(2^G, \subseteq)$ and an ordered set of descriptions. For FCAS, descriptions are pairs composed of an attribute name and an interval. For IPS, descriptions are interval vectors with fixed size. In both case, intervals are ordered with inclusion.

The first operator of the Galois connection of FCAS associates to any set of objects the set of attributes they share. Firstly, pairwise similar objects are searched for, then γ returns the maximal shared interval. With PSI, the similarity operator $\sqcap$ accomplishes the same task as it returns a description representing the similarity between its arguments: $\sqcap$ is a *kernel operator* [10,16]. Thus, this operator may handle other kind of similarities.

The second operator of the Galois connection in FCAS returns for a given description, i.e. set of pairs $(m, [a, b])$ with $m \in M$ et $a, b \in \mathbb{R}$, the maximal set of all objects that share these attributes. PSI performs operations. However, PSI does not consider a pairwise similarity involving θ. In the following, we show how it can be achieved in full compliance with the existing framework of FCA.

5.2 Similarity between Patterns

Basically, pattern structures consider the meet operator $\sqcap$ as a similarity operator [10]. Intuitively, given two objects g and h, and their respective descriptions $d = \delta(g)$ and $e = \delta(h)$ from a meet-semi-lattice, $d \sqcap e$ gives a description representing similarity between g and h. As a meet-semi-lattice is defined on the existence of a meet for any pair of elements, it follows that any two objects are similar and that their "level" of similarity depends on the level of their meet in the semi-lattice. Then, how to state that two objects are similar or not in sense of FCAS can be achieved as follows. Given $c, d \in D$ two patterns, then c and d are said to be similar iff $c \sqcap d \neq *$ where $*$ materializes the pattern that is subsumed by any other pattern. This pattern is added in D and can be interpreted as the pattern denoting "no subsumption" or "non similarity" between two patterns.

When considering patterns of type interval and remembering that any interval subsumes largest intervals containing it, the element $*$ can be introduced in association with a parameter θ as follows. Given a,b,c,$d \in \mathbb{R}$ and a parameter $\theta \in \mathbb{R}$,

$$[a,b] \sqcap_\theta [c,d] = \begin{cases} [min(a,c), max(b,d)] \text{ if } max(b,d) - min(a,c) \leq \theta \\ * \text{ otherwise,} \end{cases}$$

and

$$* \sqcap_\theta [a,b] = * \Leftrightarrow * \sqsubseteq_\theta [a,b].$$

Then, the meet-semi-lattice of intervals given in Figure 2 becomes the one given in Figure 3 when $\theta = 0.2$. In this way, we have defined a meet operator in a semi-lattice, such as the following links with FCAS hold:

$$[a,b] \sqcap_\theta [c,d] \neq * \Leftrightarrow [a,b] \simeq_\theta [c,d] \text{ and } [a,b] \sqcap_\theta [c,d] = * \Leftrightarrow [a,b] \not\simeq_\theta [c,d].$$

Operators $\sqcap$ and $\sqsubseteq$ for interval vectors use the $\sqcap_\theta$ for "constrained" intervals instead of $\sqcap$ for intervals, and formulas still hold. An example of concept is $(\{g_3, g_4, g_6\}, \langle[9,15], *, [0.5, 0.7]\rangle)$: objects in the extent are similar for the first and third attributes. In FCAS, equivalent concept is $(\{g_3, g_4, g_6\}, \{(m_1, [9,15]), (m_3, [0.5, 0.7])\})$: only shared intervals are represented, where attribute labels are inserted.

5.3 Pairwise Similarity by Means of Projections

The use of $\sqcap_\theta$ does not allow the construction of intervals whose length exceeds θ like in FCAS. However, we cannot be sure these intervals describe maximal valid sets of objects in FCAS: definition of $\mathcal{R}_v$ starts with a set of objects A and returns the maximal valid set of objects: this set contains A plus all objects similar with objects in A and pairwise similar. Then γ returns the interval shared by the resulting set of objects for a given attribute. This means that though intervals from a semi-lattice $(D, \sqcap_\theta)$ all describe valid set of objects, some of them may not be "maximal". Below, we show how to replace any interval by its "maximal" interval thanks to a so-called *projection* in a meet-semi-lattice.

"Ball of Patterns". Firstly, consider the meet-semi-lattice $(D, \sqcap_\theta)$ of interval values for a given attribute. Then, for any interval $d \in D$, we define the ball $B(d, \theta)$ as the set of intervals in D similar to d as follows.

$$B(d, \theta) = \{e \in D \mid e \simeq_\theta d\} \text{ with } e \simeq_\theta d \iff e \sqcap_\theta d \neq *$$

This ball of center d and diameter θ contains all intervals e whose meet with d is different of *, meaning that d and e are *similar*: $B([0.1, 0.1], 0.2) = \{[0.1, 0.1],$

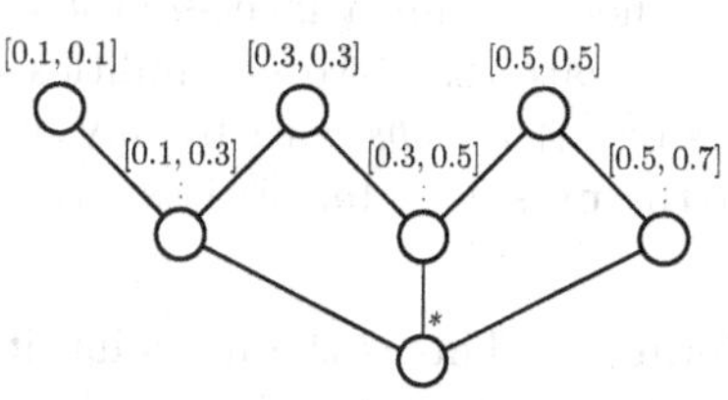

Fig. 3. A meet-semi-lattice of intervals with additional element *

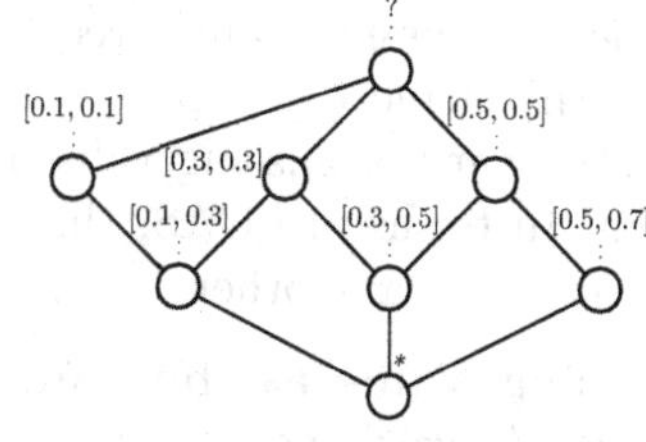

Fig. 4. A lattice of intervals with additional elements * and ?

$[0.3, 0.3]\}$. This set is linked with $\mathcal{R}$ in FCAS, for a given attribute: $B(d, \theta)$ is the set of intervals shared by objects in $\mathcal{R}(A, m)$ when $A = g$ and $m(g) = d$.

Intervals Representing Maximal Pairwise Similar Sets of Objects. Now, among this set of intervals, we should remove any pair of intervals that are not pairwise similar, i.e. computing $\mathcal{R}_v$, and build an interval with left border (resp. right border) as the minimum (resp. maximum) of all intervals, i.e. computing γ. In terms of IPS it can be done by replacing any d of the meet-semi-lattice of intervals by the meet of all intervals e from the ball $B(d, \theta)$ that are not dissimilar with another element e' of this ball, i.e. $e \sqcap_\theta e' \neq *$:

$$\psi(d) = \bigsqcap_{\theta} {}_{e \in B(d,\theta)} e \sqcap_\theta d$$
$$\text{such as } \nexists e' \in B(d, \theta) \text{ with } e \sqcap_\theta e' = *$$

In our example, $\psi([0.1, 0.1]) = [0.1, 0.1] \sqcap [0.3, 0.3] = [0.1, 0.3]$, for the third attribute and $\theta = 0.2$. In FCAS, the set returned by $\mathcal{R}_v$ is composed of objects whose attribute values respect the condition $\nexists e' \in B(d, \theta)$ with $e \sqcap e' = *$, i.e. objects are pairwise similar. Then $\bigsqcap_\theta$ returns the meet of all remaining intervals. With FCAS, we have $\gamma(g_2, m_3) = [0.1, 0.3]$ as well. In case of A is not valid w.r.t. m, remembering that any interval whose size exceeds θ is replaced by $*$, the mapping ψ returns $*$ and γ in FCAS returns $\varnothing$.

ψ is a mapping that associates to any $d \in D$ an element $\psi(d) \in (D, \sqcap_\theta)$ such that $\psi(d) \sqsubseteq d$, as $\psi(d)$ is the meet of d and all intervals similar to d and pairwise similar. The fact $\psi(d) \sqsubseteq d$ means that ψ is contractive. In sense of [10], ψ is a *projection* in the semi-lattice $(D, \sqcap_\theta)$ as also monotone and idempotent. Moreover, any projection of a complete semi-lattice $(D, \sqcap)$ is $\sqcap$-preserving, i.e. for any $d, e \in V$, $\psi(d \sqcap e) = \psi(d) \sqcap \psi(e)$ [10].

Thereby, the projection may be computed in advance, replacing each pattern by a "weaker" or "more general" pattern without loss of information. It also naturally implies better computational properties as the number of elements in the semi-lattice is reduced. Indeed, in [14] we have shown that this parameter mostly influences complexity of adapted FCA algorithms for processing pattern structures provided with an interval algebra. However, FCAS does not suggest easily such a preprocessing, and γ needs to be processed each time operators of Galois connection are calculated.

5.4 Handling Missing Values with Pattern Structures

Considering missing values requires to order them within a meet-semi-lattice of patterns or more generally within a lattice of patterns. Two possibilities are straightforward: a missing value (i) *subsumes* or (ii) *is subsumed* by any other element. In terms of FCAS, this means that the missing value (i) is *similar* or *dissimilar* with any other.

A Missing Value as the Join of All Elements. This is the most intuitive approach. As we do not know the actual value of a missing value, denoted by "?", it can be any other value: it has to subsume any element. Then we should not restrict D to a meet-semi-lattice $(D, \sqcap)$, but allow a lattice $(D, \sqcap, \sqcup)$ of patterns, such as $? \in D$. This requires some definitions: the meet $\sqcap$ is already defined except for "?", and the join $\sqcup$ has to be defined for any pair of elements. In fact, this is rather easy as we just add one element subsuming all the others in a meet-semi-lattice. Most importantly, for $d \in D$, we have: $d \sqcap ? = d \Leftrightarrow d \sqsubseteq ?$.

An example of a lattice of patterns $(D, \sqcap, \sqcup)$ is given in Figure 4: actually it results from adding "?" in the meet-semi-lattice given by Figure 3. In case of intervals, the join operator is given by

$$[a, b] \sqcup [c, d] = \begin{cases} [max(a, c), min(b, d)] \text{ iff } min(b, d) \leq max(a, c) \\ ? \text{ otherwise} \end{cases}$$

A Missing Value as the Meet of All Elements. The fact that a missing value is dissimilar with any other (except itself) is also interesting (see Section 6 for an application with real-world data). This underlines the fact that if the value is not given then it should not be considered as unknown: there is simply no information. This kind of missing value can by represented by the element $*$ introduced earlier. Indeed, * represents the dissimilarity between object descriptions and $*$ is subsumed by any other value.

Computation. In [14] we have shown how slight modifications of well-known FCA algorithms enable computation of pattern structures provided with an interval algebra. Interval vectors suggested to be implemented as arrays or vectors of intervals. With this implementation, and due to canonical order of vector dimensions, a missing value has to be materialized by * each time it is necessary, e.g. $\langle [15, 18], * \rangle$ where * is a missing value. Some data contain numerous attributes and are very sparse. Then the representation by vectors is not adequate as it leads to pattern intents containing a major proportion of * values. By contrast, FCAS suggest to IPS to consider pairs composed of an attribute name and a value, better for sparse data as representing only non-missing values.

6 An Application in Agronomy

In this section, we show how interpreting a concept lattice –designed from interval data considering a pairwise similarity of objects and missing values– may help farmers to evaluate and improve their agricultural practices.

6.1 Data and Problem Settings

Each pesticide applied to a field presents different and specific characteristics. Given these characteristics and the properties of a field, agronomists compute indicators that relate how applying the pesticide may impact on the environment. For example, the risk level for a pesticide to reach groundwater is reflected by the so-called indicator I_{gro} in [17]. This indicator takes as input variables both pesticide and field characteristics. Field characteristics are measured by experts while pesticide characteristics depend on "expert knowledge". The latter come from several information sources such as books, databases and experts in

Table 3. Characteristics of pesticide *cypermethrine*

	DT50 day	koc L/kg	ADI mg/kg.day
BUS	30	10000	
PM10	5		0.05
PM11	5		0.05
INRA			0.05
Dabène	[7,82]	[2000,160000]	0.05
ARSf	[7,82]	[5800,160000]	
ARSl	[6,60]	[5800,160000]	
Com96	[7,82]	[2000,160000]	0.05
RIVM	[61,119]	3684	
BUK	[7,70]	19433	
AGXf	[14,199]	[26492,144652]	0.05
AGXl	[31,125]	[26492,144652]	0.05

Table 4. Table resulting of the projection of Table 3

	DT50 day	koc L/kg	ADI mg/kg.day
BUS	30.0	[3684.0, 144652.0]	*
PM10	[5.0, 82.0]	*	0.05
PM11	[5.0, 82.0]	*	0.05
INRA	*	*	0.05
Dabène	[5.0, 82.0]	*	0.05
ARSf	[5.0, 82.0]	*	*
ARSl	[5.0, 82.0]	*	*
Com96	[5.0, 82.0]	*	0.05
RIVM	[30.0, 125.0]	[3684.0, 144652.0]	*
BUK	[5.0, 82.0]	[3684.0, 144652.0]	*
AGXf	*	[3684.0, 144652.0]	0.05
AGXl	[30.0, 125.0]	[3684.0, 144652.0]	0.05

agronomy. Then interesting data arise, as given by the Table 3. This table relates characteristics of the pesticide *Cypermethrin*. 12 different sources provide values for 3 different characteristics. Sometimes sources provide same values, sometimes not. Consequently, the problem for agronomists is to choose a "good" value for each characteristic, inputs of the indicator I_{gro}.

The problem statements are the following. The field characteristics are given by variables *position* = 0.5 and *leaching risk* = 0.75. The variable *leaching risk* depends on the soil characteristics. It is estimated by experts on a score between 0 and 1. It evaluates the quantity of pesticide reachable by the soil. The variable *position* is given by the farmer as a value between 0 and 1 to indicate the soil cover. A question associated to the calculation of the indicator I_{gro} can be: "*is it dangerous for the ground water to apply Cypermethrin to this field*"? To compute the indicator I_{gro}, we have to choose a value for $DT50$, koc, and ADI, from Table 3. The variable $DT50$ is the time required for the pesticide concentration under defined conditions to decrease of 50%. The variable koc is an organic-carbon constant describing the tendency of a pesticide to bind to soil particles. The variable ADI reflects chronic toxicity of the pesticide for humans.

Each row of Table 3 is an interval vector, e.g. $BUS^{\square} = \langle [30,30],[10000,10000],[0.05,0.05]\rangle$ is the description of the source (object) BUS. Then, Table 3 is a pattern structure. Missing values are replaced by the element *, meaning that missing values are considered as values dissimilar to any other. Indeed, no information is available, we consider this information as dissimilar to any other (contrasting with the use of "?"). The following experiment shows an original way of computing the input variables of the indicator I_{gro} by interpreting a concept lattice.

6.2 Method

Parameters. The choice of a θ parameter is done for each characteristic by the expert as follows: for $DT50$, $\theta = 100$, for koc, $\theta = 150\ 000$ and for ADI, $\theta = 0$. Pesticides with a $DT50$ value lower than 100 days can be considered as having weak impact on ground water quality in most temperate agronomic

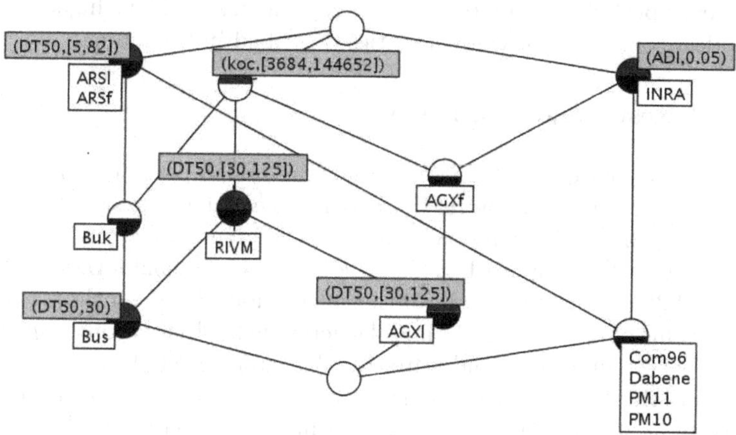

Fig. 5. Concept lattice raised from data of Table 4

conditions. Last insecticide treatments are done by the end of August and maximum leaching period begins with November rain events causing the winter leaching phenomenon. Thus, when value of $DT50$ of pesticide is 100 days or less, we can assume that most of the pesticide will be degraded in winter [18]. Then, we set $\theta = 100$ for $DT50$ and any interval whose size exceeds 100 is replaced by *. The value 150000 has been chosen as similarity threshold for *koc*. This value allows one to consider similar all values proposed by different sources ranging from 2000 and 160000. A pesticide with a *koc* value higher than 500 is hardly susceptible to leach as it is strongly retained by the soil. For ADI, it is convenient to set $\theta = 0$, meaning that all sources are reliable.

Lattice Construction and Interpretation. Projecting the pattern structure consists in replacing each value by the meet of all pairwise similar values, and does not involve computational difficulties. From Table 3, projection returns Table 4. To process the obtained pattern structure, we run the algorithm *CloseByOne* detailed in [4] and adapted for IPS in [14]. We obtain the lattice given in Figure 5 with 11 concepts. The extent of a concept is read as usual, with reduced labelling. The intent of a concept is given by a particular reduced labelling: it is read from super concepts as usual, but if two intents give different values for a same attribute, then the lowest w.r.t. $\leq$ is to consider (this is due to the interval subsumption). For example, the "most left-down" concept is $(\{BUS\}, \{(DT50, 30), (koc, [3684, 144652])\}$, and its "most left" super-concept is $\{BUS, BUK\}, \{(DT50, [5, 82]), (koc, [3684, 144652])\}$. Thereby, intents of concepts give a value for each attribute if this value is different from *, i.e. objects in extent share this value, or simply sources agreed on the given value. Then the interpretation of the lattice is the following. The higher a concept is in this hierarchy, the largest is its extent and the largest are the intervals composing its intent. However, in any concept extent, objects are similar according to the θ parameters. Then, the higher a concept is, the more sources agree on a similar value for each pesticide characteristic given by the intent. Thus, highest concepts give an ideal consensus for characteristics described in the intent.

6.3 First Result and Discussion

Direct sub-concepts of the highest concept are called co-atoms. They provide "best values" for the characteristics, as they have the largest extents. In our experiment, each co-atom gives a value for a different characteristic. Then, we use these values to compute I_{gro}, i.e. $koc = [3684, 144652]$, $ADI = 0.05$ and $DT50 = [5, 82]$, see [19] for details of the calculation. Based on these values, the computed value of I_{gro} is 10, matching the actual field observations: *cypermethrin* was never been found in ground water for this experiment [17].

The strength of the concept lattice is that, though sources of an extent do not agree on a specific characteristic, they may agree on other characteristics. More importantly, the originality of our approach w.r.t. [17] is that a value used to compute I_{gro} extracted from the lattice may not be in initial data (thanks to the interval algebra $\sqcap$). In many cases, extracting values from the lattice may not be that simple. The expert may not agree with a value from a co-atom. Then navigation through the lattice provides the expert with a decision support, materialized for each characteristic by an ordered set of values that are shared by a decreasing number of sources. Lattice-based navigation is a well known FCA application, mainly presented as a solution for interactive information retrieval tasks [9,8]. In our settings, a breadth first navigation is suited but many other methods are conceivable.

7 Conclusion

In this paper, we have shown how two interrelated classification method based on FCA (IPS and FCAS) can be combined for a better understanding and analysis of complex data. Pattern structures rely on a formalism in full compliance with FCA and with strong computational properties. FCAS brings intuitions to consider similarity and pairwise similarity of objects. It also suggests a particular implementation for IPS when data are sparse bringing interesting further research. Other directions have to be investigated. Firstly, close links have to be studied and established with possibility theory and fuzzy classification (as in [20]). Secondly, extraction of decision trees from lattices (as in [21]) and use of domain knowledge still need to be improved.

Acknowledgements

The first and the last authors were supported by the Contrat de Plan Etat-Région Lorraine: Modélisation, Information et Systèmes Numériques (2007-2013).

References

1. Barbut, M., Monjardet, B.: Ordre et Classification, Algèbre et Combinatoire. Hachette, Paris (1970)
2. Ganter, B., Wille, R.: Formal Concept Analysis. Mathematical foundations edn. Springer, Heidelberg (1999)
3. Wille, R.: Why can concept lattices support knowledge discovery in databases? J. Exp. Theor. Artif. Intell. 14(2-3), 81–92 (2002)
4. Kuznetsov, S.O., Obiedkov, S.A.: Comparing Performance of Algorithms for Generating Concept Lattices. J. Exp. Theor. Artif. Intell. 14, 189–216 (2002)
5. Valtchev, P., Missaoui, R., Godin, R.: Formal concept analysis for knowledge discovery and data mining: The new challenges. In: Eklund, P. (ed.) ICFCA 2004. LNCS (LNAI), vol. 2961, pp. 352–371. Springer, Heidelberg (2004)
6. Pensa, R.G., Leschi, C., Besson, J., Boulicaut, J.F.: Assessment of discretization techniques for relevant pattern discovery from gene expression data. In: Zaki, M.J., Morishita, S., Rigoutsos, I. (eds.) BIOKDD, pp. 24–30 (2004)
7. Kaytoue, M., Duplessis, S., Napoli, A.: Using formal concept analysis for the extraction of groups of co-expressed genes. In: An, L.T.H., Bouvry, P., Tao, P.D. (eds.) MCO. CCIS, vol. 14, pp. 439–449. Springer, Heidelberg (2008)
8. Messai, N.: Formal Concept Analysis guided by Domain Knowledge: Application to genomic resources discovery on the Web (in French). PhD Thesis in Computer Science, University Henri Poincaré – Nancy 1, France (March 2009)
9. Messai, N., Devignes, M.D., Napoli, A., Smail-Tabbone, M.: Many-valued concept lattices for conceptual clustering and information retrieval. In: Ghallab, M., et al. (eds.) Proc. of 18th European Conference on Artificial Intelligence, pp. 127–131 (2008)
10. Ganter, B., Kuznetsov, S.O.: Pattern structures and their projections. In: Delugach, H.S., Stumme, G. (eds.) ICCS 2001. LNCS (LNAI), vol. 2120, pp. 129–142. Springer, Heidelberg (2001)
11. Kuznetsov, S.O.: Galois connections in data analysis: Contributions from the soviet era and modern russian research. In: Ganter, B., Stumme, G., Wille, R. (eds.) Formal Concept Analysis. LNCS (LNAI), vol. 3626, pp. 196–225. Springer, Heidelberg (2005)
12. Chaudron, L., Maille, N.: Generalized formal concept analysis. In: Ganter, B., Mineau, G.W. (eds.) ICCS 2000. LNCS, vol. 1867, pp. 357–370. Springer, Heidelberg (2000)
13. Ferré, S., Ridoux, O.: A Logical Generalization of Formal Concept Analysis. In: Ganter, B., Mineau, G.W. (eds.) ICCS 2000. LNCS, vol. 1867, pp. 357–370. Springer, Heidelberg (2000)
14. Kaytoue, M., Duplessis, S., Kuznetsov, S.O., Napoli, A.: Two FCA-Based Methods for Mining Gene Expression Data. In: Ferré, S., Rudolph, S. (eds.) ICFCA 2009. LNCS, vol. 5548, pp. 251–266. Springer, Heidelberg (2009)
15. Serra, J.: Image Analysis and Mathematical Morphology. Academic Press, Boston (1982)
16. Raedt, L.D.: 9: Kernels and Distances for Strucutred Data. In: Logical and Relational Learning, pp. 289–324 (2008)
17. Van der Werf, H., Zimmer, C.: An indicator of pesticide environmental impact based on a fuzzy expert system. Chemosphere 36(10), 2225–2249 (1998)
18. Bockstaller, C., Girardin, P., van der Werf, H.: Use of agro-ecological indicators for the evaluation of farming systems. European Journal of Agronomy 7(1-3), 261–270 (1997)

19. Assaghir, Z., Girardin, P., Napoli, A.: Fuzzy logic approach to represent and propagate imprecision in agri-environmental indicator assessment. In: Proc. of the European Society For Fuzzy Logic And Technology Conference (2009)
20. Dubois, D., de Saint-Cyr, F.D., Prade, H.: A possibility-theoretic view of formal concept analysis. Fundamenta Informaticae 75(1-4), 195–213 (2007)
21. Guillas, S., Bertet, K., Visani, M., Ogier, J.M., Girard, N.: Some links between decision tree and dichotomic lattice. In: CLA 2008, pp. 193–205 (2008)

Author Index